Electrophilic Substitution of Heterocycles:
Quantitative Aspects

Advances in

Heterocyclic
Chemistry

Volume 47

Editorial Advisory Board

R. A. Abramovitch, *Clemson, South Carolina*
A. Albert, *Canberra, Australia*
A. T. Balaban, *Bucharest, Romania*
A. J. Boulton, *Norwich, England*
H. Dorn, *Berlin, G.D.R.*
J. Elguero, *Madrid, Spain*
S. Gronowitz, *Lund, Sweden*
T. Kametani, *Tokyo, Japan*
O. Meth-Cohn, *South Africa*
C. W. Rees, FRS, *London, England*
E. C. Taylor, *Princeton, New Jersey*
M. Tišler, *Ljubljana, Yugoslavia*
J. A. Zoltewicz, *Gainesville, Florida*

Electrophilic Substitution of Heterocycles:
Quantitative Aspects

Alan R. Katritzky, FRS
Department of Chemistry
University of Florida
Gainesville, Florida

Roger Taylor
School of Chemistry and Molecular Sciences
The University of Sussex
Falmer, Brighton
England

Advances in Heterocyclic Chemistry
Volume 47

ACADEMIC PRESS, INC.
Harcourt Brace Jovanovich, Publishers
San Diego New York Boston London Sydney Tokyo Toronto

This book is printed on acid-free paper. ∞

COPYRIGHT © 1990 BY ACADEMIC PRESS, INC.
All Rights Reserved.
No part of this publication may be reproduced or transmitted in any form or by any means, electronic or mechanical, including photocopy, recording, or any information storage and retrieval system, without permission in writing from the publisher.

ACADEMIC PRESS, INC.
San Diego, California 92101

United Kingdom Edition published by
ACADEMIC PRESS LIMITED
24-28 Oval Road, London NW1 7DX

LIBRARY OF CONGRESS CATALOG CARD NUMBER: 62-13037

ISBN 0-12-020647-1 (alk. paper)

PRINTED IN THE UNITED STATES OF AMERICA
90 91 92 93   9 8 7 6 5 4 3 2 1

Contents

PREFACE vii
DETAILED TABLE OF CONTENTS ix
CHAPTER 1. Introduction 1

Part I
Electrophilic Substitution Reactions

CHAPTER 2. Hydrogen Exchange 7
CHAPTER 3. Nitration 39
CHAPTER 4. Other Reactions 59
CHAPTER 5. Reactions Involving Formation of Carbocations at
    Side-Chain α-Positions 77

Part II
Five-Membered Heterocyclic Rings

CHAPTER 6. Reactivity of Five-Membered Rings Containing One Heteroatom 87
CHAPTER 7. Azoles 139
CHAPTER 8. Polycyclic Heteroaromatics Containing a Five-Membered Ring 181

Part III
Six-Membered Heterocyclic Rings

CHAPTER 9. Heteroaromatics Containing One Six-Membered Ring 277
CHAPTER 10. Six-Membered Rings: Electrophilic Substitution in the Azines 325
CHAPTER 11. Compounds Containing Two or More Six-Membered Rings 353
CHAPTER 12. Thiaazepines 399
REFERENCES 403

Preface

Volume 47 of *Advances in Heterocyclic Chemistry* is, unlike most volumes, a monograph and deals with the quantitative aspects of electrophilic substitution of heterocycles. It is written by Roger Taylor of the University of Sussex, Brighton, England, and your editor with one chapter contributed by Ross Grimmett of the University of Otago in New Zealand. It is hoped that this survey of the whole area of electrophilic substitution of heterocycles, covering as it does semiqualitative as well as completely quantitative aspects, will be of considerable help to workers in the field.

As is normal for volumes of our series, no subject index is included. Instead, there is a very detailed contents from which we believe it will be possible to track down most points. Of course, this volume will be indexed in Volume 51, which will be the next "index volume" of the series and will cover Volumes 46–50, just as Volume 46 covered Volumes 41–45 and Volume 40 covered Volumes 1–40.

ALAN R. KATRITZKY

Detailed Table of Contents

Chapter 1. Introduction
1. General Objectives 1
2. Significance of Mechanism in the Electrophilic Substitution of Heterocycles 2
   A. Rationalization of Experimental Results 2
   B. Guidance in Future Experimental Work 3
3. Scope and Organization of Review 3

Part I. Electrophilic Substitution Reactions

Chapter 2. Hydrogen Exchange
1. Acid-catalyzed Exchange 7
   A. Mechanism 7
   B. Exchange Conditions 10
      a. Aqueous Mineral Acids 11
      b. Organic Acids 11
   C. Steric Effects 12
   D. Hydrogen Exchange in Heteroaromatics 12
   E. Experimental Techniques 13
      a. Deuteriation 14
      b. Detritiation 15
   F. Criteria for Defining the Reacting Species 17
      a. Species Variation 17
      b. Use of Model Compounds 17
      c. Consideration of Rate Profiles 17
      d. Other Criteria 19
      e. Examples of Rate Profiles 19
   G. Standard Conditions: Choice and Procedure 27
      a. Acid-Catalyzed Hydrogen Exchange as a Quantitative Measure of Reactivity 27
      b. Justification for Selecting Standard Conditions 28
      c. Procedure for Determining Standard Rates for Deuteriation 29
      d. Reliability of k_2^o Values 34
      e. Alternative Standard Conditions 34
2. Base-Catalyzed Exchange 35
   A. Mechanism 35
   B. Exchange Conditions 36
   C. Steric Effects 36
   D. Hydrogen Exchange in Heteroaromatics 37

Chapter 3. Nitration

1. Nitration Conditions	39
2. Mechanism	40
A. The Nitrating Species	40
B. Encounter Control	40
C. Solvent Effects	41
D. Electron-Transfer Mechanism	42
E. Nitration of Bases	42
F. Ipso Attack	43
3. Experimental Techniques	44
A. The UV Technique	44
B. Calculation of Kinetics	44
C. Kinetic Complications	45
4. Criteria for Defining the Reacting Species	46
A. Survey of Possible Criteria	46
B. High-Acidity Rate Profiles	47
C. Moodie–Schofield Plots	49
D. Modified Rate Profiles	51
E. Other Types of Rate Profiles	53
F. Model Compound Studies	54
G. The Encounter Rate Criterion	54
H. Thermodynamic Parameters	55
I. Summary of Mechanistic Criteria	56
5. Standard Conditions: Choice and Procedure	56
A. Selection of Standard Conditions	56
B. Determination of Standard Rates	57
C. Alternative Procedure	58
D. Conclusions	58

Chapter 4. Other Reactions

1. Reactions Involving the Proton as Electrophile	59
A. Protiodemercuriation	59
B. Protiodeboronation	60
C. Protiodesilylation	60
D. Protiodeplumbylation	60
2. Metallation	60
A. Lithiation	60
B. Magnesiation	61
C. Mercuriation	61
D. Plumbylation	62
3. Reactions Involving Carbon Electrophiles	62
A. Alkylation	62
B. Haloalkylation and Hydroxyalkylation	63
C. Aminoalkylation	63
D. Cyanoethylation	63
E. Acylation	64
a. Formylation	64
b. Other Acylations	65
c. Alkoxycarbonylation	67

4. Reactions Involving Nitrogen and Phosphorus Electrophiles 67
   A. Nitrosation 67
   B. Diazonium Coupling 67
   C. Phosphonylation 67
5. Reactions Involving Oxygen and Sulfur Electrophiles 68
   A. Hydroxylation 68
   B. Sulfonation 68
   C. Chlorosulfonation 69
   D. Sulfenylation 70
   E. Thiocyanation (and Selenocyanation) 70
6. Halogenation 70
7. Reactions Involving Replacement of One Substituent by Another 74
   A. Lithiodehalogenation 74
   B. Iododeboronation 74
   C. Other Reactions 74

Chapter 5. Reactions Involving Formation of Carbocations at Side-Chain α-Positions

Experimental Technique 80
A. Preparative Method 80
B. Kinetic Method 80
   a. Solvolysis 80
   b. Pyrolysis 80

Part II. Five-Membered Heterocyclic Rings

Chapter 6. Reactivity of Five-Membered Rings Containing One Heteroatom

1. Acid-Catalyzed Hydrogen Exchange 87
   A. Thiophenes 87
   B. Selenophene 90
   C. Furan 90
   D. Pyrrole 90
2. Base-Catalyzed Hydrogen Exchange 93
   Furan, Thiophene, and Selenophene 93
3. Nitration 94
   A. Thiophenes 95
   B. Pyrrole 97
4. Halogenation 98
   A. Thiophene and Selenophene 98
   B. Furan and Pyrrole 101
5. Alkylation 102
   Pyrrole, N-Methylpyrrole, Furan, and Thiophene 102
6. Chloroalkylation 103
   Thiophene 103
7. Acylation 104
   A. Thiophenes 104
   B. Selenophene and Tellurophene 109
   C. Furans 109
   D. Pyrroles 111

8. Other Electrophilic Substitutions	113
A. Thiophenes	113
B. Selenophene and Tellurophene	120
C. Furans	120
D. Pyrroles	121
9. Side-Chain Reactions	122
A. Thiophene	122
B. Selenophene and Tellurophene	124
C. Furans	124
D. Pyrroles	125
10. Conclusions	125
A. Aromaticity and Relative Reactivity	126
B. Summary of Relative Rates	129
C. Sensitivity of the Five-Membered Heterocycles to Substituent Effects	132

Chapter 7. Azoles

1. Introduction	139
A. Compounds Considered	139
a. Neutral Five-Membered Rings with Two Heteroatoms	139
b. Neutral Five-Membered Rings with Three Heteroatoms	139
c. Neutral Five-Membered Rings with Four Heteroatoms	139
d. Monocationic Azoles	140
e. Anionic Azoles	140
f. Azolinones and Corresponding Thiones and Imines	140
g. N-Oxides	140
B. Reactivity Pattern	141
C. Substituent Effects	142
2. Acid-Catalyzed Hydrogen Exchange	142
A. Mechanism	142
B. Reaction at the 4-Position	143
C. Effect of Ring Nitrogen Atom on Rate	143
D. Effect of Methyl Substitution on Rate	145
E. Reactivity of Cations versus Free Bases	145
3. Base-Catalyzed Hydrogen Exchange	147
A. Introduction	147
B. Positional Reactivity Order	148
C. Substituent Effects	151
4. Nitration	154
A. Oxazoles, Thiazoles, Selenazoles and Imidazoles	154
a. Oxazoles	154
b. Thiazoles	155
c. Selenazoles	157
d. Imidazoles	157
B. Isoxazoles, Isothiazoles, Pyrazoles, and Dithiolium Ions	158
a. Isoxazoles	158
b. Isothiazoles	159
c. Pyrazoles	160
d. Dithiolium Ions	163
C. Oxadiazoles, Thiadiazoles, Triazoles, and Derivatives	164

    a. Oxadiazoles 164
    b. Thiadiazoles 164
    c. Triazoles 165
5. Halogenation 165
   A. Oxazole, Thiazole, and Imidazole 165
   B. Isoxazole, Isothiazole, and Pyrazole 167
   C. Thiadiazoles and Triazoles 170
6. Alkylation, Chloro(hydroxy)alkylation, and Acylation 170
7. Sulfonation, Sulfenylation, and Diazonium Coupling 171
8. Metallation 172
   A. Mercuriation 172
   B. Lithiation 173
9. Side-Chain Reactions 173
   A. Determination of Positional Reactivities 173
   B. Transmission of Substituent Effects 177
10. Theoretical Calculations of Reactivity 178

Chapter 8. Polycyclic Heteroaromatics Containing a Five-Membered Ring

1. General Introduction 181
2. Compounds with One Five- and One Six-Membered Ring 181
   A. Molecules Containing One Heteroatom: Benzo[b]furan, Benzo[b]thiophene, Benzo[b]selenophene, Benzo[b]tellurophene, and Indole 182
    a. Positional Reactivity Order 182
    b. Reactions 184
    c. Quantitative Aspects of the Reactivity Data 213
   B. Molecules Containing One Heteroatom: Benzo[c]selenophene, Isoindole, and Indolizine 216
    a. Reactions 216
    b. Quantitative Aspects of the Reactivity Data 218
   C. Molecules with More Than One Heteroatom in the Five-Membered Ring 220
    a. Positional Reactivity Order 221
    b. Reactions 222
   D. Molecules with Heteroatoms in Each Ring 228
    a. Positional Reactivity Order 229
    b. Reactions 230
3. Compounds with One Five- and Two Six-Membered Rings 239
   A. Molecules Containing One Heteroatom 239
    a. Positional Reactivity Order 239
    b. Reactions 242
    c. Quantitative Aspects of the Reactivity Data 248
   B. Molecules Containing Two or More Heteroatoms 251
4. Compounds with Two Five- and One Six-Membered Ring 253
   A. Acid-Catalyzed Hydrogen Exchange 255
   B. Other Reactions 257
5. Compounds with Two Five-Membered Rings 262
   A. Acid-Catalyzed Hydrogen Exchange 265
   B. Base-Catalyzed Hydrogen Exchange 268
   C. Other Reactions 269
6. Compounds with Three or More Five-Membered Rings 272
   Acid-Catalyzed Hydrogen Exchange of Dithienothiophenes 272

Part III. Six-Membered Heterocyclic Rings

Chapter 9. Heteroaromatics Containing One Six-Membered Ring

1. Introduction ... 277
   A. Compounds Considered ... 277
   B. Reactivity Patterns ... 279
      a. Pyridines ... 279
      b. Pyridine N-oxides ... 279
      c. Pyrones and Thiapyrones ... 280
      d. Pyrylium Ions ... 280
      e. Arsabenzene ... 280
      f. Phosphorins ... 280
2. Acid-Catalyzed Hydrogen Exchange ... 280
   A. Methylpyridines ... 280
   B. Aminopyridines ... 281
   C. Pyridones and Hydroxypyridines ... 283
   D. Pyridine N-Oxides ... 284
   E. Pyrones and Thiapyrones ... 286
   F. Pyrylium Ions ... 286
   G. Arsabenzene ... 287
   H. Summary of Kinetic Data ... 287
3. Base-Catalyzed Hydrogen Exchange ... 289
4. Nitration ... 292
   A. Pyridines ... 292
   B. Pyridones and Hydroxypyridines ... 297
   C. Pyridine N-Oxides ... 299
   D. 2-Pyrone ... 302
   E. Arsabenzene ... 303
   F. Summary of Kinetic Data ... 303
5. Halogenation ... 303
   A. Pyridines ... 303
   B. Pyridones and Hydroxypyridines ... 306
   C. Pyridine N-Oxides ... 307
6. Other Reactions ... 308
   A. Metallation ... 308
   B. Alkylation and Acylation ... 308
   C. Diazonium Coupling ... 309
   D. Sulfonation and Sulfenylation ... 310
   E. Demetallation ... 310
7. Side-Chain Reactions ... 311
   A. Pyrolysis of Esters ... 311
      a. Pyridine ... 311
      b. Pyridine N-Oxide ... 312
   B. Solvolysis of 1-Aryl-1-Methylethyl Chlorides ... 313
      a. Pyridine ... 313
      b. Pyridine N-Oxide ... 314
8. Transmission of Substituent Effects in Pyridine ... 314
9. Comparison of Theoretical Calculations of the Reactivity of Pyridine and Pyridine N-oxide with Observed Data ... 318

A. Pyridine Free Base	318
B. Hydrogen-Bonded Pyridine	319
C. Pyridinium Cation	320
D. Pyridine N-Oxide Free Base	321
E. Hydrogen-Bonded Pyridine N-Oxides	321
F. Protonated Pyridine N-Oxides	322
G. Pyrylium and Thiopyrylium Ions	322
H. Methyl-Substituted 2-Pyridones	322
I. Comparison of Standard Data for Nitration and Hydrogen Exchange	323

Chapter 10. Six-Membered Rings: Electrophilic Substitution in the Azines
by M. Ross Grimmett

1. Reactivity of the Monocyclic Azines	325
2. Acid-Catalyzed Hydrogen Exchange	326
A. Pyridazines	327
B. Pyrimidines	328
C. Pyrazines	330
D. Triazines	330
3. Base-Catalyzed Hydrogen Exchange	331
A. Pyridazines	331
B. Pyrimidines	333
C. Pyrazines	336
D. 1,2,4-Triazines	337
4. Nitration	337
A. Pyridazines	338
B. Pyrimidines	339
C. Pyrazines	341
D. Triazines	341
E. Borazapyridines	341
5. Halogenation	342
A. Pyridazines	342
B. Pyrimidines	342
C. Pyrazines	347
D. Triazines	348
6. Other Electrophilic Substitutions	348
A. Diazo Coupling	348
B. Nitrosation	349
C. Sulfonation	349
D. Acylation	350
E. Alkylation	350
F. Metallation	350

Chapter 11. Compounds Containing Two or More Six-Membered Rings

1. Introduction	353
A. Survey of Heterocycles Considered	353
a. Compounds Containing One Nitrogen Atom	353
b. Compounds Containing Two Nitrogen Atoms	354
c. Compounds Containing Three Nitrogen Atoms	356

d. Compounds with Four or More Nitrogen Atoms		357
e. Hydroxy Derivatives "Ones" of Compounds **11.1–11.39** with Hydroxy Group Conjugated with Nitrogen		357
f. N-Oxide Derivatives of Compounds **11.1–11.39**		357
g. Compounds Containing Other Group VB Elements		358
h. Compounds Containing Boron and Nitrogen		358
i. Benzo-Annelated Pyrylium Ions		358
j. Benzo-Annelated Pyrones		358
B. Reactivity Patterns		359
2. Acid-Catalyzed Hydrogen Exchange		361
A. Quinolines and Isoquinolines		361
B. Quinoline and Hydrogen Isoquinoline N-Oxides		365
C. Chromone and Thiachromone		365
3. Base-Catalyzed Hydrogen Exchange		365
4. Nitration		369
A. Compounds Containing One Nitrogen Atom		369
B. Compounds Containing More than One Nitrogen Atom		376
C. Xanthylium Salts		380
D. Boraza Compounds		380
E. Summary of Kinetic Data		381
5. Halogenation		382
A. Compounds Containing One Nitrogen Atom		382
B. Compounds Containing More than One Nitrogen Atom		385
C. Boraza Compounds		387
6. Other Electrophilic Substitutions		388
A. Mercuriation		388
B. Sulfonation		388
C. Miscellaneous Electrophilic Substitution		389
7. Side-Chain Reactions: Pyrolysis of 1-Arylethyl Acetates		389
8. Theoretical Calculations of Reactivity		393
A. Summary of General Methods		393
B. Quinoline and Isoquinoline		396

Chapter 12. Thiaazepines

Thiaazepines	399
References	403

CHAPTER 1

Introduction

1. General Objectives

A large and important part of the preparative chemistry of heteroaromatic compounds has been concerned with their electrophilic substitution reactions. Similarities between the chemistry of heteroaromatic compounds and benzenoid derivatives were recognized early, and reactions discovered initially in the benzene series were then applied to various heterocycles.

The classical investigations of the mechanism of aromatic electrophilic substitutions concentrated on benzenoid derivatives; thus the 1266 pages of the second edition of Ingold's definitive *Structure and Mechanism in Organic Chemistry*, written in 1969, while describing in great detail the mechanism of electrophilic substitution, barely mentioned heterocyclic chemistry. However, over the last 20 years the position has changed dramatically and several schools have made considerable headway in the detailed study of mechanism and reactivity in heteroaromatic electrophilic substitution, notably at the Universities of East Anglia, Exeter, Florida, Perugia, and Sussex, and at University College London.

The objectives of this work can be illustrated by reference to the program at the University of East Anglia over the years 1965–1980.

(1) It was first necessary to define the *species* of the heterocycle entering into reaction under any particular set of conditions. For example, basic molecules such as pyridine could react as free base or conjugate acid, whereas a potentially tautomeric compound such as 4-pyridone could react as such, or in the other tautomeric form (4-hydroxypyridine), or as the conjugate acid or base.

(2) Having defined the species reacting, the quantitative effect of the heteroatom(s) on reactivity had to be determined. This entailed a kinetic investigation which, for purposes of comparison, often needed extrapolation to standard conditions of the kinetic results (which had to be obtained under a wide variety of conditions because of the very large differences in reactivity encountered).

(3) With information available regarding the quantitative effects of the heteroatoms on the reactivity of various systems, the correlation of the effects of heteroatoms on different reactions and different substrates could be examined. Mutual interactions with substituents and other het-

eroatoms, and interactions of heteroatoms with the reagent in the transition state, were to be investigated and, if possible, explained by linear free-energy relationships (LFER), valence bond, molecular orbital (MO), or other theoretical methods.

(4) The experimental program at the University of Sussex (1970–present) of reactivity in the gas phase (involving formation of side-chain carbocations) has, in addition to the points already mentioned, demonstrated the need to take hydrogen bonding into account for both π-deficient and π-excessive heteroaromatics, showing that this can in some cases markedly alter the reactivity.

(5) Many other studies have encompassed all or some of the aspects discussed below.

2. Significance of Mechanism in the Electrophilic Substitution of Heterocycles

In addition to their own intrinsic scientific interest, the studies outlined above are of great importance in several respects.

A. Rationalization of Experimental Results

The recognition of the species which is undergoing reaction, of the quantitative effects of heteroatoms, of interactions between heteroatoms and substituents, and of the importance of hydrogen bonding have made possible, for the first time, a rational, quantitative, overall treatment of heteroaromatic reactivity patterns.

The heterocyclic literature is enormous, and a significant fraction deals with electrophilic substitution reactions of heteroaromatics. A great many authors have provided quantitative data, but the data are scattered through the literature, rarely reviewed comprehensively, and still less interpreted. Indeed, a proper interpretation is possible only by taking the wider view. This is what this book is intended to provide. It has been found possible not only to give interpretations of all of these quantitative data—in many cases for the first time—but to consider, additionally, much of the semiquantitative and qualitative work on the electrophilic substitution of heterocycles.

In our rationalization, we have relied heavily on the classical concept of aromaticity with particular emphasis on bond order and bond fixation. These concepts, together with acid–base and tautomeric equilibria and hydrogen bonding, are capable of explaining nearly all of the quantitative results. We have found MO methods less helpful: Electrophilic substitution reactions are usually carried out in condensed phases involving very strong solvent–substrate and solvent–reagent interactions, which vary considerably from ground to transition states. MO methods are still unable to cope effectively with this behavior, although this is changing rapidly.

B. GUIDANCE IN FUTURE EXPERIMENTAL WORK

The rationalizations just discussed can be used in extrapolation. Study of this book should be of considerable assistance in the optimization of experimental conditions, whether it be to improve overall yields, or to maximize the yield of one particular orientation or substitution.

The reactivity patterns disclosed in this book will be of greatest help in assessing the probability of success for new reactions, and in choosing experimental conditions likely to render such reactions successful.

3. Scope and Organization of Review

In this review we have gathered the important work on quantitative and mechanistic aspects of electrophilic aromatic reactivity of heterocycles. We have concentrated in particular on acid-catalyzed hydrogen exchange, nitration, and gas-phase elimination, these being the major efforts of our own research groups. However all other electrophilic substitution reactions are covered for completeness.

The book is divided into two parts: Part I (Chapters 2–5) is concerned with individual reactions, and Parts II and III (Chapters 6–12) with groups of related compounds.

Part I commences with hydrogen exchange, both because this is the simplest electrophilic substitution, and because the studies can be and have been extended over a far wider range of experimental conditions, and substrates, than any other electrophilic substitution. Chapter 3 deals with nitration, and Chapter 4 with other electrophilic substitutions. Chapter 5 is devoted to a study of the formation of side-chain carbocations, the results of which are of great importance in the interpretation of heteroaromatic reactivity.

Parts II (Five-Membered Heterocyclic Rings) and III (Six-Membered Heterocyclic Rings) are organized along classical lines: Monocyclic five-membered rings with one heteroatom (Chapter 6), monocyclic five-membered rings with two or more heteroatoms (Chapter 7), polycyclic compounds with five-membered rings (Chapter 8), monocyclic six-membered rings with one heteroatom (Chapter 9), monocyclic six-membered rings with two heteroatoms (Chapter 10), and polycyclic six-membered rings (Chapter 11). Little quantitative work has been reported on seven-membered or larger rings. Some of this is considered in Chapter 12.

Part I

Electrophilic Substitution Reactions

CHAPTER 2

Hydrogen Exchange

Hydrogen exchange can occur under either acid- or base-catalyzed conditions. Both can be considered electrophilic aromatic substitutions, the latter involving attack of the electrophile upon an aromatic anion, zwitterion, or ylide. The former reaction is aided by electron supply, the latter by electron withdrawal (particularly by $-I$ effects) as the rate-determining step is the initial *proton loss*. Steric hindrance, negligible in virtually all cases under acid-catalyzed conditions, appears to be of slightly greater importance under base-catalyzed conditions.

1. Acid-Catalyzed Exchange

A. MECHANISM

The mechanism of acid-catalyzed exchange has been described in very great detail elsewhere [72MI2(194)], so that only a summary of the main features together with more recent material is given here.

The reaction (in the form of deuteriation) was first shown to be an electrophilic substitution by Ingold, Wilson, and their co-workers some 50 years ago [34N(L)347; 36JCS915,1637; 38JCS28]. These workers found the order of reactivity of electrophiles to be $D_2SO_4 > D_3O^+ > DOPh > D_2O$. Shortly thereafter, Koizumi and Titani examined the reactivity of additional aromatics, including heterocycles (38BCJ95,681; 39BCJ353). Both these and subsequent studies have concentrated on two main areas, namely the determination of the mechanism of the reaction, and use of it to determine quantitative electrophilic reactivities of aromatics. In this latter respect the reaction has great advantages over other electrophilic substitutions, including (i) absence of steric hindrance; (ii) the ability to carry out studies on very small quantities of aromatic; (iii) very high kinetic accuracy; and (iv) a large rate spread due to the range of electrophiles available, including those of fairly low reactivity, which provide a reaction of quite high ρ factor.

The mechanism of the reaction was shown by Eaborn and Taylor to be (60JCS3301) an acid-catalyzed version of the S_E2 mechanism (A-S_E2), which applies to most electrophilic substitutions (Scheme 2.1). This involves a bimolecular reaction between an acid (HA) and the aromatic to give a Wheland intermediate, which then loses a hydrogen ion to give

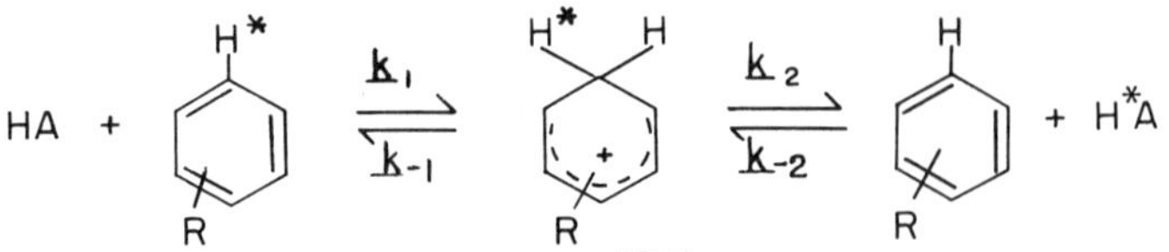

8   2. HYDROGEN EXCHANGE   [Sec. 1.A

$$HA + \underset{R}{\underset{}{\bigcirc}}\overset{H^*}{\underset{}{}} \underset{k_{-1}}{\overset{k_1}{\rightleftarrows}} \underset{R}{\underset{}{\bigcirc}}\overset{H^* \; H}{\underset{+}{}} \underset{k_{-2}}{\overset{k_2}{\rightleftarrows}} \underset{R}{\underset{}{\bigcirc}}\overset{H}{\underset{}{}} + H^*A$$

(2.1)

SCHEME 2.1. The A-S$_E$2 mechanism for acid-catalyzed hydrogen exchange.

A^-. The reaction is reversible and the profile is symmetrical about the intermediate (2.1) apart from small differences arising only from the nature of the isotopes. Bond breaking and bond making take place in essentially identical and rate-determining steps (56ACS879; 59JA5509; 60JCS3301; 61JA2877); it is this near symmetry of the reaction pathway that contributes to the low steric requirement of the reaction. The existence of a Wheland intermediate was first demonstrated by Gold and Tye, who found that anthracene in sulfuric acid gave a yellow species, attributed to 9-protonated anthracene (2.2) (52JCS2172,2184). More recently, nuclear magnetic resonance (NMR) methods have confirmed the existence of such structures in a number of cases, and even shown that the charge distribution in the benzenonium ion is as shown in structure 2.3 (58MP247; 60RTC737; 70JA2546; 71JOU1232; 72JOU1685,1808; 73JOC3212; 74BAU232, 74JA6908).

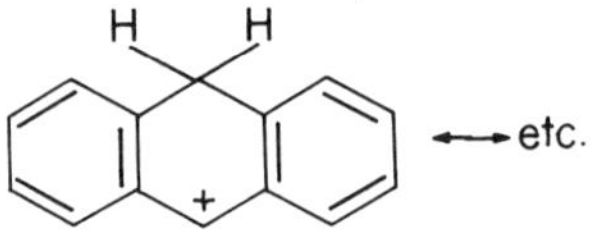

(2.2) 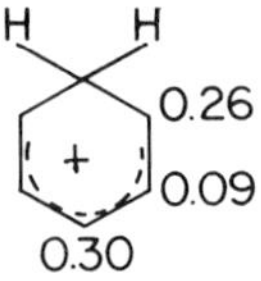   (2.3)

Early kinetic work had led to the proposal of the A-1 mechanism, i.e., one in which π-complexes are formed in a rapid pre-equilibrium, followed by rate-determining intramolecular exchange of (one form of) the intermediate into the other (55JCS3609,3619,3622; 56JCS3911). However, this was based on a linear correlation of log exchange-rate coefficient versus the acidity function $-H_0$, which was found subsequently not to hold over wider acid ranges, the slopes increasing with increasing acidity (60JCS3301); similarly, there was no correlation of exchange rates between different acids of the same H_0 value (55JCS3609). The implication drawn from the supposedly linear log k versus $-H_0$ plots was that exchange was catalyzed by specific acids (i.e., by H_3O^+ only), but later work showed that catalysis is effected by a variety of other proton-donor acidic species (general-acid catalysis). The A-1 mechanism was further based on the now-discredited Zucker–Hammett hypothesis that a water molecule cannot be covalently bound in the transition state if H_3O^+ is the catalyzing acid. This hypothesis has been demonstrated to be incorrect because similar bases can show different protonation behavior in a given acid [55JA3044; 56ACS879; 59JA5790; 60JA2965, 60JA4729, 60TL(21)12; 62JA3778, 62JA4343; 63T465; 65CC46; 66JCS(B)613; 71JA6181].

Since catalyzing acids of a wide range of strengths can be employed in hydrogen exchange, the reactivity of the electrophile could be expected to vary accordingly, thus producing a spectrum of transition states. For example, before the demise of the Zucker–Hammett postulate it was proposed that the reaction could change from the A-S_E2 to the A-1 mechanism at high acidity (59AK507; 61MI3), but no evidence to support this exists. Indeed, exchange in trifluoromethanesulfonic acid, the strongest acid (by many orders of magnitude) ever used for hydrogen exchange, showed that the selectivity of the reaction is changed little (73CC836); this rules out mechanisms involving either fast, or rate-determining, formation of π-complexes.

If transition-state structure varies with the nature of the catalyzing acid, it is a corollary that a similar variation should also be obtained for reaction of a given acid with a range of aromatics having different reactivities [65MI1(298)]. A considerable amount of work has therefore been devoted to determining the extent to which the proton transfer from the catalyzing acid to the aromatic carbon atom has taken place, using either measurement of Brönsted coefficients or of isotope effects. The Brönsted coefficient α is the slope of a plot of the logarithm of the exchange rate coefficients for a given aromatic against the pK_a value of the acid for a range of catalyzing acids. The Brönsted coefficient β is the slope of the plot of log k values for a range of aromatics against the pK_a value for protonation of the aromatic under consideration, in reaction with a given acid. Since both α and β will be 0 when no transfer has taken place, and 1.0 when transfer is complete (as in the Wheland intermediate), then reaction by the A-S_E2 mechanism should give values between 0 and 1. Experimental results appear to both confirm and contradict these expectations. For example, Thomas and Long obtained values of α of 0.61, 0.67, and 0.68 for detritiation of [1-^{3}H]azulene by anilinium ions, carboxylic acids, and dicarboxylic acid monoanions, respectively (64JA4770); that is, α increases as the catalyzing acid becomes weaker, corresponding to a later transition state. Furthermore, a smaller value (0.54) was obtained for detritiation of the more reactive [3-^{3}H]guaiazulene, corresponding to the expected earlier transition state. Detritiation of 1,3,5-trimethoxybenzene also gave α values that depended on acid strength (59JA5509; 61JA2877; 70JA6309).

Challis and Miller suggested that Brönsted coefficients may not properly represent the transition-state structure. Whereas [3-^{3}H]-indole gave β values of 0.67 and 0.75 for detritiation by hydroxonium ion and acetic acid, respectively, detritiation of [3-^{3}H]-2-methylindole gave α values that were appreciably lower, ranging from 0.46 to 0.58 [63JA2524; 72JCS(P2)1618]. One should not expect completely identical α and β values for reaction of a given aromatic, since reactions with a range of acids should give a spectrum of transition states. Thus, the observed α value will represent only the average transition-state structure. The logical conclusion is that a plot of log exchange rate versus pK_a of the catalyzing acids should be a curve, and similar arguments apply to the β values.

Although earlier work on isotope effects appeared to give a good indication of the structure of the transition state, later work has cast some doubt upon this. Because of the symmetrical nature of the hydrogen-exchange pathway, mutual compensation of trends results in small overall kinetic isotope effects. The isotope effects for the individual steps of protonation and deprotonation can also be measured and, as expected, these are larger than the overall effect. Comparison of the effects for the second (deprotonation) step of the reaction obtained with aromatics covering a 10^{13}-fold reactivity range showed a substantial (~threefold) variation, with evidence of a maximum that also coincided with zero difference in pK between the aromatic and the catalyzing acid (56JCS2743; 65MI2; 67JA1292). This maximum corresponds to the situation whereby the proton is half-transferred in the transition state (i.e., α = 0.5). However, Challis and Miller proposed that this apparent agreement is fortuitous and that the difference in isotope effects arises from proton tunneling, since in a series of indoles (though covering a much smaller reactivity range) little variation in isotope effect with reactivity was observed [72JCS(P2)1618]. This interpretation may be incorrect, and in Chapter 8 (Section 2.1.a) an alternative explanation of these results is given.

Significant variations in solvent isotope effects are also found in hydrogen exchange and k_{HX}/k_{DX} can be greater [60JCS2461; 64JCS4284; 67JCS(B)445; 70JA6309] or less [64JCS4284; 66JCS(B)613] than 1.0, depending upon the strength of the catalyzing acid and the reactivity of the aromatic.

B. Exchange Conditions

Various media have been used for exchange, ranging in acidity from aqueous solutions of ammonium ions (64JA4770) to trifluoromethanesulfonic acid (73CC836) and encompassing a 10^{28}-fold reactivity range. However, the majority of studies has involved two main conditions, as discussed in Sections 1.B.a and b.

a. *Aqueous Mineral Acids*

A variety of these has been used, but sulfuric acid has been generally preferred, largely because of availability and purity, and because many acidity function data are recorded. Nevertheless, there are two main disadvantages. The first is that sulfonation accompanies exchange, and becomes increasingly severe with increasing acidity since the rate of sulfonation increases more rapidly with increasing acidity than does the rate of hydrogen exchange (60JCS3301). It is therefore sometimes necessary to correct for sulfonation; this can be done quite easily using an ultraviolet (UV) spectroscopic method (60JCS1480). This problem can be avoided by using perchloric acid, which is of comparable acidity, but which is unfortunately hazardous especially in concentrations above 72 wt%.

The second disadvantage is that sulfuric acid is a poor solvent for many aromatics, with the solubility limit often reached well before solutions appear nonhomogeneous to the naked eye. Only very small amounts of aromatic can therefore be used, and to be valid, kinetics must be carried out using vessels only marginally larger than that needed to accommodate the sample. If this is not done, a substantial fraction of the aromatic occupies the vapor space above the sample, and exchange can occur between the two phases leading to non-first-order kinetics and anomalously low rate coefficients (60JCS3301). Fortunately, these problems arising from poor solubility do not occur with N-containing heterocycles, which, because they are either hydrogen bonded or protonated, usually dissolve readily in sulfuric acid. Consequently, many exchange data for nitrogen heterocycles have been determined in sulfuric acid, the data being extrapolated to pD = 0, and 100°C as a standard condition, as described below.

The solubility of aromatics in sulfuric acid can be significantly improved by using a cosolvent, preferably an organic acid since this is completely removed in the work-up procedure. Acetic acid is the best cosolvent (60JCS3301), and trifluoroacetic acid has also been used [74JCS(P2)394].

b. *Organic Acids*

Trifluoroacetic acid is far superior to all others because it is both a reasonably good solvent and sufficiently acidic that the exchange rates of a very wide range of aromatics can be measured at accessible temperature. The exchange (detritiation) rates at 70°C (which involves minimal or no extrapolations) constitute the largest body of rate data for any electrophilic aromatic substitution (over 350 rate coefficients being available) and this is the standard condition for exchange in all non-N-containing heterocycles.

The acidity of trifluoroacetic acid is raised dramatically by the addition of traces of mineral acids (the rate increases becoming progressively smaller for each subsequent addition of a constant amount of acid). It is therefore necessary to very carefully purify and standardize each batch of trifluoroacetic acid used. The enhanced acidity attained by mineral acid addition means that lower temperatures were used in earlier work (e.g., 61JCS2388), and, more importantly, very unreactive aromatics may be examined by this technique Trifluoromethanesulfonic acid is now the additive of choice because little is required and side reactions (cf. sulfonation with sulfuric acid, oxidation with perchloric acid) are apparently absent. Acetic acid is also used to *lower* the acidity of trifluoroacetic acid (for studying very reactive compounds); this technique has also been used to demonstrate that sulfur-containing heterocycles (and no doubt many others) are hydrogen bonded in trifluoroacetic acid (and probably in all strongly acidic media), this bonding producing a substantial reduction in reactivity.

C. Steric Effects

As already mentioned, acid-catalyzed hydrogen exchange is entirely free of steric hindrance except in the most extreme cases (i.e., those aromatic positions that are virtually completely unreactive in all other electrophilic substitutions). Thus, the central ring positions of 1,3,5-triphenylbenzene are moderately hindered [72JCS(P2)766], and rather less exchange than expected is found ortho to the extremely bulky triphenylmethyl (73CC936), as well as to the 3-pentyl and 3-hexyl, substituents [76JCS(P2)559]. The general absence of steric hindrance makes the reaction ideal for testing theoretical calculations of aromatic reactivity; it is the only reaction producing truly meaningful data in this respect.

D. Hydrogen Exchange in Heteroaromatics

Two problems may be encountered here depending upon whether N-containing heterocycles (which may react as protonated and hydrogen-bonded species) or other heterocycles (hydrogen-bonding only) are considered.

For the latter, rate data have been obtained in trifluoroacetic acid–acetic acid media, and the rate versus acidity profiles have been compared to those for compounds (e.g., alkylbenzenes) which do not hydrogen bond significantly. Exchange rates of O- and S-heterocycles relative to alkylbenzenes become progressively smaller on going to more acidic media (i.e., those containing more trifluoroacetic acid), as the heterocycle becomes increasingly deactivated by hydrogen bonding. (This is not a selectivity effect because this variation is absent in other compounds which cannot hydrogen bond.) From the rate profiles it is easy to calculate the exchange-rate coefficient that would apply in trifluoroacetic acid at 70°C if hydrogen bonding were absent. Thus far mainly five-membered sulfur-containing heterocycles have been examined and these show clearly that the sulfur atom is hydrogen bonded, since the rate reduction is directly proportional to the number of sulfur atoms in the molecules; thiaazepines are also hydrogen bonded, almost certainly at nitrogen.

For nitrogen-containing heterocycles, for which protonation is the major complication, the procedure required to obtain standardized reaction rates is more complex. The technique employed is described in detail in Section 1.F, but in general the slopes of the profiles (log rate vs. acidity function) are examined using aqueous sulfuric acid as the exchange medium. If exchange takes place on the conjugate acid of the base, then the exchange rate coefficient will increase regularly with increasing acidity. If, however, exchange occurs on the free base at acidities where the free base is a minority species, then on going to stronger acid the concentration of the free base decreases and roughly compensates for the increase in exchange rate which would otherwise occur. The overall result is thus an exchange-rate coefficient that is approximately invariant with acidity. Different substrates of the same general type will respond differently to changes in acid concentration. Thus a heterocycle containing a strongly electron-supplying substituent will protonate more readily than one without, and the equilibrium concentration of the free base will be lower, hence it is more likely to undergo hydrogen exchange via conjugate acid than will a heterocycle containing an electron-withdrawing substituent. In addition, the activation energy for reaction of a conjugate acid containing an electron-supplying substituent could be sufficiently low for reaction to be able to take place under moderate conditions, whereas electron-withdrawing substituents will raise the activation energy to the extent that reaction could preferentially take place on the small quantity of free base. From the rate–acidity profiles and rate–temperature profiles, the rate coefficients for exchange at 100°C and $pD = 0$ can be calculated.

E. Experimental Techniques

Of the six possible hydrogen-exchange reactions, the two most widely used are deuteriodeprotonation (deuteriation) and protiodetritiation (detritiation). Nowadays deuterium contents are determined by NMR, whereas tritium contents are determined by scintillation counting, which by virtue of its extreme sensitivity is the most accurate method; tritium contents have also been determined by NMR, but this requires high and potentially hazardous tritium activities.

Both deuteriation and detritiation have their advantages and disadvantages. The high sensitivity of scintillation counting permits the use of low concentrations of substrate (which renders unnecessary corrections for activity, acidity, and especially back-reaction), and facilitates examination of both sparingly soluble substrates and those which are novel to the extent that only very small quantities are currently available. The high sensitivity also means that a small proportion of a given reaction can be followed accurately. On the other hand, some aromatics are good quenching agents so that relatively high specific activities must be used to provide adequate light output, and the necessary synthesis of the specifically labeled tritium compounds often requires a considerable amount of rigorous synthetic work. Deuteriation is simpler but much less accurate, and requires unambiguous assignment of peaks to ring positions, which may be impossible for complex molecules; furthermore, peaks must be noncoincident.

a. *Deuteriation*

Deuteriation is particularly applicable to heterocycles which have high solubility in D_2SO_4 [67JCS(B)1219; 71JCS(B)2363]. The method merely involves dissolving a weighed amount (~40–80 mg) of substrate in about 0.5 ml of deuteriosulfuric acid of known concentration and heating the solution at the appropriate temperature. The reaction is generally carried out directly in a sealed NMR tube, the extent of reaction being easily evaluated by integrating the signal from the exchanging aromatic proton against a standard peak. For the latter the signal of any nonexchanging proton(s) present in the spectrum of the substrate may be used, provided the relevant peak is sufficiently resolved. Otherwise, ~15 mg of tetramethylammonium sulfate (prepared from tetramethylammonium halide and silver sulfate, and stored over P_2O_5) can be used as external standard. The tube is heated for known times in a thermostatically controlled bath, removed at intervals, cooled rapidly in an ice bath, and the NMR spectrum obtained. The averages of a ~5–10 integrals being used. At least six successive readings are used in each run, and $\ln(R_0/R_t)$ (where R_0 and R_t are the integrals for the initial and subsequent measurements, respectively) is plotted against t to give a pseudo-first-order plot. For accurate and meaningful work, it is necessary to correct for the relative molar quantities of exchangeable hydrogen in the aromatic and the acid

(61JCS247). Thus reaction of 0.5 ml of sulfuric acid with 80 mg of a heterocycle (molecular weight 100) containing five hydrogens that can exchange in the same amount of time as the hydrogen at the position under investigation will give at equilibrium ~20% of available deuterium in the aromatic, and 80% in the acid. The uncorrected pseudo-first-order kinetic plots will therefore be badly curved. With the advent of Fourier transform NMR (FT-NMR) this problem can be reduced by using lower concentrations of aromatic.

Early studies of deuteration (and protiodedeuteriation), used infrared (IR) (which is less convenient) to measure changes in the intensity of the C—H stretching frequencies. This technique was used to study the kinetics of N—H exchange in azoles [75JCS(P2)1316], the decrease in the first overtone band of the N—H stretching mode at 1.48 μm being followed.

b. *Detritiation*

The very limited solubility of some substrates in aqueous sulfuric acid media precludes the use of deuterium–protium exchange with kinetics followed by the NMR method. In these cases it is more convenient to use tritium–protium exchange, which can be studied with a much smaller substrate concentration. This is particularly appropriate for the less soluble benzenoid compounds (60JCS3301; 61JCS247; 61JCS4927). The labeled substrates are usually prepared from the bromo compounds by formation of the Grignard (or lithium) reagent, and hydrolysis of this with tritiated water (60JCS3301; 75TL435). Protiodetritiation has also been the reaction of choice for determination, with high accuracy, of rates of exchange in trifluoroacetic acid at 70°C; ~350 partial rate factors are available under this condition.

The technique (60JCS3301) involves adding sulfuric acid (260 ml) of the required concentration to a weighed amount of tritiated substrate (~25 mg). The mixture is then shaken vigorously for 10 min in a tightly stoppered long-necked conical flask (300 ml) to allow complete dissolution. In some cases, mixtures of acetic acid and sulfuric acid have been used in this method (61JCS247) and trifluoroacetic acid has also been employed to dissolve the aromatic before adding the sulfuric acid. [The current availability of tritiated water of high specific activity (5 Ci ml^{-1}) means that smaller quantities of aromatic can now be used.] Equal volumes of the solution (50 ml) are pipetted into ampoules of volume ≯53 ml, which are sealed with teflon-sleeved quickfit joints and placed in a thermostatted bath. For runs at temperatures >40°C, permanently sealed ampoules must be used, but in each case the ampoule, at the temperature of the bath, must contain a vapor space of not more than ~5% of the volume of the acid. If this is not done, the aromatic can exchange between liquid and vapor phases, and first-order kinetics are unobtainable (60JCS3301); the problem is most severe with weaker sulfuric acid media, which are poorer solvents for aromatics. Ampoules are removed at appropriate time intervals and the contents either poured into, or the ampoule broken in, 150 ml of ice/water (necessary to prevent localized heating with consequent rapid exchange) under a 20-ml layer of scintillator solution (which prevents any escape of the aromatic). If the ampoule-breaking technique is used, the ampoule must first be washed; water is adequate if, as is currently customary, polyethylene glycol is used for the heating bath medium. The water–scintillator mixture is shaken mechanically for 2 min (longer times are given in earlier papers, but this is, in fact, unnecessary), and the organic layer separated and washed with 10% sodium hydroxide (50 ml), followed by water (50 ml). Portions (10 ml) of the dried (Na_2SO_4) scintillator solution are counted in the usual way. If the extracts are counted as they are obtained, then correction for the half-life of tritium is necessary for very slow runs. However this general procedure is not recommended because of the long-term drift in scintillator counter efficiency, so that counts for a given run should, if possible, be obtained on a single day. This necessitates keeping the extracts in totally sealed containers, and in the dark (to prevent scintillator degradation), counting being delayed until the run is complete. For runs in trifluoroacetic acid, 1-ml samples are used in each ampoule, the size of which is not critical because of the much greater solubility of the aromatic in this acid. For these runs, ampoules are broken under 100 ml of 3% sodium hydroxide solution, and a single washing of the scintillator solution with 100 ml of distilled water is sufficient.

For runs carried out with >70 wt% sulfuric acid, it is necessary to correct for sulfonation, which becomes increasingly severe the stronger the acid (60JCS3301). The extent of sulfonation can be determined by using a procedure similar to that given above, except that extraction (in a completely grease-free apparatus) is carried out with spectroscopic hexane or heptane, and the UV spectrum recorded. Sulfonation produces water-soluble sulfonic acids, which are removed during the extraction process, and the residual concentration of aromatic decreases in successive extracts according to a first-order law. In runs where darkening is observed, exchange rates will appear anomalously high if a quench-correcting scintillation counter is not available. The problem can be overcome by extracting with 10 ml of the aromatic used in the run, and distilling the dried extract. Scintillator (10 ml) is then added to a known weight (which must be the same for each sample) of the distilled extract; the efficiency of

Sec. 1.F] ACID-CATALYZED EXCHANGE 17

counting will be reduced through dilution of the phosphor, but this reduction will be the same for each sample.

F. CRITERIA FOR DEFINING THE REACTING SPECIES

a. Species Variation

Exchange may occur on several species, especially among nitrogen-containing heterocycles. Species variation of the following types must be considered.

(1) Different charge types in rapid equilibrium. There may just be two, (e.g., pyridine/pyridinium cation), but there may also be more than two (e.g., 2,6-diaminopyridine/monocation/dication).

(2) Different tautomeric species (e.g., 6-chloro-2-pyridone/6-chloro-2-hydroxypyridine).

(3) Covalent hydration can be important (e.g., in certain pyrimidines exchange can occur via covalently hydrated species).

(4) Complications (1)–(3) can occur simultaneously.

(5) Base-catalyzed hydrogen exchange (e.g., in pyridinium cation, can occur in zwitterions or ylids).

Quite generally, exchange via different charge types [i.e., (1) and (5)] is distinguished by examination of rate profiles. Comparison with fixed models is needed, however, to distinguish between species of the same charge type [(2) or (3)], and this method can also be used to help distinguish different charge types.

b. Use of Model Compounds

The more direct method is to compare the rate with that for a model compound, one in which the change of species type under consideration is no longer possible. Thus, for exchange in 4-pyridone, 4-methoxypyridine serves as a model for two of the species (4-hydroxypyridine tautomer and the 4-hydroxypyridinium cation), 1-methyl-4-pyridone serves also for two of the species (4-pyridone itself and 4-hydroxypyridinium cation), whereas 1-methyl-4-methoxypyridinium cation serves as a model just for the cation (see Scheme 2.2, Section 1.F.e.iii).

c. Consideration of Rate Profiles

The appreciable basicity of many heteroaromatic compounds complicates the dependence of the rates for hydrogen exchange on the acidity, compared to that for nonbasic aromatics. The relative concentrations of the free base and the corresponding conjugate acid will depend on both the acid concentration and the pK_a of the base. Obviously, the electrophilic proton will react more slowly with the positively charged conjugate acid than with the free base. On the other hand, if the relative concentration of the conjugate acid is *much* larger than that of the free base, then the amount of exchange proceding via the conjugate acid species may nevertheless be greater.

This point is illustrated in Fig. 2.1. At pH values above the pK_a for the heteroaromatic substrate, the free base is the majority species, and as the acidity is increased within this region, so the exchange rate will increase. At acidities greater than the pK_a, the rate will become essentially invariant with increasing acidity because two factors act in opposition: as the acidity, and therefore the reactivity of the electrophile, is increased, the amount of substrate present as free base decreases. To a first approximation (the reasoning holds quantitatively only when the substrate is a true Hammett base) these two factors cancel each other, and the rate profile becomes horizontal. However, when an acidity is reached at which the concentration of the free base has become so small that despite its higher reactivity exchange proceeds almost entirely via the conjugate acid, then the rate profile shows again a unit slope.

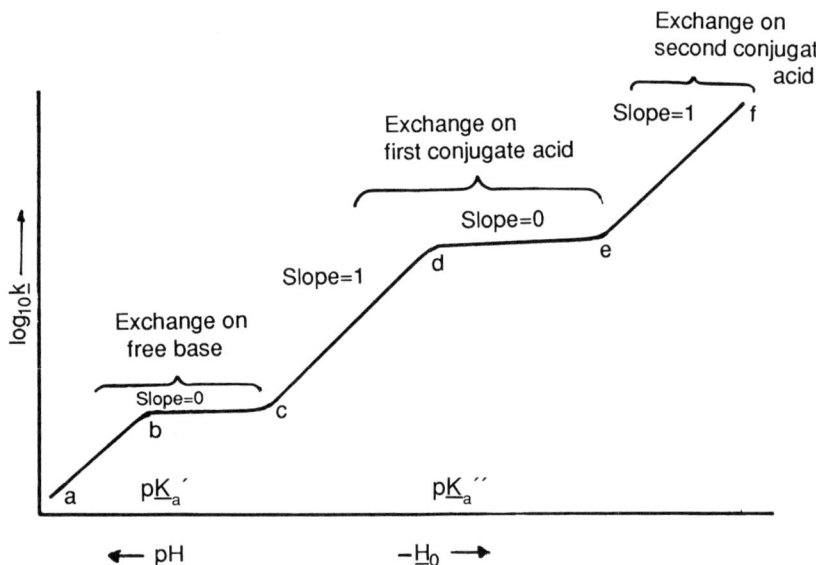

FIG. 2.1. Plot of log rate of hydrogen exchange vs. pH.

d. Other Criteria

Several other criteria are available to decide what species is reacting [67AG(E)608].

The entropy of activation should be substantially greater for reactions between two singly charged species of the same sign, than for reaction between a neutral species and one with a single charge. However, studies indicate that most of the variation in thermodynamic parameters appears to occur in the entropy term [73JCS(P2)1065], thus raising doubts on the usefulness of this approach, since the limiting value cannot be clearly defined.

The encounter rate criterion is applicable in those cases where the concentration of a free base in a strongly acid solution is so low that even if every molecular collision resulted in reaction, the calculated reaction rate would be lower than that measured. This limiting rate, the encounter rate, is given by Eq. (2.1), where η is the viscosity of the medium, k is Boltzmann's constant, N is Avogadro's number, and r_A and r_B are the radii of the ions.

$$k \text{ (encounter)} = \frac{kt(r_A + r_B)^2}{3\eta \quad r_A r_B} \frac{N}{100} \quad (2.1)$$

The relationship between activation energies and the expected effects of substituents has also been examined (63JA329). Lack of correlation indicates reaction on the conjugate acid, so that the activation energy includes the enthalpy of dissolution and the activation enthalpy of exchange. Appropriate correction for the former then leads to satisfactory Hammett plots.

e. Examples of Rate Profiles

i. *Exchange in Carbocyclic Aromatics.* With a single reacting species, exchange in carbocyclic aromatics represents the simplest case. They give plots of log rate versus $-H_0$ that are curves, concave upwards, which may be approximated to straight lines. The slopes of these increase with decreasing reactivity of the substrate, and with decreasing temperature. Some recorded slopes are 2.2 (benzene, 60JCS3301) and 1.5 (*para* position of toluene, 60JCS3301) for detritiation; and 1.22 (55°C) to 1.55 (15°C), and 1.42 (55°C) to 1.68 (15°C) for the 1- and 2-positions of naphthalene, respectively, for dedeuteration (73JA3918); Figure 2.2 shows some data that have been obtained for detritiation of [1-^{3}H]naphthalene, with an average slope of ~1.2 [74JCS(P2)394].

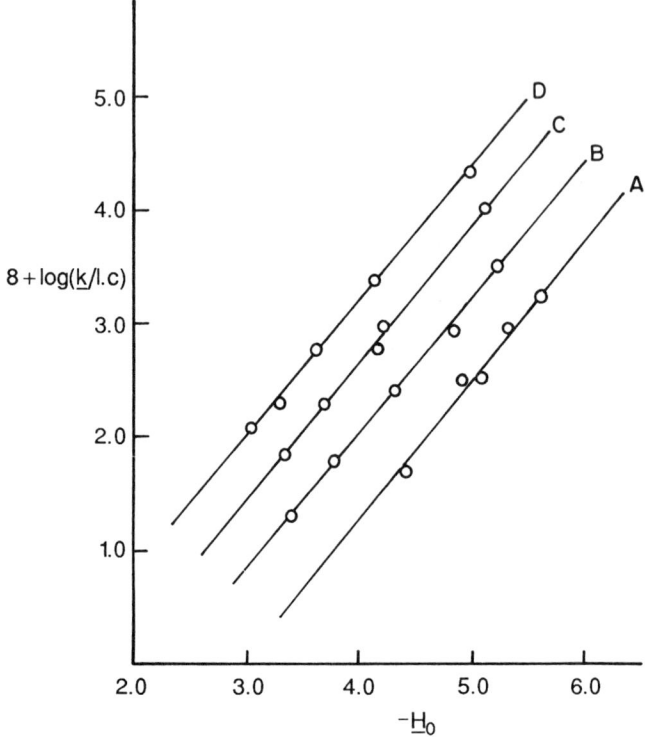

FIG. 2.2. Plots of log rate of hydrogen exchange vs. $-H_0$ for [1-^{3}H] naphthalene. A, 25°C; B, 35°C; C, 45°C; D, 55°C.

ii. *Exchange via Both Base and Conjugate Acid.* Figure 2.3 shows the idealized plot that would be obtained if reaction involves only these two species.

iii. *Exchange on Bases with More than One Protonation Site.* Heteroaromatic compounds may have several basic centers, and if this is the case then the rate profile can be more complicated, as illustrated in Fig. 2.4 [67JCS(B)1219]. As in Fig. 2.3, the unit and zero slopes are idealized, and in practice the factors of decreasing free base concentration and increasing rate do not exactly cancel, so that fractional slopes are obtained.

Figure 2.5 shows the more complicated rate profile for deuteriation of 4-aminopyridine at 107°C [67JCS(B)1219]. Below an H_0 value of about −6, exchange occurs on the first conjugate acid (ring nitrogen proton-

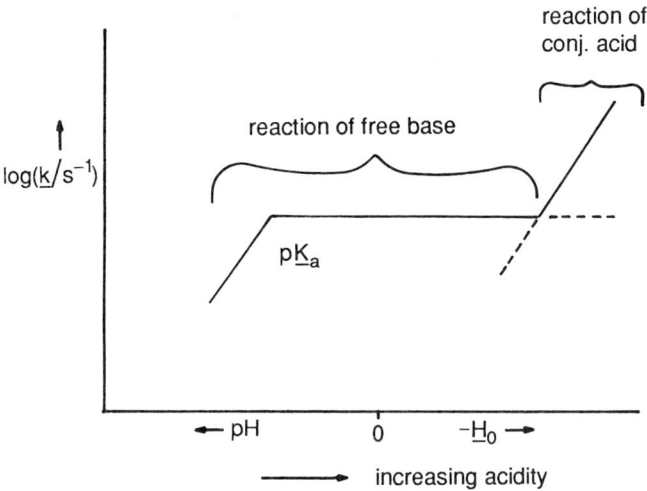

FIG. 2.3. Idealized plot of log rate of hydrogen exchange vs. $-H_0$ expected if exchange occurred only on the base and conjugate acid.

ation), which is the majority species. At higher acidity the majority species becomes the second conjugate acid (the second pK_a of 4-aminopyridine is -6.3), but exchange still takes place on the first conjugate acid, now the minority species, leading therefore to a horizontal rate profile.

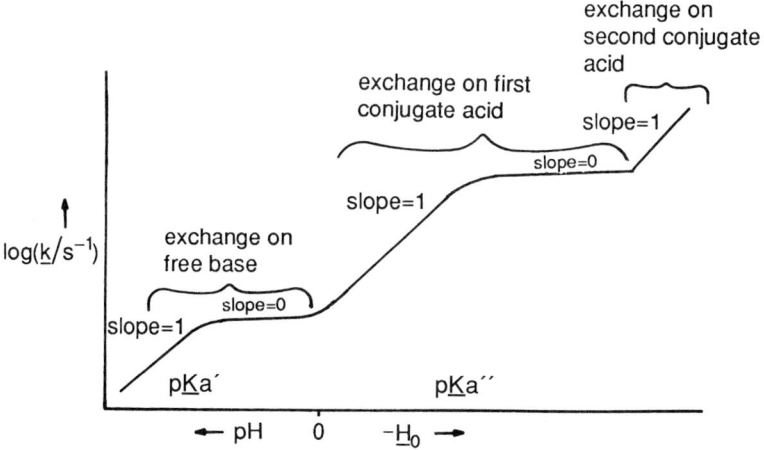

FIG. 2.4. Hydrogen exchange rate profile for heteroaromatic compound with multiple basic centers.

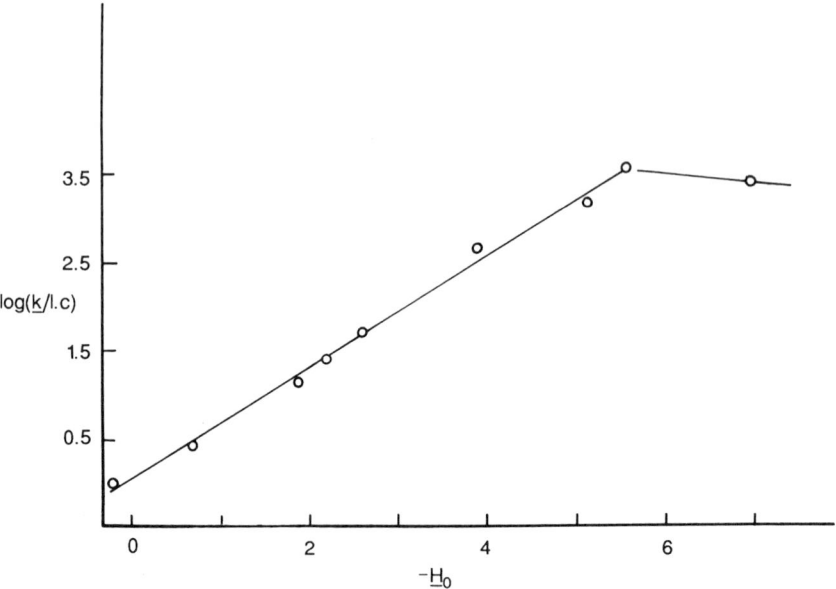

FIG. 2.5. Hydrogen exchange rate profile for deuteriation of 4-aminopyridine at 107°C.

iv. *Exchange on Free Base Only.* 4-Pyridone exists as two neutral species (4-1H-pyridone and 4-hydroxypyridine) and can give rise to a cation and an anion (Scheme 2.2), on any of which exchange could occur. The experimentally observed rate profile is shown in Fig. 2.6 [67JCS(B)1226]. Over an enormous acidity range of some fourteen H_0 units, the exchange rate changes by only a very small amount. This shows that reaction takes place on a neutral species, because over all this range

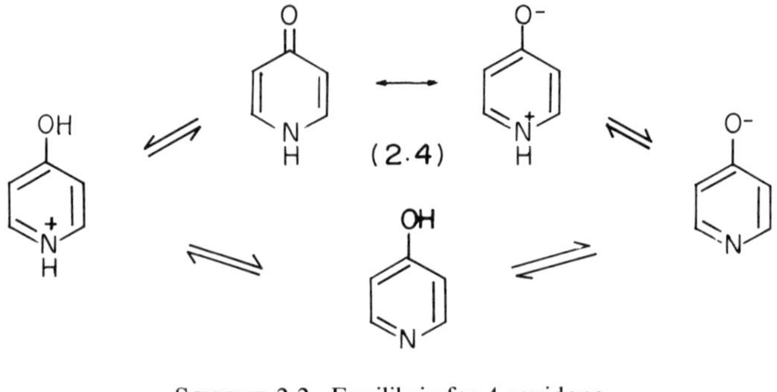

SCHEME 2.2. Equilibria for 4-pyridone.

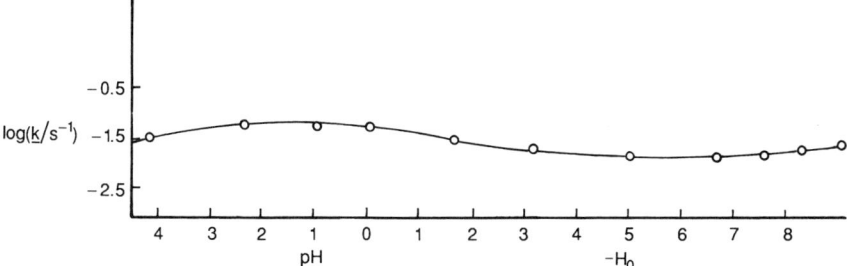

FIG. 2.6. Experimental hydrogen exchange rate profile for 4-pyridone/4-hydroxypyridine.

the majority species will be the conjugate acid. The neutral species can be identified as 4-1H-pyridone (**2.4**), because the measured rate is similar to that for 1-methyl-4-pyridone whereas 4-methoxypyridine is unreactive under these conditions.

v. *Exchange on Anion, Free Base, and Conjugate Acid.* Figure 2.7 shows the yet more complicated example of the deuteriation of 1-hydroxy-2,6-dimethyl-4-pyridone [68JCS(B)866], and here there are considerable variations in rate with acidity. An idealized representation of the experimental profile of Fig. 2.7 is given in Fig. 2.8, which shows the individual rates for exchange taking place on the anionic, neutral, and cationic forms of the compound. As the acidity increases, exchange occurs successively via the anionic, neutral, and cationic species; the solid line in Fig. 2.8 corresponds to the portions of the rate profile observed, whereas the dotted lines are theoretical extrapolations. The neutral species which undergoes exchange is the 1-hydroxypyridone (shown in Fig. 2.8) rather than the alternative tautomeric 4-hydroxypyridine 1-oxide, since the rate profile for 4-methoxy-2,6-dimethylpyridine 1-oxide given in Fig. 2.7 shows this to be very unreactive in the region

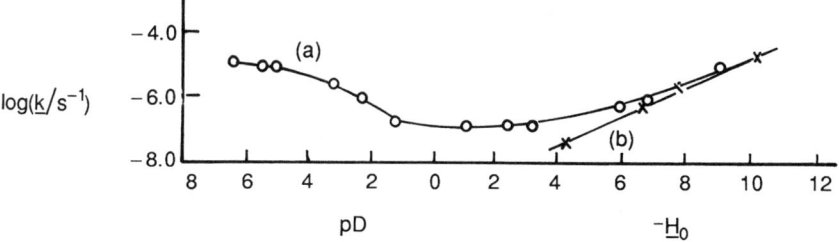

FIG. 2.7. Experimental hydrogen exchange rate profile for deuteriation of 1-hydroxy-2,6-dimethyl-4-pyridone.

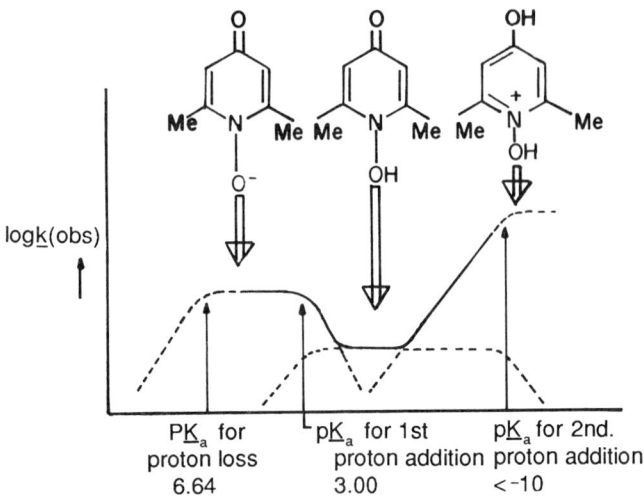

FIG. 2.8. Idealized hydrogen exchange rate profile for deuteriation of 1-hydroxy-2,6-dimethyl-4-pyridone.

where the neutral species is the principal reactant. The rate profiles for the anionic and neutral species cross *below* the proton-addition pK_a value, because of the relatively small difference between the pK_a values for the proton addition and proton loss. Consequently the observed rate actually decreases with increasing acidity in this part of the rate profile. In general, a negative unit slope indicates a reaction proceeding on a small amount of a minority species.

vi. *Profiles Showing a Base-Catalyzed Component.* Figure 2.9 shows the experimental rate profiles for deuteriation of 3,5-dimethylpyridine 1-oxide at the 2- and 4-positions [67JCS(B)1222]. For the 4-position, the rate profile shows very little change with acidity and this may confidently be attributed to a mechanism involving the free base species, where the majority species is a conjugate acid. For the 2-position this also applies at acidities greater than $H_0 = -2$. However, at weaker acidity a very pronounced increase in the rate is found, due to a change to a base-catalyzed exchange mechanism (Scheme 2.3) (see also Section 2). Base-catalyzed exchange is strongly accelerated by $-I$ effects, and so is favored here by the very strong electron withdrawal by the adjacent positive pole.

The same base-catalyzed mechanism accounts for the profile obtained in deuteriation of the 2-position of quinoline (Fig. 2.10), with the reaction occurring on the cation [71JCS(B)4]. This mechanism is found at an ex-

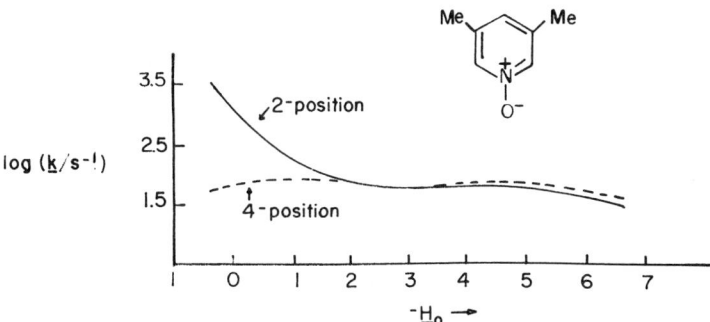

FIG. 2.9. Experimental hydrogen exchange rate profile for deuteriation of 3,5-dimethylpyridine 1-oxide at positions 2 and 4.

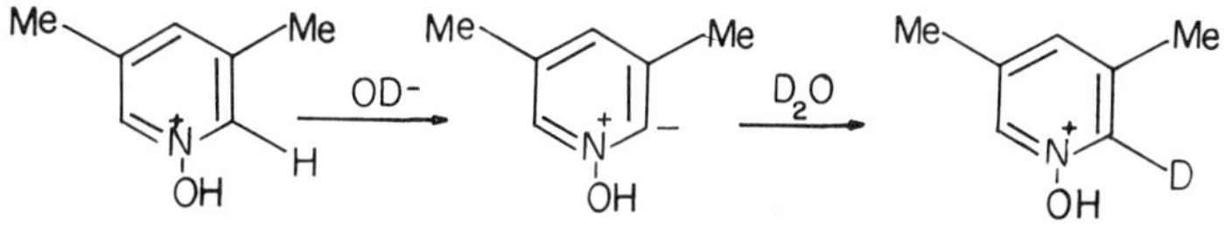

SCHEME 2.3. Base-catalyzed exchange mechanism.

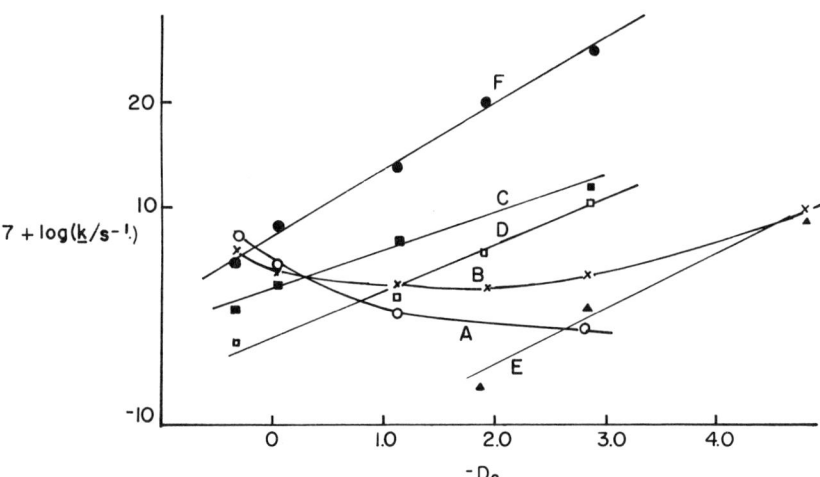

FIG. 2.10. Rate profiles at 245°C for the hydrogen exchange of quinoline. Letters on the curves indicate the position of the exchange in question: A, 2; B, 3; C, 5; D, 6; E, 7; and F, 8.

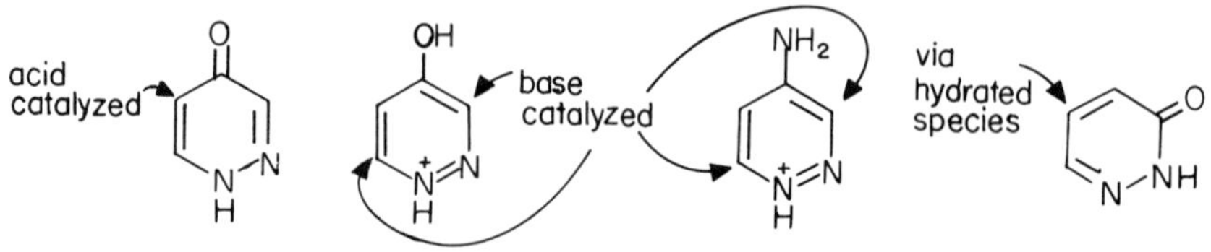

SCHEME 2.4. Hydrogen exchange of pyridazine derivatives.

ceptionally high acidity here, and this may be attributed to the high 1,2-bond order in quinoline, which strongly facilitates transmission of electronic effects (and thus the electron withdrawal by the positive pole) across the 1,2-bond. Likewise 4-pyridazinone undergoes acid-catalyzed exchange at the 5-position, whereas the protonated species and protonated 4-aminopyridazine undergo base-catalyzed exchange at the 3- and 6-positions [68JCS(B)873], because of the combined effects of the ring nitrogens, one of which is protonated (Scheme 2.4).

vii. *Exchange on a Covalent Hydrate.* Figure 2.11 shows the experimental rate profile for 2-pyrimidinone. This looks perfectly reasonable for reaction as the free base, except that the rate coefficient thus calculated for 2-pyrimidinone would be 10^4 times greater than that calculated for 2-pyridone [68JCS(B)1484]. Replacement of CH by N cannot produce rate enhancement for a normal electrophilic substitution, so the reaction is believed to occur via a small equilibrium proportion of a covalently hydrated species in equilibrium with 2-pyrimidone, as shown in Scheme 2.5. In the hydrated species with interrupted conjugation, exchange effectively takes place on a much more reactive aminoalkene. Co-

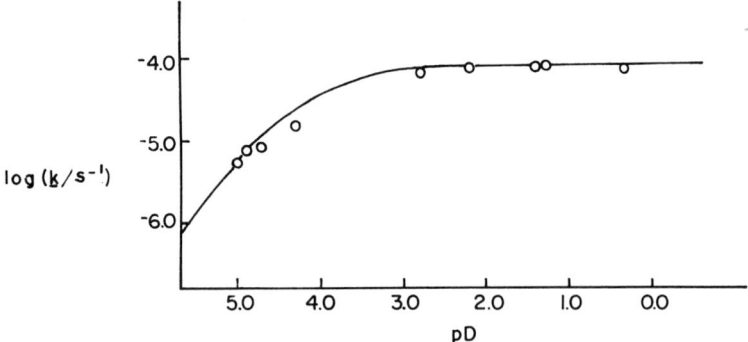

FIG. 2.11. Experimental hydrogen exchange rate profile for 2-pyrimidinone.

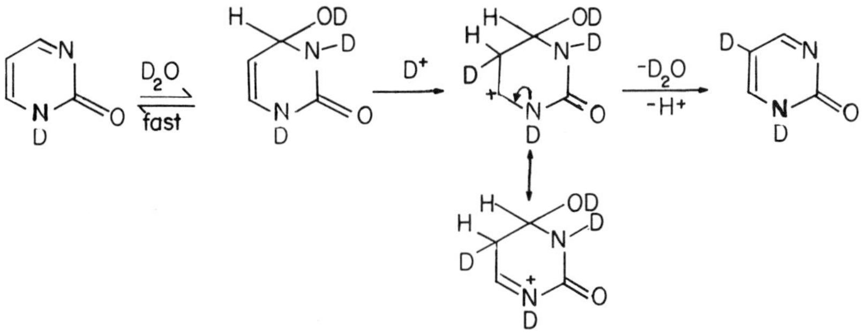

SCHEME 2.5. Exchange of 2-pyrimidinone via covalent hydration.

valent hydration of heterocycles, discovered by Albert *et al.*, is well known for polyazaheteroaromatics [65AHC(4)1, 65AHC(4)43], but the above indicates that small quantities of covalent hydrates can be kinetically significant intermediates. They have been postulated as being involved in exchange at the 3-position of quinoline (Fig. 2.10), and at the 5-position of 3-pyridazinone (Scheme 2.4).

Another possible means for distinguishing between the normal mechanism and formation of covalent hydrates when a shallow rate profile is obtained is comparison of the curve with the trend of log a_w in the same acidity region. If covalent hydration is involved in the slow step, the two curves should parallel each other closely (70CC1024).

General equations for the kinetics of acid-catalyzed hydrogen exchange of bases have been derived [68JCS(B)1484].

G. STANDARD CONDITIONS: CHOICE AND PROCEDURE

a. *Acid-Catalyzed Hydrogen Exchange as a Quantitative Measure of Reactivity*

Acid-catalyzed hydrogen exchange has important advantages as a quantitative measure of heteroaromatic reativity.

(1) Deuteriation and detritiation are complementary techniques.
(2) Rates may be measured over very wide acidity and temperature ranges.
(3) Steric hindrance is absent.
(4) Aqueous acids may be used in which the acidity function behavior is much better understood than for any other solvent.

(5) By choosing the appropriate acidity and temperature range, a very large range of heteroaromatic compounds of widely differing reactivity can be made to undergo acid-catalyzed exchange.

The mechanism of heteroaromatic hydrogen exchange can therefore be determined for a wide variety of substrates in the manner described in the previous sections; in particular, the nature of the species undergoing reaction may be elucidated. The next stage is to compare rates, and this can be done provided all the kinetic data can be extrapolated to the same standard conditions of acidity and temperature. These relative rates will, however, as in the case of electrophilic substitution of all heterocycles in solution (especially π-deficient heterocycles in protic solvents), be subject to the modifying effect of hydrogen bonding, correction for which has so far been applied in only a relatively few cases.

b. *Justification for Selecting Standard Conditions*

For detritiation the standard conditions are trifluoroacetic acid at 70°C. This acid is virtually the strongest organic acid, readily available, reasonably inexpensive, and easily purified; the acidity can be altered in both directions by the addition of other acids and the solvent property increased by addition of small quantities of carbon tetrachloride or chloroform. The effects of these modifications can be monitored easily with standard compounds, since there are more rate data available for detritiation than for any other electrophilic aromatic substitution. The temperature of 70°C was chosen for both practical advantages (the pressure in sealed ampoules is not too high) and because the exchange rates of a wide range of aromatics (including benzene itself) can then be measured directly. This condition has been used exclusively for determining the reactivities of π-excessive heterocycles.

For deuteriation the standard conditions (exchange in sulfuric acid) were chosen as pH = 0 and 100°C for the following reasons [67JCS (B)1226; 68JCS(B)866].

(1) Acidity: Measuring rate coefficients at pH = 0 is the best available means of converting pseudo-first-order rate coefficients into second-order rate coefficients (k_2^o) defined by Eq. (2.2), since the value of [H$^+$] is 1 mol liter^{-1}. Moreover, all the acidity functions merge near pH = 0.

$$\text{Rate} = k_2^o[\text{substrate}][\text{H}^+] \tag{2.2}$$

(2) Temperature: 100°C was chosen to minimize extrapolations, as most rates have been measured within the range 20–180°C.

c. Procedure for Determining Standard Rates for Deuteriation

The procedure for determining k_2° (100°C) at pH = 0 is complex and requires the following steps [73JCS(P2)1065].

(1) The value of the acidity function at T (°C), the effect of dissolved substrate on this value, and the effect of using D_2SO_4 instead of H_2SO_4 must be determined.

(2) The rate versus acidity profile must be constructed and extrapolated to pH = 0.

(3) The variation of rate with temperature must be measured.

(4) Correction must be made for the concentration of minority species, assumptions being required regarding protonation behavior of substrate pK_a with temperature.

(5) Correction for isotope effects when comparing exchange rates involving different hydrogen isotopes.

Each of these steps is discussed in detail in Sections 1.G.c.i–v.

i. Determination of k_1 (stoich) at a Particular Acidity and Temperature (step 1). This requires the following steps.

(1) Standardization of the acid by normal titration.

(2) Corrections to the acidity to allow for protonation of the substrate; these corrections can be substantial, i.e., 5 wt% or more. If y g of a substrate S of equivalent weight E are taken in z g of acid of w wt%, then the new wt% of acid is given by $(Ewz - 10008y)/(Ez - 100.08y)$ for D_2SO_4, or $(Ewz - 9808y)/(Ez - 98.08y)$ for H_2SO_4 [73JCS(P2)1675; 74JCS(P2)399]. Use of figures reported in [73JCS(P2)1065] (i.e., 5004 for D_2SO_4 and 4904 for H_2SO_4) applies only to acidities of pH > 0, whereas at H_0 < 0 sulfuric acid should be treated as a monobasic acid.

(3) The acidity function for solutions of D_2SO_4 has to be determined. Wyatt elucidated the effect of using D_2SO_4 instead of H_2SO_4 and showed that K_a for BH^+ in H_2SO_4–H_2O is about twice as great as that for BD^+ in D_2SO_4–D_2O [JCS(B)1570]. His values for D_0 against wt% D_2SO_4 at 25°C are slightly more negative than H_0 (up to 92 wt% acid), but most (if not all) of the difference can be accounted for by the difference in molecular weights. Hence this graph may be used to obtain values of H_0.

(4) The acidity function is temperature corrected. H_0 has been measured in the temperature range 25–90°C and from these data it was deduced that the variation of acidity with temperature for a given concentration of sulfuric acid is given by Eq. (2.3), where K is the proportionality coefficient for the variation of the acidity function with temperature. For

each acidity there is a particular K value, these being obtained from plots of H_0 values against the corresponding $1/T$ values for a given acidity; intermediate values of K can be obtained by interpolation from tabulated values (69JA6654).

$$H_0(T°C) = H_0(25°C) + K(298.15 - T)/298.15T \qquad (2.3)$$

(5) Corrections for salt effects. In general the kinetic salt effects on the exchange rates are rather larger than are the corresponding thermodynamic effects. However there appears to be considerable variation in this effect with substrate, and possibly with acidity, but not with temperature (75G539). Accordingly, no such correction is made, since the extrapolated log k_2^o should not be in error through kinetic salt effects by more than 0.3 log units. Thus, the effect of dissolved substrates on acidity functions can be serious (75JA760).

ii. *Determination of k_1 (stoich) T (°C) at pH = 0 (step 2).*
The determination of a rate coefficient at pH = 0 for a reaction which has been investigated only at higher acidities requires the construction and extrapolation of a rate profile. Acid-catalyzed hydrogen-exchange reactions give plots of log rate versus H_0, which may be approximated to straight lines. Extrapolation is then straightforward and introduces only relatively small errors; where alternative data are available, extrapolation is carried out independently on each data set.

iii. *Determination of k_1 (stoich) (100°C) at pH = 0 (step 3).*
Three methods are available for this, and that which is selected depends upon the data available.

(1) The activation parameters at a given acidity may be determined. Because of the variation of acidity with temperature, activation parameters determined using solutions of the same wt% of sulfuric acid at different temperatures give only "apparent values"—indeed the Arrhenius plots will be curved, as has been found for exchange in trifluoroacetic acid [78JCS(P2)751]. Therefore, to find true activation parameters referring to a definite H_0 value, the most satisfactory procedure is to construct two or more rate profiles at different temperatures, and then to use interpolated rates which refer to the same H_0 value. The derived thermodynamic parameters were given the notation $\Delta \overline{H}^{\ddagger}$ and $\Delta \overline{S}^{\ddagger}$ [75JCS(P2)1600], and were preferred to E_a and log A because of the sounder theoretical basis of the Eyring equation (51JA5628).

(2) A less accurate variation of the above requires just a rate profile and an individual point at a different temperature (i.e., for a single wt%

sulfuric acid). From the rate coefficient of the individual run, and the interpolated value on the rate profile corresponding to the same H_0 value, the activation parameters are calculated by the Eyring equation. If more than one rate coeffcient is available for a single wt% sulfuric acid, the following procedure is adopted: (a) the corrected H_0 values are calculated for the individual rates at the various temperatures, but constant wt% sulfuric acid; (b) an average H_0 (e.g., x is selected from those just determined; (c) by making the approximation that the rate profile slope does not change with temperature, the rate coefficients at the H_0 values of x at the various temperatures are calculated; (d) the activation parameters at the H_0 value of x are calculated using these rate coefficients.

(3) A standard enthalpy of activation has been used to simplify the procedure in many cases where the available data consist of a rate profile at one temperature, together with rate coefficients at various temperatures for a single wt% sulfuric acid (Fig. 2.12). The relationships between activation parameters at constant wt% acid and constant H_0 were deduced as Eqs. (2.4) and (2.5), where K is defined by Eq. (2.3); the differences in each set of thermodynamic parameters are of course small [75JCS(P2)1600; 77JHC893]. In Eqs. (2.4) and (2.5), and in the following equations of this section, m is the slope of log $([BH^+]/[B])/-H_0$ and depends on the acidity function followed (i.e., 1.0 for H_0, ~0.65 for H_A, ~1.9 for H_R, etc.). A standard temperature of 25°C was used.

$$\Delta \overline{H}^\ddagger - \Delta H^\ddagger = 2.303 RmK \qquad (2.4)$$
$$\Delta S^\ddagger - \Delta S^\ddagger = 2.303 RmK/n \qquad (2.5)$$

Provided that the rate versus acidity profile is not curved, a consequence of Eq. (2.4) is the nonvariance of $\Delta \overline{H}^\ddagger - \Delta H^\ddagger$ with acidity. Furthermore, if the rate versus acidity profiles at different temperatures are parallel, then $\Delta H^\ddagger$ will be acidity invariant, and so too will $\Delta \overline{H}^\ddagger$. This appeared to be the case for hydrogen exchange [73JCS(P2)1065] and also for nitration [75JCS(P2)1600] so that most of the variation in rate was

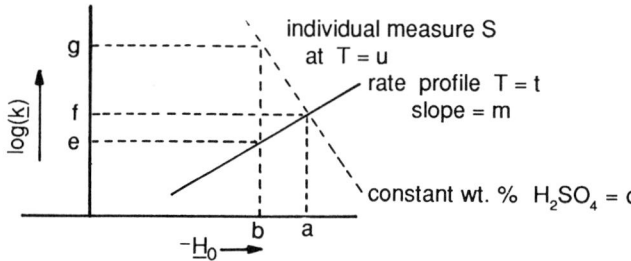

FIG. 2.12.

considered to derive from the entropy term. Consequently, for the hydrogen exchange data, an average value of 30 kcal mol^{-1} was used for $\Delta \overline{H}^{\ddagger}$ for all compounds.

However, other workers find that the variation in rate with acidity is smaller the higher the temperature, so that $\Delta \overline{H}^{\ddagger}$ is not constant (72TL2191; 73JA3918) and this is certainly true for detritiation in trifluoroacetic acid [78JCS(P2)751]. Moreover, a variation in $\Delta \overline{H}^{\ddagger}$ with acidity would be expected as a consequence of the reactivity–selectivity principle, and also by virtue of the fact that hydrogen exchange is not sterically hindered (i.e., $\Delta \overline{S}^{\ddagger}$, rather than $\Delta \overline{H}^{\ddagger}$ should be constant). The foregoing assumption may not therefore be strictly correct, but it has been shown that an error in $\Delta \overline{H}^{\ddagger}$ as high as 10 kcal mol^{-1} would lead to a relatively small error of 0.4 in log f (where f is the partial rate factor) when derived from an extrapolation of 40°C (72TL2191).

iv. *Determination of k_2° [100°C at pH = 0 (step 4)].* A number of further corrections may be required in this determination [Eq. (2.6)].

$$k_2^{\circ} = k_1 \text{ (stoich) [stoich]/[minor species]} \quad (2.6)$$

(1) Correction must be made for the minority species. For reactions that proceed on a majority species, k_1(stoich) = k_2°, but for reactions that proceed on a minority species (e.g., on a free base below the pK_a value) a correction given by Eq. (2.6) must be made. As is evident from Fig. 2.13, $k_2^{\circ} > k_1$(stoich) for a base with p$K_a > 0$, but $k_2^{\circ} < k_1$(stoich) for bases with p$K_a < 0$. As all bases in the pH region show Hammett behavior, the correction in the first case is addition of pK_a log units to k_1(stoich). In the second case, the correction is m·pK_a log units (the pK_a value is now negative).

(2) Correction must be made for the pK_a of the deuterated solvent. Bases are stronger in deuterium-containing acidic media than in the analogous protium-containing one. For primary amine protonation in the pH range, Δ pK_a is ~0.55 units, and for weaker bases, ~0.35 units (60JA15). A standard value of 0.4 was therefore used in all cases.

(3) The pK_a values used above must be corrected to 100°C. For the protonation of primary aromatic amines, Eqs. (2.7) and (2.8) apply (68AJC939; 70JA1567). Combining these equations gives Eq. (2.9), which can be used directly for correcting pK_a values.

$$R \ln K_a = -\Delta G_{25}^{\circ}/298.15 + \Delta H_{25}(T - 298.15)/T \quad (2.7)$$

$$pK_a(25°C) = 0.88 \Delta H_{25} - 1.97 \quad (2.8)$$

$$pK_a(100°C) = 0.83 \, pK_a(25°) - 0.33 \quad (2.9)$$

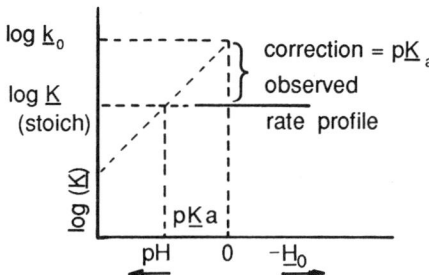

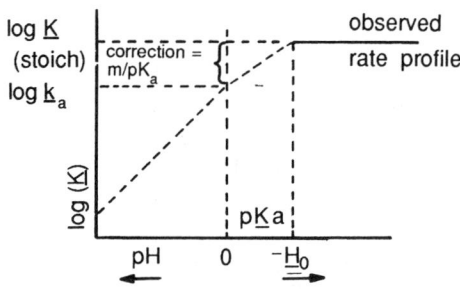

Fig. 2.13. (a) Base with $pK_a > 0$. (b) Base with $pK_a < 0$.

From relationships analogous to Eq. (2.8) for protonation at the pyridine nitrogen atom, and at any oxygen atom, Eqs. (2.10) and (2.11), respectively, may be derived. Slightly different equations are used if the pK_a value at 20°C is available [73JCS(P2)1065].

$$pK_a(100°C) = 0.82\ pK_a(25°C) + 0.09 \quad (2.10)$$
$$pK_a(100°C) = 0.76\ pK_a(25°C) + 0.01 \quad (2.11)$$

v. *Correction for Isotope Effects (step 5)*. Much of the hydrogen-exchange data for heterocycles have been obtained using deuteriation in D_2SO_4, whereas other rate data (and in particular those for benzene) have been measured using detritiation in H_2SO_4. For the direct comparison of rate coefficients it is necessary to consider the isotopic effects on the exchange rates [72MI2(199)]. Because the exchange pathway is not completely symmetrical, the value of the isotope effect varies according to the reactivity of the substrate (since the position and hence the symmetry of the transition state will vary in consequence) and likewise according to the strength of the exchanging acid. For the only case where the rate coefficients for deuteriation and protiodetritiation have been compared directly (for 1,3,5-trimethoxybenzene in aqueous perchloric acid at 24.6°C), k(T-H)/k(H-D) was 0.45 (62JA3976; 67JA4411). Evidence from protiodedeuteriation versus protiodetritiation indicates that the values will be larger the less reactive the aromatic substrate [72MI2(199)] and in the present work a value of 0.55 was used.

d. *Reliability of* k_2° *Values*

It is appropriate to estimate the error involved in such a complex procedure. The simplest empirical approach is to compare standard rate coefficients calculated for the same compound from different sets of data (at different temperatures or from different areas of the rate profiles). Most of the variation lies within 0.3 log units from the mean value, and it has therefore been assumed that the maximum error should be 0.35 log units [78JCS(P2)861], this being subject to the qualification noted above regarding the assumption of a constant $\Delta \overline{H}^{\ddagger}$ value.

Support for the reliability of the standard rate coefficients so defined is given by the results obtained by applying the standard procedures to the few data for exchange (of benzene derivatives) in mineral acids other than sulfuric acid: The extrapolated standard rate coefficients agree within a factor of 5 or better [73JCS(P2)1077].

e. *Alternative Standard Conditions*

An alternative standard procedure has been developed, and this is aimed at producing standard rate coefficients for hydrogen exchange, under the same conditions as those selected for nitration ($H_0 = -6.6$, $T = 25°C$). These alternative standard rate coefficients, defined as $k_2^{\circ'}$, are calculated as follows [78JCS(P2)613].

For substrates reacting as conjugate acids it is sufficient to extrapolate the linear arm of the rate profile to $H_0 = -6.6$, and subtract from log k_2° the value of 4.42, which is the logarithm of the rate difference between 100 and 25°C for a reaction of $\Delta \overline{H}^{\ddagger} = 30$ kcal mol^{-1}. (An error of ± 5 kcal mol^{-1} in this will produce a maximum error of ± 0.8 in this correction factor.)

For reactions on the free base, the calculation is not as simple, since the pK_a value should be taken at 25°C, and therefore cannot be directly derived from log k_2°, which itself was derived using pK_a values at 100°C. The procedure suggested required the following steps, starting from the value of log k_1 (stoich) (100°C) at pH = 0.

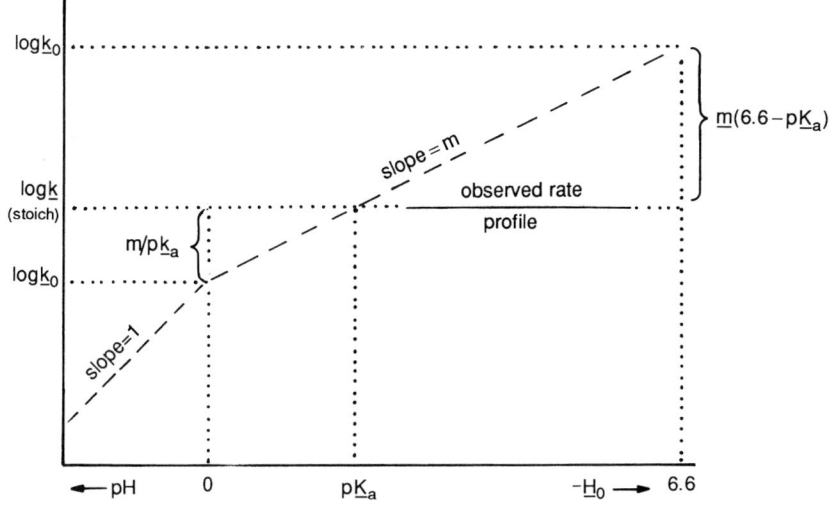

FIG. 2.14. Minority species hydrogen exchange rate correction for $pK_a < 0$.

(1) Acidity extrapolation [Eq. (2.12)].

$$\log k_1 \text{ (stoich)} + 6.6 \, d(\log k)/d(-H_0) = A \quad (2.12)$$

(2) Temperature extrapolation [Eq. (2.13)].

$$A - 4.42 = B \quad (2.13)$$

(3) Minority species correction, using Eq. (2.14) for $pK_a < 0$ and Eq. (2.15) for $pK_a > 0$.

$$\log k_2^{\circ\prime} = B + (pK_a + 6.6)m \quad (2.14)$$
$$\log k_2^{\circ\prime} = B + pK_a + 6.6m \quad (2.15)$$

Figure 2.14 illustrates the minority species correction for the case where $pK_a < 0$.

2. Base-Catalyzed Exchange

A. Mechanism

Strong bases will remove sufficiently activated protons from aromatics in a rate-determining step [Eq. (2.16)]. This is followed by rapid abstraction of a proton from the solvent [Eq. (2.17)]. The reaction pathway is symmetrical,

$$\text{ArH}^* + \text{MB} \underset{\text{fast}}{\overset{\text{slow}}{\rightleftharpoons}} \text{Ar}^-\text{M}^+ + \text{BH}^* \qquad (2.16)$$

$$\text{Ar}^-\text{M}^+ + \text{BH} \rightleftharpoons \text{ArH} + \text{MB} \qquad (2.17)$$

except for isotopic differences, so that the slow and fast steps become, respectively, fast and slow in the reverse direction. The intermediate carbanion, zwitterion, or ylid has the electron pair in a σ orbital and these cannot therefore be directly involved in conjugation in the ring. It is therefore stabilized largely by inductive electron withdrawal, so that exchange is particularly rapid adjacent to −I groups, such as positive poles.

B. Exchange Conditions

Studies of base-catalyzed exchange (which have been fewer than for acid-catalyzed exchange) have involved deuteriation, protiodedeuteriation, and protiodetritiation. Reagents have included potassamide in liquid ammonia, potassium *t*-butoxide in *t*-butanol, lithium and cesium cyclohexylamide in cyclohexylamine, and sodium methoxide in methanol [72MI2(266)]. Dedeuteriation is ~1.5–2.5 times faster than detritiation, and the methods for following these reactions are essentially those used in the acid-catalyzed exchange. In some cases, the reaction conditions are more rigorous (e.g., air must be excluded to prevent side reactions).

C. Steric Effects

Since the base must attack the ring hydrogens in the plane of the ring, but away from it, one would expect steric effects to be minimal, and certainly no more than for acid-catalyzed exchange. This appears to be the case for attack by amide, but the bulkier cyclohexylamide shows evidence of some steric hindrance [72MI2(266)].

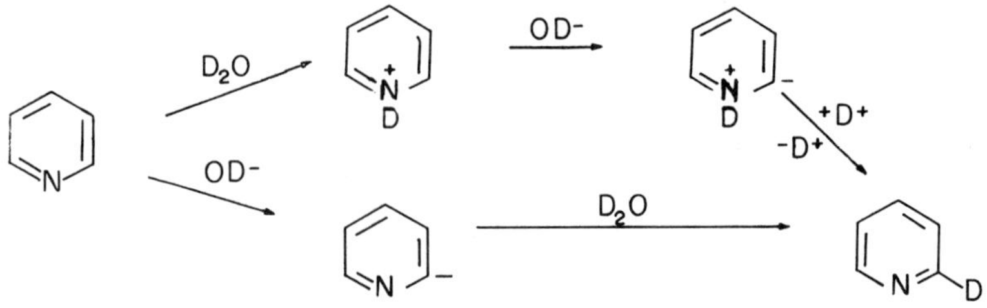

SCHEME 2.6. Base-catalyzed exchange mechanism for pyridine involving free base and conjugate acid.

D. Hydrogen Exchange in Heteroaromatics

The spread of rates in base-catalyzed exchange is smaller than under acid-catalyzed conditions, and this makes direct comparison of reactions easier, so that extrapolative techniques have not been necessary.

Examples have already been noted in Section 1, where the mechanism for exchange of heterocycles changes from acid-catalyzed to base-catalyzed as the medium becomes neutral, then basic. There is also evidence that in some near neutral or weakly alkaline media two mechanisms can take place for neutral π-deficient heterocycles (e.g., for pyridine, Scheme 2.6) (67JA3358). The mechanism involving the conjugate acid (likely to be present in very low concentrations) is favored by the very strong electron withdrawal by the positive pole. The evidence for the dual mechanisms comes from a change in the relative exchange rates at the 2-, 3-, and 4- positions with changing basicity. A similar mechanism is believed to account for exchange at the 2-position of quinoline [71JCS(B)4].

CHAPTER 3

Nitration

Nitration, the most widely studied electrophilic aromatic substitution, has been reviewed many times [59MI; 65MI1(61); 69MI1; 71ACR240, 71ACR248, 71MI1; 72MI2(10); 80MI1]. Only a brief overview is therefore presented here, in order to put heteroaromatic nitration in the context of current knowledge.

1. Nitration Conditions

Nitration of many substrates by nitric acid alone is very slow, and is also largely heterogeneous. The reaction rate can be greatly increased by the addition of sulfuric acid, which increases the concentration of nitronium ion, the nitrating species, and also removes water, the reaction by-product. Solutions of nitrates in sulfuric acid may also be used. Nitrous acid present in nitric acid can cause rate acceleration as a result of nitration via nitrosation; often this can be suppressed by the addition of urea, but in some cases more effective nitrous traps (e.g., sulfanilic acid) are necessary. Nitric acid (or nitric acid/sulfuric acid) in organic solvents increases homogeneity, which reduces the tendency for dinitration (caused by the nitro products being often more soluble in nitric acid than are the starting aromatics). The lowering of medium polarity, and dilution of the nitrating reagent, can also give less reactive nitrating conditions, which is important for some reactive heterocycles (e.g., thiophene). Acetic acid and acetic anhydride are the most widely used cosolvents, and nitromethane, sulfolane, acetonitrile, and carbon tetrachloride have also been employed. Acetic anhydride as a solvent for preformed acyl nitrates is a popular nitrating mixture. Acetic acid has also been used as a solvent for cupric nitrate. Nitronium salts in organic solvents have been used for aromatics (e.g., aroyl halides) that are readily hydrolyzed, but cost precludes their general use. Lastly, alkyl nitrates in sodium ethoxide fulfill the requirement for very mild nitrating conditions required, for example, for heteroaromatic conjugate bases.

2. Mechanism

A. THE NITRATING SPECIES

Until the 1960s, the nitration of aromatic compounds by solutions of nitric acid in sulfuric acid or other mineral acids, as well as in organic solvents, was confidently discussed in terms of the attack by the nitronium ion, NO_2^+. Support for this hypothesis came from spectroscopy, cryoscopic measurements, and the comparison between the rate of nitration and the rate of ^{18}O exchange between nitric acid and the media. Accordingly the generally accepted mechanism of aromatic nitration involved the steps outlined in Equations (3.1)–(3.4). Depending on the conditions and the aromatic substrate, either of

$$HNO_3 + H^+ \rightleftharpoons H_2NO_3^+ \text{ (fast)} \quad (3.1)$$
$$H_2NO_3^+ \rightleftharpoons NO_2^+ + H_2O \quad (3.2)$$
$$NO_2^+ + ArH \rightleftharpoons [ArHNO_2]^+ \quad (3.3)$$
$$[ArHNO_2]^+ \rightleftharpoons ArNO_2 + H^+ \quad (3.4)$$

Eqs. (3.2), (3.3), or even (3.4) may be rate-determining. Obviously the use of nitration rates in discussing relative reactivities is of little value unless the same step in Equations (3.1)–(3.4) is rate-determining for each substrate. If Eq. (3.2) is rate-determining (and this is frequently the case for nitration in organic solvents where heterolysis is slow because of the relatively low polarity of the medium) then nitration is zero*th* order for fairly reactive aromatics. However, for unreactive aromatics, the mechanism changes over to the most common condition for nitration: Namely, Eq. (3.3) is rate-determining, so that the reaction is first-order in both nitronium ion (or its precursor) and aromatic. The last step, Eq. (3.4), is fast except in the case of nitration of some substituted 1,3,5-tri-*t*-butylbenzenes (but not the parent compound), where the exceptional steric hindrance causes the reverse step of Eq. (3.3) to become fast relative to the forward step of Eq. (3.4) (60ACS219; 66JA1569; 68JA2105).

B. ENCOUNTER CONTROL

A more recently recognized kinetic complication is that, for all aromatics more reactive than toluene, Eq. (3.3) is encounter limited, so that all these substrates react at the same rate [68JCS(B)800]; in 68.3 wt% H_2SO_4 this is ~40 times that of benzene (67CC352). A problem here is that although there is no intermolecular selectivity under these conditions, the intramolecular selectivity typical of electrophilic substitutions remains (71ACR240). This necessitates the postulation of the intermediacy of a species of unidentified structure (possibly a π-complex) [71ACR240; 72JA7448; 75JCS(P2)648] prior to the formation of the Wheland intermediate. Encounter control may be responsible for the variation of isomer distribution with medium acidity [71JCS(B)2443; 77JCS(P2)1693], and it is an important factor to be considered when measuring the reactivity of π-excessive heterocycles, but not for most π-deficient heterocycles, which are less reactive than benzene unless they contain strongly electron-donating substituents (e.g., for 2,4,6-trimethoxypyridine).

C. Solvent Effects

Another complication has been the observation for a number of aromatic substrates of rate ratios widely differing according to the nature of the solvent. This has led to the belief that the nitronium ion might not be the electrophile in some media, and also that π-complexes may be involved in the rate-determining step (62JA3687; 65AJC1377). It is now recognized that these anomalous data arose from the high rates of nitration in certain media, such that in competitive experiments, mixing control was taking place. That is, inadequate speed of mixing produced attenuated relative substrate (but not positional) nitration rates [70JCS(B)797].

Nitration by nitric acid in acetic anhydride presents problems of a different kind since it gives markedly different ortho/para isomer ratios for nitration of some substrates including heterocyclic ones from those found for other nitrations of the same substrates. The relative nitration rates depend upon the length of time and the temperature at which the nitric acid and acetic anhydride mixture is allowed to stand before being used for nitration, and these in turn depend upon the nitric acid concentration; this is due to removal of water from the nitric acid through acid-catalyzed hydrolysis of acetic anhydride, and the formation of acetyl nitrate [66JCS(B)727]. The nitration rate also depends upon the purity of the acetic anhydride [77JCS(P2)1361], probably due to the presence of traces of nitrate ion, which in turn implies nitronium ion (or at least a cation) as the electrophile (58JA5329). Since acetoxylation accompanies nitration in this medium (and nitration is strongly catalyzed by added sulfuric acid, 64JCS3691) protonated acetyl nitrate was thought to be the electrophilic species responsible for both reactions, a view supported by calculations (69T5777). However, acetoxylation is now known to be produced by addition-elimination involving acetyl nitrate (70CC641, 70TL2793) so the nature of the electrophile remains in doubt; it is possible that protonated acetyl nitrate (which is also nitronium ion solvated by acetic acid) is the precursor for the nitronium ion. The current evidence seems irreconcilable with, on the one hand, both nitric acid–sulfuric acid and nitric acid–acetic anhydride giving fairly similar toluene/benzene rate ratios [71JCS(B)1256], and similar ortho/para ratios for a range of alkylbenzenes (72RTC831); whereas against this argument for a common electrophile must be set the fact that substrates like anilides, ethers, and biphenyl give abnormally high ortho/para ratios with the latter medium. Meaningful analysis is further complicated by reports that biphenyl also gives abnormally high ratios in nitric acid–sulfuric acid (76TL771), and the ratios are substantially influenced by both nitrosation and the heterogeneity of the reaction mixture (72TL1755).

D. Electron-Transfer Mechanism

This mechanism requires the electron pair to be transferred from the aromatic to the electrophile in a stepwise fashion, so that a radical cation (formed from the aromatic) occurs along the reaction pathway. Some evidence for this was provided by an apparent correlation of logarithms of relative nitration rates with ionization potentials of the aromatics (73T579), and radical cations can be shown to be present in solutions of phenols under conditions in which nitration occurs (75JOU1883). Perrin has argued that a one-electron transfer mechanism applies to nitrations which occur at the encounter rate, since electron transfers from reactive aromatics to nitronium ion are exothermic, and exothermic electron transfer should be encounter controlled (77JA5516). However, it has been shown that such conclusions cannot be generally true, and that it is false to assume that the retention of intramolecular selectivity under conditions of encounter control is inexplicable by the normal nitration mechanism [80MI1(109)]. More recently, chemically induced dynamic nuclear polarization (CIDNP) measurements have indicated the presence of radical cations in nitrous acid-catalyzed nitration of phenols, amines, and mesitylene [84JCS(P2)1659,1667], and in the rearrangement of the initially formed *ipso* Wheland intermediate in nitration of durene to give 3-nitrodurene in trifluoroacetic acid [85JCS(P2)1227].

E. Nitration of Bases

As in hydrogen exchange of bases, complications arise here since reaction may take place either on the free base or the conjugate acid; in general, bases containing electron-withdrawing substituents will tend to react as the former, and vice versa. The analysis of rate versus acidity profiles, however, is not always straightforward, because the important effect of hydrogen bonding must be considered. In addition, there is now firm evidence that amino-substituted aromatics containing electron-withdrawing substituents can undergo initial rate-determining attack by nitronium ion upon the amino group. Subsequent fast rearrangement leads to the C-nitro products, and this overall mechanism is probably the normal one for anilines which nitrate via the neutral molecule (72CC641).

F. *Ipso* Attack

The mechanism of nitration is complicated by the importance of *ipso* attack (i.e., attack of the reagent at the position occupied by the substituent). (*Ipso* attack is well known in other electrophilic substitutions.) The possible fates of the *ipso* Wheland intermediate (**3.1**) are: (1) capture by a nucleophile; (2) rearrangement by 1,2- or 1,3-migration of the nitro group, with loss of a proton; (3) similar migration of the group X; (4) loss of X^+ (i.e., *ipso* substitution); (5) loss of a proton or related group from a substituent remote from the *ipso* position; (6) return to starting materials.

Capture by a nucleophile generally leads to addition products (e.g., *ipso*-nitrocyclohexadienyl acetates obtained in Ac_2O) (73CC300; 74CJC3960). Rearrangement by 1,2-migration may be repeated giving rise to meta products, and there are a number of examples of this formal 1,3-rearrangement (e.g., 77CC301). 1,2-Migration of X rarely occurs (72BCJ2534), but *ipso* substitution has been the longest recognized consequence of *ipso* attack [80MI1(198)]. Three types of loss from remote positions are common: deprotonation of hydroxyl, demethylation of methoxyl (both with formation of the ketone), and deprotonation of methyl (76ACR287). The rules governing the competition for the various processes undergone by the Wheland intermediate are not yet well understood. The importance in recognizing the possibility of *ipso* attack is that since there are two routes to the production of a given isomer, then it is not possible to translate isomer distributions into a direct measure of positional reactivity and selectivity. *Ipso* attack takes place at the position of highest electron density in an aromatic (75MI3), and applies mainly therefore to homocyclic substrates containing at least one substituent that is electron supplying; only a few substituted heteroaromatic compounds may undergo *ipso* attack.

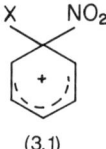

(3.1)

3. Experimental Techniques

A. The UV Technique

Nitration kinetics are usually followed by measuring the absorption of the nitro product by UV spectroscopy [67JCS(B)1204; 68JCS(B)800; 76JCS(P2)1135]. The wavelength chosen for measuring the optical density is usually close to $\lambda_{\max}$ for the nitro product.

The method of carrying out kinetic runs depends upon the rate of nitration and the temperature. For relatively slow reactions at or near room temperature, the easiest procedure is to carry out the reaction in a volumetric flask to which are added a weighed amount of substrate (~40 mg), sulfuric acid of known composition nearly to the mark, and, finally, a measured excess (~0.5 g) of ~70% nitric acid. Urea (one-tenth of the molar concentration of nitric acid) must also be added to prevent nitrosation. The flask is then placed in a thermostatted bath and, at appropriate intervals, portions are removed, quenched, and the optical density measured. For kinetic runs above 50°C the procedure differs only in that aliquots are sealed in Pyrex tubes before thermostatting.

For compounds that nitrate rapidly at room temperature, two methods may be used. The substrate may be dissolved in sulfuric acid of appropriate strength and introduced into one limb of a two-limbed flask. A solution of nitric acid in sulfuric acid is introduced into the other limb, the flask is thermostatted, and the reaction started by vigorously shaking the flask. Aliquots are then withdrawn and quenched as before. However, by far the most convenient method is to carry out the kinetics directly in the thermostatted spectrophotometric cell and this would now be the method of choice in view of the automatic monitoring facilities currently available on spectrophotometers. For very fast reactions, stopped-flow techniques have been used.

B. Calculation of Kinetics

The concentration x_t of the nitrated product formed after time t is obtained from Eq. (3.5). Pseudo-first-order rate coefficients are found from Eq. (3.6) [by plotting log $(D_\infty - D_t)$ against t] and second-order rate coefficients are calculated from Eq. (3.7), where b is the concentration of nitric acid, a is the initial concentration of the substrate, and D_0, D_t, and D_∞ are, respectively, the optical densities initially, after time t, and after complete reaction. (In some cases there is no initial absorption at $\lambda_{\max}$ so D_0 is zero.) Second-order rate coefficients, which are obtained from the

$$x_t = a(D_t - D_0)/(D_\infty - D_0) \tag{3.5}$$

$$k_1 = \frac{2.303}{t} \log \frac{(D_\infty - D_t)}{D_\infty - D_0} \tag{3.6}$$

$$k_2 = \frac{2.303}{(b-a)t} \log \frac{a(b - x_t)}{b(a - x_t)} \tag{3.7}$$

pseudo-first-order rate coefficients through division by the stoichiometric concentration of nitric acid, should be corrected for variations in densities of the measured solutions if the reaction is not run at room temperature.

C. KINETIC COMPLICATIONS

In most studies the reaction mixtures were quenched by dilution with water. When dealing with heteroaromatic substrates care must be taken to ensure that only one species (either free base or conjugate acid) of both reactant and product is present in the resulting acid solution. In some cases, on diluting with water the pH of the resultant solution would fall close to the pK_a values for the substrate and/or its nitro derivative. In these cases, to avoid measuring absorptions arising from both the protonated and nonprotonated forms, the reaction mixtures are quenched with sodium hydroxide solution, so that the resulting pH is far from the pK_a value; in the reference cell a solution of sodium hydroxide of comparable concentration should be used.

Because of the limited solubilities of nonbasic aromatics in aqueous acids, the effect of acetic acid as a cosolvent has also been investigated (77JOC2511). Addition of $\sim 10^{-3}$ M AcOH has been shown to increase appreciably the solubility of aromatic substrates without affecting the rate coefficients.

When kinetics have been complicated by partial decomposition of the reaction product (e.g., 3,5-dimethoxy-2,6-dinitropyridine), the rate coefficients for the second nitration could be obtained by taking measurements at the isosbestic point of the mono- and dinitro-compounds, and also at another wavelength if there is a large extinction coefficient difference between the two compounds [67JCS(B)1211]. This method assumes that the decomposition product does not absorb at the wavelength concerned. The concentration x_t of product formed after time t was calculated from Eq. (3.8), where the subscript i refers to optical densities at the isosbestic point.

$$x_t = \frac{a}{D_\infty}(D_t - D_0)\frac{D_{0i}(D_t - D_0) + D_\infty(D_{0i} - D_{ti})}{D_{0i}D_t - D_{ti}D_0} \tag{3.8}$$

For one compound (3,5-dimethoxy-2-nitropyridine 1-oxide), the UV method could not be used since the dinitro product precipitated from solution even at the concentrations used in UV spectroscopy. The kinetics were therefore followed by NMR [67JCS(B)1213].

A further complication occurs if two isomeric nitro products are formed, as from, for example, 2-pyridone [72JCS(P2)1953] and 1-methyl-2-methylaminopyridinium perchlorate [72JCS(P2)1950]. In such cases, the rate coefficients for the individual positions as well as the proportions of the two isomeric products can be determined by measuring the optical densities at two distinct and carefully selected wavelengths.

4. Criteria for Defining the Reacting Species

A. SURVEY OF POSSIBLE CRITERIA

Several criteria have been employed to determine the reactive species in the nitration of N-heteroaromatics. The problem is to decide the following. (1) Which of the one or more species present is the one undergoing predominant reaction in a given medium. (2) What is the dependence of the rate of nitration of each of these species upon both acidity and temperature. (3) What measure of acidity (if any is needed) is appropriate for correlating with nitration rates; this is an important factor if data are to be extrapolated to a "standard" condition. (4) How does the activity of the nitronium ion vary with the acidity of the medium. This is particularly important at high acidity where a decrease in nitration rate with increasing acidity may be due to reaction taking place predominantly on the conjugate acid or to a decrease in nitronium ion activity, or both. The problem is a difficult one and has been reviewed in considerable detail [71MI1; 80MI1(147)] so that a condensed overview is given here.

The main criteria which have been used are listed below and then discussed in detail.

(1) Analysis of rate versus acidity profiles. This is a long established and reliable technique and has been used previously in aromatic chemistry to determine, for example, which forms of phenols and carboxylic acids are the reacting species in electrophilic substitution.

(2) Comparison of the rate of reaction of the compound under investigation with that of a fixed derivative, in which the possibility of ionization or tautomerism is wholly or partially eliminated.

(3) Comparison of the observed and calculated encounter rates.

(4) Determination of the thermodynamic parameters.

Various possible *rate profile* approaches have been employed. The *high-acidity rate profiles* refer to reactions studied in 85–98 wt% sulfuric acid, and for these the sulfuric acid concentration, and subsequently H_0, were used as the measure of acidity [63JCS4204; 67JCS(B)1204]. The *low-acidity rate profiles* refer to studies carried out in 40–85 wt% sulfuric acid, the acidity being measured in terms of the function (H_R + log a_{H_2O})(64MI2). The *modified rate profiles* refer to the same acidity range, expressed arbitrarily as H_0, but the rate coefficients are corrected for the nitronium ion concentration [68JCS(B)1477]. A systematic survey of rate–acidity profiles using H_0 as the measure of acidity has shown this approach to be of wide generality [75JCS(P2)1600]. Finally, a method has been introduced to cover the acid range 40–98 wt% H_2SO_4, whereby nitration rates are found to give excellent correlations with the activity coefficient ratios M_c of the solutes [84JCS(P2)1163].

Use of the H_0 acidity function in rate versus acidity profiles is purely arbitrary and is not meant to imply that the concentration of either the substrate or nitronium ion is a function of it. It is the most suitable acidity function, having been the most studied, with values available over a range of temperatures (69JA6654).

B. High-Acidity Rate Profiles

The high-acidity rate profiles exhibit a maximum at ~88 wt% H_2SO_4 (H_0 = −9 at 25°C, but lower at higher temperatures) wherein falls the pK_a of nitric acid. The slopes obtained for nitration proceeding via either the free base or the conjugate acid form of the substrate are quite different (Fig. 3.1). Since variations of the rate coefficient depend upon the concomitant variation of the acidities of the reactants (substrate and nitronium ion), a quite different slope of the rate profile is expected in the two cases. For reaction of a conjugate acid, where initially the stoichiometric concentration of the substrate is present throughout the acidity range, the observed rate increase (with increasing acidity) up to the maximum is due to the increasing amount of nitronium ion. Thereafter, it should level off, and eventually slightly decrease, as does nitronium ion activity (Fig. 3.2). For free-base reactions, although the variation in nitronium ion activity with acidity is the same, the concentration of free base decreases due to increased protonation. Consequently, the slope of the rate profile up to the rate maximum is less positive than that for the conjugate acid, and becomes large and negative at higher acidity; in this latter region the average value of the slope is −0.1 for conjugate acids and −0.5 for free bases (Fig. 3.2).

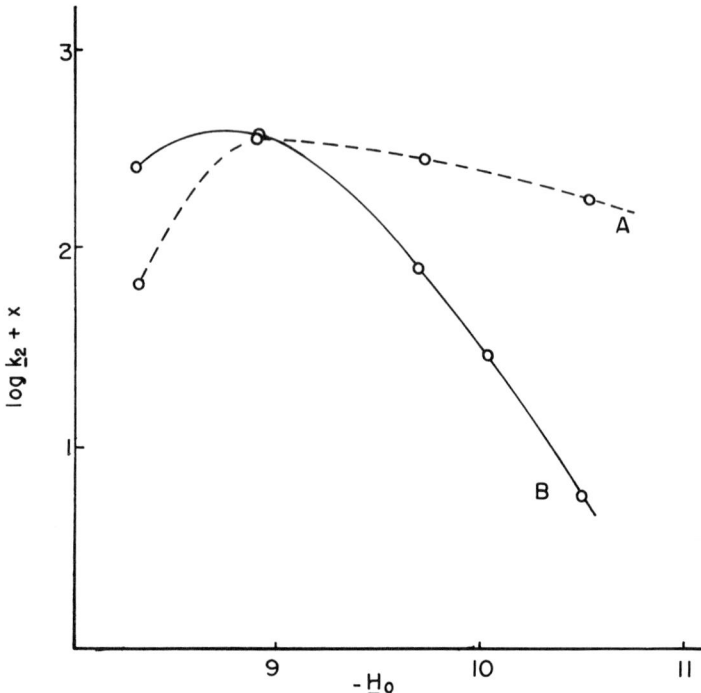

FIG. 3.1. Form of plots of log k_2(obs.) versus H_0 for (A) majority species and (B) free base minority species.

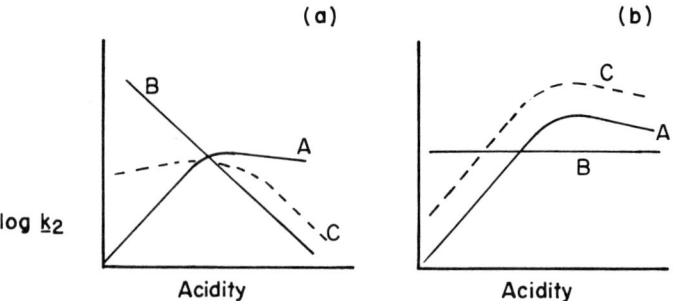

FIG. 3.2. (a) Relation of (A) majority and (B) free base minority species nitration rates to (C) the decreasing amount of free base. (b) As before but with the minority species rate corrected to constant concentration.

A further analysis of rate profiles obtained for free base nitrations is possible. If allowance is made for the decrease in the concentration of the reactive species, the resulting rate profile should have a slope similar to that for a conjugate acid. The *free base rate coefficient* $k_2(\text{fb})$ is defined by Eq. (3.9) and can be calculated by Eq. (3.10), in which m is the slope of the substrate protonation correlation.

$$k_2(\text{obs.})[\text{substrate}] = k_2(\text{fb})[\text{free base}] \tag{3.9}$$

$$\log k_2(\text{fb}) = \log k_2(\text{obs.}) + pK_a(\text{obs.}) + pK_a - mH_0 \tag{3.10}$$

C. Moodie–Schofield Plots

For compounds which undergo nitration at lower acidity, another useful criterion was proposed by Moodie and Schofield (64MI2). This again depends upon the difference between a free base and a conjugate acid in its concentration dependence with acidity. Westheimer and Kharasch (46JA1871) had shown that the degree of protonation of nitric acid was correlated with the degree of protonation of triarylmethanols (which describes the H_R acidity function). Accordingly, Deno and Stein (56JA578) found a linear relationship between the rate of nitration of several aromatic compounds and the acidity function H_R in the region 40–60 wt% H_2SO_4, although deviations were observed at higher acidity. Moodie and Schofield noted that nitric acid is monohydrated in the range 60–85 wt%, and therefore good straight lines of slopes close to unity (for ~15 wt% acidity range) could be obtained for the nitration of a large number of substrates in the overall acidity range 40–85 wt% H_2SO_4, according to Eq. (3.11) (64MI2). It is now recognized that the assumption that hydrated nitric acid is involved is unnecessary [80MI1(36,148)].

$$\log k_2(\text{obs.}) = -m(H_R + \log a_{H_2O}) \tag{3.11}$$

For nonbasic aromatics, and for heteroaromatics undergoing nitration as conjugate acids, a unit slope is therefore expected, whereas free-base minority species should give smaller slopes convertible to unity if calculated free-base rate coefficients are used (Fig. 3.3). A survey showed that compounds undergoing nitration via the majority species had an average slope of 1.00 [80MI1(151)], whereas those nitrated via the minority species had slopes in the region 0.3–0.9 [80MI1(153)].

Although the theoretical values for the slopes are often obtained from kinetic data at room temperature, this is not the case (and the recalculation of the free-base rate coefficients also fails) for data obtained at higher temperature. This discrepancy may be attributed to the variation of acidity function with temperature. In a redetermination of the variation of H_R

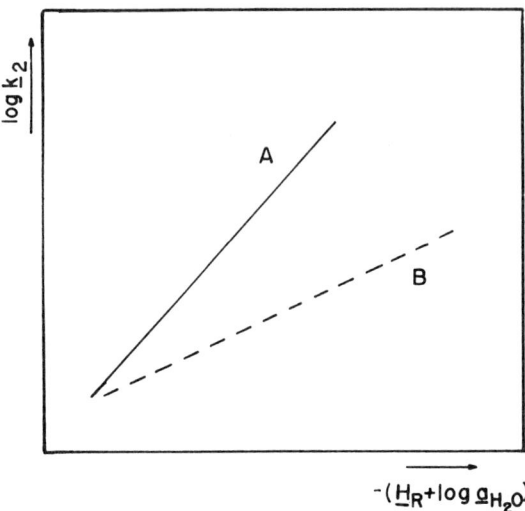

FIG. 3.3. Plots of log K_2(obs.) versus (H_R + log a_{H_2O}) for A, conjugate acids and B, free base minority species.

with wt% H_2SO_4, the temperature dependence [Eq. (3.12)] was shown to be similar to that for H_0, though the proportionality coefficient A is always smaller at each concentration (75JA760). Vapor-pressure measurements have provided values for a_{H_2O} (39JA2370; 61JA4956), but the temperature variation of these is known only for <40 wt% H_2SO_4 (35JA27). In this range, the dependence of log a_{H_2O} upon temperature is given by Eq. (3.13) [75JCS(P2)1600], where T is the absolute temperature (in K).

$$H_R(T) = A/T + B \qquad (3.12)$$
$$\log a_{H_2O}(T°) = \log a_{H_2O}(298.15) + (T - 298.15)B \qquad (3.13)$$

In Eq. (3.13), B is a constant for a given wt% H_2SO_4; a plot of B against H_0 is a smooth curve, and a third-degree polynomial gives log B as a function of H_0. However, this implies that B reaches a maximum at $H_0 = -2.8$ (45 wt% H_2SO_4 at 25°C) and thereafter decreases with increasing H_0. If this is correct, then at high acidity a_{H_2O} would decrease with increasing temperature, in contrast to its behavior at low acidity, which seems unlikely. Moodie–Schofield plots cannot therefore be accurately corrected for temperature at present.

For benzene, the slope of the Moodie–Schofield plot is 1.0 in the (narrow) range 63–68 wt% H_2SO_4, but the slope increases to 1.2 at higher acidity [68JCS(B)800]. This may diminish the utility of a comparison of the rate of nitration of a given substrate with that of benzene as a means of studying aromatic reactivity, for the comparison would be dependent on the acidity (as is the case for hydrogen exchange). The curvature cannot be due to the rate of nitration at the higher acidities studied being only six times less than the encounter rate [75JCS(P2)1600], because any encounter control of the rate would diminish the slope of the rate–acidity plot, and calculations have more recently confirmed this [77JCS(P2)1693]. Rate profiles for compounds more reactive than benzene, such as toluene, biphenyl, xylenes, mesitylene, and naphthalene, are linear (over small ranges of acidity for each compound) in the overall range 56–80 wt% H_2SO_4. By contrast, an investigation using a wide acidity range [77JCS(P2)845; 77JOC2511] confirmed that the profiles for benzene and the halogenobenzenes are curved (as are also the corresponding plots against H_R). Plots against H_0 were linear (see also Section 5) with slopes of ~2.4 (converging toward higher acidity). In this work it was shown that these correlations must be empirical and have limited value as mechanistic criteria for nitration. Mechanistically, it is not unreasonable for rate versus acidity plots to converge toward higher acidity (as they do in hydrogen exchange), since the higher activity of the nitronium ion in the more concentrated acids should lead to a smaller spread of rates, according to the reactivity-selectivity principle. Certainly the precise unit slopes of some rate–acidity profiles are now recognized as rather fortuitous in view of the different protonation behavior of various bases in a given acid.

D. MODIFIED RATE PROFILES

Katritzky and Tarhan introduced another mechanistic criterion [68JCS(B)1477] that refers to the same acidity region discussed in Section 3, but permits observation of a mechanistic changeover. The so-called modified rate profile consists of correcting the observed rate coefficients for the acidity dependence of the nitronium ion concentration, according to Eqs. (3.14) and (3.15). A new rate coefficient k_2^* is defined in which $[HNO_3]_{st}$ refers to the stoichiometric quantity of nitric acid in the reaction mixture. Values of $\log[HNO_3]_{st}/[NO_2^+]$ are computed assuming that nitric acid is half-protonated in 88 wt% H_2SO_4 and that its protonation follows ($-H_R = \log a_{H_2O}$).

$$k_2(\text{obs.})[HNO_3]_{st}[ArH] = k_2^*[NO_2^+][ArH] \tag{3.14}$$

Hence:

$$\log k_2^* = \log k_3(\text{obs.}) + \log[HNO_3]_{st}/[NO_2^+] \tag{3.15}$$

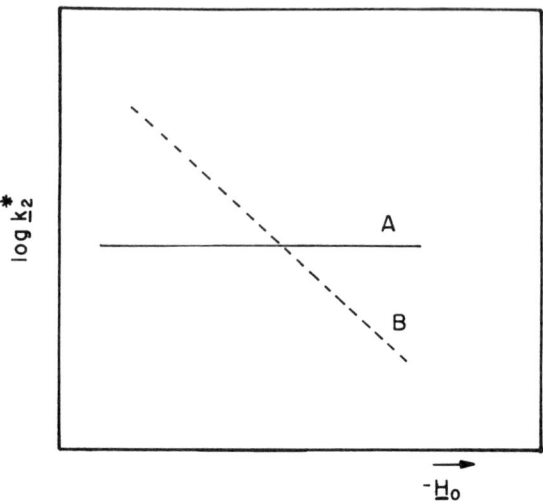

FIG. 3.4. Plots of log k_2^* versus $-H_0$ for reactions of (A) conjugate acids and (B) free base minority species.

Thus plots of log k_2^* versus $-H_0$ for reaction of conjugate acids should yield a line of zero slope, whereas compounds reacting via the free base should give straight lines with a unit negative slope (Fig. 3.4). Other slopes may be obtained, and this reflects the proportionality between the acidity function followed and $-H_0$. The method was first applied to the kinetic data for 4-pyridone, which was thereby shown to react as the free base at lower acidities and as the conjugate acid at higher ones (Fig. 3.5).

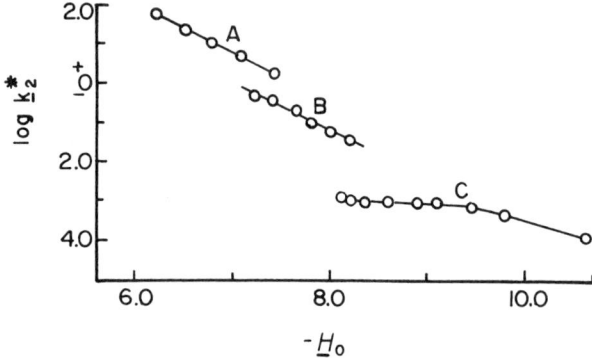

FIG. 3.5. Plots of log k_2 versus H_0 for 4-pyridone at (A) 157.5°C, (B) 133.5°C, and (C) 86°C.

The method can be applied only at or near room temperature because although the values of H_0 up to 90°C are well known, the variation of the pK_a of nitric acid with temperature, and indeed the precise acidity function followed by nitric acid, are not known; attempts to determine these factors have failed [75JCS(P2)1600].

E. OTHER TYPES OF RATE PROFILES

An alternative simple criterion makes use of H_0 values. Statistical analysis of 12 data sets for linear plots of log k_2(obs.) against $-H_0$ (<86 wt% H_2SO_4) showed correlation coefficients slightly better than those for the corresponding plots against $-H_R$ or against $(-H_R + \log a_{H_2O})$ [75JCS(P2)1600]. Hence the slope of the $d[\log k_2(\text{obs.})]/d(-H_0)$ plots was proposed as an alternative mechanistic criterion. With only three exceptions, among 131 data sets, reactions of majority species exhibit slopes greater than 1.7 (mean value 2.24 ± 0.27), whereas slopes for all minority species reactions are less than 1.7 (mean value 1.23 ± 0.24). Compounds undergoing reaction with rate profile slopes close to 1.7 have to be investigated using other criteria.

The most recent procedure, not yet widely applied, provides a linear description of rate profiles for substrates reacting as free bases, throughout the whole of the acidity range 40–98 wt% H_2SO_4. A function, log Z, of the observed rate coefficient [given by Eq. (3.16)] is plotted against M_c, the activity coefficient ratio of the solutes; M_c is given by Eq. (3.17), when subscripts B = base, A = acid, and C = conjugate acid [84JCS(P2)1163].

$$Z = k_2(\text{obs.}) \left[1 + \frac{[NO_3^-]}{[HNO_3]} + \frac{[NO_2^+]}{[HNO_3]} \right] \frac{a_{H_2O}}{[H^+]} \tag{3.16}$$

$$M_c = -\log \frac{f_B f_A}{f_C}$$

Another criterion suggested [71JCS(B)2454] for identifying the nature of the reacting species was comparison of the behavior of a base on nitration in acetic anhydride with that on nitration in sulfuric acid. Applied to weak bases the method seemed to work well, but subsequent work by the same authors showed that it fails with stronger bases, such as most heterocycles, since they form salts in acetic anhydride–nitric acid solutions, as well as in sulfuric acid; application of this method to some relatively weak bases may also be vitiated by side reactions [72JCS(P2)1654].

The whole question of reliability of rate profile interpretation is still under discussion. Moodie and Schofield have criticized the utility of rate–acidity profiles, noting that for weak bases, especially those containing oxygen functions, the occurrence of hydrogen bonding can render the criterion ambiguous or anomalous, and the dichotomy of free base and cation may be too simple [77JCS(P2)1693]. Nitration of substrates that show kinetic ambiguities or anomalies can be accompanied by marked acidity dependence of product compositions. Hence they have proposed a further tentative criterion, which is to compare the variation of rate with acidity for each substrate with that of benzene. If the ratio of activity coefficients were unity, such plots would be rectilinear with unit slope. Statistical analysis on a large number of data suggested that values of the slope ratio $d[\log k_2^{ArH}(obs.)]/d[\log k_2^{PhH}(obs.)] > 0.85$ are characteristic of minority species reactions. Again, for compounds giving slopes close to this value the rate profile is not reliable on its own, and other criteria must be investigated.

F. Model Compound Studies

The most effective single criterion of whether free base or conjugate acid nitrates is the comparison of a model compound in which the possibilities of prototropic and tautomeric equilibria have been eliminated. For example, the model compound for a substitued pyridinium cation is its N-methyl cation. If the nitration of the pyridine proceeds via the conjugate acid, the rate profile will have the same shape as for the N-methyl model compound. The individual rates of nitration for the model compound may or may not be exactly the same as for the pyridine itself, depending upon the effect of the methyl group upon the reactivity. If, on the other hand, nitration of the pyridine takes place on the free base, then the N-methyl cation will not be nitrated under the same conditions.

This method is therefore the most reliable for determining the species reacting, provided that the model compound does not decompose under the experimental conditions.

G. The Encounter Rate Criterion

As in the case of hydrogen exchange discussed in chapter 2 (Section 1.F), it is possible to apply the Smoluchowski equation [Eq. (2.1)] to calculate the encounter rate. For nitration of heteroaromatic substrates, Ridd took a fixed value of 6 for the ratio of the ions (63JCS4204), thus simplifying the expression to Eq. (3.17). Since values of η at various temperatures are well known, the calculation of $k(enc.)$ at various acidities and temperatures is easy. If a nitration takes place via the free base the calculated k_2(fb) value (= k_2(obs.)$[S]_{stoich}/[S]$) should not be greater than k(enc.) at any acidity, whereas for conjugate acid reactions the calculated k_2(fb) value will be greater than k(enc.). However, if the observed second-order rate coefficient is less than k(enc.), the reaction *may* proceed via the free base, so this method affords evidence for conjugate acid nitrations only.

$$k(\text{enc.}) = 8RT/300\eta \qquad (3.17)$$

H. Thermodynamic Parameters

The activation parameters have been determined for many nitrations by means of the Arrhenius and Eyring equations, but their values cannot be predicted with much certainty. The usefulness of these parameters as mechanistic criteria is limited by the wide range of values observed for both free base and conjugate acid reactions. [Their precise interpretation is impossible due to lack of knowledge of all the factors governing their size (67MI6).] However, the relative magnitudes for free-base and conjugate-acid nitrations can provide evidence for one particular mechanism if used in conjunction with other criteria. In general, $\Delta S^{\ddagger}$ values for conjugate acid nitrations are more negative than those for free-base nitrations. Frost and Pearson (61MI2) predicted that the electrostatic contribution to the activation entropy is -10 e.u. for reactions between two ions of like charge and zero for reactions between an ion and a neutral molecule. The $\Delta S^{\ddagger}$ value for conjugate acid and free base nitrations is generally about -20, and -10 e.u., respectively (equivalent to log A being 8.4 and 10.6 sec^{-1}), leading to the corresponding activation energies of 13–17 and 20–26 kcal mol^{-1} (69TH1). (However, this would mean that the conjugate acid > free base reactivity order would reverse at sufficiently high temperature.)

A subsequent survey using the Eyring equation showed that there is a large spread in the distribution of the activation entropies (the mean value is -16 ± 9 e.u.), whereas the heats of activation lie within a smaller range (17.6 ± 4.0 kcal mol^{-1}). Moreover the average value of $\Delta H^{\ddagger}$ is larger for reactions taking place on a minority species (21.7 ± 4.0 kcal mol^{-1}) than for those taking place on a majority species (16.1 ± 2.8 kcal mol^{-1}) [75JCS(P2)1600].

As noted in Chapter 2 (Section 1.G.c), parameters determined from solutions of identical wt% H_2SO_4 at different temperatures are only apparent values, and the parameters ($\Delta \overline{H}^{\ddagger}$, $\Delta \overline{S}^{\ddagger}$) should be calculated using solutions of the same H_0. This approach [75J(P2)1600] indicated that most of the variation occurs in the entropy term, indicating strong solvent interactions, whereas $\Delta\overline{H}^{\ddagger}$ values appear roughly constant for all compounds and in any acidity region (mean value 34.9 ± 3.9 kcal mol^{-1}). Furthermore, the mean value for minority species reactions is close to the overall average, thus indicating that parameters at constant H_0 are not good mechanistic criteria. Nor can a difference be observed between the activation entropies for the two types of species. The average $\Delta\overline{S}^{\ddagger}$ is highly positive (42 ± 14 e.u.), presumably because the transition state is less solvated than the ground state.

I. Summary of Mechanistic Criteria

The above mechanistic criteria almost invariably allow a firm conclusion to be made concerning the species reacting in heteroaromatic nitration. Model compound studies appear to provide the best evidence, and the rate profile approach can be very successful, whereas activation parameters and the encounter rate criterion give, by themselves, less useful information. However, use of a single criterion can often give equivocal evidence, and a firm conclusion frequently requires several criteria used together.

5. Standard Conditions: Choice and Procedure

A. Selection of Standard Conditions

As in hydrogen exchange (Chapter 2, Section 1.G), evaluation of standard nitration rates permits comparison of substrate reactivities under the same conditions. Previous comparisons between nitration rates of compounds of widely differing reactivity have required a stepwise procedure (71MI1), which carries the implicit assumption that partial rate factors do not vary appreciably with acidity.

The definition of standard conditions is more difficult for nitration than for hydrogen exchange because of uncertainty in the NO_2^+ activity variation, and the peculiar behavior of benzene itself. The standard conditions chosen were 25°C and H_0 −6.6 (i.e., 75 wt% H_2SO_4 at 25°C) [75JCS(P2)1600]. The choice of 25°C was made because kinetics for many nitrations have been followed at this temperature, and most in the range 0–100°C. The standard acidity of H_0 −6.6 minimizes the extrapolations needed for the rate profiles of many substrates and is close to the range of "normal" behavior found for benzene.

B. Determination of Standard Rates

The method for obtaining k_2° values requires four consecutive steps [75JCS(P2)1600].

(1) *Determination of $k_2(obs.)$ at the Experimental Acidity and Temperature.* Knowledge of the acidity function of the acid and its variation with temperature is required. The standardization of the acid is usually done by titration. The low substrate concentration required for UV analysis obviates correction of the acidity to allow for losses due to protonation of the substrate, or salt effects.

The variation of H_0 with wt% H_2SO_4 at 25°C is used to obtain the H_0 value of the acid used. The variation of H_0 with temperature at any given acid concentration is given by Eq. (2.4). The appropriate values of K are found from a graph constructed from the acidity variation (69JA6654).

(2) *Determination of $k_2(obs.)$ at T (°C) and H_0 -6.6.* The determination of a rate coefficient at a given acidity for a reaction which has been investigated in a different range of acidity requires construction and extrapolation of a rate profile. Reasons have been given above (Section 4.C) for using a plot of log $k_2(T°C)$ against $-H_0$; this type of rate profile is the only one that can, at present, be correctly constructed at the experimental temperature.

(3) *Determination of $k_2(obs.)$ at 25°C and H_0 -6.6.* Since we are referring to constant acidity, the rate coefficients for 25°C are calculated from Eq. (3.18), using the statistical average value of 35 kcal mol^{-1} for $\Delta \overline{H}^\ddagger$.

$$\log k_2(T°C) = \log k_2(25°C) - \frac{\Delta \overline{H}^\ddagger}{4.574}\left(\frac{1}{273.15 + T°C} - \frac{1}{298.15}\right) \quad (3.18)$$

(4) *Determination of Standard Rate Coefficients (k_2° at 25°C and H_0 -6.6).* For reactions proceeding as a majority species $k_2^\circ = k_2$. For reactions proceeding on a minority species (i.e., as the free base below the pK_a value) the correction provided by Eq. (3.19) [or more easily Eq. (3.20)] must be made, where $H_0^{1/2}$ is the half protonation value for the particular substrate, and m is the slope of log[BH$^+$]/[B] against H_0.

$$\log k_2^\circ = \log k_2(obs.) + pK_a + 6.6m \quad (3.19)$$

$$\log k_2^\circ = \log k_2(obs.) + m(H_0^{1/2} - H_0) \quad (3.20)$$

C. ALTERNATIVE PROCEDURE

If sufficient rate data are unavailable for acidities below $H_0\ -8.5$, extrapolated rate coefficients may be obtained from the high-acidity region of the rate profiles by a different procedure. This latter may also be used for the extrapolation of data for compounds which show a mechanistic changeover [75JCS(P2)1600].

For reaction of majority species at 25°C, in which no such change in mechanism occurs, the difference ($\Delta \log k_2$) between the values at H_0 -6.6 (75 wt% H_2SO_4) and -9.66 (94 wt% H_2SO_4) lies within a narrow range. It is an experimental fact that rate profiles at temperatures higher than 25°C continue to show a maximum near 88 wt% H_2SO_4 (i.e., at progressively less negative H_0 values) as the temperature increases. Little variation (with no clear trend) was found between $\Delta \log k_2$ and temperature and the mean value was found to be close to that for 25°C. The overall average for $\Delta \log k_2$ is 4.0 ± 0.7.

The alternative procedure, which works only for conjugate acid nitrations, is as follows. Values of $\log k_2$(obs.) at T°C are obtained from the rate profile at $H_0(T)$ corresponding to 94 wt% H_2SO_4. Subtraction of 4.0 from this value gives an estimate of $\log k_2$(obs.) at H_0 (T°C) corresponding to 75 wt% H_2SO_4. These values are then converted to those appropriate for 25°C by making a temperature extrapolation at constant wt% H_2SO_4 [i.e., using Eq. (3.18)] with $\Delta H^{\ddagger}$ (the average value for which is 17.6 kcal mol^{-1}) in place of $\Delta \overline{H}^{\ddagger}$.

D. CONCLUSIONS

The data available for ~130 compounds were processed in this way to derive the standard rate coefficients [75J(P2)1600]. Note that when nitration occurs at more than one position, the slope of the rate profile refers to the overall reaction. Standard rate coefficients for nitrations at the individual positions are then obtained using the isomer distribution at the measured acidity nearest to 75 wt% H_2SO_4. When nitration occurs at two or more equivalent positions, the calculated $\log k_2^\circ$ values refer to overall reactivity, and must therefore be statistically corrected.

The error involved in the standard procedure is estimated on the basis of standard rate coefficients obtained from different data sets for the same compound to be 0.2 log units. The error involved in the alternative procedure is expected to be much larger, although fair agreement between the two methods was found in the only two cases where both applications were possible.

CHAPTER 4

Other Reactions

Various other reactions have been used to determine the quantitative electrophilic reactivities of heteroaromatics, and rather more have been used to determine the most reactive sites in a given molecule. For those reactions in which hydrogen is replaced, the method most generally used consists of determining the positions of reaction through gas–liquid chromatography (GLC) and NMR analysis of reaction products, the overall reactivities of the molecules relative to benzene (or some other standard) being determined by the competition method. This is satisfactory except for accurate determination of partial rate factors at sites of low reactivity where, particularly in reactions with large ρ factors, the percentage of a given product isomer may be less than 1%. In a few reactions a substituent is replaced by hydrogen (more recently referred to as *ipso* substitutions), the rate of this replacement relative to replacement of the same substituent in benzene (usually measured spectroscopically) giving the partial rate factor directly. This method is very accurate and is akin to hydrogen exchange in that the reaction site is specifically "labeled."

For some reactions, particularly those in which the reverse reaction is known to occur readily, it is necessary to determine if the product isomer distribution is due to thermodynamic rather than kinetic control. Products which are sterically hindered are particularly susceptible to rearrangement to less hindered isomers. It is also necessary to determine whether reaction occurs on the free base or conjugate acid, though for many of these reactions the conditions are considerably less acidic than those used in nitration or hydrogen exchange.

In the following account, references are given to some examples of the uses of particular reagents and methods; these are described in greater detail in subsequent chapters.

1. Reactions Involving the Proton as Electrophile

A. PROTIODEMERCURIATION

Protiodemercuriation involves acid cleavage either of ArHg from ArHgAr or of HgCl from ArHgCl [72MI2(278)]. The latter procedure has been used to determine the quantitative reactivities of furan, thiophene, and selenophene (65AJC1513).

B. PROTIODEBORONATION

Protiodeboronation is the process whereby the $B(OH)_2$ group is cleaved from an aromatic ring by acid [72MI2(287)]. Quantitative kinetic data for heteroaromatics have been obtained only for furan and thiophene (65AJC1513,65AJC1521; 75JHC195).

C. PROTIODESILYLATION

This reaction involves cleavage of SiR_3 groups (R usually methyl) from aromatic rings, either (1) by acid (acid-catalyzed desilylation), in which the proton attacks the ring in the first and rate-determining step; or (2) by base (base-catalyzed desilylation), in which the aromatic anion is produced by loss of the silyl group, and the proton reacts with the anion in the second and fast step.

Acid-catalyzed desilylation has been used to determine the quantitative reactivities of thiophene [56JCS4858; 59JCS2299; 61JCS4921; 70JCS(B)1364], of a range of substituted thiophenes [81JCS(P2)931], and of benzo[b]thiophene, dibenzofuran, dibenzothiophene, and N-ethylcarbazole (61JCS4921).

Base-catalyzed desilylation of thiophenes has been used to examine both the mechanism of the reaction, and the effects of substituents [76JCS(P2)925; 81JOM(204)153]. In this work the reactivities of the 2-positions of furan, benzo[b]furan, and benzo[b]thiophene were also examined, as were base-catalyzed degermylation (cleavage of GeR_3) and destannylation (cleavage of SnR_3).

D. PROTIODEPLUMBYLATION

Protiodeplumbylation (cleavage of PbR_3 groups from aromatics) was used in the first demonstration that furan was more reactive than thiophene toward electrophiles (32RTC1054).

2. Metallation

A. LITHIATION

Replacement of hydrogen in an aromatic ring by lithium is an electrophilic substitution, entry of lithium taking place in the second and fast step of the reaction; the reaction mechanism parallels that for base-catalyzed hydrogen exchange. Since the most acidic hydrogen is replaced, the substitution pattern differs from other electrophilic substitutions. Removal of the proton from the ring in the first step was shown to be rate-determining in a study of the kinetic isotope effect in lithiation of thiophene (55AK343). Though there are numerous examples of this reaction, relatively few have involved quantitative reactivity data, interest centering upon the position of lithiation, particularly in substituted heteroaromatics. Much work has been carried out in this direction with thiophenes [53JA3697; 54AK361, 54JOC70; 58AK269,295; 59BAU1925, 59JGU2003; 60AK309, 60AK363, 60BAU1700; 62AK(18)513; 71JOC1053; 77JCS (P1)887]. The relative reactivities of furan, thiophene, and selenophene, and the position of lithiation of tellurophene, have also been examined [71CHE938; 76ACS(B)605].

The positions of ring lithiation of the following compounds have likewise been determined: isothiazole (64JCS446,3114), 1-phenylpyrazole (58JA6271), 1-substituted imidazoles (71BAU1429; 73JOC3762; 77JH C517), 4-methylisothiazole (70CJC2006), 4-aryl-2-methylthiazoles (74JOC 1192), 1-methyltetrazole (71CJC2139), 2-$3H$-thiazolthiones (80S800), 2-(2-thienyl)thiazole (74BSF2099), benzo[*b*]furan, benzo[*b*]thiophene, benzo-[*b*]selenophene, indole [72CHE13; 73CHE953; 77JCS(P1)887; 84JCS (P1)2839], thieno[3,2-*b*]pyridine (84JHC785), furo[3,2-*c*]pyridine (83T 1777), thieno[2,3-*b*]pyridine (74JHC355), thieno[2,3-*b*]pyrazine (80JHC 1019), imidazo[1,2-*a*]pyridine (72JHC1157; 83S987), pyrrolo-[1,2-*a*]pyrazine, imidazo[1,2-*a*]pyrimidine, 1,2,4-triazolo[1,2-*a*]pyrimidine (72JHC 1157), dibenzofuran, dibenz *N*-substituted carbazoles (38JOC120; 39 JA951; 40JA2606; 41JA2479; 43JA1729; 45JA877; 54JA5775), pyridine (40 JA446; 48JA1037), and pyrazines (68JOC1333; 71RTC513; 74JOC3598).

B. MAGNESIATION

The reaction between aliphatic Grignard reagents and aromatics containing an acidic C—H bond results in replacement of hydrogen by MgX in the same way that hydrogen is replaced by Li. This is called *magnesiation*, and like lithiation, is limited mainly to heteroaromatics since these have acidic hydrogens arising from neighboring heteroatoms. The relative reactivities of benzothiazole, 1-methylimidazoles, and 1-methylbenzimidazole in this reaction have been studied (69JGU1816).

C. MERCURIATION

The principle methods of mercuriation involve reaction of an aromatic with either mercuric acetate (both in the presence and absence of a protic acid), mercuric chloride, or mercuric perchlorate [72MI2(186)]. The former reactions are the more common and are sometimes referred to as acetoxymercuriation and chloromercuriation, respectively. Mercuriation has been used to determine the positions of substitution in indole, methylindoles, benzo[b]selenophene (70CHE254; 71CHE1401; 72CHE18), benzo[c]selenophene [79CA(91)73755], 9-ethylcarbazole and dibenzofuran (36JOC146), pyridine (23CB2223; 32JCS1263; 37USP2085063), pyridine 1-oxide (58RTC340; 62RTC124), quinoline and isoquinoline (31JPJ542), and quinoline 1-oxide (53YZ823; 62RTC124; 69CPB906).

D. PLUMBYLATION

Reaction of aromatics with lead tetracarboxylates leads to substitution of a tri(acyloxy)lead group into the aromatic. This reaction takes place more readily if the acyl group contains electron-withdrawing groups such as halogen. The reaction with thiophene has been examined [58DOK(123)295; 74TL853].

3. Reactions Involving Carbon Electrophiles

A. ALKYLATION

Alkylation is brought about with either an alkyl halide and a Lewis acid catalyst, or with an alkene (or alcohol) in the presence of a protic acid. Complications include alkylation of nitrogen heteroatoms, coordination of the catalyst with heteroatoms, protonation at the heteroatom or elsewhere in the ring (which can give rise in the latter case to ring opening), polysubstitution, and rearrangement of the products, with respect to both the position of the alkyl group and within the alkyl group.

As a consequence, there have been relatively few quantitative studies of the alkylation of heterocycles, and these have mainly involved the introduction of secondary or tertiary alkyl groups, since less severe catalysis is then required. Alkylation under these conditions has been used to determine positional reactivities for furan and thiophene (Chapter 6, Section 5), benzo[b]furan (69BAU2446; 71CHE953), benzo[b]thiophene [56JOC584; 62JOC2026; 66JOC3093; 72JCS(P1)414], alkylindoles (72TL5277; 84JHC1485), and various pyrimidines (74JOC587). Single reports describe the ethylation of dibenzofuran (50USP2500732), the pyridylethylation of barbituric acid (62JOC174), and the cyanoethylation of indoles (56ZOB557) (the last two methods use vinylpyridine and vinyl cyanide, respectively, without a catalyst). The methylation of quinoline (61BRP845562) and of an activated pyrimidine (62JCS3172) have been described.

The relative reactivities of indole, 2- and N-methylindole, pyrrole, thiophene, and furan have been determined in alkylation by a benzenonium ion coordinated to iron tricarbonyl (73CC540). The effects of methyl substituents in pyrrole were determined in alkylation by 4-(N,N-dimethylamino)benzaldehyde [76JCS(P2)696]. In neither of these methods, nor in the alkylation of indole by aziridinium tetrafluoroborate [67AG(E)178], nor in self-alkylation of a λ^5-phosphorinyl tetrafluoroborate [73AG(E)753], is a catalyst required.

B. Haloalkylation and Hydroxyalkylation

Haloalkylation involves a more reactive electrophile than in alkylation and so weaker Lewis acid catalysts such as zinc chloride may be used. Moreover the entering substituent contains an electron-withdrawing group so the tendency for further reaction is diminished. Both these advantages have resulted in haloalkylation being more widely studied than alkylation. The most common reaction is chloromethylation using formaldehyde and hydrogen chloride (the hydroxymethyl compound is the initial product with chlorine replacing hydroxyl in a subsequent fast step); bis(chloromethyl) ether/protic acid has also been used.

These reagents were employed to determine isomer yields in chloromethylation of various thiophenes (e.g., 42JA477; 47JA1549; 60CCC1058; 62CCC372, 62MI2; 71CHE1265), 1,2,5-thiadiazole (74CC585), 2,1,3-benzothiadiazoles (64JGU2491), dibenzofuran [73CA(78)84153], dibenzothiophene (76BAU2609), and uracils (60JA991), and reactivity orders in isoxazole and derivatives (57ZOB3210; 58ZOB2376).

Isomer yields have been determined in the chloroethylation of some thiophenes (62MI2). Hydroxymethylation has been used to determine the position of substitution in indole (51G613).

C. Aminoalkylation

Aminoalkylation has advantages similar to those of haloalkylation. Aminomethylation has been used to determine positional reactivities in hydroxypyridine N-oxides (72JOU416; 74BAU2023), hydroxy-2-pyridones (71BAU2222), and pyrimidines (60JA991; 79BAU633; 82CHE297; 83CHE1008).

D. Cyanoethylation

The electron-withdrawing ability of the cyano group facilitates alkylation by a primary alkyl group through effectively increasing the reactivity of the electrophile. Cyanoethylation has been used to determine the isomer yields for indole and 2-methylindole (56ZOB557).

E. ACYLATION

The electrophile in acylation is generally more reactive than that in alkylation, and the electron-withdrawing property of the substituent means that further substitution is retarded. In consequence, acylation has been widely studied, and under a variety of conditions. Those used in formylation differ from those employed in most other acylations, so formylation is considered separately.

a. *Formylation*

This is generally carried out with $POCl_3$/DMF (Vilsmeier formylation involving the electrophile [Me_2NC^+HCl]$PO_2Cl_2^-$) and has been used in a large number of studies of the positions of substitution in thiophene and derivatives (Chapter 6 Section 7.A), pyrroles, and various alkylpyrroles [70JCS(C)2563]. It has also been used to determine the relative effects of 1- and 3-methyl substitution in pyrrole [84JCS(P2)1179, 84JCS(P2)1607]; substituent effects in indoles [72CC427; 77CA(86)171800]; the position of substitution in various alkyl- and arylpyrazoles (57JCS3314; 59JCS1819; 61JCS2733; 73JOU840), pyrimidines (65M1567; 71CPB215,1216, 71JHC 445; 73CPB260), indolizine and 5-methylimidazo[1,5-*a*]pyridine (75JHC379), dibenzothiophene (69AJC1963), and benzo[*d,e*]cinnolines (81JOU2183); and the relative reactivities of thiophene, thieno[3,2-*b*]thiophene, selenolo[3,2-*b*]thiophene, and selenolo[3,2-*b*]selenophene [80CS(15)206].

N-Methyl-*N*-phenylformamide has been used as an alternative amide in determination of the relative reactivities toward formylation of thiophene and selenophene (73JGU871).

Phosgene ($COCl_2$) has also been used instead of phosphoryl chloride, and this gives rise to a more reactive electrophile. Thus kinetics are first order in all the aromatics (thiophenes) examined [74JCS(P2)1610], whereas with phosphoryl chloride kinetics are first order only for reaction with thiophenes of moderate reactivity; with reactive aromatics (2-methoxythiophene) the reaction becomes zeroth order in aromatic because formation of the electrophile is then rate-determining [72JCS(P2)2070]. Formylation with phosgene has been used to determine the relative reactivities of thiophene and selenophene [73JCS(P2)2097], and the activation effects of a 2-methyl group in selenophene and tellurophene (77G339).

b. *Other Acylations*

Acylations are usually carried out either with an acid anhydride in the presence of a protic or Lewis acid, or with acyl halide in the presence of a Lewis acid. Use of a Lewis acid may lead to problems arising from coordination of the catalyst with heteroatoms in the substrate, and/or the ketonic product which has a greater electron density on the carbonyl oxygen than in the starting reagents. Steric hindrance to acylation is considerable and also dependent upon the size of the Lewis acid. Hindrance is greater toward acetylation than toward benzoylation because the degree of coordination between the acylating agent and catalyst differs [72MI2(181)]; acylation of substituted thiophenes demonstrates both features (73T413). For reaction with heteroaromatics, anyhdrides are the favored reagents. Vilsmeier–Haack acetylation uses $POCl_3/N,N$-dimethylacetamide (the electrophile here being considerably less reactive than that in formylation). This reagent has been employed to study acetylation of thiophene [74JCS(P2)1610], and positional reactivities together with a range of substituent effects in indole and carbazole [77JCS(P2) 1284].

Acetylation has been used to determine the relative reactivities of benzene, dibenzofuran, dibenzothiophene, thieno[3,2-*b*]selenophene [80CS(15)206], tellurophene, selenophene, thiophene, and furan (including partial rate factors for the last two molecules) [67T1739; 69AC(R)787, 69T4599; 70JCS(B)1153; 71TL3833; 73JCS(P2)2097]; the reactivity of furan is however abnormally low because of coordination with the Lewis acid catalyst. Coordination also adversely affects the reactivity of pyrrole (67CJC897, 67MI4), but acylation can be achieved without a catalyst in view of the high reactivity of pyrrole (67MI4).

Various acylations have been used to determine the effects of substituents in thiophene (47JA3093; 55JCS21; 55JA4066; 67T1739; 70ACS99; 72ACS1851; 73CHE447, 73T413; 75CJC1; 77TL389), furan [73BSF1760; 72CR(C)(275)49], and indole (47JCS1631; 63JOC2262); and the isomer ratios in arsabenzene (78TL2537); indole (72CC77); benzimidazoles (77LA145); dibenzothiophene (38JA2628); carbazoles [35JCS741; 63CA(58)2422c,d; 74CA(81)168829; 75CA(82)16654; 76T2595]; thieno[3,2-*f*]quinoline [70JCS(C)2334]; thieno[2,3-*b*][1]thiophenes [71JCS(C)463, 71JCS(C)1308];4*H*-furo[3,2-*b*]indole (78JHC123); 1*H*,5*H*-pyrrolo[2,3-*f*]indole and 3*H*,6*H*-pyrrolo[3,2-*c*]indole (83CHE871); pyrrolo[1,2-*a*]benzimidazole [67CHE723; 69CA(70)68249]; selenolo[2,3-*b*][1]benzothiophene, thieno[2,3-*b*][1]benzoselenophene [76CA(84)135510], 1-methyl[1]benzofuro[3,2-*b*]pyrrole, and its thieno-, selenolo-, and [2,3-*b*]analogues (83JHC61); 2,3-dimethylthienofurobenzene (65BSF1473); 2-arylcyclohepta[4,5]pyrrolo[1,2-*a*]imidazole [84JCR(S)390]; and 6-aminouracils [23CB2482; 58LA(612)173; 72JOC578]. Isomer ratios and partial rate factors for benzo[*b*]furan and benzo[*b*]thiophene [64JCS173; 71JCS-(B)79; 73JCS(P2)1250; 78JCS(P2)1053; 80CA(92)215176], and substituent effects in both molecules [52JA766,2185; 60T(10)215; 61BSF1534, 61JOC359,363; 64BSF1525; 65BSF1473, 65NKZ99,637,643, 65NKZ1067; 67BRP1058468; 67JCS(C)2084, 67M2039, 67NKZ751; 70AHC(11)327, 70BCJ3496, 70BSF3601; 78JCR(S)10; 79JHC1029; 82JHC279; 84JHC-177], have likewise been determined.

The differences in steric hindrance in five-membered rings compared to benzene have been demonstrated in acetylations of alkylthiophenes (71T4667).

Benzoylation of *N*-arylpyrazoles takes place sufficiently readily that no catalyst is required (1889G128; 26CB611; 60ZOB203). Benzoylation has been used to determine partial rate factors for dibenzofuran (77NKK1518), and the position of substitution in imidazoles (77LA159; 78S675) and dibenzofurans (54JA6407; 72NKK387; 73NKK1505; 74NKK1708; 80S139).

Chloroacetylation has been found to be *slower* than acetylation for reaction of activated thiophenes, but behaves normally (i.e., is faster than acetylation) in reaction with deactivated thiophenes. The former result arises from coordination of the $HAlCl_4$ co-product with the aromatics (75JOU412). Various haloacylations of indoles have been reported (73T971; 84H241).

Trifluoroacetylation with trifluoroacetic anhydride is sufficiently rapid that no catalyst is required; acetyl trifluoroacetate may also be used but is accompanied by acetylation, which tends to be the main reaction. The former reagent has therefore been used to determine partial rate factors (relative to thiophene) for a range of substituted thiophenes [72JCS(P2)71]; the relative reactivities of thiophene, furan, pyrrole, tellurophene, and selenophene, [69T4599; 71CC1441; 73JCS(P2)2097; 77JCS(P2)1284]; the effects of 2-methyl substituents in the latter two molecules (77G339); and partial rate factors (relative to either furan or pyrrole) for a range of substituted furans and pyrroles, respectively [72JCS(P2)71; 73BSF1760]. Positional reactivities in indole (76T2595) and 6-methylpyrrolo[2,1-*b*]thiazole [66JCS(C)1908] have been determined using trifluoroacetic anhydride, and in furan and thiophene using acetyl trifluoroacetate [70JCS(B)1153; 71TL3833].

Cyanoacetylation also involves a more reactive electrophile, and this reaction has been used to examine reactivity patterns in indoles (80CB3675).

c. *Alkoxycarbonylation*

Reaction of imidazole with chloroformates in the presence of triethylamine produces 2-carboxylates (78S675).

4. Reactions Involving Nitrogen and Phosphorus Electrophiles

A. NITROSATION

The products of nitrosation tend to be unstable and to rapidly oxidize to the nitro derivative, and the electrophile is rather unreactive. This has limited studies mainly to a few reactive azoles: a kinetic study has been made of nitrosation of a range of substituted indoles [73JCS(P2)918], and the position of nitrosation determined in 2-phenyl-3-substituted indoles [84JCS(P2)165], 6-methylpyrrolo[2,1-*b*]thiazole [66JCS(C)1908], and 6-(2'-furyl)imidazo[2,1-*b*]thiazole (72CHE1223). The reaction with pyrimidines containing very strong substituent activation has also been examined [44JCS315; 54YZ674; 58LA(612)158; 60MI1; 64JCS1001; 66LA(691)142; 71TL851; 78H247; 82CPB3392].

B. DIAZONIUM COUPLING

This reaction involves an unreactive electrophile, due to delocalization of the positive charge from nitrogen into the aryl ring. Consequently, there are reports of diazonium coupling only for reactive heteroaromatics, the reactivity being enhanced in some cases by the presence of strongly electron-releasing groups.

A detailed kinetic study has been made of the coupling of phenyldiazonium ion with a wide range of methylpyrroles, which showed that the neutral pyrrole species was involved [77JCS(P2)1452]; likewise, reaction of *p*-nitrophenyldiazonium ion with indoles also involved the neutral molecule (57JCS2398). Studies of the position of diazonium coupling have also been carried out with 6-phenylimidazo[2,1-*b*]thiazole [77HC(30)1], 5-benzyl-3-hydroxypyridine 1-oxide (74BAU2023), pyridazine-3,4-1H,2H-diones (70JPR591), activated pyrimidines (83AJC1659), 3-hydroxyquinoline, and 4-hydroxyisoquinoline (72BAU452).

C. PHOSPHONYLATION

A diamide of methylphosphonic acid [MeP(O)(NBu$_2$)$_2$, a phosphonoamidate] has been shown to lose the methyl group and substitute into indole (80JGU618); there are no other reports of this reaction.

5. Reactions Involving Oxygen and Sulfur Electrophiles

A. HYDROXYLATION

The only example of the quantitative study of the electrophilic hydroxylation of a heteroaromatic concerns quinoline [54JBC(208)741].

B. SULFONATION

Sulfonation is an important electrophilic substitution, especially from the preparative viewpoint. Not only are many products important per se, but the bulky sulfonic acid group is valuable for blocking sites in a molecule. This obliges further substitution to take place elsewhere, the location depending upon the electronic directing effect of the group. The sulfonic acid group is then readily removed by reaction with hot dilute acid (protiodesulfonation), the ease of this reaction being to a considerable extent due to the weak C—S bond, itself arising from steric interactions between the group and the aromatic ring. However, the bulk of the sulfonic acid group also means that substitution patterns in sulfonation are very poor indicators of the intrinsic site reactivities, and there are therefore few quantitative reactivity data for heteroaromatics. Moreover, steric hindrance between the sulfonic acid group and the aromatic ring causes many products to be thermodynamically unstable, so that they tend to rearrange to less hindered products; this is especially true for compounds with the sulfonic acid group in α-naphthalene-like positions.

A variety of electrophiles are probably involved in sulfonation by sulfuric acid, but the principal ones are HSO_3^+ in more weakly acidic media, and SO_3 in stronger media, especially oleum; these electrophiles may also be solvated by one or more sulfuric acid molecules. These are very reactive electrophiles so that all classes of heteroaromatics should, in principle, undergo substitution. However, the strong acid conditions frequently result in protonation of the heteroatoms, thereby drastically reducing the reactivity of the heteroaromatic. One way to avoid this is to use SO_3 in an aprotic solvent, pyridine being especially suitable since it removes any sulfuric acid produced by the presence of adventitious water; the reagent under these conditions is $Py:SO_3$. Coordination of the SO_3 to substrate heteroatoms may occur, but such coordination is usually less strong than to pyridine. Sulfonation may also be accomplished using chlorosulfonic acid, or, less commonly, fluorosulfonic acid.

The strength of the electrophiles in sulfonation is such that other sub-

stituents are sometimes cleaved, examples being found in sulfodebutylation of 3-*t*-butylthiophene (56BAU627) and sulfodeacylation of 5-alkyl-2-acylthiophenes (1886CB660,1886CB2623; 1896CB2560).
There have been many reports of the isomer yields in sulfonation of thiophene and derivatives [24LA(437)14; 33LA(501)174; 48MI2; 51ZOB1524; 52JGU189; 53JGU263, 53MI1; 54CB1184; 56BAU627; 60BSF793; 80URP707916], likewise in pyridine, its *N*-oxide, and derivatives (42CB1108; 43JA2233; 53JA3865; 55JA2902; 58RTC963; 59RTC586; 66JA986; 75CHE745); and various quinolines [1870LA(155)311; 1882CB683,1882CB1979; 1887CB731; 1888JPR258; 54USP2689850; 61USP2950283; 72BAU404,72BAU406]. The quinolines in particular show thermodynamic stability effects.

The positions of sulfonation have also been examined for the following compounds: furan (49ZOB531; 54JOC894), benzo[*b*]furan (53CA10519c), dibenzofuran (49JA1593), selenophene (64JGU1814,2201), benzo[*b*]thiophene [34CR(198)2260], dibenzothiophene 1,1-dioxide [79JCS(P2)224], pyrroles (49ZOB538,1365,2118; 51ZOB281), indoles (51JGU1415; 73T669), carbazole (49CA6205f; 50MI1), phenyloxazoles (53GEP869490; 55GEP926249; 63JCS1363), dialkyloxazoles (40G1,11), isoxazole (59ZOB535), methylisothiazoles [63AG(E)714; 65JCS7283], 2,4-dimethylselenazole (48YZ195), imidazoles (24JCS919; 27JCS2711; 57JA2188), pyrazoles [1894LA(279)217], benzoxazolone (41JA879), 2-alkylbenzimidazoles (75ZPK2241), indazole (50BSF466), 2,1,3-benzothiadiazoles [64CA(60)10670g; 64JGU1265], naphtho[1,2-*c*]-2,1,3-oxadiazole (73CHE1331), amino- and hydroxypyrimidines (54JCS4206, 54JGU2212; 56JA401; 59JA5166; 61JOC3863; 63JOC1994), acridizinium ion (66JOC565), phenanthridin-6-one (57JA5479), quinoxaline-2,3-dione and methyl derivatives (64JAP26975; 66BRP1043042), 1,10-phenanthroline (61AC867), phenazine (50G651), coumarins (23CB480; 28JIC433; 57JIC3545; 57JOC884), and 2-methylchromones (56JOC1104).

C. CHLOROSULFONATION

Chlorosulfonic acid produces both sulfonation and chlorosulfonation, the balance depending upon conditions and the nature of the aromatic. The first-formed sulfonyl chloride may then react further to give, as a final product, the diaryl sulfone. Reports have been limited to reactions of thiophenes containing electron-withdrawing substituents [34LA(512)136; 55JA3410; 60BSF793; 81JMC959, 81PS111], some phenyloxazoles (53GEP869490; 55GEP926249; 63JCS1363), and dimethylbenzo[*b*]thiophenes [74CA(81)105154; 76CA(84)89931].

D. SULFENYLATION

Sulfenylation involves substitution of the SR group into an aromatic ring and is accomplished with a sulfenyl halide RSX, with X usually chlorine. The most common variation employs sulfur dichloride as reagent, the products being a diaryl sulfide. The electrophile is rather unreactive, and studies with heteroaromatics have been limited to reaction of SCl_2 with 2-pyridone (64ACS269), and CF_3SCl with N-methyl-2-thiomethylimidazole (75JHC597) and with substituted indolizines (81JFC67).

E. THIOCYANATION (AND SELENOCYANATION)

Thiocyanation is very closely related to sulfenylation, the group SCN being introduced into the aromatic ring, usually by reaction with thiocyanogen $[S(CN)_2]$. The electrophile here is also rather unreactive and there are studies only of thiocyanation of substituted indoles (60JA2742) and carbazoles [73CA(78)111049; 77CA(87)151948; 78CA(88)152349].

Selenocyanation has been used only in the reaction with indole (63ACS268).

6. Halogenation

Halogenation has been carried out using a wider range of reaction conditions than any other electrophilic substitution and most of these have been used in studies of the reactivities of heteroaromatics. Conditions may be divided roughly into those in which the halogen is virtually fully polarized (positive halogenation), and those in which it is slightly polarized (molecular halogenation). The former conditions are found in acidified hypohalous acids, and probably in mixtures of halogens and Friedel–Crafts catalysts. Molecular halogenation applies in mixtures of halogen and various solvents, and in some of these there may be a substantial degree of coordination between the halogen and solvent. The spread of reactivities in positive halogenation is much less than in molecular halogenation. The order of reaction in halogen may be greater than 1.0 for molecular bromination [an example being the bromination of substituted benzo[*b*]furans (74BCJ1267)], and this order may differ according to the reactivity of the aromatic substrate, so it is necessary to check for this when determining relative reactivities. Halogenation may be accompanied by addition, particularly for those heteroaromatics (e.g., furan) (67BCJ130) that have highly localized bonds. The *vic*-dihalo products thus obtained may then undergo elimination of hydrogen halide to give the normal substitution product, albeit via a different mechanism. Side-chain substitution also frequently accompanies ring halogenation [e.g., in chlorination of 2,3-dimethylbenzo[b]thiophene) [68JCS(B)397; 76JCS(P2)266]]. In some instances halodehalogenation occurs, and the displaced halogen enters another position of the heteroaromatic [e.g., in chlorination of 3-bromobenzo[b]thiophene (73IJS233)].

The positional reactivities may differ substantially for each halogen in view of their differing sizes. For example, 1,2-disubstitution is likely to be more prevalent in chlorination than in bromination. A common method of reducing disubstitution is to use an N-haloamide in a polar acidic solvent. This produces a small equilibrium concentration of halogen electrophile and prevents a large excess of halogen and hence disubstitution. Examples are the bromination of pyrrole (81JOC2221), 3-phenylindole (74T2123), and cyclazines (72ACS624; 73ACS2421) with N-bromosuccinimide (NBS), and of thiophene with N-bromoamides [19LA(426)61; 44LA(556)1; 51MI2; 52MI2]. The differing selectivities of molecular chlorination and molecular bromination also mean that the latter gives a higher relative yield of the most abundant isomer when mixtures of products are obtained [e.g., in halogenation of benzo[b]thiophene [71JCS(B)79]].

Other molecular brominating reagents used for heterocycles include dioxane dibromide (54JGU1251; 73CHE95; 78JHC123), potassium bromate/hydrogen bromide (82BAU2104), 2,4,4,6-tetrabromocyclohexa-2,5-dienone [72JCS(P2)2567; 73JCS(P1)68], 1,3-dibromo-5,5-dimethylhydantoin (51MI1), bromine alone (46JA453; 71BAU2108; 80JHC1399), bromine in dimethylformanide (DMF) [72JCS(P2)2567], pyridine [67KKZ63; 77CS(12)1], carbon disulfide (47JA1920), nitrobenzene (73JHC153, 73JHC409), chloroform [61CPB414; 71BAU2108; 73JOU2216; 80JCR(S)201], carbon tetrachloride (68JOC1384; 83JOC1064), water (47JPJ87; 60JOC1916; 63JCS1276; 67CPB1411; 78CPB3884), alcohols [74CA(80)3443; 75CPB923], hydrogen bromide (63JCS2203; 64JCS2760; 65JOC526), aqueous sodium methoxide (54ZOB488), or acetic acid [54JCS4142; 64TL2093; 65RTC1101, 65T945; 66ACS1448; 70JCS(C)2334, 70CPB1680; 72ACS624; 73ACS2421). Ethyl bromide in dimethyl sulfoxide (DMSO) (65T945) probably involves the intermediate formation of Me_2SBr_2. Bromine in thionyl chloride or sulfur monochloride has also been used, but the mechanism may not be a normal electrophilic substitution (60JA4430). For chlorination of unreactive heteroaromatics, (e.g., pyrazines and pyrimidines), both $SOCl_2$/DMF and $POCl_3$/PCl_5 have been used [67JCS(C)1922; 72JCS(P1)2004].

Positive halogenation (e.g., HOBr, 90% aqueous dioxane) has been used to determine the relative reactivities of thiophene and selenophene [70JCS(B)43], of substituted pyridines (74JOC3481; 76JOC93), and (using $Cl_2/AlCl_3$) the reactivities of furans and thiophenes with substituents containing carbonyl groups (70BAU2592; 71BSF238; 72CHE541; 73BAU2666; 76JHC393), though here the product distribution depends substantially on the extent of coordination between the aluminum halide and the substituent. Chlorine in the presence of iron has been used to determine the positions of substitution in dibenzofuran (55JOC657). A positive bromination species is present in a mixture of bromine–silver sulfate–sulfuric acid (71BAU2687). This has been used to show that bromination of 2,5-diphenyloxazole under these conditions occurs on the free base (71JOU1835), and to determine the position of substitution in pyridine 1-oxide (62T227), quinoline (60JCS561), isoquinoline (59MI1), quinoxaline-2,3-diones (62JCS1170), and benzo[c]cinnoline [79JCS (P1)1503]. Fairly polar electrophiles are probably involved in bromination with cyanogen bromide [22LA(430)79], or with bromine in sulfuric acid/ oleum (62RTC864, 62T227). The latter have been used to examine the reactivity of bromopyridine 1-oxides (62RTC864), methylpyridines (65RTC951), 1,5-naphthyridines (63RC1589), 2,6-dimethylpyridines (70RC779), aminopyridines [70JCS(B)117], and benzo[1,2-d:4,5-d']diimidazole (52JGU1069), and the relative reactivities of 3,5-dihydroxy-1,2,4-triazine and uracil (76JOC4004). Likewise Cl_2/H_2SO_4 has been used for the chlorination of 2-aminopyridine (76JOC93).

Molecular bromination and/or chlorination has been used to determine the reactivities of thiophene (65T843), selenophene [70JCS(B)43], furan (70TL1389), and pyrrole [68JCS(B)392]; rates or partial rate factors (relative to thiophene) of a range of substituted thiophenes [70JCS(B)848; 71AHC(13)235; 74CC333; 81IJC(19A)1183], thieno[3,2-b]thiophene, selenolo[3,2-b]thiophene, and selenolo[3,2-b]selenophene [80CS(15)206]; and partial rate factors (relative to benzene) of benzimidazole and indazole [78J(P2)865], and of N-acetylcarbazole [66JCS(B)521]. They have also been used to determine the positions of substitution of various substituted thiophenes (67JOC463; 68JOC2902); furans and pyrroles (65JCS459); 2,1,3-benzothiodiazole (63JGU223); benzotriazole (75JOU889,902); naphtho[1,2-c]-1,2,5-oxa- and -thiadiazole (73CHE1331); pyrrolo[1,2-a]quinoline [67JCS(C)1164; 68JCS(C)2848]; thieno[2,3-b][1]benzothiophene and thieno[3,2-b][1]benzothiophene [71JCS(C)1308]; selenolo[2,3-b][1]-benzothiophene, thieno[2,3-b][1]benzoselenophene, and their [3,2-b] isomers [76CA(84)135510]; 1-benzoyl-4H-furo[3,2-b]indole (78JHC 123); benzothieno[3,2-d]pyrimidine (80JHC1399); thieno[2,3-b]quinoline [80JCR(S)201]; thieno[3,2-f]quinoline [70JCS(C)2334]; s-triazolo[3,4-a]iso-quinoline [74CA(80)3443]; naphtho[1,2-d]imidazole (54ZOB488); 1-benzoyl-4H-furo[3,2-b]indole (78JHC123); benzo[1,2-d:3,4-d']diimidazole (73CHE95); dithienobenzene [77CS(12)97]; 2-arylcyclohepta[4,5]pyrrolo[1,2-a]imidazole [84JCR(S)390]; 1H-thieno- and selenolo-[3,2-d]pyrazoles (73JOU2216); 2H-thieno[4,3-b]pyrazole [77CS(12)1]; 2-quinoxalinone (63MI2); and various pyrimidinones (79JOC3256; 81JOC4172).

Molecular bromination has been used to determine the positional reactivities in oxazole [59LA(626)83,92], thiazole (34RTC77), imidazole [28JPR(118)33], isoxazole (60G356), isothiazole [63AG(E)714; 64CI(M)207; 72AHC(14)17], pyrazole [66AHC(6)391], 1,2-benzoisothiazole [80JCR(S)197], as well as furazane (74JHC813), 2- and 4-pyridone, 3-hydroxypyridine (50RTC1281; 82JA4142; 83CJC2556), and various of their derivatives. Among these the phenyl derivatives [and also those of thiophene (67JOC463; 68JOC2902; 72ACS1851), 1,2,3-triazole (63CJC2380)], and pyrrolo[2,1-b]thiazole [77HC(30)1]] have attracted considerable interest, partly because the relative reactivities of the phenyl and heterocyclic rings give a very rough indication of the relative electrophilic reactivities of the heteroaromatic to that of benzene. This method has also been used to determine the qualitative relative reactivities of thiophene and pyrazole, through bromination of thienylpyrazoles [80CS(15)102], and of imidazole and imidazo[1,2-a]pyridine (which reacts as the neutral species) (74AJC2349). Bromination has been used to determine the effect of silver complexing upon the relative reactivities of each ring in 4-chloromethyl-2-phenylthiazole (74ACH381).

Iodination involves a less reactive electrophile, with the transition state being commonly regarded as being nearer to the Wheland intermediate and the second step of the reaction becoming partially rate determining. A substrate kinetic isotope effect is therefore usually observed (e.g., in iodination of indole) [68AC(R)1435].

Iodination can be complicated by the hydrogen iodide co-product, which, being the strongest of the halogen acids, can cause protolytic cleavage of heteroaromatic rings. This may be avoided by adding mercuric oxide, which removes the acid as it is formed. $I_2/HIO_3/HOAc$, ICl, and I_2 in 50% aqueous HOAc have also been used in various studies [1884CB1558; 43OSC357; 60LA(634)84; 71MI2; 82IJC(A)417], the latter conditions in determination of partial rate factors (relative to thiophene) for iodination of 5-halothiophenes [82IJC(A)417].

Iodination has been used to assess the effect of Ni(II) complexation upon the reactivities of pyrazole and imidazole (66JA5537; 72JA2460), which curiously are opposite in effect. Iodination has been used to determine the relative reactivities of pyrazole and its anion (conjugate base) (67JA6218). Under alkaline conditions, pyrazole reacts as the anion which is so reactive that it may be tribrominated [55LA(593)179;

74FRP2193823]. Results for iodination of substituted indoles [69AC(R)799] reveal that substituent effects are poorly transmitted between the benzenoid and heterocyclic rings. Iodination has been used to determine the position of substitution in 2-benzyl-2-hydroxypyridine and its N-oxide (75CHE478), in 3-hydroxypyridine 1-oxides (66RC1875; 74BAU2023), and in amino- and hydroxypyrimidines (62CCC2550; 84S252).

Fluorination of aromatics with fluorine usually occurs explosively fast. However, pyrimidin-2($1H$)-ones have been successively fluorinated by fluorine in acetic or hydrofluoric acids (77CCC2694), and 5-substituted barbiturates by perchloryl fluoride (79JOU357).

7. Reactions Involving Replacement of One Substituent by Another

A. Lithiodehalogenation

The replacement of halogen in aromatics by lithium is an electrophilic substitution, but, like lithiation, it involves entry of lithium in a fast step, so that the normal substituent effects of electrophilic substitution do not apply. Relative rates of lithiodehalogenation have been determined for some alkylbromothiophenes [74CS(5)217], the positional reactivity orders determined in pyrimidine (65ACS1741), and site reactivities examined for bromopyridines (46JA103; 51JOC1485; 52JA6289; 55RTC1003), 5-halopyrimidines (56JA2136; 59T225; 65ACS1741), and pyrazines (65JHC209).

B. Iododeboronation

Rates of cleavage by iodide of $B(OH)_2$ from thiophene have been determined (65AJC1527); there are no other reports of this reaction for heteroaromatics.

C. Other Reactions

There have been reports for the following reactions on thiophene, but none lead to rate or relative rate data: alkyldelithiation, acyldelithiation, carboxydelithiation, silyldelithiation, acyldemagnesiation, acyldezincation, acyldecadmiation, acyldemercuration, halodemercuration, halodethalliation, bromodealkylation, nitrodeacylation, sulfodealkylation, bromodeacylation, mercuridecarboxylation, halodenitration, mercuridesulfonylation, acyldehalogenation, nitrodehalogenation, and sulfodeiodination [86HC(44,2)1]. In addition, there have been reports of nitrodeacylation accompanying nitration of 4*H*-pyrrolo[1,2-*a*]benzimidazole [71CA(74)76367], and acyldesilylation accompanying acetylation (but not chloroacetylation) of 4-trimethylsilylindole (84JOC4409).

CHAPTER 5

Reactions Involving Formation of Carbocations at Side Chain α-Positions

The reactivity of an aromatic ring toward an electrophilic reagent is usually measured by the ease with which it stabilizes the transition state arising from electrophilic attack [e.g., Eq. (5.1)]. An alternative quantita-

$$\text{MeO}-\!\!\!\bigcirc\!\!\!-\text{H} \xrightarrow{Z^+} \text{MeO}-\!\!\!\bigcirc\!\!\!\!\!\!{}^{Z}_{H} \longleftrightarrow \text{MeO}{=}\!\!\!\bigcirc\!\!\!\!\!\!{}^{+Z}_{H} \text{ etc.} \longrightarrow \text{products} \quad (5.1)$$

tive measure is provided by the degree to which it stabilizes a carbocation formed at the α-position of a side chain [Eq. (5.2)]. This latter method

$$\text{MeO}-\!\!\!\bigcirc\!\!\!-\overset{X}{\underset{|}{\text{C}}}- \xrightarrow{-X^-} \text{MeO}-\!\!\!\bigcirc\!\!\!\!-\overset{+}{\text{C}}- \longleftrightarrow \text{MeO}{=}\!\!\!\bigcirc\!\!\!={\text{C}}\diagdown\text{ etc.} \longrightarrow \text{products} \quad (5.2)$$

was introduced by H. C. Brown (58JA4979), who used, as a standard reaction, the S_N1 solvolysis (in aqueous acetone) of 2-aryl-2-chloropropanes [Eq. (5.3)]. The substituent reactivity parameters determined in this reaction were designated σ^+ and were found to apply to a wide range of electrophilic substitutions (63AP035).

$$\text{ArCMe}_2\text{Cl} \underset{\text{slow}}{\rightleftharpoons} \text{ArCMe}_2{}^+ + \text{Cl}^- \xrightarrow[\text{OH}^-]{\text{fast}} \text{ArCMe}_2\text{OH} \quad (5.3)$$

The method can clearly measure not only the effect of a substituent upon the reactivity of benzene as outlined above, but also the reactivity of a *different aromatic system,* by placing it instead of the substituted benzene in, for example, Eq. (5.2). This method was introduced by Taylor (62JCS4881), who used, however, a different reaction, the pyrolysis of 1-arylethyl acetates (described below): This was the first application to the determination of heterocyclic reactivities.

The solvolysis reaction has certain disadvantages. First, the 2-aryl-2-chloropropanes are often too reactive to be prepared in the pure state (because of easy thermal elimination of hydrogen chloride or self-condensation) and so have to be generated *in situ* from the corresponding alcohol and hydrogen chloride. Second, the reaction is carried out in a strongly solvating medium, and differential steric hindrance to solvation at various

5. REACTIONS AT SIDE CHAIN α-POSITIONS

positions in the ring can cause the substituent to *appear* to be less electron releasing than is in fact the case [75JCS(P2)1463; 77JCS(P2)678; 78TL267; 82JCS(P2)181, 82JCS(P2)187]. Thirdly, the opportunity for hydrogen bonding may affect the reactivity of, in particular, nitrogen-containing heterocycles. The importance of steric hindrance to solvation is as yet relatively unexplored with regard to this type of heterocyclic reactivity, since few substituent effects have been determined via solvolysis. By contrast, hydrogen bonding is known to be a factor affecting the relative reactivity of heterocyles (many of which are, of course, water soluble for this very reason) in other reactions.

A reaction which overcomes the above deficiencies in the solvolysis technique is the pyrolysis of 1-arylethyl acetates [Eq. (5.4)]. This takes

$$\text{Ar-CH-CH}_2 \quad \xrightarrow{\Delta} \quad \text{Ar-CH}\cdots\text{CH}_2^{\delta+} \quad \longrightarrow \quad \text{ArCH=CH}_2 + \text{MeCO}_2\text{H} \quad (5.4)$$

place in the gas phase, so there are no complications from solvation or hydrogen bonding. The reaction is semiconcerted, and proceeds via the partial formation of a carbocation at the side-chain α-position, and indeed gives an excellent correlation with σ^+ values (Fig. 5.1). Deviations in Fig. 5.1 are in each case due to solvation factors affecting the corresponding solution data. Other esters may be used for determining heterocyclic reactivities (e.g., carbonates) [79JCS(P2)624], but the appropriate ρ factor for the ester type must first be determined. The *solvolysis* of esters [e.g., acetates (69JA7381) or *p*-nitrobenzoates (72JOC2615)] has also been used for determining the electrophilic reactivity of heterocycles, and again the ρ factor for reaction of substituted benzenes must first be determined. Solvolysis of esters is preferable to solvolysis of the the corresponding chlorides, since the former compounds are more stable; it is important, however, to be sure that the mechanism remains S_N1 for all the compounds in the series.

In some cases to be described in the appropriate chapters, the σ^+ values determined from the gas-phase and solution reactions differ (as indeed they do between different electrophilic substitutions carried out in solution). The factors outlined above are at least partially responsible, but, in addition, the polarizability of the aromatic needs to be considered. Thus, the extent of resonance between the aromatic and the positively charged center (be it an electrophile or side-chain cation) depends upon the magnitude of the charge. Yukawa and Tsuno (59BCJ971) introduced the most successful means [Eq. (5.5)] of taking this into account. Subse-

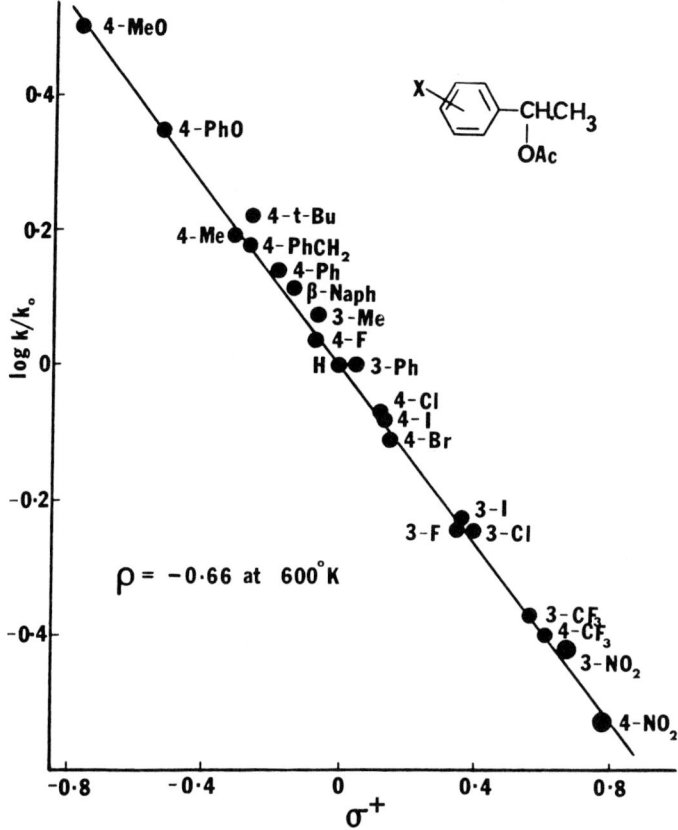

FIG. 5.1 Linear free-energy relationship for pyrolysis of 1-arylethyl acetates.

quently, the more rigorous Eq. (5.6) was introduced (66BCJ2274); Eq. (5.6) differs from Eq. (5.5) only in that the magnitude of the resonance parameter r' is greater.

$$\log k/k_0 = \rho[\sigma + r(\sigma^+ - \sigma)] \quad (5.5)$$

$$\log k/k_0 = \rho[\sigma° + r'(\sigma^+ - \sigma°)] \quad (5.6)$$

For solvolysis of 2-aryl-2-chloropropanes, r is 1.0 by definition, and this value appears to be approximately correct for pyrolysis of 1-arylethyl acetates. Although the magnitude of the charge is smaller in the latter reaction, resonance stabilization of this charge by the aromatic is proportionally greater because of the absence of the solvent (63T937). Nevertheless, for some very polarizable heterocycles, differences in reactivity between the gas phase and solution due to polarizability effects is evident, and is referred to in the appropriate chapters.

Experimental Technique

A. Preparative Method

Both the solvolysis and pyrolysis reactions have one substantial disadvantage, which is the need to synthesize each of the required heterocyclic derivatives, a process usually requiring many steps. The magnitude of the synthetic task in some cases is such that, by comparison, carrying out the kinetic studies is a trivial exercise. The general synthetic routes require the preparation of either the 1-arylethyl alcohol or the 1-aryl-1-methylethyl alcohol, and these are then esterified with acetic anhydride, or reacted with hydrogen chloride as appropriate. The key intermediates (and the ones difficult to prepare) are the aryl halide, the ketone, or the aldehyde; the subsequent conversion of these to the corresponding alcohol is usually straightforward.

B. Kinetic Method

a. *Solvolysis*

Progress of reaction has often been followed by titrating the liberated acid (e.g., 58JA4979). This is unsatisfactory, particularly with heterocycles, because they will be protonated by this acid, leading to a fall-off in reaction rate. More recent methods have therefore used automatic correction to a constant pH (e.g., 72JOC2615), the progress of reaction being determined by the extent of compensation required.

b. *Pyrolysis*

This technique has been used in the majority of studies of heterocycle reactivity for the advantageous reasons described above, yet it is less familiar to the majority of chemists. A full description is therefore given here.

The reaction has a stoichiometry of 2.0 (or 3.0 if carbonates are used, since the alkyl- or arylcarbonic acid instantly decarboxylates). Thus, if the reaction is studied under constant volume conditions, the pressure will double during the reaction. If a pressure-sensitive device is built into the wall of the reactor, then the progress of the reaction can be monitored.

The esters are high-boiling and cannot be studied using the traditional glass apparatus since there are of necessity cold zones in the transfer lines and valves (and upon which both the products and reactants condense). Moreover, such an apparatus requires all of the compound to be transferred into the pyrolysis vessel (it must all arrive simultaneously) by flash distillation, which is impossible. The problem was overcome by using a stainless steel reaction vessel, the whole of which (including the valves and pressure detector) is held at the temperature of the reaction [68JCS(B)1397; 71JCS(B)255; 86JCS(P2)1581]; the apparatus is shown schematically in Fig. 5.2.

The reactor consists of an ~185-ml stainless steel cylinder, enclosed at one end by a stainless steel diaphragm equipped with a platinum contact on the outside, and at the other by two stainless steel valves. One of these permits evacuation, the other permits direct injection of the ester into the vessel through a silicone rubber septum and using a long-needle hypodermic syringe. The septum is enclosed in a perspex dome, through which nitrogen is passed so that there is no possibility of oxygen (which activates the reactor surface) entering the reactor. Nitrogen is also circulated within the barrel of the "gas-tight" syringe to counter any failure in this respect. The reactor rests in a close-fitting aluminum block with capacity

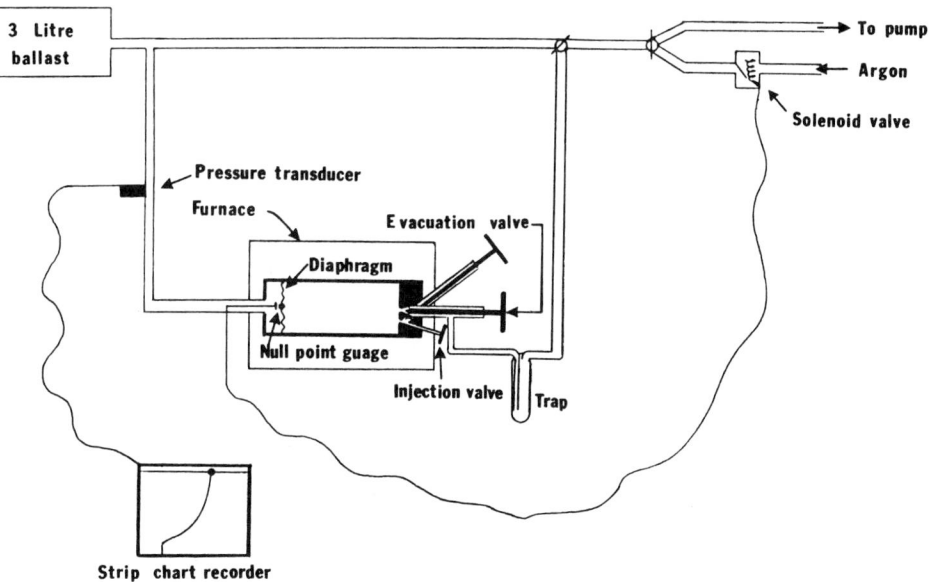

FIG. 5.2 Stainless steel pyrolysis reaction vessel.

of ~1 ft³, and which is fitted with dual thermocouples to record temperature, and stainless-steel heater rods operated by a platinum-resistance thermometer and temperature controller; the temperature can be maintained to better than ±0.1°C over several days. This is achieved through the high mass of the aluminum block, and the heavy and close fitting insulation which surrounds the furnace. The valve ends which pass through the insulation are water cooled to permit handling.

The kinetic method consists of evacuating the reactor (injection valve in the open position) and then closing the evacuation valve. The ester is injected through the system directly into the reactor, the hypodermic needle is partly withdrawn, and the injection valve moved to the closed position; this process takes ~10 sec overall. Pressure build-up in the reactor causes the diaphragm (sensitive to ~0.02 mm pressure change) to deflect, thereby closing an electrical contact. This activates a solenoid valve, which admits oxygen-free argon to the balancing line thus opening the platinum contacts. Further pressure increases within the reactor cause the contacts to close and the process repeats until the reaction is complete. Pressure in the balancing gas line is monitored by a pressure transducer coupled to a strip chart recorder, which therefore draws the first-order kinetic plot. At reaction completion the products are pumped out through a cold trap so that they may be examined as necessary. The method gives kinetics markedly superior to other systems (first-order linearity to >95% of reaction is common) and it is rapid; 20 kinetic runs during a day can be carried out under favorable conditions.

The main disadvantage of pyrolysis kinetics is the possibility of incursion of surface-catalyzed reactions. This can be minimized by prepyrolysis of 3-butenoic acid and, most recently, by coating the reactor surface with iridium-free gold [86JCS(P2)1581]. The traditional method of ascertaining the presence of surface catalysis is to change either the surface-to-volume ratio of the reactor, or the quantity of reagent in a run. A major disadvantage in this procedure has been the tendency to describe small variations in rate (if any) that are observed as being insignificant or within experimental error. Consequently, some kinetics originally described as wholly homogeneous have subsequently been shown to be partially heterogeneous (e.g., 71JHC1101). The best test is to obtain the Arrhenius plot, which should be *completely linear* with no deviations from the mean line by more than ±1%, if surface catalysis is absent. If it is present, then scattered plots are found (the scatter being worse at low temperature, where surface catalysis is more prevalent). Kinetics should therefore be carried out at a minimum temperature range of 50°C (equivalent to 25°C at room temperature) and until a minimum linear rate set is obtained with deviations within the above limit. Since high temperature cannot be measured accurately (in contrast to temperature *differences*) comparisons of rates with literature data are unsatisfactory. Thus, for each study, the rate coefficients for a standard compound should be determined. For 1-arylethyl acetates this is of course 1-phenylethyl acetate. The ρ factor for pyrolysis of 1-arylethyl acetates is -0.66 at 600 K and -0.63 at 625 K. Thus, division of the log k_{rel} value determined from the Arrhenius plot by the ρ factor for the appropriate temperature gives the σ^+ value. The positional reactivites of 40 heteroaromatics have been determined by this method.

Part II

Five-Membered Heterocyclic Rings

CHAPTER 6

Reactivity of Five-Membered Rings Containing One Heteroatom

The reactivity of five-membered rings containing one heteroatom has been reviewed previously in part or in whole in *Advances in Heterocyclic Chemistry*. For example, the electrophilic substitution of thiophenes [63AHC(1)1], furan [66AHC(7)377], selenophenes [70AHC(12)1], pyrroles [70AHC(11)383], and five-membered heteroaromatics [71AHC(13)235] have all been reviewed. Electrophilic substitution in thiophene has also been comprehensively surveyed in *The Chemistry of Heterocyclic Compounds* [86HC(442)1]. This chapter brings the subject completely up-to-date.

1. Acid-Catalyzed Hydrogen Exchange

An important development in the reactivity of five-membered heterocycles is the discovery, through hydrogen-exchange studies, that hydrogen bonding reduces the reactivity to a small but quite significant extent. This is true for thiophene [82JCS(P2)295] and will be more so for furan and especially for pyrrole (where bonding may be so strong that the reduction in rate approaches that expected for N-protonation). Though the effects of hydrogen bonding have not yet been determined in other reactions, it is clear that all data obtained in strong acids will be affected.

A. THIOPHENES

Quantitative data have been reported by several groups using different experimental conditions.

Protiodetritiation in trifluoroacetic acid (TFA) at 70°C was studied by Baker *et al.* [72JCS(P2)97], the rates relative to a position in benzene being 9.2×10^7 and 7.84×10^4. After correction for hydrogen bonding [82JCS(P2)295], the corresponding σ^+ values were -0.913 and -0.560, for the 2- and 3-positions, respectively. Butler and Eaborn also reported rates of 2-detritiation of some 5-substituted thiophenes using a range of TFA–HOAc mixtures at 25°C (Table 6.1), the data giving a Hammett correlation with $\rho^+ = -7.2$. (Errors introduced by the assumption that both

TABLE 6.1
RELATIVE RATES OF 2-SUBSTITUTION OF THIOPHENES
RC$_4$H$_3$S IN ACID-CATALYZED HYDROGEN EXCHANGE

R	Dedeuteriation[a]	Detritiation[b]
5-OMe	>10^{6c}	
3-OMe	>10^{6c}	
5-SMe	1270	
3-SMe	5500	
4-SMe	0.2	
5-Me	200	204
3-Me	305	
4-Me	10	
5-t-Bu		232
5-(5'-Me-thien-2-yl)		144
5-thien-2-yl		42.2
5-Ph	26	15.5
3-Ph	59	
4-Ph	0.25	
4,5-Ph$_2$	7.5	
3,5-Ph$_2$	1800	
3,4-Ph$_2$	14	
H	1	1
5-Cl		0.17
5-Br		0.13
5-I		0.19

[a] 68JGU1933, 68JGU1944; 71JGU2314.
[b] 68JCS(B)370.
[c] As a measure of substituent effects these values will be at least 10-fold too small because methoxy is strongly hydrogen bonded in trifluoroacetic acid [72JCS(P2)97].

the ρ factor and the extent of hydrogen bonding are the same in each medium are unlikely to give an error of more than 0.1 in the ρ$^+$ value.)
Mixtures of TFA and HOAc were also used by Shatenshtein and co-workers for dedeuteriation at 25°C (Table 6.1) and for deuteriation which gave a 2:3-position rate ratio of 3000 (70JGU1609). They also measured (68JGU1933; 70JGU1609) the effects of some substituents on relative rates of 3-exchange as follows: 4-Me (35), 4-MeS (61), 2-MeS (1.4 × 10^4), 5-MeS (1.0), 2,5-Et$_2$ (8000). The substituent effects in detritiation and dedeuteriation show close similarity [in general, in hydrogen exchange at the same temperature log f(detrit.) = 0.96 log f(dedeut.)]. The higher activation by the SMe, Me, and Ph groups across the 2,3- compared to the 2,5-positions demonstrates the high 2,3-bond order in thiophene. Likewise,

the very low 3,4-bond order causes the 3,4-interactions to be much smaller than the corresponding 2,3-interactions for the same substituent, the 2,3 : 3,4 activation ratios being 8.7 (Me) and 90 (MeS). *Meta* interactions (in which conjugation is effectively absent) are even smaller, with deactivation observed in some cases for Ph and SMe. The results also indicate that demand for electron availability by resonance at the less reactive β-position is greater that at the α-position, as theory demands. Thus, 2-SMe activates the 3-position more than the reverse interaction; likewise, 4-SMe deactivates the 2-position but no deactivation is found in the reverse interaction.

Dedeuteriation of the 2-position of thiophenechromium tricarbonyl showed that the $Cr(CO)_3$ ligand is approximately twice as electron withdrawing as an α-chloro substituent (76MI1).

Hydrogen exchange of thiophene with mineral acids has also been studied. Partial rate factors (sulfuric acid at 25°C) were determined for the 2- and 3-positions as 8.7×10^6 and 8.3×10^3 (dedeuteriation) and 5.3×10^6 and 5.7×10^3 (detritiation), respectively (70AK89); these values may be depressed somewhat by hydrogen bonding. The α/β reactivity ratio of ~1000 was confirmed by Butler and Hendry [70JCS(B)852]. They also showed that log k was approximately linearly related to $-H_0$, the slope being least for the more reactive compound as is generally observed for benzenoid derivatives (60JCS3301), and that the ratio increased with decreasing temperature (as required by the Arrhenius relationship). Further, they confirmed (see Chapter 2, Section 1.A) that different acids of identical $-H_0$ value give different exchange rates. Using aqueous perchloric acid as the exchange medium, Butler and Alexander [77JCS(P2)1998] showed that a 5-methyl substituent activated the 2-position 194-fold, in good agreement with the data in Table 6.1. However, dimethylthiophenes were unexpectedly unreactive (giving anomalous 4- and 3-methyl substituent effects if additivity was assumed); this is almost certainly due to reaction of the less soluble dimethylthiophenes being heterogeneous (significantly, the Arrhenius plots showed marked scatter). Hydrogen bonding, which would be greater in the dimethyl compounds, provides an alternative explanation [86HC(44,2)1].

The surprisingly large α : β exchange ratio of 1950 at 20°C for deuteriation in aqueous sulfuric acid-methanol (67ZC58) was based on measurement at one acidity only so it is difficult to judge if the effect is real, indicating H_2OMe^+ to be a less reactive and more selective electrophile than H_3O^+, or due to some other cause. It may be significant that exchange of furan under these conditions showed marked medium effects [72JPR(314)603].

B. SELENOPHENE

Dedeuteriation of selenophene by sulfuric acid in aqueous methanol at 20°C gave the approximate partial rate factors 2.1×10^8 and $<4 \times 10^4$ for the 2- and 3-positions, respectively [72JPR(314)603]. Under these conditions, the Hammett ρ factor for the reaction was determined as approximately -8.4. Direct measurement of the exchange rates relative to that in benzene was not possible, so there may be some error in the corresponding σ^+ values of -0.95 and <-0.53. These σ^+ values indicate that selenophene is more reactive than thiophene, as do other studies described below (Section 7.B).

C. FURAN

Very few quantitative data exist for acid-catalyzed exchange of furans, because of the high sensitivity to acid media, which results successively in hydrogen bonding, substantial protonation, and decomposition. The deuteriation rate in sulfuric acid aqueous methanol is *less* than that for thiophene [72JPR(314)603], contrary to their known relative reactivities toward other electrophilic substitutions. However, the slope of the rate versus acidity profile was anomalous compared to the slopes for the other five-membered heterocycles, and the rate relative to that of thiophene decreased with increasing acidity; this behavior is attributable to hydrogen bonding and/or protonation which would increase with acidity.

In perchloric acid (4.6–7.2 M), hydrogen exchange of furan is only seven times faster than the decomposition rate [70JPR(312)882]. This factor becomes 200 for 2-methylfuran (73ACS153), which permitted determination of its relative exchange rates at 100°C as 1.0 (detritiation), 1.31 (deuteriodetritiation), 1.47 (dedeuteriation), and 3.22 (deuteriation); the activation energies (70–100°C) vary inversely with the reactivity ratios. Decomposition involves both polymer-forming reactions, and hydrolytic cleavage (via rate-determining protonation at the β-position); the former is the more rapid (73ACS153).

The 2- : 3-position reactivity ratio for furan for deuteriation in acetic acid/trifluoroacetic acid (2 : 1) was determined as 500 (72MI1), but decomposition severely affects the accuracy of the β-exchange rate, and hence of the reactivity ratio.

D. PYRROLE

Dedeuteriation of pyrrole in aqueous methanol containing 0.5% sulfuric acid at 20°C gave $\log f_{2,3} \simeq 15$, which, with the k_2/k_3 rate ratio of 0.74 for deuteriation in aqueous hydrochloric acid, indicates $\sigma_2^+ -1.70$ and $\sigma_3^+ = -1.71$ [72JPR(314)603]. However, very large extrapolations of rate data are required to obtain these results. Bean and Wilkinson found that deuteriation by DOAc in aqueous dioxan (buffered with potassium acetate) gave k_2/k_3 ratios of 3.0–1.01 (the smaller values being obtained in more acidic media); N-methylpyrrole shows a higher reactivity at the α-position under comparable conditions. For pyrrole in aqueous dioxan-trifluoroacetic acid, the k_2/k_3 values were 0.97–1.04, the smaller values being obtained in the less acidic media [71CC421; 78JCS(P2)72].

Values of $\log f_2$ and $\log f_3$ under standard conditions were calculated as 10.3 and 10.1, respectively [73JCS(P2)1072]; the corresponding values for N-methylpyrrole were 10.6 and 10.3 [73JCS(P2)1675]. These agree within an order of magnitude with those noted above, after allowing for the temperature difference; however, in a more recent paper [78JCS(P2)72], the values were revised downward by 0.6 units. It has been argued that the values are anomalously low [73JCS(P2)1072], since $\log f_3$ for hydrogen exchange at the 3-position of indole was calculated as 15.4, whereas in acetylation the reactivity of the 3-position of indole is comparable to that of the 2-position of pyrrole (72CC427). This analysis may be incorrect because benzo annelation strongly stabilizes the canonicals for 3-substitution of indole (and in benzene itself produces $\sim 10^3$ increase in hydrogen exchange rate on going to the 1-position of naphthalene). Moreover, acetylation of the 3-position of indole will be almost as sterically hindered as it is at the 1-position of naphthalene. The situation is also complicated by the calculation by Bean and Wilkinson of $\log f_3$ for indole in hydrogen exchange as only 9 [78JCS(P2)72].

Bean and Wilkinson have reported deuteriation rates for several alkylpyrroles in buffered aqueous dioxan-acetic acid at 25°C [78JCS(P2)72]; under these conditions, neither 2-formyl- nor 2-nitropyrrole underwent exchange. Partial rate factors are given in Table 6.2. These show a number of interesting features.

(1) N-Alkyl groups activate very weakly, because their hyperconjugative effects cannot operate from the 1- to any other position.

(2) Activation across the 2,3-bond is much higher than across the 3,4-bond, because of the differences in bond order [cf. the effects of a 2-methyl substituent in naphthalene [68JCS(B)1112]].

(3) Alkyl groups activate more strongly between the 2- and 3-positions than between the 2- and 5-positions. This is general for five-membered heterocycles and reflects the unfavorable bond fixation in the resonance structure for 5-substitution [86HC(44,2)1]. The partial rate factors are overall rather lower than benzene, which may arise from the fact that the

TABLE 6.2
PARTIAL RATE FACTORS RELATIVE TO PYRROLE FOR
DEUTERIATION OF ALKYLPYRROLES C_4H_4NR

R	Position of exchange			
	2	3	4	5
1-Me	2.6	1.6	1.6	2.6
1-t-Bu	3.1	2.4	2.4	3.1
2-Me	—	230	8	75
2-Et	—	100	5	80
2-t-Bu	—	(not measured)	15	40
3-Me	170	—	30	8
3-Et	100	—	20	5
3-t-Bu	60	—	15	7
1,2-Me$_2$	—	440	30	360

transition state for pyrrole hydrogen exchange is nearer to the ground state than that for benzene exchange [65MI1(143)].

(4) The alkyl groups give the highest "Baker–Nathan" activation order (Me > Et > t-Bu) yet found for a chemical reaction. This is because a larger proportion of the transition-state charge resides at the site bearing the alkyl group (and in some cases also adjacent to it) than is the case in benzene; for example, for exchange at the 3-position of 2-alkylpyrroles, the canonicals are given by **6.1** and **6.2**. Steric hindrance to solvation [which gives rise to the Baker–Nathan order [77JCS(P2)678]] is, therefore, very considerable, leading to the observed results [85JCR(S)318].

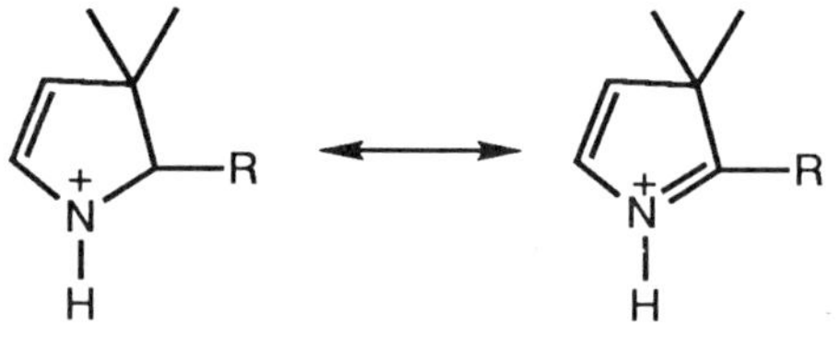

(6.1)   (6.2)

(5) The methyl substituent effects are approximately additive, confirming that (as in benzene) steric effects toward hydrogen exchange are effectively absent.

Alexander and Butler showed that the ratio of exchange rates at the 2-position of 1,3,4-trimethylpyrrole versus the 3-position of 1,2,5-trimethylpyrrole in aqueous phosphate buffer at 25°C was ~4 (for reaction with undissociated acid). This reactivity difference is close to that found for N-methylpyrrole itself, and therefore confirms that the activating effect of the methyl group across the 2,3-bond is the same in each direction [80JCS(P2)110].

Muir and Whiting measured the relative rates of C—H and N—H deuteriation in both pyrrole and indole in aqueous dioxan containing either perchloric or trichloroacetic acids [75JCS(P2)1316; 76JCS(P2)388]. In general, exchange in indole was faster than in pyrrole by factors of 10 (C—H exchange) and 180 (N—H exchange). In pyrrole, N—H exchange was 10- to 15-fold faster than C—H exchange. The k_2/k_3 rate ratio also varied with conditions, being 0.85 ($HClO_4$) and 2.7 (CCl_3CO_2H), probably reflecting a fine balance at the 2-position between the inductive withdrawing, and the resonance donor effects ($-I$ and $+M$), respectively, of nitrogen. The transition state under conditions involving a less reactive electrophile (CCl_3CO_2H) will require greater stabilization by the resonance donor ($+M$) effect. Both C—H and N—H exchange showed general acid catalysis (and hence occur via the A-S_E2 mechanism) but displayed different medium dependence. Different numbers of water molecules appeared to be involved in the respective transition states in an unknown way. It may be relevant that in contrast to C—H exchange, the transition-state charge in N—H exchange cannot be delocalized away from the nitrogen atom.

Reactivities, relative to benzene, in gas-phase protonation of five-membered heterocycles by $^3HeT^+$ are pyrrole, 30; N-methylpyrrole, 6; furan, 0.7; and thiophene, 0.5 (84JA37). The low substrate selectivities are consistent with the anticipated high reactivity of the unsolvated electrophile. Demands for conjugative electron release by the heteroatom will be small, thus accounting for the low reactivity of furan and thiophene (see Section 5).

2. Base-Catalyzed Hydrogen Exchange

FURAN, THIOPHENE, AND SELENOPHENE

This reaction generates a carbanion in a σ orbital in the rate-determining step (Chapter 2, Section 2.A) and so does not directly involve the π cloud. Inductive effects are therefore of primary, and conjugative effects only of secondary, importance. The 2- : 3-position rate ratio for deuteriation of thiophene in DMSO at 25°C is 2.5×10^5 (64MI3, 64MI4; 66MI2), but no direct comparison with benzene is available; thiophene is likely to

TABLE 6.3

RELATIVE RATES OF SUBSTITUTION OF THIOPHENES RC_4H_3S
IN BASE-CATALYZED DEUTERATION

(A) 2-Substitution

R =	5-Me	4-Me	3-Me	H	5-MeS	4-MeS	3-MeS	5-MeO	3-MeO
$k_{rel.}$ =	0.12	0.18	0.06	1.0	50	30	>100	0.39	>100

(B) 3-Substitution

R =	5-MeS	2-MeS	3-Mes	H
$k_{rel.}$ =	10	70	130	1.0

be the more reactive. The reaction gives a small kinetic isotope effect ($k_D : k_T$) of 1.2 due to the symmetry of the reaction pathway. The original interpretation (68JGU1938, 68MI3) that the C—H bond is not broken in a rate-determining step is incorrect. The relative reactivities for furan, thiophene, and selenophene are 1 : 500 : 700 for the 2-positions, and 1 : 1 : 7 at the 3-positions (71JGU1945; 72MI1). Substituent effects in thiophene (Table 6.3) are produced by a subtle balance of the inductive and conjugative effects (64MI3, 64MI4; 66MI2; 68JGU1933, 68MI4). The deactivation of all positions by methyl is straightforward. By contrast, 5- and 3-methoxy groups deactivate and activate, respectively, the 2-positions, due to a strong resonance donor ($+M$), weak inductive acceptor ($-I$) combination in the former case, and a stronger resonance donor ($+M$), much stronger inductive acceptor ($-I$) combination in the latter. For the methylthio group, the different balance of these effects can explain all of the observed results. In addition, there could be some $p \rightarrow d$ π-electron withdrawal ($-M$ effect), but if this were significant one would expect 3-MeS to activate the 2-position more strongly than the 4-position due to the differences in bond order; this is not observed.

As in thiophene, both 3- and 5-methyl groups in furan deactivate the α-position ~10-fold (66CHE643).

3. Nitration

A. THIOPHENES

Three main conditions have been used for nitrating thiophenes.

(1) Nitric acid in various forms: nitric acid-sulfuric acid (cosolvents are frequently used to improve solubility); nitric acid in acetic acid (which

gives better solubility of the aromatic, reducing the possibility of dinitration); nitric acid in trifluoroacetic acid.

(2) Acyl nitrates, usually produced from nitric acid or cupric nitrate in acetic anhydride.

(3) Nitronium salts.

For thiophene itself, conditions of type (1) tend to produce very fast reactions. However, work has shown that the explosively fast reaction with nitric acid-acetic acid, due to nitrosation, can be largely suppressed by using urea [71JCS(B)102] and sulfanilic acid would probably be even better in this respect [cf. 77JCS(P2)248, 77JCS(P2)1693]. Conditions (1) have therefore been confined mainly to deactivated thiophenes, with conditions (2) being preferred for activated thiophenes. It is believed that acetic anhydride suppresses side reactions preceeding via nitrosation under conditions (2) [71JCS(B)102].

Use of nitric acid-acetic anhydride gives an 80% yield of nitrothiophenes; the 2-:3-product ratio was 6.2 with benzoyl nitrate, which also gave a small secondary inverse isotope effect (k_T/k_H = 1.14) [62AK(19)499]. Even higher overall yields have been obtained using nitronium salts (56JCS4257). The kinetics of nitration of thiophene by nitric acid-sulfuric acid in sulfolane or nitromethane [68JCS(B)800] showed that the reaction is governed by the rate at which the reagents can encounter each other and not by their true reactivities; the encounter rate is ~300–400 times that of the rate of nitration of benzene. Nitration of 2-methylthiophene, thiophene, and 2-chlorothiophene by nitric acid in acetic acid occurs at or just below the encounter rate for the first two compounds, hence the substituent effects are anomalously small [71JCS(B)102]. This is also shown by the "standard" nitration rate for thiophene being only 5.6 times greater than that for benzene [75JCS(P2)1600], which is clearly much too small.

The effects of substituents invariably reflect the 2,3-, 2,5-, and 3,4-conjugative interactions, and the high 2:3-bond order, superimposed upon the intrinsic 2 > 3 positional reactivity order. Thus, with cupric nitrate in acetic anhydride, the 3-Me-, 3-AcOCH$_2$-, and 3-HO$_2$CCH:CH substituents gave mainly 2- and 5-substitution [57JA3800; 80CS(15)20]. With the same reagent, 2-phenylthiophene gave the 3- and 5-derivatives in the ratio 0.67:1, whereas 3-phenylthiophene gave the 2- and 5-derivatives in the ratio 9:1 (58JGU1288; 67ACS2823). Both sets of data, in combination with the intrinsic 2:3 rate ratio of ~6, demonstrate the high 2:3-bond order. Interestingly, just as nitric acid-acetic anhydride gives anomalously high ortho/para ratios in the nitration of biphenyl [66JCS(B)727], so nitration of 2,2'-bithienyl (**6.3**) with nitric acid-acetic acid produced a 3-:5-product ratio of 0.31, which becomes 1.12 with nitric acid-acetic anhydride. Likewise, for 2,3'-bithienyl (**6.4**), the 3:5 ratios are correspondingly 0.46 and 1.5 (71JHC849; 74JHC1017). 2-Methyl- and 2-methoxythiophene gave 5-:3-product ratios of 2.3 (60AK563) and 1.5 (53JA3697), respectively, the lower ratio in the latter case reflecting the effect of the high 2:3-bond order upon the larger resonance donor ($+M$) effect for methoxy.

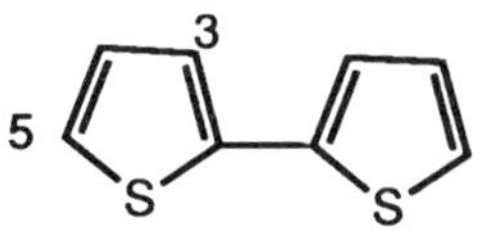

(6.3) 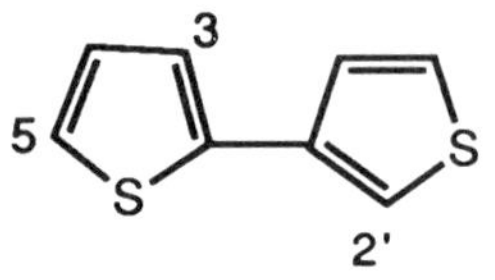 (6.4)

Nitric acid and acetic anhydride have been used to show that the 2-SMe (56JCS4114), 2-CH$_2$Cl [61CR(252)2419], 2-CH$_2$OAc [61CR(252)2419; 62BCJ1420], 2-Cl (52JA2965), 2-Br (35JA1763; 53JA3517), 2-I (43CB419) and 2-CH$_2$Ph substituents all direct into the 5-position. The yields of 5-, 4-, and 3-nitro products for 2-benzylthiophene were 87, 3, and 10%, respectively (72JHC849), which incidentally shows the greater electrophilic reactivity of thiophene compared to benzene. Yields of the 4-isomers in nitration of thiophenes with the following deactivating substituents at the 2-position were CH(OAc)$_2$, 14; CO$_2$H, 31; CN, 43; COCH$_3$, 52; SO$_2$Cl, 78; NO$_2$, 80–85% (63BSF1651; 72USP3707480). These yields reflect the increasing deactivation of the 3- and 5-positions by the mesomerically withdrawing (-M) effect of the substituents along this series. For nitration by nitric acid in trifluoroacetic acid, deactivating 2-substituents gave slightly different results, the 5-:4-product ratio being 0.77 (CHO), 1.7 (CN), and 0.98 (NO$_2$) (68ACS2754). Insignificant 3-substitution accords with the high 2:3-bond order. Likewise, these substituents at the 3-position gave 80–90% of 5-substitution. The remaining product arose from approximately equal amounts of 2- and 4-substitution: the high and low bond orders for the 2:3- and 3:4-bonds, respectively, creates a large differential in deactivation, which cancels out the intrinsically greater reactivity of the α- compared to the β-site.

Nitration of 2-nitro- and 2-formylthiophene by nitropicolinium fluoroborate gave 5-:4-product ratios of 0.61 and 1.0, respectively, whereas with nitric acid-sulfuric acid, 2-acetyl- and 2-formylthiophenes gave corresponding ratios of 1.0 and 4.0 (71BAU1142; 71JOU1803). These results suggest that, in sulfuric acid, nitration occurs on species in which the carbonyl groups are protonated, especially since the proportion of the 4-isomers increased with increasing acidity. Nitration of 2-nitrothiophene with nitric acid-sulfuric acid gave 56 and 44% of 4- and 5-dinitro derivatives, respectively (62AK527).

Lastly, nitration of 2-substituted thiophenes by nitric acid-acetic anhydride was said to give a Hammett correlation with $\rho = -6.7$ (82CHE127). However, this is invalid since σ values were used instead of σ^+, it was assumed that all substitution occurred at the 5-position (certainly incorrect), and the correlation line missed the origin by a very large amount.

B. PYRROLE

The nitration of pyrroles by a variety of reagents over a range of temperatures was studied by Morgan and Murray (66T57; 70T5101; 71T245), who found fewest side reactions with nitric acid-acetic anhydride. These conditions gave a temperature-independent 2- : 3-product ratio of ~4 and partial rate factors $f_2 = 1.3 \times 10^5, f_3 = 3 \times 10^4$. These values are anomalously low compared to (say) hydrogen exchange and very probably refer to a strongly hydrogen-bonded species; partial protonation is also expected to decrease the available concentration of neutral species.

For the nitration of 1-, 2-, or 3-acylpyrroles (Pyr·COR, R = H, Me, OH, or OMe), overall yields were ~25, 45, and 55%, respectively. The effects of the acyl groups on isomer ratios were exactly as expected: 1-Acyl groups produced equal amounts of 2- and 3-substitution, and 2-COMe gave a 4 : 5-product ratio of ~2, whereas this became 1 for the more electron-withdrawing 2-CO_2Me substituent, and 3-acyl groups gave mainly 5-substitution. Only the 5-isomer was detected in the nitration of 3-methoxycarbonylpyrrole (41RTC650) and 1-methyl-3-nitropyrrole (59BAU1258).

Nitration of 2-methylpyrrole by fuming nitric acid and acetic anhydride at low temperature gave a 12–15% yield of mononitro products, the 5- : 3-product ratio being 5.7 (70JHC399). Nitration of 1-benzylpyrrole gave 60% of substitution at the 3-position, much greater than for pyrrole or 1-methylpyrrole (67CJC2227), suggesting that a steric effect operates here.

Nitration of 2- and 3-nitropyrrole and their 1-methyl analogues by nitric acid-sulfuric acid at 25°C showed the nitro groups to deactivate by factors of 1.5×10^5 and 2.6×10^4, respectively; this is less than that in benzene, which may reasonably be attributed to the higher reactivity of pyrrole. The 1-methyl compounds were only 1.5 times more reactive (75CHE571), which is consistent with the results obtained in hydrogen exchange (see Section 1.D.).

4. Halogenation

A. THIOPHENE AND SELENOPHENE

The high degree of bond fixation in thiophene is such that addition readily occurs, subsequent dechlorination and dehydrochlorination leading to substitution products. However, addition can be minimized by using polar solvents (e.g., aqueous acetic acid) and exclusion of light; selenophene probably behaves similarly. The high selectivity for molecular chlorination, and the relatively weak deactivation by the chloro substituent, result in ready dichlorination, which can be minimized by using high thiophene/chlorine reagent ratios. An ingenius method for minimizing dichlorination involves passing chlorine over refluxing thiophene. Chlorothiophene thus formed does not reflux and so does not come into contact with chlorine, and this gives the highest yield (72%) of any monochlorination method (78GEP2749235).

Addition is not a problem accompanying molecular bromination, but dibromination is. Maximum yields (67–78%) of 2-bromothiophene can be obtained by using dilute thiophene solution (45JA2092; 53YZ1023). Likewise, yields of up to 71% can be obtained with N-bromoamides [19LA(426)61; 44LA(566)1; 51MI2; 52MI2], which produce dilute equilibrium concentrations of bromine. Other methods give yields of 45% [cyanogen bromide; 22LA(430)79], 60% [potassium bromate/hydrogen bromide (82BAU2104)], and 100% [dioxan dibromide (54JGU1251)].

Molecular chlorination and bromination of thiophene give only 1 and 0.2% of the 3-isomer, respectively, among the monohalo products [70JCS(B)1153], due to the high selectivities of the reactions. However, if halogenation is carried out at very high temperature (up to 750°C), the proportion of the 3-isomer increases greatly and is dominant in bromination (53JA3517). This suggests that the 3-isomer is thermodynamically more stable, due possibly to repulsions between the sulfur d orbitals and the α-halogen. The rates of molecular halogenation of thiophene relative to benzene are 1.3×10^7 (Cl_2) and 1.7×10^9 (Br_2) (65T843), from which we calculate $f_2 = 3.86 \times 10^7$ (Cl_2) and 5.09×10^9 (Br_2); $f_3 = 3.9 \times 10^5$ (Cl_2) and 1.02×10^7 (Br_2). The f_3 values are probably inaccurate since 3-substitution is such a minor reaction; the f_2 values lead to σ^+ values of -0.76 and -0.81, respectively. In the perchloric acid-catalyzed positive bromination of thiophene by hypobromous acid in 90% aqueous dioxan at 25°C [70JCS(B)1153], k(thiophene) : k(benzene) was 6.3×10^4, with only ~0.8% of 3-bromination detected. This yields $f_2 = 1.88 \times 10^5$ and $f_3 = 1.5 \times 10^3$, with $\sigma_2^+ = -0.85$. In these reactions, selenophene was more reactive than thiophene by factors of 47.5 (Br_2), 6.5 (Cl_2), and 4.5 (Br^+)

[70JCS(B)43], which is qualitatively consistent with the reaction selectivities. The first two factors afforded σ_2^+ values of -0.94 and -0.84, respectively. No addition was detected and in each bromination the proportion of the 3-isomer was less than 1%.

The high α/β selectivity in molecular halogenation means that 2-substituents give mainly 5-substitution; 3-substituents give 2- and 5-substitution. Bromination of 2-methyl- (68JOC2902), 2-ethyl- (52BSF713), 2-*t*-butyl-, and 2-octylthiophenes (1886CB644) takes place almost exclusively at the 5-position, especially when the alkyl group is large. 2-Phenylthiophene gave 96% 5- and 3% 3-bromo derivatives (72ACS1851), and 2-chlorothiophene forms 98% of 2,5-dichlorothiophene (50USP2492644). Bromination of 2-iodothiophene gave 88% of the 5-bromo derivative, the reaction being accompanied by deiodination products [37LA(527)237]. Conjugatively electron-withdrawing substituents such as 2-CHO or 2-COCH$_3$ can outweigh the high α/β selectivity, but the extent to which this occurs depends very much on the amount of catalyst (usually AlCl$_3$) necessary for halogenation of these less reactive compounds. An excess of catalyst, or a substituent more readily coordinated to AlCl$_3$, will give a greater proportion of the 4-isomer. Thus, AlCl$_3$-catalyzed chlorination of 2-formylthiophene or 2-thienoic acid gave 74 and 48% of 4-substitution, respectively (76JHC393), showing the former substituent to coordinate more easily as expected. 2-Benzoylthiophene gives mainly the 5-isomer if only catalytic quantities of AlCl$_3$ are used, but with excess AlCl$_3$ the 4- : 5-product ratio becomes 6 (70BAU2592). Likewise, bromination of 2-formylthiophene with *N*-bromosuccinimide gives 70% of the 5-bromo derivative, but with bromine in sulfuric acid (which protonates the substituent) the 4- : 5-product ratio of mono-substituted products is ~ 1.0 (accompanied by 50% dibromination) (71BAU2687). The low 3 : 4-bond order prevents resonance donor ($+M$) substituents at the 3-position from substantially activating the 4-position, so that 3-methyl-2-phenylthiophene and 2-methyl-3-phenylthiophene gave 95 and 93% of the 5-bromo derivatives, respectively [76JCS(P1)2355].

3-*t*-Butylthiothiophene brominates 57% in the 2-position (73BAU2233), the combination of the resonance donor ($+M$) effect and high 2 : 3-bond order more than balancing steric hindrance; similar reasoning accounts for the almost exclusive 2-bromination of 3-methylthiophene (68JOC2902) and 2-halogenation of 3-acetaminothiophene (54JA2447). 3-Iodothiophene gave 90% of the 2-bromo derivative (together with deiodination products) [37LA(527)237]. 3-Phenylthiophene gave both 2- *and* 5-bromo derivatives (only the former would be expected from consideration of electronic effects), but this is due to the greater thermodynamic stability of the 5-isomer as a result of steric hindrance in the 2-isomer (67JOC463;

68JOC2902). Resonance acceptor ($-M$) substituents at the 3-position strongly deactivate the 2-position because of the high 2 : 3-bond order (especially if the substituent is coordinated to a Lewis acid), and thus the $AlCl_3$-catalyzed bromination of 3-acetylthiophene goes 74–80% into the 5-position (75BSF2334; 76T1403). The low 3 : 4-bond order means that large resonance acceptor ($-M$) deactivation of the 4-position does not take place and hence 3-formylthiophene gave 83% of the 4,5-dichloro derivative (73M1599).

Iodination of thiophene has been carried out with (1) iodine/mercuric oxide (to remove HI and thereby prevent protolytic cleavage of the thiophene ring); (2) iodine/iodic acid/acetic acid, and (3) iodine monochloride. In each case, 70–80% yields of 2-iodothiophene have been obtained [1884CB1558; 43OSC357; 60LA(634)84; 71MI2; 82IJC(A)417]. 2-Iodothiophene iodinates largely in the 5-position (1884CB1558), while 3-iodothio-

TABLE 6.4
PARTIAL RATE FACTORS (RELATIVE TO THIOPHENE) FOR 5-HALOGENATION OF SUBSTITUTED THIOPHENES AT 25°C

		Bromination (Br_2)		
Substituent	Chlorination (Cl_2)[a]	HOAc[b]	85% aq. HOAc[c]	Iodination (ICl)[d]
2-Et	—	—	402	—
2-Me	—	631	392	—
(2,5-Me_2)[e]	—	350	—	—
3-Me	—	1020[f]	89	—
2-t-Bu	—	—	291	—
2-CH_2Ph	—	300	—	—
2-Ph[g]	—	—	74	—
2-F	—	—	1.45	—
2-Cl	0.36	0.52	0.081	0.23
2-Br	0.48	0.38	0.070	0.21
(3-Br)[f]	—	—	0.035	—
2-I	0.53	0.94	0.14	0.72
2-CO_2H	1.1×10^{-4}	3.3×10^{-5}	—	—
2-CO_2Et	1.38×10^{-4}	1.1×10^{-5}	—	—
2-NO_2	1.1×10^{-6}	—	—	—

[a] In HOAc; 70JCS(B)848; 71AHC(13)235; 74CC333.
[b] 71AHC(13)235.
[c] 70JCS(B)848; 81IJC(19A)1183.
[d] 82IJC(A)417. Data are for 26°C in 50% aqueous HOAc.
[e] Reaction at the 3-position.
[f] This must refer to reaction at the 2-position.
[g] Only the thiophene ring brominates.

phene iodinates in the 2- and 4-positions in 93 and 7% yields, respectively (63AK191).
Partial rate factors for the 5-halogenation of substituted thiophenes are gathered in Table 6.4.

B. FURAN AND PYRROLE

Both furan and pyrrole are rather unstable under halogenation conditions but introduction of an electron-withdrawing substituent stablilizes the rings and permits determination of their reactivities. Thus molecular bromination of the 2-methoxycarbonyl derivatives of furan, thiophene, and pyrrole in acetic acid at 25°C gave relative reactivities as thiophene, 1.0; furan, 120; pyrrole, 5.9 × 10^8 [68JCS(B)392]. Only monobromo derivatives were obtained: 100% 5-bromo for 5-methoxycarbonylthiophene and -furan, and 23% 5-bromo and 77% 4-bromo for the corresponding pyrrole (cf. 65JCS459). The relative rates should be regarded only as semiquantitative estimates for the parent ring systems, since the deactivation by the substituent may vary between the systems. A similar approach indicated furan to be 50-fold more reactive than thiophene in chlorination (70TL1389). Bromination of 3-acetyl- and 3-methoxycarbonylpyrrole occurs exclusively in the 5-position (67CJC897).
Bromination of pyrrole with bromine gives mainly the thermodynamically more stable 3-bromo derivative, which is produced from the 2-isomer, formed initially under kinetic control, through isomerization catalyzed by the reaction product HBr. If N-bromosuccinimide is used for bromination, no HBr is produced and the 2-isomer dominates (81JOC2221). Bromination of furan gave cis- and trans-2,5,-dibromo-2,5-dihydrofurans (isolated as the dimethoxy derivatives) in a ratio of ~3 : 1 (67BCJ130). A spectroscopic study carried out at low temperature (−50°C) confirmed this ratio and also showed that 20% of trans-2,3-dibromo-2,3-dihydrofuran was formed (75CC875). It is not known whether these adducts (which demonstrate the competition between 1,4- and 1,2-addition) are intermediates along the pathway for bromine substitution, or whether they are produced in a separate equilibrium as is the case for bromination of benzo[b]furan.
The effect of complexing between the carbonyl group and aluminum chloride also shows up here in that the $AlCl_3$-catalyzed bromination of 2-acetylfuran gave equal amounts of 4- and 5-bromo product (with the 4,5-dibromo isomer as the main product). 2-Formylfuran (more easily coordinated, presumably due to lower steric hindrance) gave mainly the 4-isomer (71BSF238; 72CHE541; 73BAU2666). Halogen exchange also occurs for both 2-formylfuran and 2-methoxycarbonylfuran (and also 2-formyl-

thiophene), the normal 4,5-dibromo derivative being accompanied by the 4-bromo-5-chloro product [71BSF238; 73JCS(P1)1766]. Iodination of 2-formylfuran gave 32% of the 5-iodo derivative (55AK87).

Bromination of 1-benzylpyrrole gives mainly the 3-substituted product (as in nitration) although polysubstitution occurs very easily (67CJC2227).

5. Alkylation

PYRROLE, N-METHYLPYRROLE, FURAN, AND THIOPHENE

An unusual alkylating electrophile, which is also very selective due to the delocalization of charge, is **6.5**; this gives the following relative rates of formation of **6.5a** in acetonitrile at 20°C: pyrrole, 5×10^5; furan, 3×10^3; thiophene, 1. Indole had the same reactivity as pyrrole, of interest because there are few comparative data for these molecules (73CC540).

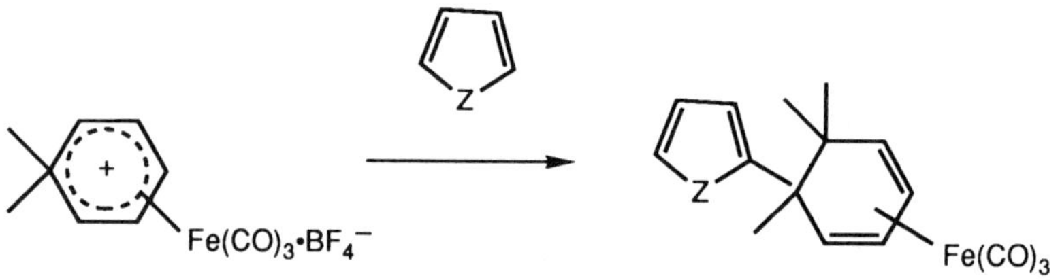

However, alkylation in solution usually requires strongly acidic conditions, so data are mainly confined to thiophene, and even this tends to polymerize, though i-propylation of 3-acetyl- and 3-methoxycarbonylpyrrole has been reported, exclusive 5-substitution being obtained (67CJC897). The problem of poly-, as distinct from dialkylation, is less than in benzene because of the difference in the 2-:3-position reactivity ratio; consequently, reaction tends to cease at the 2,5-dialkyl stage. 2-t-Butylation of thiophene with 2-methylpropene and sulfuric acid gave a 48% yield (with 17% 2,5-di-t-butylation) (50BP625173); phosphoric acid–Kieselguhr in an autoclave gave 73% (67BAU1120). In the latter work, the 2-:3-product and hence rate ratio was 3 for thiophene, and 11.5 for furan [which gave an 84% overall yield (67BAU1561, 67MI5)] at low temperature, but the product ratio decreased in both cases with increasing

TABLE 6.5
REACTIVITY RELATIVE TO TOLUENE, AND 2:3-REACTIVITY RATIOS FOR GAS-PHASE ALKYLATION WITH R^+

R^+	Pyrrole	N-Methylpyrrole	Furan	Thiophene
MeF^+Me	~0.6	1.4	1.7	1.1
$t\text{-}Bu^+$	1.0	2.2	5.2	1.0
2- : 3-Rate ratio for reaction with $t\text{-}Bu^+$	0.29	0.89	10.0	4.0

temperature, the lower thermodynamic stability of the 2-isomer due probably to steric hindrance. 2-Alkylation with alkyl halide/aluminum chloride gave a 2-: 3-rate ratio of 4.9 for formation of *t*-butylthiophenes and 1.5 for the formation of *i*-propylthiophenes (77JOU329). Alkylation with alcohols and aluminum silicate catalyst gave corresponding ratios of 2.0 and 1.5 (66MI1). These results demonstrate the low selectivity of alkylation (as found in benzene chemistry) and also the general observation [72MI2(149)] that *t*-butylation, in which there is greater opportunity for delocalizing the positive charge in the electrophile, is more selective than *i*-propylation. Selectivity should, in both cases, be less in the gas phase since the electrophiles will be unsolvated, and indeed methylation with $CH_3F^+CH_3$ and *t*-butylation with $t\text{-}Bu^+$ (Table 6.5) show very little substrate selectivity, though the positional selectivity is in some cases quite high. However, it is believed that these gas-phase reactions may involve reversible formation of an electrostatic adduct, and thus the mechanism differs significantly from that which applies in solution [82JA7084, 82JA7091; 83JCS(P2)1491].

Other data for alkylation of substituted pyrroles are noted in Section 8.D.

6. Chloroalkylation

THIOPHENE

The most common chloroalkylation reaction, chloromethylation, is usually carried out with formaldehyde and hydrogen chloride, the electrophile being $^+CH_2OH$ (OH is replaced by Cl in a subsequent fast step). Lewis acids, especially zinc chloride, are sometimes used as cocatalysts.

The reaction favors 2-substitution of thiophene, yields being 40–62% [42JA477; 49OS(29)31; 50USP2527680]. Halogens in the 2-position direct chloromethylation into the 5-position (70–80%) (47JA1549) as do acetyl, vinyl, ethoxycarbonyl (60CCC1058), propionyl (62MI2), and ethynyl groups (62CCC372). In the presence of excess aluminum chloride, 2-acetylthiophene gave a 4- : 5-product ratio of 4.3, due to increased electron withdrawal at the 5-position by the coordinated carbonyl group; 2-formylthiophene behaves similarly (73JOU1542). Sulfuric acid-catalyzed chloromethylation of 2-acetylthiophene by bis(chloromethyl) ether showed an increase in the 4- : 5-product ratio with increasing acidity for a similar reason (71CHE1265).

Chloroethylation (using acetaldehyde instead of formaldehyde) involves a less reactive electrophile (the positive charge being better delocalized by the methyl group). Reaction of 2-methoxycarbonylthiophene with $ZnCl_2$/HCl and either formaldehyde or acetaldehyde gave the 5-chloromethyl and 5-(1-chloroethyl) derivatives in 72 and 32% yields, respectively (62MI2).

7. Acylation

Vilsmeier–Haack formylation (using DMF/$POCl_3$), acetylation, and trifluoroacetylation have been studied. The reagent in trifluoroacetylation (the acid anhydride) is sufficiently electrophilic that no catalyst is required, an advantage when rings are susceptible to cleavage. Use of acetyl trifluoroacetate gives mixed acetylation and trifluoroacetylation, the proportion of the latter increasing with increasing reactivity of the aromatic. However, for compounds of the reactivity of 2-methylthiophene, trifluoroacetylation constitutes only ~1.4% of the total reaction (67MI3). The reliability of the relative rate data depends upon there being no complexing of the aromatics with the Lewis acid catalysts. Such complexing is so strong for pyrrole that it is very unreactive in the presence of these, and there is strong evidence (Section 7.C) that the reactivity of furan is also affected by complexing. As acylation deactivates and stabilizes the five-membered rings, the reaction is attractive for kinetic studies.

A. THIOPHENES

Formylation of thiophene goes 70% into the 2-position, and the 2-Me, -Et, -Pr, -Cl, -Br (48JOC635; 49JOC405,638), -SMe (54JCS237), -SEt (60BAU1700,1705), and -cyclohexyl (80CHE339) derivatives formylate 45–85% in the 5-position, whereas for 2-arylthiophenes the yields were 80–100% (72ACS1851; 73CCC1809; 74CHE136). For 2-bromothiophene, nucleophilic substitution of bromine by chlorine occurred as a side reaction. For 3-substituted thiophenes, the 2-:5-product ratios are 16 [Ph (70ACS99)] and 7 [Me, Br [73JCS(P1)2327]], but steric hindrance causes 3-t-butylthiophene to give 85% of 5-substitution (55JCS21).

Kinetic investigation of the formylation of thiophene and of 2- and 3-methylthiophene in 1,2-dichloroethane showed the reactions to be first order in each of aromatic, DMF, and $POCl_3$. However, for the much more reactive 2-methoxythiophene, the reaction is second order overall, and zeroth order in aromatic. This is consistent with a slow pre-equilibrium to form the electrophile $[(Me_2NC^+HCl)PO_2Cl_2^-]$, which reacts with the aromatic to give the Wheland intermediate in a step which is rate-determining except when the aromatic is very reactive, whereupon this step becomes relatively fast. Finally, the intermediate decomposes to products in a fast step [72JCS(P2)2070].

Use of $COCl_2$ instead of $POCl_3$ gives a less reactive electrophile $[Me_2NC^+HCl]Cl^-$ (because the negative charge is not delocalized in the anion). Thus, kinetics are first-order in aromatic for all thiophenes, thereby permitting meaningful comparisons of reactivity (Table 6.6). These data correlate approximately with σ^+-values, giving $\rho = -7.3$. A more limited data set for formylation with DMF–$POCl_3$ (Table 6.6) was said to give a ρ factor of -6.5, consistent with the latter reagent being less selective (since it is more electrophilic). However, in this work σ

TABLE 6.6
PARTIAL RATE FACTORS RELATIVE TO THIOPHENE FOR 5-ACYLATION
OF SUBSTITUTED THIOPHENES

Substituent	Formylation[a]	Formylation[b]	Acetylation[c]	Trifluoroacetylation[d]
2-OMe	1.0×10^6	—	—	1.8×10^6
2-SMe	—	—	—	5200
2-t-Bu	398	80	—	540
2-Et	217	34	—	520
2-Me	196	—	17	380
2-Ph	178	—	—	110
3-Me	7	—	—	—
2-Cl	—	0.14	0.071	0.58
2-Br	—	—	—	0.46
2-I	—	0.51	—	—

[a] DMF–$COCl_2$ in $CHCl_3$ at 30°C [74JCS(P2)1610].
[b] DMF–$POCl_3$ in $CHCl_3$ at 80°C (80CHE230).
[c] Ac_2O–$SnCl_4$ in 1,2-dichloroethane (1,2-DCE) at 25°C (67T1739).
[d] $(CF_3CO)_2O$ in 1,2-DCE at 75°C [72JCS(P2)71].

instead of σ^+ values were incorrectly used, and the difference in temperature was neglected. Use of σ^+ gives $\rho = -5.1$ at 80°C, which is equivalent to -6.0 at 30°C, the temperature used in the other study.

Many quantitative data are available for acetylation and benzoylation of thiophene under a variety of conditions, including use of acid chlorides or anhydrides with aluminum chloride or stannic chloride catalysts, acids or anhydrides with an inorganic acid catalyst, acids with phosphorus pentoxide, and N,N-dimethylacetamide with phosphorus oxychloride. With regard to the two principal catalysts, stannic chloride needs to approach the acylating reagent more closely because its coordinating ability is weaker than aluminum chloride. This, coupled with its size, makes acylation with stannic chloride more hindered than with aluminum chloride as catalyst. The latter tends to decompose thiophene, though this can be minimized by premixing the acyl halide and catalyst, in which case up to 90% yield of 2-acetylthiophene can be obtained (1886CB636). On the other hand, the ketone products are much more strongly coordinated to aluminum chloride, thereby deactivating the ring so that disubstitution is least with this catalyst. Acylation of thiophene by an equimolar mixture of acetyl and benzoyl chlorides gave 88% of 2-acetylthiophene, showing that acetylation is the faster reaction (70JOU2531).

Steric hindrance is greater for acetylation than for benzoylation (delocalization of charge in the former means that the electrophile tends to be the polarized complex $RCO^+AlCl_4^-$ rather than the free acylium ion [72MI2(181)]). The higher steric requirement of acetylation is consistent with the relative yields of **6.6**, which are 60% (R = Me, 48RTC309) and 72% (R = Ph, 78CJC1970).

(6.6)   (6.7)   (6.8)

Acetylation of 5-methyl-2-phenylthiophene (in which the 3-position is the most activated but also the most hindered) gives 84% of **6.7** (R = Me, Ar = Ph) if $AlCl_3$ is the catalyst, but 77% of **6.8** (R = Me, Ar = Ph) if $SnCl_4$ is used. For acetylation with bulkier acyl chlorides, the amount of 3-substitution (to give **6.7**) was always less with $SnCl_4$ than with $AlCl_3$ as catalyst. However, stannic chloride-catalyzed benzoylation, which is less hindered, gave 75% of **6.7** (R = Ph, Ar = Ph) (73T413). Acetylation is less hindered than formylation (the bulk of the $DMF-COCl_2$ electrophile is self-evident), and so in contrast to the above results, 5-methyl-2-phenylthiophene will not formylate at all [72CR(C)(275)49]. The steric requirement of acetylation is also evident in reaction of 2-phenylthiophene, which gives slightly more of the 3-isomer (0.1–1.0%) than does formylation (0.05%) (72ACS1851). In *apparent* conflict with this is the lower 2-:5-rate ratio of 2.3 (compared to 16 in formylation) for acetylation of 3-phenylthiophene (70ACS99), but here the much higher selectivity of formylation causes a very large differential ortho- versus meta- phenyl substituent effect. The effect of steric hindrance shows up in many other data. For example, the 2-SMe and 2-OMe substituents give 5-:3-product ratios of >10 (73CHE447), while 3-alkylthiophenes gave the following percentages of 2- and 5-acetyl derivatives, respectively: 74, 18 (Me); 31, 48 (*i*-Pr); —, 82 (*t*-Bu) (47JA3093, 55JA4066, 55JCS21). The lower steric hindrance in benzoylation is shown by the higher 2-:5-product ratio of 6 for 3-methylthiophene (75CJC1) (cf. 4 for acetylation). Steric hindrance is likely to be the reason for both 2- and 3-*t*-butylthiophenes giving the same yield (82%) of the 5-isomer in acetylation (55JCS21).

Steric hindrance in five-membered rings should be less than in benzene because of the different angles between adjacent C-substituent bonds (60° in benzene, 72° for a regular pentagon), which produces a greater distance between adjacent substituents in the former. The importance of this effect is demonstrated, for example, for stannic chloride acetylation of methyl and *t*-butyl derivatives of benzene and thiophene. Partial rate factors, relative to the unsubstituted aromatic, are shown in Scheme 6.1 (71T4667).

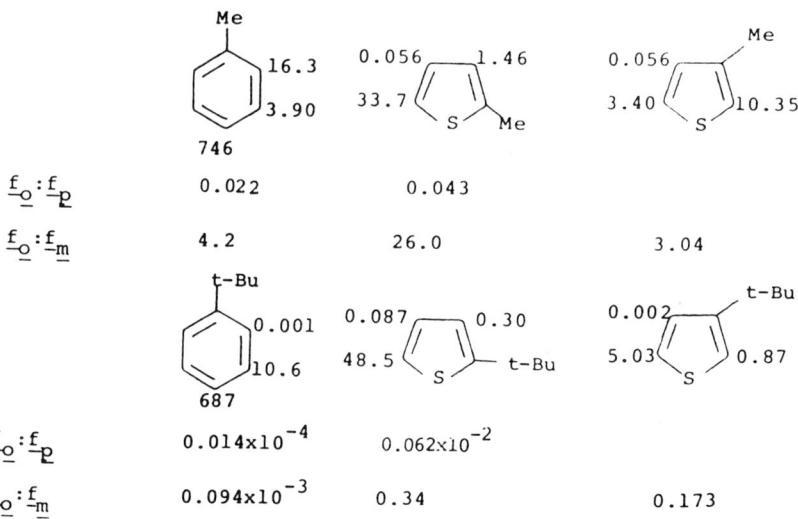

SCHEME 6.1. Partial factors for stannic chloride-catalyzed acetylation.

The ratio $f_o:f_p(Me)/f_o:f_p(t\text{-Bu})$ is 15,700 for the alkylbenzenes but only 69 for the 2-alkylthiophenes. Likewise the $f_o:f_m(Me)/f_o:f_m(t\text{-Bu})$ ratios are 44,700 (alkylbenzenes), 76.4 (2-alkylthiophenes), and 17.6 (3-alkylthiophenes). These results are free from any bond-order effects (which would equally affect the activation by both methyl and t-butyl substituents). They indicate that the 2-t-butyl group produces more hindrance than the 3-t-butyl group, possibly through buttressing between it and the sulfur d orbitals. This view is reinforced by results for electrophilic substitution of benzo[b]thiophene (Chapter 8, Section 2.A.b.i).

The relative reactivity ratio of thiophene to benzene in stannic chloride-catalyzed acetylation by acetic anhydride in 1,2-dichloroethane at 25°C, determined from the relative reactivity ratio of thiophene to anisole (3.1) (67T1739) and of anisole to benzene (2.9 × 10^5), is 9 × 10^5 [70JCS(B)1153]. The benzene : anisole ratio is for aluminum chloride-catalyzed acetylation by acetyl chloride (62JA1658) but this exhibits similar selectivity to the stannic chloride-catalyzed acetylation [70JCS(B)1153]. The amount of 3-substitution at 25°C in acetylation by acetic anhydride/stannic chloride, by acetic anhydride/iodine (which shows similar selectivity but lower reactivity, 67T1739), and by acetyl trifluoroacetate, and in benzoylation by benzoic anhydride/stannic chloride, was very small (0.3–0.7%), leading to a 2-:3-rate ratio of ~200. The ratio decreased slightly with increasing temperature (reactivity–selectivity effect), but the amount of 3-isomer was less than 1.5% at 75°C [70JCS(B)1153]. From these data, $f_2 = 2.7 \times 10^6$ and $f_3 = 1.35 \times 10^4$, hence $\sigma_2^+ = -0.71$ and $\sigma_3^+ = -0.45$.

Partial rate factors (relative to thiophene) for acetylation and trifluoroacetylation at the 5-position of 2-substituted thiophenes (Table 6.6) show that the trifluoroacetylation is slightly more selective than formylation (with $\rho = -7.4$), whereas the acetylation is much less selective. In acetylation (in benzene at 20°C), 2-(2-thiophenyl)thiophene and 3-(3-thiophenyl)thiophene are ~50-fold more reactive than thiophene (77TL389).

Acetylation has also been carried out with N,N-dimethylacetamide/phosphoryl chloride, and is ~5 × 10^3 times slower than formylation with DMF/phosphoryl chloride [74JCS(P2)1610]. This finding is consistent with greater delocalization of the charge in the electrophile for the acetylation compared to the formylation.

Aluminum chloride-catalyzed acylation may be accompanied by formation of σ complexes (i.e., thiophenes protonated by the very strong acid co-product $HAlCl_4$) (75JOU412). This will be more serious under any of the following conditions: (1) the thiophenes contain electron-supplying groups, as they will then be better bases; (2) solvents that do not contain nitro groups are used, because aluminum chloride is then not coordinated to a nitro group and $HAlCl_4$ formation is facilitated; (3) the halide RCOCl contains an electron-withdrawing group in R, which lowers the ability of aluminum chloride to coordinate at oxygen in the acyl halide or the ketone product, again increasing the concentration of $HAlCl_4$; (4) excess aluminum chloride is used.

These features explain why chloroacetylation of activated thiophenes is *slower* than acetylation, whereas the converse order should apply, as it does for deactivated thiophenes.

B. SELENOPHENE AND TELLUROPHENE

The reactivity of selenophene relative to thiophene has been determined as 2.28 (acetylation), 3.64 (formylation in phosgene), and 7.33 (trifluoroacetylation), under the conditions given in Table 6.6. The corresponding values for tellurophene are 7.55, 36.8, and 46.4, respectively [71CC1441; 73JCS(P2)2097]; both compounds are less reactive than furan (Section 7.C). The reactivity ratio of selenophene to thiophene was also confirmed (semirigorously) in formylation (with N-methyl-N-phenylformamide/phosphoryl chloride) and acetylation (with acetic anhydride/ phosphoric acid) of **6.9**; the rate ratios for substitution in the α-positions were 4.0 and 3.75, respectively. Competitive formylation of selenophene and thiophene under these formylation conditions gave a rate ratio of 4.9 (73JGU871).

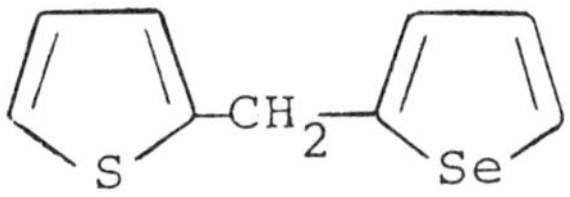

(6.9)

The activating effects of a 2-methyl group in selenophene and tellurophene are, respectively, similar and slightly greater than that in thiophene. The factors are 300 and 620 in formylation, and 280 and 500 in trifluoroacetylation (77G339).

C. FURANS

Furan is more reactive than thiophene by factors of 11.9 (acetylation), 107 (formylation in phosgene), and 140 (trifluoroacetylation), under the conditions given in Table 6.6; iodine-catalyzed acetylation gave a rate ratio of 9.3 [67T1739; 69T4599; 73JCS(P2)2097]. Accurate 2- : 3-rate ratios have been determined in acetylation as 6800 and 800 (Ac_2O, $SnCl_4$, 25 and 75°C), and 6000 (acetyl trifluoroacetate, 75°C) (71TL3833). Partial rate factors for the 2- and 3-positions in acetylation have been calculated as 3.2×10^7 and 4.7×10^3, respectively (70TL1389; 71TL3833). These lead to σ^+ values substantially less than for other reactions and also low relative to thiophene. It is reasonably certain that the reactivity of furan is lowered through coordination with the Lewis acid (significantly, pyrrole is virtually unreactive in the presence of such catalysts). Moreover, 2-methylfuran failed to give a quantitative yield in stannic chloride-catalyzed acetylation, and this is believed to be due to complex formation. The 2-methyl substituent activates the 5-position by factors of 880 and 1700 in formylation and trifluoroacetylation, respectively (69T4599; 77G339) showing furan to be more selective to substituent effects than other five-membered heterocycles; this has been confirmed by other data [86HC(44,2)1]. The nucleophilicity of 2-methylfuran is high enough to give 18.5% of trifluoroacetylated product in the reaction with acetyl trifluoroacetate (67MI3) (cf. 2-methylthiophene, Section 7.A).

Partial rate factors for trifluoroacetylation at the 5-position of 2-substituted furans in 1,2-dichloroethane at 75°C [Table 6.7, [72JCS(P2)71]] correlate approximately with σ^+ values, with $\rho = -10$. The effects of para substituents on the rate of acetylation and trifluoroacetylation of 2-aryl-5-methylfurans have been studied. In the latter reaction, the relative rates of substitution at the 3-position (to give **6.10**) were H, 1.0; Cl, 0.43; Me, 5.3; and OMe, 35.4, which correlated with the Yukawa–Tsuno equation, $\rho = -2.6$, $r = 0.74$. For acetylation, the corresponding relative rates for 3-substitution were 1.0, 0.58, 3.4, and 12.5 ($\rho = -1.76$, $r = 0.74$), and for 4-substitution (to give **6.11**), 1.0, 0.9, 1.7, and 1.4 (73BSF1760). For $SnCl_4$-catalyzed acetylation of 5-methyl-2-phenylfuran, the 3- : 4-product ratio was 1.5, larger than for the corresponding thiophene [72CR(C)(275)49], presumably because the higher 2 : 3-bond order in furan outweighs any steric considerations.

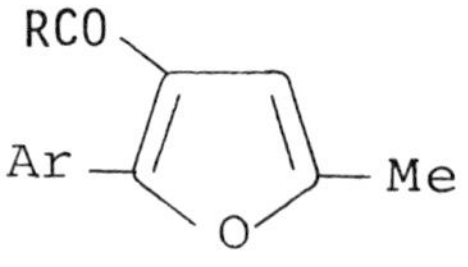

(6.10) 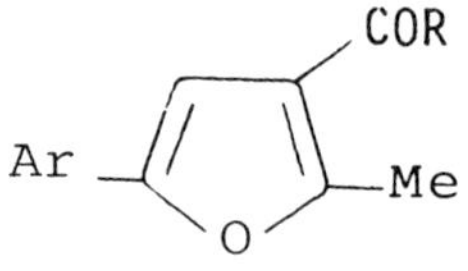   (6.11)

(R = CF_3 or CH_3)

TABLE 6.7
PARTIAL RATE FACTORS (RELATIVE TO FURAN) FOR TRIFLUOROACETYLATION
OF THE 5-POSITION OF 2-SUBSTITUTED FURANS

Substituent	SMe	Me	Et	t-Bu	Ph	Cl	Br
f	17000	1700	1400	860	300	0.071	0.036

D. PYRROLES

Vilsmeier–Haack formylations of pyrrole and of 1-alkylpyrroles take place in good yields, but the 2- : 3-product ratio depends markedly on the size of the alkyl group as follows: Me, >99; Et, 11.5; i-Pr, 1.9; t-Bu, 0.07; PhCH$_2$, 6.2; Ph, 9 [70JCS(C)2563]. These results show again that formylation is very sterically hindered. This steric hindrance has been erroneously estimated as producing a 2000-fold rate reduction, since 1-methylpyrrole undergoes Vilsmeier–Haack formylation at the 2-position 100 times less readily than pyrrole, whereas a 3-methyl substituent activated the 2-position ~20-fold [84JCS(P2)1179, 84JCS(P2)1607]. This calculation assumes that a methyl group will activate equally across the 1,2- and 2,3-bonds, but because of bond fixation this will not be the case. The bond fixation effect means that para substituents in the 1-phenyl group cannot conjugate with either reaction center, so they produce very little effect on the 2- : 3-product ratio. By contrast, conjugatively electron-withdrawing substituents (COMe, COPh, CO$_2$Et) on nitrogen produce exclusively 2-formylation (in quite high yield). This is because juxtaposition of like charges makes the canonicals **6.12b** and **6.13c** very unstable and **6.12a** and **6.13b** fairly unstable.

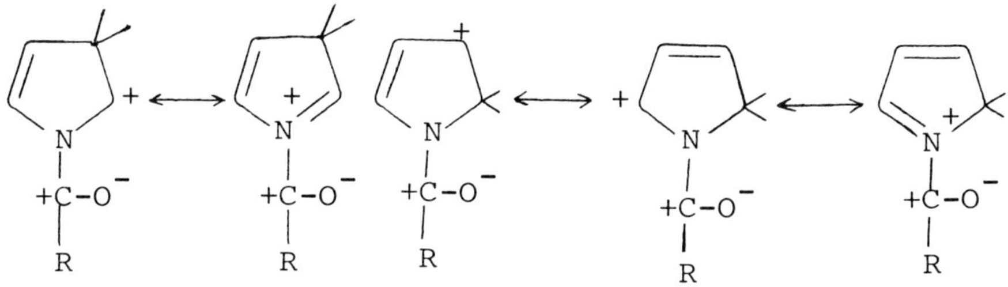

(6.12a)   (6.12b)   (6.13a)   (6.13b)   (6.13c)

Substitution at the 3-position is therefore favored by the relative stability of the remaining canonical (**6.13a**). Moreover, the 1-acyl groups (unlike the 1-alkyl groups) will be coplanar with the ring so that approach of the electrophile (perpendicular to the plane of the ring) will not be appreciably hindered [70JCS(C)2563]. The high steric hindrance to formylation is also shown by the formation of 20% of **6.14** and 80% of **6.15**. For the corresponding stannic chloride-catalyzed acetylation, the percentages become 55 and 45, respectively [72CR(C)275)49]. Formylation of 2-methoxycarbonylpyrrole gives mainly the 5-isomer [cf. acetylation, below (67CJC897)].

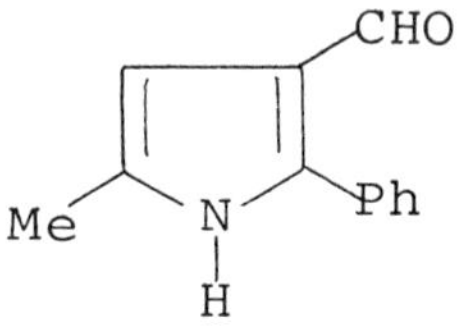

(6.14) 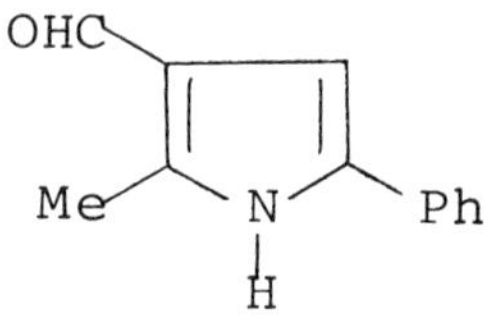 (6.15)

Acetylation of pyrrole is difficult because if forms a 2 : 1 complex with stannic chloride (29CB226). Hence, under the conditions used for the other five-membered rings (i.e., acetic anhydride in the presence of one hundredth molar equivalent of stannic chloride or iodine), no reaction occurs, and only 20% acetylation is obtained if the molar proportion of the catalyst is reduced 10-fold. The effect of complex formation also shows up in the inhibition of stannic chloride-catalyzed acetylation of furan or thiophene, on addition of pyrrole (67MI4). Catalyzed acetylation of 2-cyano-, 2-formyl-, or 2-methoxycarbonylpyrrole gives mainly 4-substitution (67CJC897) indicating that the catalyst must also be coordinated with the substrate; 1-methyl-3-nitropyrrole acetylates only in the 5-position (57CJC21).

The high reactivity of pyrrole means that acetylation can be achieved without any catalyst, a mixture of 2- and 1-acetyl derivatives being obtained (50% yield), in an initial ratio of 6 which decreases to 1.7 as reaction proceeds. 1-Acetylation probably takes place on the pyrrole anion (the electrophile would be attracted by the high charge density), since the 2- : 1-product ratios are 10 and 1 in the presence of acetic acid and sodium acetate, respectively. The presence of triethylamine increases the anion concentration and removes the acetic acid co-product, so giving 80% of 1-acetylpyrrole. For the maximum yield (85%) of 2-acetylpyrrole, the best method involves Vilsmeier–Haack acetylation with N,N-dimethylacetamide (67MI4). Bis(1-methylpyrrole-2-yl) is three times more reactive than 1-methylpyrrole in acetylation by $MeCOCl/SnCl_4$ (77TL389).

Pyrrole is 5.3×10^7 times more reactive than thiophene toward trifluoroacetylation in 1,2-dichloroethane at 75°C. This very high reactivity is due to reaction on the neutral molecule rather than the pyrrole anion, since 1-methylpyrrole is even more reactive by a factor of 1.9 (69T4599); similar results are obtained in Vilsmeier formylation and acetylation [72CC427; 77JCS(P2)1284].

Partial rate factors, relative to pyrrole, for trifluoroacetylation at the 5-position of some 2-alkylpyrroles are: Me, 23.8; Et, 24.8; t-Bu, 24.8 [72JCS(P2)71]. Thus, the activating effect of a 2-alkyl group is far lower than that found in furan and thiophene, and this is confirmed in Vilsmeier formylation and acetylation of pyrroles where 2-methyl activates by factors of 15 and 42, respectively [77G339, 77JCS(P2)1284].

In acylation by acetyl trifluoroacetate, the ratio of trifluoroacetylation to acetylation increases with increasing nucleophilicity of the aromatic (70G556), the percentage of trifluoroacetylation being 50% for pyrrole and 92% for 2-methylpyrrole (67MI3). Curious is the very large difference in the 2- : 3-product ratios for pyrrole which are 6 (acetylation) and >1000 (trifluoroacetylation) [70JCS(B)1153]; in acylation by fluorinated immonium salts, the 2- : 3-product ratio (for formation of pyr·COCHClF derivatives) is almost 1.0 (75CC956).

8. Other Electrophilic Substitutions

The reactivities of the five-membered rings and derivatives have been determined in a number of other reactions, including those where the group replaced is not a proton. These latter have the advantage that the reactivities of individual positions relative to a proton in benzene can be determined directly.

A. THIOPHENES

Many thiophenes have been lithiated en route to other derivatives. The lithiation reaction [Eqs (6.1) and (6.2)] is related to base-catalyzed hydrogen exchange, but differs from it in the relative rates of steps (6.1) and (6.2), the latter being obviously faster and non-rate-determining for lithiation.

$$\text{ArH} + \text{R}^- \rightleftharpoons \text{Ar}^- + \text{RH} \qquad (6.1)$$
$$\text{Ar}^- + \text{Li}^+ \to \text{ArLi} \qquad (6.2)$$

Lithiation shows a kinetic isotope effect, with a $k_H : k_T$ ratio of 5.9 (55AK343). Lithiation occurs initially at the 2-position (the carbanion there being better stabilized by electron withdrawal), and then at the 5-position [77JCS(P1)887]. The preference for α-substitution is such that both electron-supplying and electron-withdrawing substituents at the 2-position (e.g., Me, OMe, SR) direct into the 5-position [53JA3697; 59BAU1925, 59JGU2003; 60BAU1700; 62AK(18)513]; SMe was shown to activate (59JGU3592). At the 3-position, inductive acceptor $(-I)$ substituents (OMe, SMe, Br) direct into the 2-position, inductive acceptor/resonance donor $(-I, +M)$ aryl groups direct into the 2- and 5-positions, whereas inductive and resonance donor $(+I, +M)$ alkyl groups direct into the 5-position (54AK361, 54JOC70; 58AK269,295; 60AK309, 60AK363, 60JA1447; 71JOC1053).

Halogens can also be replaced by lithium in an analogous reaction. In lithiodebromination of 3-alkyl-2,5-dibromothiophenes (**6.16**) with *n*-butyllithium, the ratio of 3-alkyl-2-bromo-/3-alkyl-5-bromothiophenes decreased from 4.26 (R = methyl) to 0.75 (R = *t*-butyl) [74CS(5)217], showing very nicely the effect of steric acceleration; as R is made larger, the 2-bromo group is preferentially removed to relieve ground state strain.

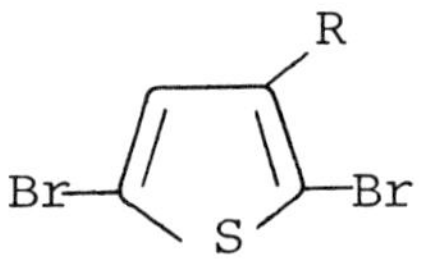

(6.16)

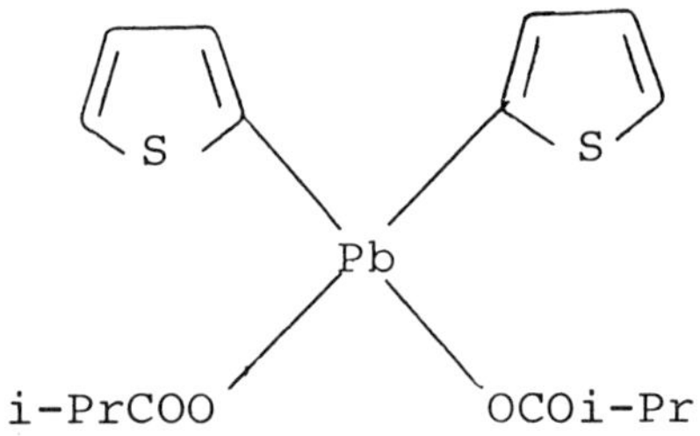

(6.17)

Another reaction of considerable preparative value for thiophenes is mercuriation (with mercuric chloride or mercuric acetate), the lower selectivity here (especially with the acetate) giving significant yields of β-substituted derivatives. These arylmercurial derivatives are readily converted into other compounds. Mercuric chloride substitutes in the 2- and then the 5-position [1892LA(267)172; 14LA(403)50], and the OR and SR substituents direct into the 5- and then the 3-position [32LA(495)166].

A number of examples of α- and β-substitution with mercuric acetate are known [86HC(44,2)1, Table 8]. Mercuriation of thiophene with mercuric acetate was claimed to give a high 2-:3-positional selectivity [70JCS(B)1153], though no quantitative data are available. The reactivity ratio of thiophene to benzene was also high ($\sim 10^5$) (57MI1) leading to a partial rate factor $f_2 = 1.85 \times 10^6$, and hence $\sigma_2^+ = -1.565$ (since ρ for mercuriation is -4.0). In this work, partial rate factors (relative to the 2-position of thiophene) were also determined for 5-substituents as follows: Cl, 0.26; MeCO, 0.0028; EtCO$_2$, 0.0034.

The value for σ_2^+ in mercuriation is completely out of line with other data, which led to the proposal [71AHC(13)235] that mercuriation of thiophene involved coordination at sulfur. Although such coordination would not be unreasonable, this explanation must be incorrect for the following reasons.

(1) By the principle of microscopic reversibility, coordination must be involved in the reverse reaction, protiodemercuriation, and indeed this reaction (ethanolic HCl, 70°C) gives a partial rate factor $f_2 = 1720$, yielding an anomalously high value for σ_2^+ of -1.125 (65AJC1513). However, protiodemercuriation of furan also yields anomalously high σ^+ values, viz $\sigma_2^+ = -1.25$, $\sigma_3^+ = -0.75$ (65AJC1513). The difference, $\sigma_2^+ - \sigma_3^+$, of 0.5 is the same as that observed in a range of reactions of furans, and since a coordination step is most improbable for 3-substitution, it must also be ruled out for 2-substitution of furan.

(2) Although coordination is evidently not involved in protiodemercuriation of the 2-position of furan, the difference σ_2^+(furan) $- \sigma_2^+$(thiophene) for protiodemercuriation is 0.1, the same as found in a range of other reactions. Clearly, coordination cannot be involved in the pathway for demercuriation of the 2-position of thiophene, and therefore it cannot be involved in mercuriation.

This leaves unexplained the anomalously high reactivities for both mercuriation and protiodemercuriation. For the latter, it may be relevant that with thiophene and substituted thiophenes kinetics were second order for only 20% of reaction, and there were marked differences in entropies of activation for the heterocyclic and phenyl derivatives. Even for benzenoid compounds the reaction is not well behaved as (1) the Hammett correlation shows marked curvature, suggesting that reactive compounds undergo a different mechanism to unreactive ones; (2) the reaction is sensitive to oxygen; and (3) it is sensitive to added chloride ion, due possibly to salt effects and reaction via hydrogen chloride ion pairs (65AJC1521). A reexamination of both mercuriation and protiodemercuriation of thiophene would seem desirable.

While plumbylation with lead tetraacetate does not occur for thiophene, the presence of electron-withdrawing groups in the acid moiety produces reaction. Thus, with one ligand replaced by dichloroacetate, a 63% yield of 2-thiophenelead triacetate is obtained (74TL853). Another probable driving force for the reaction is relief of steric hindrance in the lead reagent, since plumbylation with lead tetra(2-methylpropanoate) gave di(2-thiophenyl)lead bis(2-methylpropanoate) (**6.17**) [58DOK(123) 295].

Thiophene is sulfonated 76% in the 2-position, with only a small amount of the 3-isomer produced [24LA(437)14; 33LA(501)174]. The yield of the 2-isomer can be boosted to 86% using pyridine sulfur trioxide complex (52JGU189). Sulfonation of the 2-sulfonic acid gave the 2,5-disulfonic acid (48MI2; 53M11), but the more electron-withdrawing 2-SO_2Cl substituent directs sulfonation into the 4-position [24LA(437)14; 60BSF793]. Chlorosulfonation of the 2-sulfonyl chloride gave a 4 : 5-product ratio of 2.0 (81PS111). 2-Methyl- (**6.18**) and 2-*t*-butylthiophene (**6.19**) are sulfonated in the 5-position (56BAU627; 80URP707916), but 3-*t*-butylthiophene (**6.20**) undergoes sulfode-*t*-butylation (56BAU627), presumably because the site most strongly (electronically) activated toward electrophilic substitution happens to be that most sterically hindered, so the ipso activation of the 3-position by the *t*-butyl group becomes of overriding importance. 2,5-Dimethylthiophene (**6.21**) and 2,5-di-*t*-butylthiophene (**6.22**) are sulfonated in the 3-position (75–95% and 78% yields, respectively) (53JGU263). The result for 2,5-di-*t*-butylthiophene is particularly surprising in view of the behavior of 3-*t*-butylthiophene. The effect of steric hindrance is also shown by the positions of sulfonation of **6.23** and **6.24** (53JGU263); the free site in **6.24** is so crowded that sulfode-*t*-butylation of the 3-position is preferred.

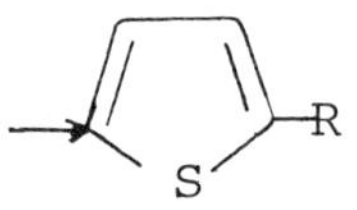

(6.18, R = Me
6.19, R = t-Bu)

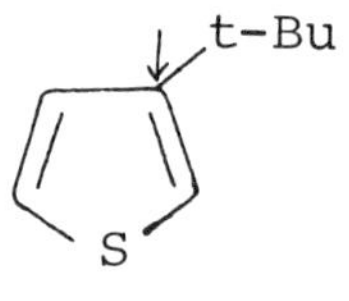

(6.20)

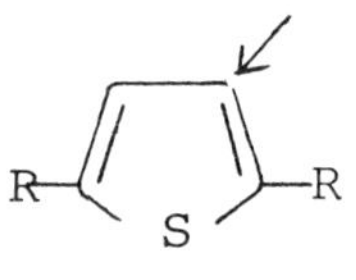

(6.21, R = Me
6.22, R = t-Bu)

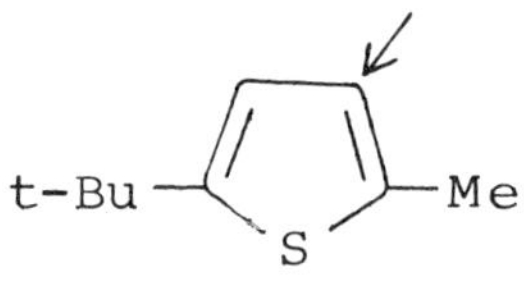

(6.23)

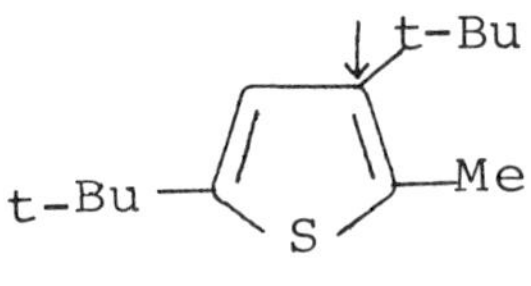

(6.24)

2-Chloro-, 2-bromo-, and 2-iodothiophenes are sulfonated in the 5-position in 95, 90, and 77% yield, respectively (51ZOB1524). 2-Chlorothiophene is chlorosulfonated 48% in the 5-position (a surprisingly low yield), the comparable yield for 2-bromothiophene (in the presence of PCl_5) is 96% [34LA(512)136; 60BSF793; 81JMC959]. Steric hindrance appears to be responsible for the 5-substitution (in 62% yield) for chlorosulfonylation of 2,5-dibromothiophene (in the presence of PCl_5), whereas 2,5-dichlorothiophene gave 71% of the 4-derivative [34A(512)136; 55JA3410; 81JMC959]. The strong activation by the NHAc substituent across the 2,3-bond is such that, whereas room-temperature sulfonation of 2-acetylaminothiophene gives the 5-sulfonic acid, at 100°C an 89% yield of the 3,5-disulfonic acid is produced (54CB1184). 2-Acylthiophenes are sulfonated in the 5-position, but the greater electron withdrawal by the 2-nitro group at the 5-position is such that 2-nitrothiophene is sulfonated in the 4-position (60BSF793). 5-Alkyl-2-acylthiophenes not only are sulfonated in the expected 4-position (strongly activated by the 5-alkyl group) but surprisingly undergo sulfodeacylation at the 2-position (1886CB660, 1886CB2623; 1896CB2560).

Partial rate factors for the protiodeboronation of thiophene [acid cleavage of the $B(OH)_2$ group] have been determined as $f_2 = 5.5 \times 10^5$, $f_3 = 5.5 \times 10^3$ (65AJC1521; 75JHC195); these give corresponding σ^+ values ($\rho = -5.0$) of -1.16 and -0.755, respectively. Like the values obtained in protiodemercuriation, these are anomalously high, and Brown *et al.* have suggested that the differing coordinating ability of boron in the phenyl and heterocyclic compounds is probably responsible. Moreover, the ρ factor for the reaction is not accurately known, being very dependent upon the reaction conditions, as is the mechanism [72MI2(287)]. However, the difference in the positional reactivities, given by $\sigma_2^+ - \sigma_3^+$ (0.405), is close to that obtained from a number of other reactions.

Partial rate factors have also been determined in protiodesilylation (acid cleavage of the $SiMe_3$ group) as $f_2 = 4060$, $f_3 = 97$ [56JCS4858; 59JCS2299; 61JCS4921; 70JCS(B)1364] yielding corresponding σ^+ values ($\rho = -4.6$) of -0.785 and -0.43. Eaborn and Seconi have used the reaction to provide a set of data for substituted thiophenes (Table 6.8), which

TABLE 6.8

Partial Rate Factors (Relative to the 2-Position of Thiophene) for Protiodesilylation at 50°C

Substituent	5-OMe	5-Me	5-Cl	5-Br	3-Br	4-Br	5-NO_2	3-NO_2
f	5900	36	0.128	0.098	0.055	0.0065	9.1×10^{-7}	7.1×10^{-7}

covers a greater reactivity range than for any other reaction and are, therefore, particularly meaningful [81JCS(P2)931]. Noteworthy features are the following.

(1) The 3-Br substituent is less deactivating relative to the 5-Br substituent ($f_{3\text{-Br}}/f_{5\text{-Br}} = 0.56$) than is the case for the corresponding interaction in benzene ($f_{2\text{-Br}}/f_{4\text{-Br}} = 0.24$) [72MI2(335)]. This may reasonably be attributed to better transmission of the $+M$ effect across the 2,3-bond in thiophene, due to the higher bond order.

(2) Substituents produce a *larger* effect than they do in benzene, indicating that the transmission of electronic effects is greater in thiophene. This is in contrast with the results obtained in hydrogen exchange and *apparent* results obtained in a number of other reactions.

As is discussed in detail later (Section 10.C), an important factor, overlooked in previous analyses, is the greater ability of resonance donor and acceptor ($+M$ and $-M$) substituents to become coplanar with the thiophene ring (due to the greater bond angles external to the ring). This is demonstrated in desilylation, in which the nitro group deactivates $\sim 10^3$ times more than in benzene; however, a quantitative measure of the effect needs to take into account that in the strong acid medium used for desilylation of the nitrothiophenes, hydrogen bonding would certainly have been significant. Protiodesilylation of thiophene was also used in the first demonstration that $SiMe_3$ group is cleaved more readily than the $SiPh_3$ group from aromatic rings (49JA2066). Protiodeplumbylation was used in the first demonstration that furan was more reactive than thiophene toward electrophiles through acid cleavage of **6.25**, from which furan was removed more readily than was thiophene (32RTC1054).

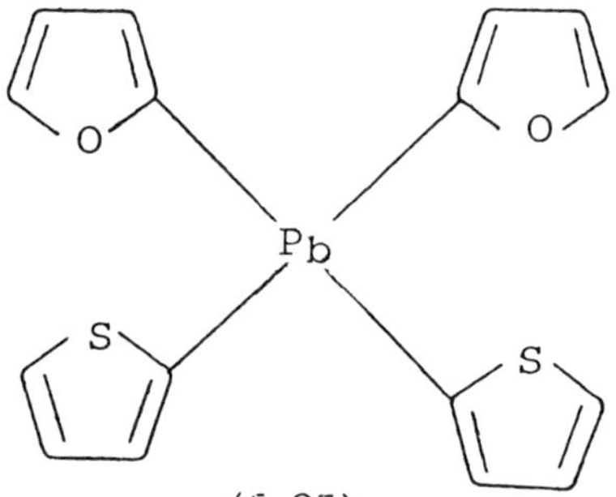

(6.25)

Rates of base-catalyzed protiodesilyation of substituted 2-trimethylsilylthiophenes have been measured along with the solvent isotope effects

(Table 6.9) [81JOM(204)153]. The latter were measured in terms of the product isotope effect (PIE), and the rate isotope effect (RIE), obtained by carrying out the reaction in 1 : 1 MeOH–MeOD, and either MeOH or MeOD at constant base concentration, respectively. In this reaction, electron-withdrawing substituents increase the reaction rate and there is a parallel between the rates and energies of deprotonation of the substituted thiophenes, calculated by the *ab initio* method (STO-3G level). This result, and the magnitudes of the respective isotope effects, were shown to be consistent with formation of aryl anions in the rate-determining step of the reaction. It is noteworthy that comparison of the rates of these desilylations with those of the corresponding benzene derivatives showed the nitro substituent to activate more in the thiophenes; correlation of the substituent effects against $\sigma°$ values showed that both the NO_2 and COPh substituents were slightly more activating than expected. Again it seems probable that greater coplanarity between the substituent and the thiophene ring permits greater electron withdrawal by these electron acceptor ($-M$) substituents than is the case in benzene.

Brown *et al.* studied iododeboronation of thiophene [cleavage of $B(OH)_2$ by potassium iodide in methanol]. The partial rate factors at 25°C were $f_2 = 9.7 \times 10^3$ and $f_3 = 7.0 \times 10^2$. In this reaction there was a marked difference of entropy of activation between the phenyl compounds on the one hand and the thiophenyl compounds on the other, lead-

TABLE 6.9
RELATIVE RATES AND ISOTOPE EFFECTS IN CLEAVAGE OF SUBSTITUTED TRIMETHYLSILYLTHIOPHENES[a]

Substituents	$k_{rel.}$	RIE	PIE
5-NMe_2	0.04		
5-OMe	0.64	0.50	
5-Me	0.20		
H	1.0		1.0
5-Cl	95	0.47	1.2
5-Br	130	0.45	1.2
4-Br	284	0.48	
3-Br	1460	0.48	
4,5-Br_2	1.56×10^4		
5-COPh	5000	0.43	1.1
5-CN	1.29×10^5	0.45	
5-NO_2	8.75×10^5	0.42	1.1
3-NO_2	1.75×10^6	0.50	1.2

[a] By NaOH/NaOMe at 50°C.

ing to the suggestion that differential solvation effects were responsible (65AJC1527). A similar factor is probably the cause of the very low log f_2/log f_3 ratio (1.40), compared to other reactions.

The unique combination of substituent and site reactivity evident in five-membered heterocycles shows up in the nitration of 5-methylthiophene-2-carboxylic acid (**6.26**). Nitration takes place at the 4-position accompanied by nitrodecarboxylation at the 2-position (32RTC1134). Despite the presence of the carboxyl group, the 2-position is the most activated site; moreover, in the species reacting the carboxy group is probably present as the much less deactivating anion. Likewise, 27% nitrodebromination occurred in nitration of 2-bromo-5-cyclopropylthiophene (**6.27**) (81CHE152), due to the very strong activation of the 2-position (cyclopropyl is the strongest electron-releasing alkyl group). Similar arguments account for nitrodebromination at the 3- and 5-positions of 2-acetylamino-3,5-dibromothiophene (**6.28**) (47JA1173).

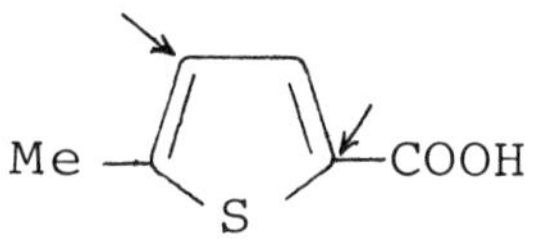

(6.26) 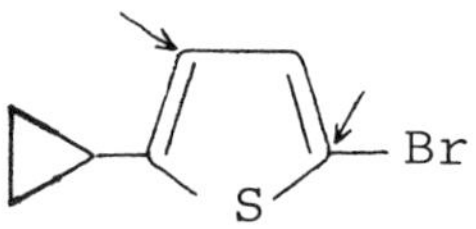   (6.27) 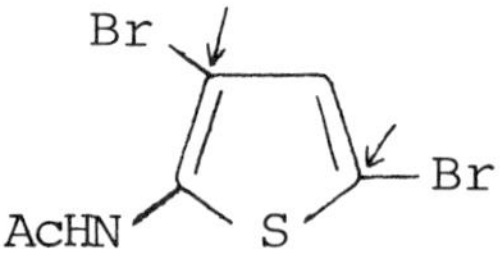   (6.28)

B. Selenophene and Tellurophene

Selenophene is 1.5 times more reactive than thiophene toward lithiation by Ph$_3$CLi. Selenophene (71CHE938) and tellurophene [76ACS(B)605] are preferentially lithiated in the 2-position. Selenophene sulfonates at the 2-position, and 2-methyl-, 2-carboxy-, and 2-formylselenophene sulfonate at the 5-position (64JGU1814,2201). Towards protiodemercuriation at the 2-position, selenophene is either slightly more or less reactive than furan, depending on the reaction temperature (65AJC1513).

C. Furans

Furan is more reactive than thiophene toward lithiation by n-BuLi [77JCS(P1)887].

Partial rate factors have been determined for protiodemercuriation, protiodesilylation, and protiodeboronation of the corresponding furan derivatives. The values for f_2 and f_3, respectively, are 4000, 150 (demercuriation, 65AJC1513); 14,600, 117 [desilylation, 61JCS4921; 70JCS(B)1364]; 9.1 × 10^5, 7050 (deboronation, 75JHC195; 76JHC1265). All three sets of data show excellent consistency relative to the corresponding thiophene data in that the 2- and 3-positions of furan are the more reactive in every case. The corresponding σ^+ values are -1.255, -0.76 (demercuriation); -0.905, -0.45 (desilylation); and -1.19, -0.77 (deboronation). However, as with thiophene, only the desilylation values are consistent with those obtained generally. The other reactivities relative to benzene appear anomalously high for the reasons noted for thiophene in Section 8.A. Furan and 2-methylfuran sulfonate in the 2(5)-position, but when both α-positions are blocked (as in 2,5-dimethylfuran) sulfonation takes place at the 3(4)-positions (49ZOB531; 54JOC894).

D. Pyrroles

The high basicity and consequent electrophilic reactivity that hinder the determination of the reactivity of pyrrole relative to other aromatics facilitates reaction with unreactive electrophiles. This was utilized by Butler and co-workers, who measured partial rate factors for methyl substituents (Table 6.10) in coupling with benzenediazonium ions [77JCS(P2)1452] and in alkylation by 4-(N,N-dimethylamino)benzaldehyde [76JCS(P2)696]. The selectivities of both reactions are remarkably similar. Reaction with p-X-benzenediazonium ions (X = OMe, CN, NO_2,

TABLE 6.10

Partial Rate Factors (Relative to the 2-Position of Pyrrole) for Diazonium Coupling and Alkylation of Pyrroles at 25°C

Substituent	Diazonium coupling[a]	Alkylation[b]
1-Me	3.7	4.1
2-Me	41	39
1,2-Me$_2$	133	154
2,3-Me$_2$	78	293
2,4-Me$_2$	1970	2230
2,5-Me$_2$	4.8	—
3,4-Me$_2$	128	393
1,2,3-Me$_3$	—	771
1,2,5-Me$_3$	3.6	98
2,3,5-Me$_3$	0.26	—
3-Et-2,4-Me$_2$	26,000	5560

[a] In 0.05 M HCl.
[b] In 1.5 M HCl.

SO$_3$H) indicated that reaction occurred via an A-S$_E$2 mechanism and the neutral pyrrole species. Assuming additivity in diazonium coupling, the activating effects of a 3-, 4-, and 5-methyl group can be calculated as 48, 1.9, and 41, respectively, so once again the 2,3-interaction is greater than the 2,5-interaction. Comparison of the results for the 2,5- versus the 3,4-dimethyl compounds shows the α-position to be ~25 times more reactive than the β-position. The anomalously low reactivities of the 1,2,5- and 2,3,5-trimethyl compounds are almost certainly due to steric hindrance, but the reactivity of the 3-ethyl-2,4-dimethyl compound is inexplicably high (additivity predicts a value ~10-fold lower).

Pyrrole is sulfonated in high yield at the 2-position by SO$_3$-pyridine (49ZOB538) and 5-sulfonation is preferred for most substituted pyrroles unless both α-positions are blocked (49ZOB538,1365,2118; 51ZOB281).

9. Side-Chain Reactions

Measurement of rates of reactions in which a carbocation is formed in the side chain of an aromatic leads to a direct measure of the electrophilic reactivity at the aromatic ring. This method, first used for five-membered heterocycles by Taylor using pyrolysis of 1-arylethyl acetates [68JCS(B)1397], has been extended to solvolysis of 1-arylethyl acetates by Hill and co-workers (see, e.g., 69JA7381), and to solvolysis of 1-arylethyl p-nitrobenzoates by Noyce and co-workers (see, e.g., 70JOC1718). The advantages of the method are that the ρ factors are fairly small so that rate comparisons are possible under nearly identical conditions. This is particularly true for pyrolysis, which has the additional advantage of the absence of solvent, so that the results are unaffected either by protonation or by hydrogen bonding.

A. Thiophene

Taylor reported pyrolysis rates for 1-(2'- and -3'-thiophenyl)ethyl acetates at temperatures between 318 and 379°C [68JCS(B)1397]. Comparison with data for 1-phenylethyl acetate gave log k/k_0 values at 600 k of 0.524 and 0.251, respectively, and hence $\sigma_2^+ = -0.795$ and $\sigma_3^+ = -0.38$ ($\rho = -0.66$).

Solvolysis of the same compounds in 30% aqueous ethanol at 25°C by Hill et al. (69JA7381) gave k/k_0 values of 5.4 × 10^4 and 480, respectively, hence $\sigma_2^+ = -0.83$ and $\sigma_2^+ = -0.47$ ($\rho = -5.7$). 5-Methyl and 5-bromo substituents altered the reactivity of the 2-position by factors of 74 and

0.18, respectively. The factor for Me is greater than predicted for a para substituent in benzene for a reaction of this ρ factor (30.4 and 0.20, respectively), suggesting that substituent effects are transmitted better in thiophene (but see below). Solvolysis of the 1-arylethyl *p*-nitrobenzoates in 80% aqueous ethanol gave rate coefficients for the 2-position at 25 and 45°C, and for the 3-position at 75°C (70JOC1718). Extrapolation of the former data to 75°C gave $k_2/k_3 = 60$, which is equivalent (Arrhenius equation) to 120 at 25°C. Using the value of f_2 at 25°C (2.27×10^{-6} sec^{-1}) and the value for 1-phenylethyl *p*-nitrobenzoate of 4.1×10^{-11} sec^{-1} [there is some uncertainty in this value, which is interpolated from data for the corresponding chloride (69JOC1008)] gives $f_2 = 5.5 \times 10^4$ and hence $f_3 = 460$, leading to $\sigma^+_2 = -0.815$ and $\sigma^+_3 = -0.460$. The ρ value is taken as -5.8 (72TL3893), an estimate based on the value for the chloride (68JA418), itself determined by assuming that $k_{rel.}$ values are independent of both solvent composition and temperature, which is not rigorously correct. Thus, of the parameters for nitrobenzoate solvolysis, the most accurate is probably the k_2/k_3 value.

In this reaction, partial rate factors (relative to reaction at the 2-position) were determined (Table 6.11) (72JOC2615), and if the points for the CO_2Et substituents are excluded (5-CO_2Et is particularly deviant), a very good correlation with σ^+ is obtained (ρ = -6.7), and again the effects of

TABLE 6.11

Rate Factors (Relative to the 2-Position of the Heterocycle) for Solvolysis of Substituted 1-(2-Aryl)ethyl *p*-Nitrobenzoates at 25°C[a]

Substituent	Thiophenes	Furans
5-OMe	1.61×10^5	—
5-Cyclopropyl	503	—
5-Me	81	214
5-Ph	15.4	—
5-Br	0.137	0.271
5-CO_2Et	1.21×10^{-4}	3.70×10^{-4} [b]
5-NO_2	—	1.45×10^{-6} [b]
4-Br	1.87×10^{-3}	4.76×10^{-3}
4-CO_2Et	7.05×10^{-4}	7.67×10^{-4}
4,5-Br_2	6.93×10^{-4}	—
4-SMe	—	0.25[c]

[a] Data from (72JOC2615).
[b] From solvolysis of the corresponding 1-arylethyl chlorides.
[c] Corrected from 45°C.

substituents are greater than in benzene. The enhanced deactivation by the CO_2Et substituents parallel exactly the behavior of the NO_2 substituent in desilylation, and the same explanation seems applicable (see Section 8.A).

The reactivity of the 2-position of thiophenes has been determined in three other reactions proceeding via formation of a carbocation at the side-chain α-position. In the solvolysis of 1-arylethyl chlorides, f_2 = 16,070, hence $\sigma_2^+ = -0.78$ (ρ = -5.4) [75JCS(P2)551]. In the isomerization of cis-1-aryl-2-phenylethenes in aqueous sulfuric acid, f_2 = 350, hence $\sigma_2^+ = -0.77$ (ρ = -3.3) (68JA4633; 70JOC1718). For the rearrangement of 1-arylbut-2-en-1-ols, f_2 = 35, giving $\sigma_2^+ = -0.595$ (ρ calculated as -2.6 at 25°C from the relative rate of the p-methoxyphenyl compound) (52JCS1528).

B. Selenophene and Tellurophene

In the pyrolysis of 1-arylethyl acetates, $\log f_2$ for selenophene was 0.563 at 600 K, hence $\sigma_2^+ = -0.855$ (86UP1), in very good agreement with the value of -0.885 determined from solvolysis of the same ester for which f_2 was 1.09×10^5 (77G339). Pyrolysis of 1-(2-tellurophenyl)ethyl acetate was affected by surface catalysis so no rate data are available. However, in the solvolysis, f_2 was 4.76×10^3 giving, $\sigma_2^+ = -0.995$ (72G534); this is the only reaction in which tellurophene is more reactive than furan. Unfortunately, in this work the experimental conditions were not verified by the usual procedure of measuring the rate of a standard compound. A 5-methyl substituent increased solvolysis by a factor of 23 in selenophene and 12 in tellurophene (77G339), both values being substantially less than in thiophene.

C. Furans

The first quantitative measure of the electrophilic reactivity of both positions of furan was obtained through pyrolysis of 1-arylethyl acetates, which gave $\log f_2 = 0.588$ and $\log f_3 = 0.274$; hence $\sigma_2^+ = -0.88$, and $\sigma_3^+ = -0.415$ [68JCS(B)1397]. This showed that the reactivity order is 2-furan > 2-thiophene >> 3-furan > 3-thiophene, and that the difference in reactivity is greater at the 2- than at the 3-position. This order has since been confirmed in other reactions. Thus, in solvolysis of the same esters, $f_2 = 2.1 \times 10^5$ and $f_3 = 670$; hence $\sigma_2^+ = -0.93$ and $\sigma_3^+ = -0.50$ [cf. thiophene above (69JA7381)]. In this reaction, a 5-methyl substituent activated the 2-position 164-fold; this is the highest activation among the five-membered heterocycles studied. From rates of solvolysis of 1-(2-furanyl)ethyl p-nitrobenzoates at 25 and 45°C, and of the corresponding 3-furanyl ester at 75°C (72JOC2620, 72JOC2623), the value of k_2/k_3 can be deduced as 315 at 75°C, which is equivalent to 830 at 25°C. Since at 25°C $k_2 = 1.17 \times 10^{-5}$ sec^{-1} (cf. 4.1×10^{-11} sec^{-1} for 1-phenylethyl p-nitrobenzoate), this gives $f_2 = 2.85 \times 10^5$ and $f_3 = 378$; hence $\sigma_2^+ = -0.94$, and $\sigma_3^+ = -0.445$.

The effects of 4- and 5-substituents on rates of solvolysis at the furan 2-position have been measured (Table 6.11) (69JOC1008; 72JOC2623), and in addition a 5-ethyl substituent activates the 3-position 8.9-fold at 75°C (equivalent to 12.8-fold at 25°C). These data give a much poorer correlation with σ^+ values than do the corresponding thiophene data, and it should be noted that although the methyl group activates more than in thiophene, all the other substituents (including the bromo substituents for which coplanarity factors do not apply) deactivate *less* than in thiophenes. The claim (72TL3893) that the ρ factor is greater in the furan series has, therefore, no foundation. In the solvolysis of the corresponding 2-aryl-2-propyl p-nitrobenzoates at 25°C, the 5-methyl substituent activated the 2-position sixfold (72JOC2620), but this seems anomalously low, despite the higher reactivity of these esters.

Finally, in the rearrangement of 1-arylbut-2-en-1-ols, f_2 was 90 leading to $\sigma_2^+ = -0.75$, which is substantially lower than in other reactions, as was also the case for thiophene (52JCS4158).

D. Pyrroles

From solvolysis of 1-[2-(1-methylpyrrolyl)]ethyl acetates, f_2 was estimated as 5.76×10^{10}. There is uncertainty in this value due to the need to measure the solvolysis rate in 100% ethanol and to extrapolate to 30% ethanol (69JA7381), but this result gives $\sigma_2^+ = -1.89$, which seems of the correct magnitude.

The activation by a 5-methyl group on the 2-position was calculated as 170-fold (77G339) from unpublished data of Noyce *et al.* for solvolysis of 1-arylethyl p-nitrobenzoates. This indicates a sensitivity to substituent effects greater than either those of furan or thiophene, despite the far higher reactivity of pyrrole.

10. Conclusions

The experimental results described in previous sections are summarized here to indicate the relative behavior of five-membered heteroaromatics toward electrophilic substitution. Although the origin of the observed reactivity order can be readily explained, the cause of the different sensitivities of the ring systems to substituent effects remains in part obscure. A brief discussion directed toward clarifying these sensitivity differences is presented here.

A. Aromaticity and Relative Reactivity

When considering the reactivity of the "π-excessive" five-membered heterocycles, it is important to distinguish between ground-state and transition-state properties. The ground state stabilities of these compounds will be directly related to their aromaticity, and there is general agreement from a variety of techniques that the order is benzene > thiophene > pyrrole > selenophene > tellurophene > furan, with pyrrole nearer to furan than to thiophene [60MI2; 61JSP58; 62JSP124; 65CC160; 66MI3; 70AHC(12)1; 74AHC(17)255, 74JCS(P2)332; 76T1767]. However, although this aromaticity order correctly predicts all these heterocycles to be more reactive than benzene, the order within the heterocyclic sequence must be modified to take account of a more important factor, the polarizability of the heteroatom. A nitrogen atom with a lone pair is much the most polarizable of all these heteroatoms, and this accounts for the very high reactivity of aniline and its derivatives toward electrophilic substitution. Strong electron release from the nitrogen into the ring thus explains why pyrrole is the most reactive compound (though in reactions of very low demand for resonance, it is conceivable that it could be less reactive than furan, selenophene, or tellurophene).

Among the Group 6A elements, there is evidence from the difference ($\sigma_p^+ - \sigma_p$) that sulfur is more polarizable than oxygen [69JCS(B)21; 71JCS(B)1450], indicating that the difference in electronegativity is here more important than the size of the heteroatom and carbon p orbitals. Hence, the polarizability order could be Te > Se > S > O. At present there is insufficient evidence to confirm this tentative proposal (and indeed in Group 7 the polarizability order is F > Cl > Br > I so that for the halogens, the size factor appears to be the more important). Nevertheless, with this polarizability order it is possible to construct a plot (Fig. 6.1) of reactivity versus electron demand (proportional to the reaction ρ factor), which provides some of the observed reactivity orders; in this plot the slopes of the lines are proportional to the heteroatom polarizabilities. Thus, under appropriate conditions, tellurophene (and even selenophene) may become more reactive than furan.

The existence of these differential polarizabilities means (as indeed it does in benzenoid compounds) that a single σ^+ parameter will not accu-

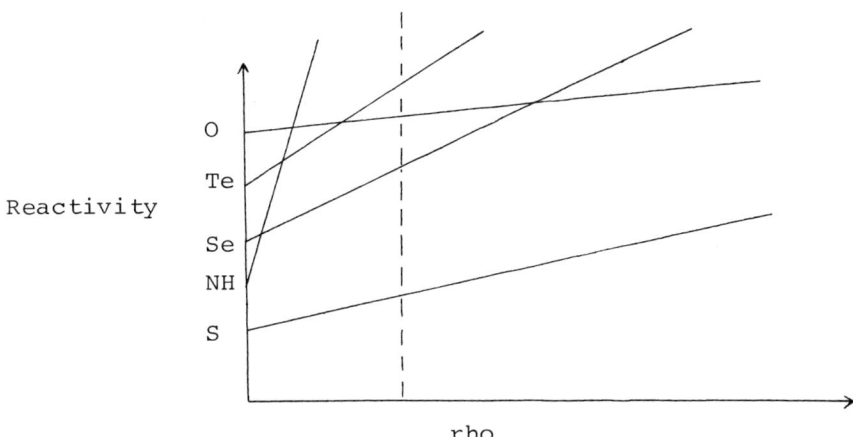

FIG. 6.1. Plots of reactivity versus electron demand for a series of five-membered heterocycles with one heteroatom (indicated on the plots).

rately describe the quantitative reactivities of the heterocycles under all conditions, but rather that more negative values will be required under conditions of higher electron demand [increased demand for the resonance donor polarization $+M$ (or polarizability $+E$) effect], and, indeed, precisely this is observed.

The existence of polarizability thus renders impossible any unique scale of electrophilic reactivity. The two most common theoretical measures, namely π-electron densities and localization energies, correspond to transition states approximating the ground state and the Wheland intermediate, respectively, whereas the transition state (the precise structure of which is unknown), lies somewhere in between. π Densities, which relate to a situation where inductive effects are dominant, will tend to predict a relatively low 2- : 3-rate ratio since all of the heteroatoms are inductive acceptors $(-I)$. By contrast, since π electrons are delocalized from the heteroatoms more to the 2- than to the 3-position, localization energies will predict a high 2- : 3-rate ratio. The importance of these factors becomes particularly evident in consideration of the substitution of benzo derivatives of these molecules (Chapter 8).

Additional problems in theoretical calculations are (1) selection of the coulomb and resonance integral parameters; (2) whether an auxiliary inductive parameter be used for the α-carbons; (3) whether d orbitals, etc., be taken into account for S, Se, and Te; and (4) what type of calculations to use.

The coulomb integral α_X is a measure of the heteroatom electronegativity, so too high a value reduces the reactivity of the 2-position. Values of α_X/α_C are usually chosen as ~2.5, 1.5, and 1.0 for O, NH, and S, respectively, with corresponding values of β_{CX}/β_{CC} of 0.7, 1.0, and 0.6 [68JCS(B)1397; 71AJC2679]. In general, the calculations seem less successful as electronegativity of the heteroatom increases.

π Densities are less successful than localization energies for predicting observed reactivities and can give, for example, either 2 > 3 or 3 > 2 positional reactivity orders for thiophene [86HC(44,2)1] and pyrrole (67M254). Localization energies can give reasonable results for a given molecule [e.g., the greater reactivity difference in the 2- and 3- positional reactivities in furan compared to thiophene and pyrrole [68JCS(B)1397; 71AJC2679]], but they are less good at predicting relative reactivities from one molecule to another where different heteroatoms are involved, [e.g., the order for 2-substitution is predicted to be Te > Se > S > O [73JCS(P2)2097]]. On the other hand, where a single heteroatom is involved (e.g., S,) then calculations, particularly Hückel calculations, are very good though here it is probable that the similarities in the electronegativities of carbon and sulfur are primarily responsible [82JCS(P2)295, 82JCS(P2)301; 83JCS(P2)813].

The valence bond method of predicting positional reactivities should also be considered. The method is incapable of predicting differences in reactivity between molecules containing different heteroatoms, but is very good at predicting reactivities within a given heterocycle. Thus for α- and β-substitution, the canonical forms are as shown in Scheme 6.2. For α-substitution there are three canonical forms, whereas for β-substitution there are only two. The charge from the electrophile is therefore

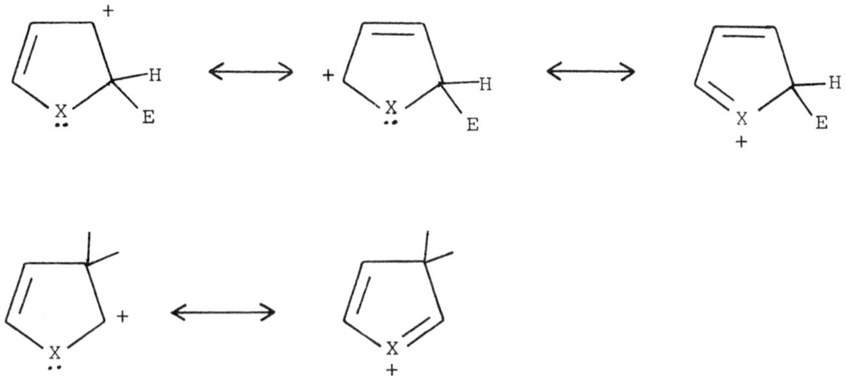

SCHEME 6.2. Canonicals for α- and β-substitution of five-membered heterocycles.

better delocalized in the transition state for α-substitution, which is therefore favored. The valence bond method is also very satisfactory for predicting reactivities of benzo derivatives of five-membered heterocycles and this is considered in Chapter 8.

B. SUMMARY OF RELATIVE RATES

The previously described reactivity data for the five-membered heterocycles are gathered (in terms of σ^+ values) in Table 6.12; no data are given for nitration because the rates are encounter controlled and meaningless in terms of electronic effects. Among the other data, those for mercuriation, protiodemercuriation, and protiodeboronation are doubtful, and other qualifying aspects are noted in the table footnotes. The following main features are noteworthy.

(1) The order of reactivities of the 2-positions appear to be pyrrole > furan > tellurophene > selenophene > thiophene. The order for furan and tellurophene was also obtained in trifluoroacetylation (for which no σ^+ values are obtainable). However, in solvolysis of 1-arylethyl acetates, tellurophene is more reactive than furan.

(2) In all four reactions in which the reactivities of both positions in furan and thiophene have been determined under the same conditions (pyrolysis of 1-arylethyl acetates, solvolysis of 1-arylethyl acetates, protiodesilylation, and protiodeboronation), the reactivity order is the same: 2-furan > 2-thiophene > 3-furan > 3-thiophene. In two other reactions, the order 3-thiophene > 3-furan *appears* to be obtained. However, in the solvolysis of 1-arylethyl *p*-nitrobenzoates, these esters were measured neither relative to each other nor to a standard compound (nor by the same worker), and the conclusion is almost certainly in error. Likewise in acetylation, the low reactivity of the 3-position of furan (which also shows up in the 2-isomer) is attributable to coordination with the Lewis acid catalyst.

(3) The *accurate* data for Table 6.12 correspond to the region to the right of the dotted line in Fig. 6.1. But other reactivity orders are theoretically possible, and may in due course be observed.

The magnitude of the σ^+ values varies for a number of reasons.

(1) Between the reactions, there are differential demands for resonance. This may, in particular, account for the low values in reaction 2, which should have the lowest demand for resonance. [The ρ factor for reaction 1 is smaller, but refers to a high temperature and is equivalent to

TABLE 6.12
Reactivity Parameters ($-\sigma^+$) for Five-Membered Heterocycles

Reaction	ρ	Pyrrole 2	Pyrrole 3	Furan 2	Furan 3	Thiophene 2	Thiophene 3	Selenophene 2	Selenophene 3	Tellurophene 2
1. Pyrolysis of 1-arylethyl acetates	−0.66	—	—	0.88	0.415	0.795	0.38	0.855	—	—
2. Rearrangement of 1-aryl-but-2-enols	−2.6	—	—	0.75	—	0.595	—	—	—	—
3. Isomerization of cis-1-aryl-2-phenylethenes	−3.3	—	—	—	—	0.77	—	—	—	—
4. Protiodemercuriation	−2.87	—	—	(1.255)[a]	(0.76)[a]	(1.125)[a]	—	—	—	—
5. Mercuriation	−4.0	—	—	—	—	(1.565)[a]	—	—	—	—
6. Protiodesilylation	−4.6	—	—	0.905	0.45	0.785	0.43	—	—	—
7. Iododeboronation	−4.8	—	—	—	—	0.83	0.59	—	—	—
8. Protiodeboronation	−5.0	—	—	(1.19)[a]	(0.77)[a]	(1.16)[a]	(0.755)[a]	—	—	—
9. Solvolysis of 1-arylethyl chlorides	−5.4	—	—	—	—	0.78	—	—	—	—
10. Solvolysis of 1-arylethyl acetates	−5.7	1.89[b]	—	0.93	0.50	0.83	0.47	0.885	—	0.995
11. Solvolysis of 1-arylethyl p-nitrobenzoates	−5.8	—	—	0.94	0.445	0.815	0.46	—	—	—
12. Positive bromination	−6.2	—	—	—	—	0.85	0.51[c]	0.955	—	—
13. Hydrogen exchange, aq. H_2SO_4	−7.7	—	—	—	—	0.873	0.488	—	—	—
TFA	−8.75	—	—	—	—	0.913[d]	0.560[d]	—	<0.53	—
MeOH–H_2SO_4	−8.4	1.70	1.71	0.82	0.405	0.71	0.45	0.95	—	—
14. Acetylation	−9.2	—	—	—	—	0.76	0.56[c]	~0.74[c]	—	~0.795[c]
15. Molecular chlorination	−10.0	—	—	—	—	0.76	0.56[c]	0.84	—	—
16. Molecular bromination	−12.0	1.54[f]	—	0.98[f]	—	0.81	0.58[c]	0.94	—	—

[a] See text, Sections 8.A, 8.C.
[b] For 1-methylpyrrole.
[c] These values are inaccurate and could be ~0.1 σ units smaller, but not larger.
[d] These are the only values for solution reactions, which are corrected for hydrogen bonding.
[e] 2:3-Rate ratio not determined.

−1.3 at 25°C; further, the demand for resonance in the reaction is exalted relative to the ρ factor, because of the absence of solvent (63T937).]

(2) Errors may be introduced by use of overlap techniques; only the data for reaction 1 are entirely free of this approximation.

(3) Reactions carried out in media that are good proton donors will be affected by hydrogen bonding; only the solution data for hydrogen exchange are corrected for this. Hydrogen bonding will reduce the reactivity of the heterocycle and this is clearly shown by the hydrogen exchange (TFA) values for thiophene (which are corrected for hydrogen bonding) being higher than in any other solution reactions.

Nevertheless, it is approximately true that in reactions of average demand for resonance, σ_2^+(furan) − σ_2^+(thiophene) is 0.1, σ_2^+ − σ_3^+(furan) is 0.45, and σ_2^+ − σ_3^+(thiophene) is 0.40. Reasonably consistent values of σ_2^+ for furan, thiophene, selenophene, and tellurophene are −0.91, −0.81, −0.88, and −0.90, respectively; the "best" σ_3^+ values for furan and thiophene are −0.45, and −0.43, respectively. No data are available for the relative reactivities of the 3-positions of selenophene or tellurophene. The 3-position of pyrrole is slightly more reactive than the 2-position in hydrogen exchange, but the reactivities are so close that under different demands for resonance, the reactivity order can be inverted.

A further point seems relevant here. In Vilsmeier–Haack formylation, the activation parameters (Table 6.13) were determined for furan, thiophene, selenophene, and tellurophene [73JCS(P2)2097]. A good linear correlation was obtained between the activation enthalpies and the resonance energies of the four rings, indicating that the differences in ground-state energy play a fundamental role in determining the relative reactivities at the α-position. (This result may, however, be merely fortuitous because it would predict pyrrole to be intermediate in reactivity between thiophene and selenophene.) The constant activation entropies (within experimental error) suggest that the transition states in each case lie in a

TABLE 6.13
ACTIVATION PARAMETERS FOR VILSMEIER–HAACK FORMYLATION IN $CHCl_3$

Parameter	Heterocycle			
	Furan	Thiophene	Selenophene	Tellurophene
Heat of activation ($\Delta H^\ddagger$, kcal mol^{-1})	14.7 ± 1.1	17.1 ± 0.8	16.5 ± 0.7	15.5 ± 1.2
Entropy of activation ($\Delta S^\ddagger$, cal mol^{-1} K^{-1})	−27.7 ± 1.8	−29.5 ± 1.1	−28.8 ± 1.0	−27.5 ± 2.1

similar position along the reaction coordinate. Moreover, the high negative entropes of activation are consistent with a high degree of formation of a new bond (i.e., the transition states resemble the σ-complex). This is consistent with the high ρ factor for formylation, estimated as > -10.

C. Sensitivity of the Five-Membered Heterocycles to Substituent Effects

For electrophilic substitutions, the Hammett ρ factor is generally regarded as a measure of the reactivity of the electrophile (and consequently of the position of the transition state along the reaction coordinate) [65MI1(298)]. Compare (say) the ρ factor for substituted benzenes with that for substituted methylbenzenes. Since the ring system transmitting the effects is the same in each case, it is reasonable to interpret the observation of a smaller ρ factor for the more reactive compounds in terms of differing reactivities of the aromatic substrates (benzene and methylbenzene), and this is a manifestation of the reactivity–selectivity principle first described by Norman and Taylor [65MI1(298)]. For the five-membered heterocycles, however, such a conclusion may not necessarily be justified because, in addition to the differing reactivities of the parent heterocycle, there will also be differing degrees of conjugation within the molecules that will affect their ability to transmit substituent effects of which the ρ factor is also a measure. It may then turn out (see below) that the most reactive system does not necessarily produce the smallest ρ factor. This should not be regarded as a breakdown of the reactivity–selectivity principle, because the latter is meant to relate to aromatics with like π systems, though it has sometimes incorrectly been rigorously applied elsewhere.

In considering the results obtained to date, some general points need to be considered.

(1) The substituent effects show that conjugative effects are transmitted strongly, and since they will be poorly transmitted by the heteroatom flanked by substantially single bonds, the carbon chain is principally involved and the extent of conjugation in this will be particularly important. It seems reasonable to assume that the greater the degree of bond fixation in the heterocycle, then the more difficult it will be to place a double bond across the 3,4-positions, as required for conjugation between the 2- and 5-positions. On this basis, the ease of transmission ought to parallel the order of aromaticity (i.e., benzene > thiophene > pyrrole > selenophene > tellurophene > furan). In the majority of reactions (though not in proti-

odesilylation, one of the best documented), thiophene is less effective at transmitting substituent effects than is benzene.

(2) Any direct field component of substituent effects will be better transmitted in the heterocycles than in benzene, because even in thiophene the C-2—C-3, C-2—C-4, and C-2—C-5 distances will be shorter than their counterparts in benzene; on this basis furan and pyrrole should be the best transmitters among the heterocycles. Such effects should be less significant in side-chain reactions where the charge to be stabilized is one atom further away.

(3) Resonance donor ($+M$) and acceptor ($-M$) substituents may produce substantially larger effects than they do in benzene because they are able to become more coplanar with the heteroaromatic ring. A proper analysis requires a range of substituent types to detect such anomalies. These need, in turn, to be considered carefully because hydrogen bonding may reduce the reactivities of more reactive compounds relative to less reactive ones, so producing attenuated ρ values.

(4) The ability of side chains to become coplanar with the heteroaromatic ring will be greater than in benzene, so that for reactions involving side-chain carbocations, the heterocycle may appear to be a better transmitter than benzene.

It is thus probably impossible at present to interpret unambiguously the magnitude of ρ, and this difficulty is compounded by the paucity of data. Some ρ factors for electrophilic substitutions of benzene and thiophene are gathered in Table 6.14; only the acetylation, detritiation, and protio-

TABLE 6.14
HAMMETT ρ FACTORS FOR ELECTROPHILIC SUBSTITUTION OF SUBSTITUTED BENZENES AND THIOPHENES

Reaction	Benzenes	Thiophenes
Bromination	-12.1	-10.0
Chlorination	-10.0	~ -7.0
Acetylation	-9.1	-5.6
Detritiation	-8.8	-6.5
Protiodesilylation	-5.3	-6.25
Mercuration[a]	-4.0	~ -4.7

[a] The ρ factor of -5.7 claimed for mercuration of substituted selenophenes was based on a few results for electron-withdrawing substituents and did not include selenophene itself (66CHE686).

desilylation factors for thiophene are accurate. For the other reactions of thiophene the factors are based mainly on data for a few electron-withdrawing substituents, largely of the resonance acceptor $(-M)$ type. If these are excluded, the ρ factors for halogenation of thiophene would be much smaller than in benzene, whereas that for mercuriation (based simply on the value for the 2-bromo compound) would be the same.

The general trend seems to be for thiophene to give smaller ρ factors, consistent either with it being more reactive, or that it transmits substituent effects less readily. However, the relative values of the (well-documented) ρ factors for desilylation are puzzling and confirms that a single overall explanation will not suffice. It is probably relevant that in this reaction the demand for resonance stabilization of the transition state is substantially lower.

In an attempt to shed further light on the overall problem, Clementi and Marino have collated the effects of a 5-methyl substituent upon the reactivity of the 2-positions of the heterocycles in a number of reactions (described in the previous sections) [77CS(11)87], and these are gathered in Table 6.15. (In the original publication the values were presented in terms of approximate ρ factors, calculated from $\log k_{rel.}/\sigma^+_{p\text{-Me}}$; note that in this paper there are errors in the column headings, footnotes, and ρ factors, which account for discrepancies from Table 6.15.)

Consideration of the results for electrophilic substitution relies heavily upon the (uncertain) ρ factor for trifluoroacetylation. Clementi and Ma-

TABLE 6.15

ACTIVATING EFFECT OF A 5-METHYL SUBSTITUENT AT THE 2-POSITION IN ELECTROPHILIC SUBSTITUTIONS AND SIDE-CHAIN SOLVOLYSIS

Aromatic	Trifluoro acetylation[a]	Formylation[b]	Solvolysis of 1-arylethyl acetates	Solvolysis of 1-arylethyl p-nitrobenzoates
Furan	1700	880	164	214
Thiophene	380	196	74	81
Selenophene	280	300	23	—
Tellurophene	500	620	12	—
Pyrrole	24	15	—	170
(Benzene)[c]	>30,000[d]	>5000	~59	~64

[a] In 1,2-dichloroethane at 75°C.
[b] Vilsmeier–Haack formylation in $CHCl_3$ at 30°C.
[c] Effect of a para-substituent in benzene.
[d] Not measured, but calculated from the ρ value itself determined by comparison of the relative rates of the heterocycles with their σ^+ values; the error in the substituent effect may be 100%.

rino gave the value as -10.2, based upon a plot of σ^+ for the heterocycles against the log k_{rel} values. This ρ factor depends largely on the accuracy of the σ^+ value for pyrrole (which Table 6.12 shows to vary widely) and the accuracy of log k_{rel} for pyrrole, which in this reaction is 10^6 times more reactive than the other heterocycles. The importance of this is seen in the fact that if only the σ^+ values for the other heterocycles are used, ρ becomes -15.0 (or -19.9 if the σ^+ values quoted in this review are employed). A value in the latter region is more likely to be correct because (1) use of the same technique with the acetylation data predicts ρ to be -10, which is very close to the accepted value of -9.2; and (2) the spread of rates among the five-membered heterocycles is much greater in trifluoroacetylation than in acetylation, so the ρ factor must be considerably higher than the value of -9.2 for acetylation. Thus the order of transmission of substituent effects is benzene > furan > tellurophene > thiophene > selenophene >> pyrrole, being slightly different to formylation in that the positions of thiophene and selenophene are reversed. The results confirm that more than one factor is at work because if the small substituent effect in pyrrole reflected solely its high reactivity, then the next smallest effect should be observed in furan, the opposite being the case.

The question of whether more reactive compounds should give smaller ρ factors by virtue of their greater reactivity is relevant to the seeming incompatibility of the Hammond postulate (more reactive compounds have transition states more like the ground state and less reactive ones more like the products) and the Hammett equation. This implies that a set of σ^+ values derived from any one reaction should give a curve when plotted against the log k_{rel} values for any other, if the transition states for the reactions differ widely. Such curvature may be obscured by the need to use overlap techniques for reactions of high ρ factor, by the variation in resonance between reactions, or because the above reasoning is flawed (90MI1). The variation in transition-state structure for reaction of a given group of substituted benzenes (say) with a range of electrophiles is represented in Fig. 6.2 [74MI1(219)], in which left to right represents reaction with electrophiles of decreasing reactivity and increasing selectivity. (The Hammond postulate requires that the lines shown as straight might need to be slightly curved relative to each other). Likewise, for reactions of a given set of derivatives of a range of substrates of differing reactivities (e.g., pyrrole, furans, benzenes) with a given electrophile, the corresponding variations in transition state structure are given by Fig. 6.3, in which left to right represents aromatics of decreasing reactivity. Figure 6.3 has been represented elsewhere with the slope of the lines increasing with later transition states [77CS(11)87]. Whether or not this is

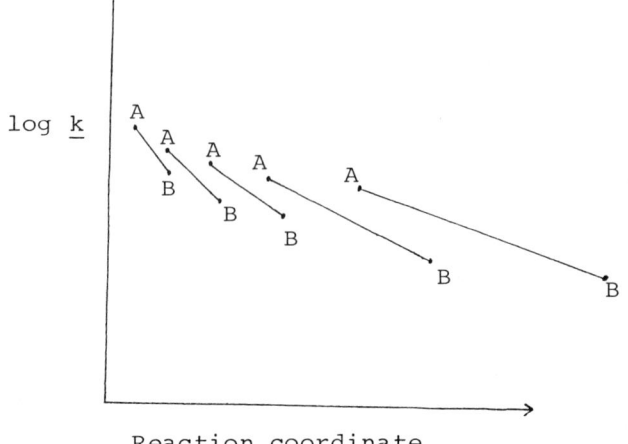

FIG. 6.2. Variation in transition-state structure with electrophile reactivity in reactions of substituted benzenes. A, Activated benzene (e.g., anisole); B, deactivated benzene (e.g., nitrobenzene).

more appropriate (to Fig. 6.3 and also to Fig. 6.2) depends upon whether the spread of transition-state structures is greater for a reaction with a higher ρ factor, or not; at present this cannot be properly decided.

For reactions involving side-chain carbocations, the order of transmission of substituent effects is quite different to the electrophilic substitutions, being furan > pyrrole > thiophene > benzene > tellurophene.

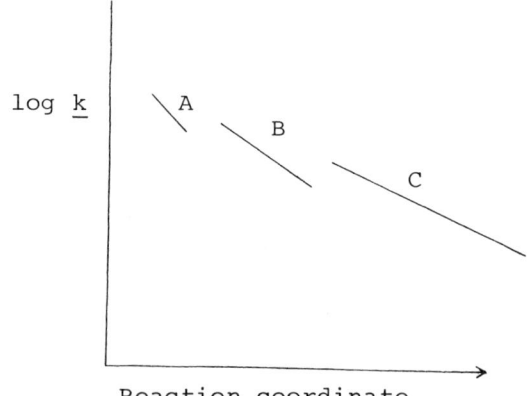

FIG. 6.3. Variation in transition-state structure with aromatic reactivity in reaction of a given electrophile. A, Pyrroles; B, furans; C, benzenes.

Here again, one can devise reasons which appear to explain the results, but they may not necessarily be the correct ones. For example, calculation of the difference in charge at a given site between the ground-state molecule and $ArCH_2^+$ indicates that substituent effects should be larger in furan than in thiophene (72TL3889), but corresponding calculations are not available for all of the molecules. However, it may also be relevant that the observed sequence is close to if not exactly that in which a —CMe_2^+ side chain can achieve coplanarity with the aromatic ring; the steric hindrance to the sp^2 carbon becoming coplanar with the aromatic ring will be smaller for a five-membered ring than for that of benzene, but will increase within the series as the size of the heteroatom becomes larger. A straightforward test of this possibility would be to compare the rate of the 5-methyl and unsubstituted compounds, in solvolysis of the 2-(2-aryl)-2-propyl acetates. If there is a regular departure from a linear free-energy correlation between these data and those for 1-arylethyl acetates in Table 6.15, with increasing size of the heteroatom, it would show that steric effects inhibit coplanarity in ArC^+Me_2 and possibly therefore in $ArCHMe^+$ as well.

CHAPTER 7

Azoles

1. Introduction

A. Compounds Considered

The most important compounds considered to fall within the scope of this chapter are listed in the following subsections a–g, but reactivity data are as yet by no means available for all of them.

a. *Neutral Five-Membered Rings with Two Heteroatoms*

7.1	Z	7.2
Pyrazole	NH	Imidazole
Isoxazole	O	Oxazole
Isothiazole	S	Thiazole
Isoselenazole	Se	Selenazole

b. *Neutral Five-Membered Rings with Three Heteroatoms*

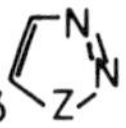

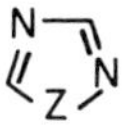

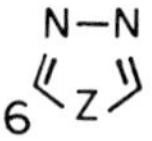

1,2,3	1,2,4	1,2,5	1,3,4
Z = NH	Triazoles	Z = S	Thiadiazoles
Z = O	Oxadiazoles	Z = Se	Selenadiazoles

c. *Neutral Five-Membered Rings with Four Heteroatoms*

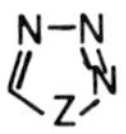

1,2,3,4		1,2,3,5	
Z = NH	Tetrazoles	Z = S	Thiatriazoles
Z = O	Oxatriazoles	Z = Se	Selenatriazoles

d. Monocationic Azoles

Ions such as isoxazolium and pyrazolium are produced by protonation of a multiply bonded nitrogen in the above species, and are therefore present under acidic conditions. Similar cations can be formed by alkylation.

e. Anionic Azoles

These ions (**7.9–7.13**) are produced via proton loss from an NH group and are therefore found under basic conditions.

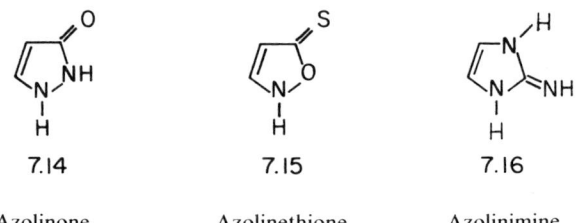

7.9   7.10   7.11   7.12   7.13

f. Azolinones and Corresponding Thiones and Imines

There is a very large number of these. Examples of each main type are shown in structures **7.14–7.16**.

7.14 Azolinone   7.15 Azolinethione   7.16 Azolinimine

g. N-Oxides

In principle there is also a very large number of these; two examples are shown in structures **7.17** and **7.18**.

7.17   7.18

B. REACTIVITY PATTERN

The reactivities of these molecules show the following general features.

(1) Neutral multiply bonded ("pyridine-like") nitrogen atoms in the ring deactivate the ring as, for example, in pyridines. Such deactivation is much increased in corresponding positively charged compounds. Hence, compounds are much less reactive in reactions carried out under strongly acidic conditions where they exist in the protonated form. Hydrogen exchange, nitration, and sulfonation are usually carried out under strongly acidic conditions, whereas halogenation and mercuriation usually are not. The effect of N-oxide formation is more subtle: the $:N^+(O^-)$ group is more effective both as an electron acceptor and as an electron donor than the (neutral) nitrogen.

The greater the number of multiply bonded nitrogens in the ring, the lower is the reactivity. This occurs to such an extent that electrophilic substitution is unknown in azoles containing four heteroatoms. As the number of heteroatoms in the ring increases, it becomes less probable that reaction will take place upon the conjugate acid species. Not only is the conjugate acid too unreactive, but the lower overall electron availability on any particular nitrogen, and hence the lower basicity, will increase the concentration of free base species.

Each multiple bonded pyridine-like nitrogen atom deactivates all positions especially those α and γ to it.

(2) The effect of NH, O, and S groups in the ring is in each case activating, in the order NH > O > S, just as occurs in pyrrole, furan, and thiophene. Formation of the anion from a compound containing an NH group (e.g., in pyrazole and imidazole) brings about a substantial increase in reactivity as expected. Each of these groups activates positions α and β to it, the former the more strongly.

(3) The effect of multiply bonded nitrogen will be modified (compared to its effect in pyridine) by any bond fixation present in the molecule. Thus, the nitrogen in thiazole deactivates the 2-position much more strongly than the 4-position, since the mesomeric electron-withdrawing $(-M)$ effect of nitrogen operates more effectively across regions of high π-bond density.

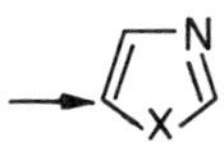

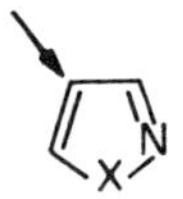

The overall result is that compounds of type **7.19** are more reactive than those of type **7.20**, the preferred substitution site being as indicated, the positional reactivity order being 5 > 4 > 2 in **7.19** and 4 > 5 > 3 in **7.20**. In imidazole, the reactivities of the 4- and 5-positions are both expected, and found, to be fairly similar.

The above positional reactivity orders are predicted for pyrazole and imidazole by CNDO/2 calculations [78JCS(P2)865].

C. Substituent Effects

The effects of substituents at ring carbon atoms on electrophilic substitution rates and orientations qualitatively parallel those in benzene, but again they will be affected by bond fixation, which in particular will modify the mesomeric (resonance) effect. Thus a substituent will generally produce a smaller effect across the 3,4-bond of an azole, and a bigger one across the 2,3-bond (in either direction), compared to benzene. This clear picture applies to neutral species, but is complicated in the cationic species formed by protonation. An activating substituent will increase the concentration of protonated species so that reaction via conjugate acid becomes more significant; a deactivating substituent shows the opposite effect. Additionally, the distinction between the two nitrogen atoms, and thus between the 2,3- and the 3,4-bonds, disappears in protonated pyrazoles and imidazoles.

For these reasons, directly observed relative rate coefficients do not necessarily give a true picture of the substituent effects, and an attenuated Hammett ρ factor may result. To obtain the true substituent effects, the procedure given in Section 9.B must be adopted.

2. Acid-Catalyzed Hydrogen Exchange

A. Mechanism

The rate coefficients for deuteration in aqueous D_2SO_4 have been determined for a wide range of azoles and their methyl derivatives [71JCS(B)2365; 73JCS(P2)1675; 74JCS(P2)399]. In general, the compounds reacted as the free bases with a changeover to reaction on the conjugate acid at higher acidity for the more reactive compounds. As expected, this changeover point occurred at a lower acid concentration the higher the temperature, since increasing temperature makes the electrophile more reactive and also affects the acidity function. When reaction took place via the conjugate acid, this was frequently confirmed by the deuteration rates for the corresponding N-methyl derivative, which in every case reacted at a closely similar (but slightly faster) rate. Correction of the rate data to standard conditions (100°C, pD = O) gave the partial rate factors given in Scheme 7.1 in terms of their log values; temperatures of rate measurements are shown in parentheses.

B. REACTION AT THE 4-POSITION

The 4-positions are the most reactive, as theory requires. The order of reactivity at the 4-positions (1-methylpyrazole > isoxazole > isothiazole) exactly parallels that for the corresponding positions in 1-methylpyrrole, furan, and thiophene (cf. Chapter 6). The order for isoxazole and isothiazole were stated orginally [74JCS(P2)399] to be different from that in furan and thiophene, but this conclusion was based on an incorrect literature analysis of the reactivity of these latter two molecules. The problem arises because the data for hydrogen exchange of furan in sulfuric acid are not corrected for the substantial protonation and for hydrogen bonding which occurs [72MI2(262)]. The relative reactivities given in more recent reviews [79AHC(25)147; 84MI13] do not therefore reflect the true situation.

C. EFFECT OF RING NITROGEN ATOM ON RATE

Comparison of the data for isothiazole and thiophene show the meta nitrogen to produce a rather small deactivation effect, with log $k_{rel.}$ = 3.6 − 5.0 (i.e., −1.4). Since ρ under these conditions is −7.5, this gives σ^+ for meta nitrogen as ~0.2. This is substantially lower than in the effect of uncharged nitrogen at the 3-position of pyridines (cf. Chapter 9) and reflects a phenomenon for which there is increasing evidence that the transmission of electronic effects to meta sites in molecules with substantial bond fixation is attenuated compared to that in those without bond fixation (e.g., benzene). The same feature is evident in low meta partial rate factors in hydrogen exchange of substituted naphthalenes [68JCS(B)1112]; it accounts for the 4-position of isoquinoline being the most reactive (Chapter 11); and it is reflected in low meta-methyl substituent effects in the azoles (see Section 2.D). For N-methylpyrazole versus N-methylpyrrole, the effect is even smaller, σ^+ being (9.8 − 10.3)/−7.5 = 0.07. From these two molecules the deactivating effect of the nitrogen upon the 5-position may be calculated in σ^+ terms as (5.6 − 10.6)/−7.5 = 0.67, again smaller than the value for the corresponding effect at the 4-position in pyridine (cf. Chapter 9).

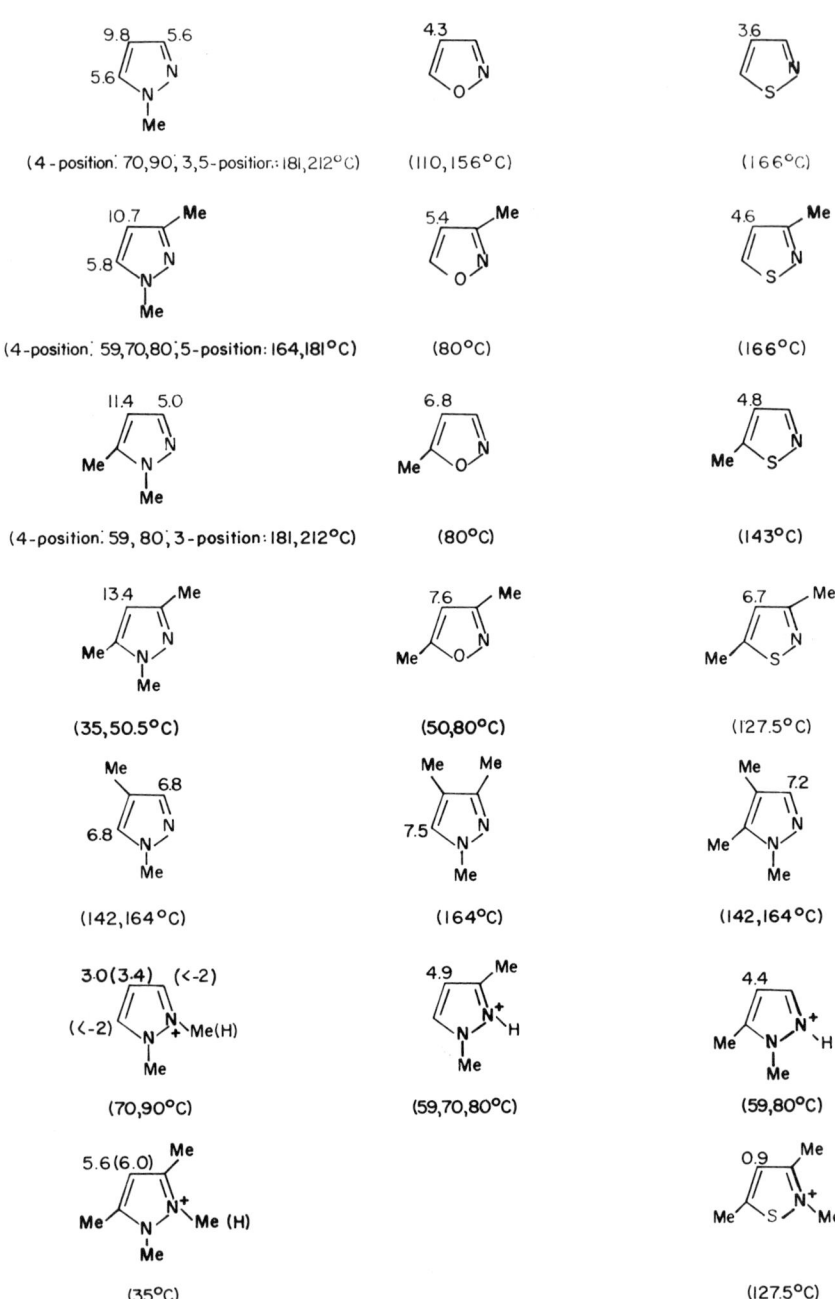

SCHEME 7.1. Log partial rate factors for acid-catalyzed hydrogen exchange. Reaction temperatures are given in parentheses.

D. EFFECT OF METHYL SUBSTITUTION ON RATE

The effects of an ortho-methyl substituent are summarized in Table 7.1. The majority of the results demonstrates the differences in the bond orders within a given molecule, which causes activation across the 4,5-bond to be greater than that across the 3,4-bond. Curiously the combined effects of two methyl groups are not additive, nor are the discrepancies between observed and calculated reactivities in the same direction: dimethylisoxazole is less reactive than predicted (the expected result), whereas the other dimethylazoles and 1,2-dimethylpyrazolium are more reactive than predicted. There are no obvious explanations for these discrepancies other than the fact that large extrapolations of rate data are required in this work. This latter may account for the fact that activation by the ortho-methyl group in 1-methylpyrazolium is in the opposite direction expected from the bond-order difference.

The effects of meta-methyl substituents are summarized in Table 7.2. Again there are significant discrepancies (deactivation being apparently observed in a number of cases), which probably reflect the experimental difficulties. A contributing factor may, however, be steric hindrance to solvation of the methyl derivatives, which is probably particularly important in these highly reactive molecules. Overall, the meta-methyl activation is insignificant, which is consistent with the weak meta-deactivating effects of nitrogen noted in Section 2.C.

E. REACTIVITY OF CATIONS VERSUS FREE BASES

Cations are much less reactive than the free bases, the reactivity difference being $10^{7.4}$ between 1,3,5-trimethylpyrazolium ion and its free base, and $10^{5.9}$ between 3,5-dimethylisothiazolium ion and 3,5-dimethylisothia-

TABLE 7.1

LOGARITHMS OF THE ACTIVATING EFFECTS OF AN ORTHO-METHYL SUBSTITUENT UPON THE REACTIVITY OF THE 4-POSITION OF AZOLES[a]

	Compound				
Substituent	1-Methyl-pyrazole	1-Methyl-pyrazolium	Isoxazole	Isothiazole	1,2-Dimethyl-pyrazolium
3-Methyl	0.9	1.5	1.1	1.0	0.4
5-Methyl	1.6	1.0	2.5	1.2	0.4
3,5-Dimethyl	3.6	2.6	3.3	3.2	2.6

[a] [73JCS(P2)1675; 74JCS(P2)399].

TABLE 7.2
LOGARITHMS OF THE EFFECTS OF META-METHYL SUBSTITUENTS IN AZOLES

Methylazole (A)	Parent compound (B)	Position	log k_A - log k_B
1,3-Dimethylpyrazole	1-Methylpyrazole	5	0.2
1,5-Dimethylpyrazole	1-Methylpyrazole	3	−0.6
1,3,4-Trimethylpyrazole	1,4-Dimethylpyrazole	5	0.7
1,4,5-Trimethylpyrazole	1,4-Dimethylpyrazole	3	0.4
1,2-Dimethylpyrazolium ion	1-Methylpyrazolium ion	4	−0.4
1,2,3,5-Tetramethylpyrazolium ion	1,3,5-Trimethylpyrazolium ion	4	−0.4

zole. The differences in reactivities between like positions in the cations and free base are, for N-methylpyrazole, larger at the 3,5-positions (log k_{rel} > 7.6) than at the 4-position (log k_{rel} = 6.4).

The relative positional deuteration rates have been determined for N-methylimidazolium ion in D_2SO_4 at 163°C (positions in parentheses) as 1 (2), 73 (4), 120 (5) (74JOC2398), revised to 1 (2), 30 (5), 50 (4) (79JOC4240). This shows that the 4-position is the most reactive, whereas for the free base, the 5-position is predicted to be the most reactive.

A particularly interesting result concerns the reactivity of imidazoles containing fluorine at the 4- or 2-position (Scheme 7.2) (79JOC4240). Because of the strong inductive electron-withdrawing $(-I)$ effect of ortho-fluorine, the nitrogen at the 3-position is not protonated so that these compounds undergo reactions on the free bases. As a consequence they appear to be more reactive than the corresponding nonfluorinated compounds. The reactivity is enhanced by the electron-donor mesomeric effect $(+M)$ of fluorine, which is particularly effective across the 4,5-bond with its high bond order. [In the case of 2-fluoro-4-methylimidazole the $(+M)$ effect of methyl is similarly enhanced.] In the corresponding nitro- or trifluoromethyl compounds, although protonation does not occur, the strong $(-M)$ deactivation results in exchange being extremely slow. The

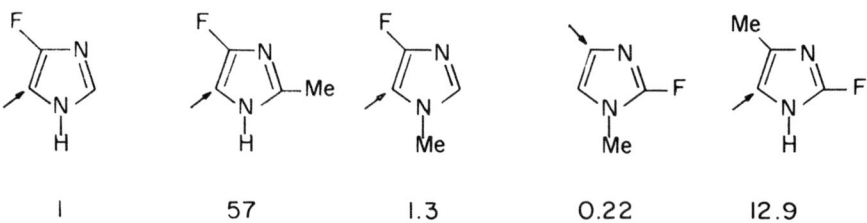

SCHEME 7.2. Relative rates of deuteration of fluoroimidazoles at 50°C (79JOC4240).

results show methyl to activate 57-fold across the 2,5-position and only 1.3-fold across the 1,5-bond (consistent with the very low 1,5-bond order).

3. Base-Catalyzed Hydrogen Exchange

A. Introduction

Base-catalyzed exchange involves rate-determining removal of a proton to give a carbanion intermediate, Ar^-, which will be strongly stabilized by positive centers adjacent to the reaction site. These are therefore particularly effective in accelerating base-catalyzed exchange while electron supply reduces the reaction rate. We have described above how in acid-catalyzed exchange reaction may occur via the (more reactive) free base rather than the (more abundant) conjugate acid. As the acidity of the medium is increased, reaction will ultimately take place only on the conjugate acid as it will effectively be the only species present. A parallel situation exists for base-catalyzed exchange: at acidity somewhat weaker than corresponding to the pK_a, both the free base and the conjugate acid will be present. The latter is now the minor, but the much more reactive, component, so reaction will, in many cases, preferentially take place upon the conjugate acid by way of ylide intermediates. In more basic reaction conditions, this mechanism will be superceded by one involving reaction upon the free base, since this will be the only species present in appreciable concentration. The reaction rate initially increases with increasing base concentration, but then levels off as this is compensated by a decrease in the concentration of the conjugate acid.

All these aspects are nicely demonstrated by exchange data for azoles. In addition, a particular feature for compounds **7.21** and **7.22** with X = NH, is that the rates *decrease* at high basicities due to formation of unreactive anions. In six-membered rings an adjacent lone pair on a neutral nitrogen inhibits the free-base mechanism because of unfavorable juxtaposition of the lone pairs in the σ orbitals of the carbanion and nitrogen (78JOC3565). However, this effect is smaller in five-membered rings (**7.23**) than in six-membered rings because the angle between the adjacent lone pairs is greater.

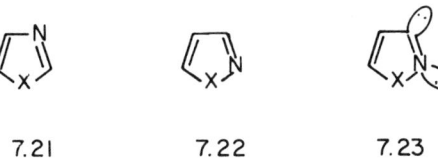

    7.21   7.22   7.23

B. POSITIONAL REACTIVITY ORDER

For azoles with 1,3-heteroatoms (**7.21**), the $-I$ effect of the heteroatom X should activate both the 2- and the 5-positions making these more reactive that the 4-position. Moreover, the 2-position is adjacent to nitrogen so the overall reactivity order should be 2 > 5 > 4 and this has been observed for oxazole [69JCS(B)270] and thiazole [72JA5759; 78CI(M)348]. For imidazole, however, there will be little difference in the electronegativity of X(—NH—) and =N—, so the reactivity of the 4- and 5-positions will be closely similar (and may in fact be largely dependent upon the lengths of the N—C-3 or N—C-5 bonds). Thus the reactivity order in 1-methylimidazole is 2 > 4 ≅ 5, with relative rates of 54,500 : 1.6 : 1. The variation in rate with basicity shows reaction here to involve the ylide intermediate (**7.24**) for reaction at the 2-position, and carbanion for the 4- and 5-positions [65CI(L)1728; 66LA(695)55; 70JOC1141; 74JOC2398]. The related 1,3-dithiolinium cation **7.25** also underwent exchange (at 34°C) at the 2-position (in D_2O–CF_3CO_2D in which OD^- is probably the effective reagent) [65AG(E)435]. Rates of exchange at the 2-position of imidazole and L-histidine are almost identical (73JA6139).

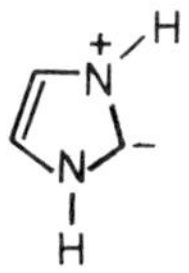

7.24 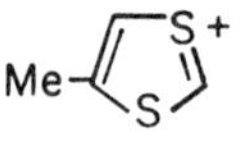   **7.25** 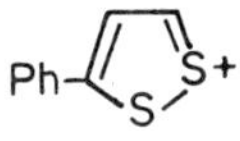   **7.26**

For azoles with heteroatoms in the 1,2-positions (**7.22**), reasoning similar to that above predicts the reactivity order 5 > 3 > 4, and the highest reactivity of the 5-position has been confirmed for isothiazole (66JA4263; 69JHC199) and pyrazole (70JOC1146).

The 2-position in azoles with 1,3-heteroatoms should be more reactive than the 5-position of azoles with 1,2-heteroatoms, but for reactions of the free base this turns out not to be the case because of the adjacent lone pair effect illustrated by the relative reaction rates in Scheme 7.3. Thus the 2-position of thiazole is 7 times less reactive than the 5-position of isothiazole. The same reasoning accounts for the 3-position of isothiazole being less reactive than the 4-position of thiazole. The former should be the more reactive since the electron-withdrawing effect of nitrogen should be greater across the bond of higher order, and the fact that it is not more reactive suggests that the effect of the adjacent lone pair is more severe across the shorter C-3—N bond in isothiazole. For reaction of the azol-

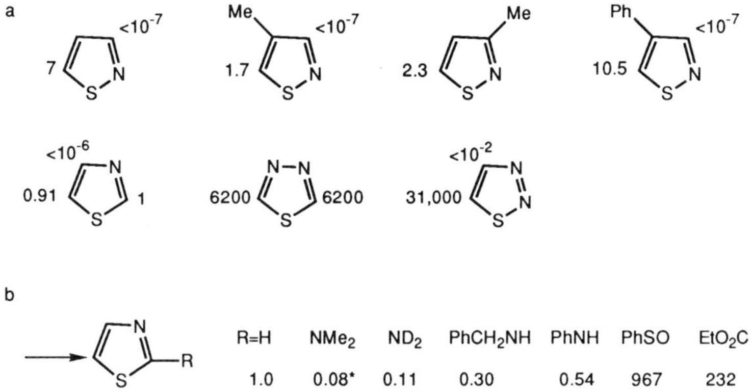

SCHEME 7.3. a, Relative free base deuteriation rates by NaOMe/MeOD at 31°C (66JAY265). b, Relative deuteriation rates by NaOMe/MeOD/DMSO) at 30.5°C [79JCS(P2)1145]. *232 for the 3-methylthiazolium iodide.

ium ions, the adjacent lone-pair effect no longer applies and the order 2-position (1,3-azoles) > 5-position(1,2-azoles) is observed (see below).

Comparison of the reactivity of the 2-position of thiazole with that of the 3-position of isothiazole gives $k_2 : k_3$ rate ratio of $\sim 10^7$, which is 10- to 20-fold greater than that which applies in thiophenes (Chapter 6, Section 2).

The reactivity order for the azoles with different heteroatoms (X) should be O > S = NH, and this has been confirmed in deuteriation of the azolium ion at $\sim 33°C$ (Scheme 7.4) (64JA1865; 69JA1113). The rather

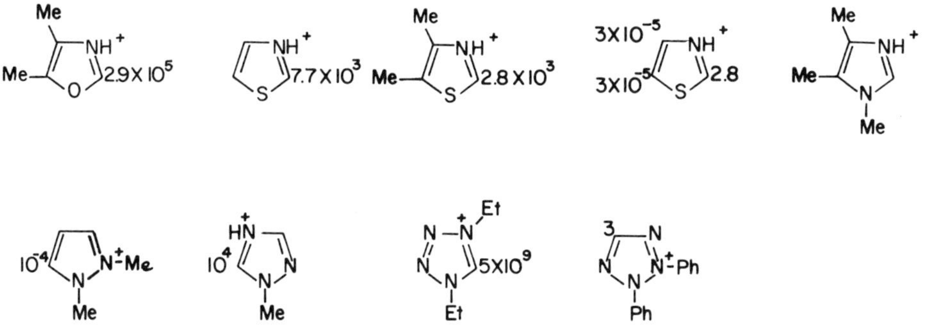

SCHEME 7.4. Relative rates of deuteriation of azolium ions at $\sim 33°C$ (64JA1865; 69JA1113).

large reactivity difference between the sulfur- and nitrogen-containing compounds has been attributed to additional electron withdrawal into the empty d orbitals of sulfur (63JA4044).

Substitution has been shown to occur on protonated species by the technique subsequently employed in acid-catalyzed exchange (i.e., by comparison of the exchange rates of the free base with those of the corresponding quaternary compounds). Thus, thiazole (**7.27**) and 3-ethylthiazolinium iodide (**7.28**) at low pH gave closely similar exchange rates (66JA4263). At high pH the rate difference becomes very large because of the higher reactivity of the quaternary compound (**7.28**) compared to that of the free base (**7.27**), which cannot exist as the conjugate acid under these conditions. A quantitative measure of these reactivity differences is shown by 4,5-disubstituted oxazolium methiodide **7.29**, which exchanges ~2000 times faster than the corresponding free base [66LA(695)55; 64TL845]. A larger difference of ~10^6 in rate coefficient applies between the isothiazolium cation (**7.30**) and the isothiazole free base (66JA4263).

7.27   7.28   7.29   7.30

This is consistent with the reactivity–selectivity principle, which requires a larger reactivity difference in the less reactive isothiazole. A value of 2900 applies between 2-N,N-dimethylaminothiazole and its 3-methylthiazolium iodide (Scheme 7.3) and this would be larger after correction for the effect of the N-methyl group. Kinetic studies on 1,2,4-triazole (74JOC2934) and 1-methyltetrazole (72JA5759) have each demonstrated the intermediacy of ylides in the exchange. The rate–acidity profile in the latter study, covering the acidity range $H_0 = -2$ to pH 7, clearly demonstrated that the principal exchanging species alters with reaction conditions.

Data for reaction of the azolium ions show that reaction takes place α to NR considerably faster than β to NR, paralleling the situation in thiazole versus isothiazole for exchange in the free bases (above), and the $\alpha : \beta$-reactivity ratio in thiophene (Chapter 6, Section 2.A). This is indicated by the 10^5 greater reactivity of 1-methyltetrazole (**7.31**) at 31°C compared to 2-methyltetrazole (**7.32**) (70MI1); similar reactivity differences are found between the 3- and 5-position of **7.33** and **7.34** and between the 4- and 5-positions of **7.35** and **7.36** (70MI1). These differences probably reflect the effect of the electronegative NR group activating the α-posi-

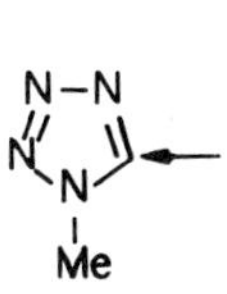

7.31 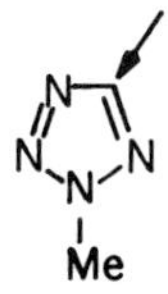   7.32 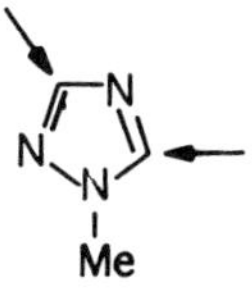   7.33 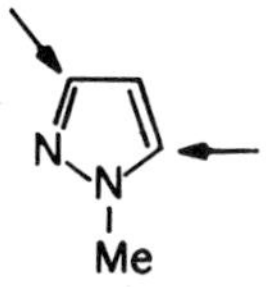   7.34 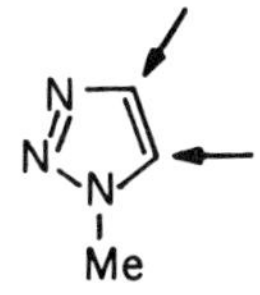   7.35 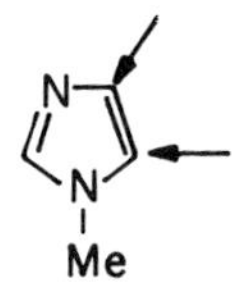   7.36

tion, but could not be satisfactorly explained by CNDO/2 calculations (73T3469).

C. Substituent Effects

The effects of electron-supplying substituents attached to the ring carbon atoms are generally small and differ little for reactions of either the free base or quaternary ion. (By contrast, nitrogen replacing CH in the ring has a very large effect.) For example, data in Scheme 7.3 show that ortho- and meta-methyl groups deactivate 4- and 2.3-fold, respectively; para-amino deactivates 10-fold; whereas ortho-phenyl activates 1.5-fold. Data for the thiadiazoles show that, by contrast, activation factors for the substitution of ring CH by N are 6800 (ortho), 34,000 (para, i.e., 2,5) and 4500–6200 (meta); the comparatively low ortho activation again is attributable to the adjacent lone-pair effect. Electron-withdrawing substituents (Scheme 7.3.) also produce substantial activation. Meta-methyl produces a fivefold deactivation in 3-methylthiazolium ion, the 3,4-dimethyl compound being five times less reactive (63B1298), and the combined deactivating effects of 3,4-dimethyl substituents is 2.8-fold in both thiazolium and imidazolium ions (Scheme 7.4).

Substituent effects in the more reactive oxazole system appear to be smaller (Table 7.3) which is to be expected by the reactivity–selectivity principle. The $k_{rel.}$ value (0.95) [69JCS(B)270] for 2-deuteriation of 1-methyl-1,2,4-triazole (**7.33**) was also determined. Thus, assuming that the activation by the 5-nitrogen upon the 2-position is the same as in thiadiazole,

TABLE 7.3
Relative Rates of 2-Deuteriation of Substituted Oxazoles at 33°C[a]

Substituent	H	4-Me	5-Ph	5-(p-ClC$_6$H$_4$)	5-(p-MeOC$_6$H$_4$)	5-Me-4-Ph
$k_{rel.}$	1	0.71	1.91	2.82	1.01	0.61

[a] [69JCS(B)270].

this makes the N—Me group 3.6×10^4 times less activating than O, in reasonable agreement with the value that may be determined from Scheme 7.4. Another report indicated that the effects of 4,5-di-*n*-propyl- and 4,5-diphenyl substituents on exchange at the 2-position of the oxazolium ion was negligble (64TL845). These small effects of phenyl in both oxazole and thiazole suggest that the 3-deuteration observed in 5-phenyl-1,2-dithiolium cation (**7.26**) in D_2O–CF_3CO_2D at 74°C [66AG(E)513] reflects the 3-position being more reactive then the 4-position, as theory requires. The higher exchange temperature required, compared to that for **7.25**, suggests that the 3-position of 1,2-dithiolium cation is less reactive than the 2-position of 1,3-dithiolium cation, again as expected.

Relative deuteration rates at 31°C of thiazolium and isothiazolium cations and their derivatives (Scheme 7.5) (66JA4263) show that the reactivity of the 2-position of the former (here unaffected by the adjacent lone-pair effect) is now much more reactive than the 5-position of the latter (cf. Scheme 7.3 for the free bases). Likewise (Scheme 7.4), the 2-position of imidazolium ion is $\sim 10^5$–10^7 times more reactive than the 5-position of pyrazolium ion. The data in Scheme 7.5 also demonstrate a 2.5-fold deactivation by ortho-methyl and a 10-fold activation by ortho-phenyl. However, the effects of replacing ring —CH= by —N= are much smaller than those given in Scheme 7.3, the activation factors being 2000 (ortho), 3000 to >40 (meta), and 5000 (para, or 2,5). By contrast, the results given in Scheme 7.4 suggest a meta activation factor of 5×10^4, and 10^4 for the para interaction. That these factors should be larger than in the more reactive sulfur analogues is not unexpected, as a consequence of the reactivity–selectivity effect. However, the anomalous meta versus para activation order may reflect uncertainties in the kinetic data.

Many substituent effects have been determined in imidazole. Rate coefficients were measured for deuteration at 50°C of imidazolium ion and

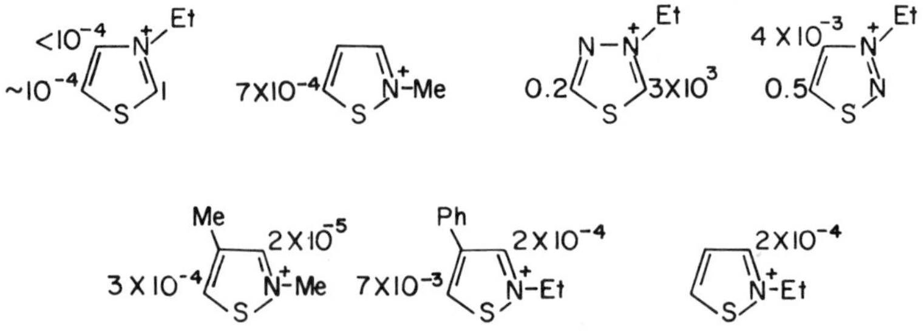

SCHEME 7.5. Relative deuteration rates at 31°C (66JA4263).

the 1-, 2-, and 4-methyl-, and 1,2-, 1,4-, and 1,5-dimethyl derivatives. In the parent compound, exchange at the 2-position is 10^3 times faster than at the 4- and 5-positions (78JOC3565). Exchange at the 5-position by the free base mechanism was ~10-fold slower than exchange at the 2-position via the ylide mechanism for the conjugate acid. Methyl substituents produced small and rather random effects; *activation* of the 5-position by 4-methyl (free base mechanism) and by 1-methyl (ylide mechanism) is particularly difficult to explain. Substantial activity effects are produced by nitro and fluoro substituents (the former being the more effective) for both the free base and quaternary ion mechanism as shown in Scheme 7.6 (relative rates in parentheses) (78JOC3570). The two sets of data are not directly comparable since they relate to a different "standard" rates, but in general they demonstrate the adjacent lone-pair effect (i.e., positions adjacent to ring nitrogen are substantially more reactive in reactions of the quaternary ion than in reactions of the free base). Some features are, however, difficult to explain (e.g., the greater activation of the 2-position by a 5-fluoro compared to a 4-fluoro substituent). Unlike the case of the corresponding nitro compounds (where the same pattern is observed), no electron-withdrawing ($-M$) effect can operate between a 5-fluorine and the 2-position.

Finally, for 5-exchange of 1-aryltetrazoles, electron-supplying substituents in the aryl ring decreased, and electron-withdrawing substituents in the aryl ring increased the reaction rate, as expected. The ρ factor at 30°C is 1.4 for reaction of the 4-ethyltetrazolium ion via the ylide mechanism (in 9 M CF_3CO_2D) and 1.3 for reaction of the free base (in piperidine–MeOD–DMF) (69TL3377).

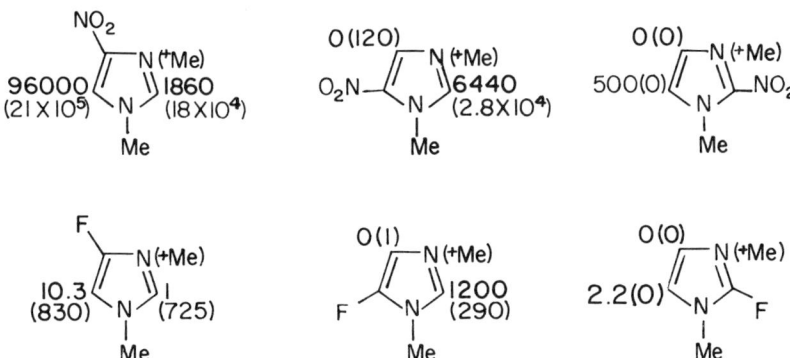

SCHEME 7.6. Relative rates of base-catalyzed deuteriation of imidazoles (and imidazolium ions) at 50°C (78JOC3570).

4. Nitration

Much of the work on nitration of azoles has concerned the phenyl derivatives. These substitute in the phenyl ring at positions that depend on the reaction conditions and the relationship of the phenyl ring to the heteroatoms. Preferential nitration in the phenyl ring does not necessarily mean that the azole is intrinsically less reactive than benzene, because the phenyl ring may be strongly activated by one of the heteroatoms whereas the azole is relatively weakly activated by the phenyl group. The combination of bond-order effects and relationship to the heteroatom means that the preferred sites of substitution are as shown by arrows in structures **7.37–7.42**. The effect of protonation is discussed later.

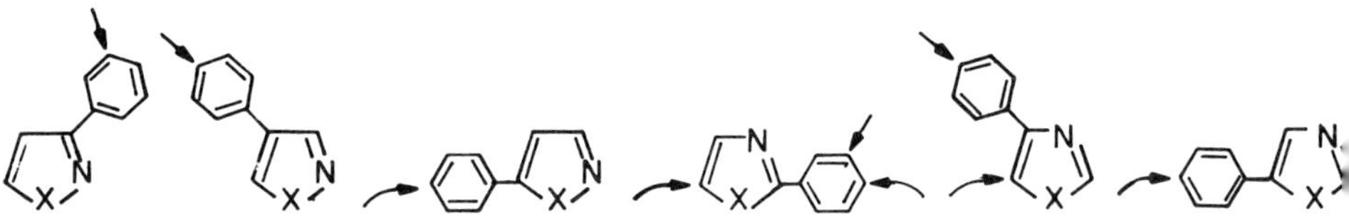

7.37   7.38   7.39   7.40   7.41   7.42

A. Oxazoles, Thiazoles, Selenazoles, and Imidazoles

a. *Oxazoles*

Oxazole itself has not been nitrated. For the reasons given above, the nitration positions for phenyl derivatives are 2-methyl-4,5-diphenyloxazole [phenyl rings, para in each (63JCS1363)], 2-phenyloxazole [phenyl ring, meta, para (42JA2444; 85CS(25)295)], and a mixture of meta and para nitration in the phenyl ring in 4-chloromethyl-2-phenyloxazole [72AP(305)509]. If there is an electron-supplying group at the 2-position, 4-phenyloxazoles are nitrated in the 5-position of the oxazole ring (97% for 2-NMe$_2$, 91% for 2-Me) (55CA296e; 59CB1944); 5-Bromo- or 5-iodooxazoles with alkyl or aryl groups at the 2- or 4-position will nitrodehalogenate (the iodo compounds are the more reactive), since the 5-position is the most activated site (81JHC885). With *N*-nitropicolinium tetrafluoroborate, 2-phenyloxazole is nitrated seven times faster than 2-phenylthiazole [85CS(25)295].

b. *Thiazoles*

Except for the benzyl derivative (which is nitrated para in the phenyl ring), thiazoles with electron-supplying substituents at the 2-position are nitrated at the 5-position [42JPJ105; 45PIA(A)343; 50HCA306; 52MI1; 67CR(C)(264)1652]. By contrast, the mesomeric electron-withdrawing ($-M$) effect of the 2-nitro substituent directs nitration into the 4-position (80%) of 2-nitrothiazole (62JOC2282). The result for 2-benzylthiazole indicates that the thiazole ring is here less reactive than the benzene ring; protonation of the thiazole ring causes this. 4-Alkyl groups direct nitration into the 5-position of the thiazole ring [39JPJ462; 40JPJ433; 47HCA2110; 52MI1; 53PI(A)758; 70BSF3155], and 5-alkyl groups into the 4-position (52MI1; 53PI(A)758; 70BSF3155), but 2-amino-4-phenylthiazole gives a mixture of para- and 5-nitro derivatives (59JOC187). Nitration in 2-, 4-, and 5-phenylthiazoles occurs para in the phenyl ring (Table 7.4) (71BSF4310), which indicates that there is no strong electron withdrawal from any of the heterocyclic ring positions (cf. isothiazole, below). It is probable that some 5-substitution accompanies nitration of the 2-phenyl isomer. 4,5-Diphenylthiazole is also nitrated at the para positions of each phenyl ring (39JPJ462).

The greater reactivity of thiophene compared to thiazole has been demonstrated by the positions of nitration in **7.43** and **7.44** (74BSF2099).

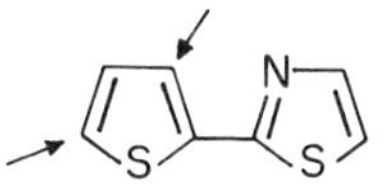

7.43 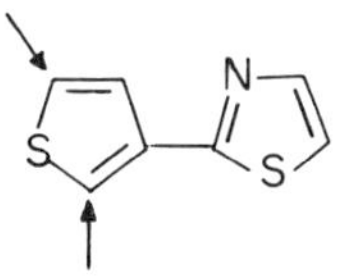   7.44

Nitration rates relative to a single position of benzene ($= 1$) for alkylthiazole and -isothiazole derivatives are shown in Scheme 7.7. Thiophene is 2.83×10^5 times more reactive than 4-methylthiazole [68CR(C)(266)714; 69JHC575; 72BSF162]. Since the activating effect of an ortho-methyl group across a 4,5-bond may be estimated as ~60-fold (i.e., 50% greater than in benzene), this indicates that the 2-position of thiophene is 1.7×10^6 times more reactive than the 5-position of thiazole. Taking ρ for nitration as -6.2 (58JA4979), this gives a $\sigma^+_{m\text{-}N}$ value of ~1.0. The magnitude of this value is consistent with it corresponding to an m-NH^+ group (cf. Chapter 9), which is an indication that nitration takes place on the protonated species.

TABLE 7.4
ISOMER DISTRIBUTION OBTAINED ON NITRATION IN THE
PHENYL RING OF PHENYLTHIAZOLES AND
PHENYLISOTHIAZOLES[a]

	Isomer ratio (%)		
	Ortho	Meta	Para
2-Phenylthiazole	3	8	89
4-Phenylthiazole	7	4	89
5-Phenylthiazole	15	2	83
3-Phenylisothiazole	25	65	10
4-Phenylisothiazole	25 (15)[b]	3	72 (81)[b]
5-Phenylisothiazole	20	19	61

[a] [71BSF4310; 85CS(25)295].
[b] Values in parentheses from 69JHC841.

Nitration of conjugate acid species is confirmed by rate versus acidity profiles for alkylthiazoles, alkylthiazolium ions, and 2-thiazolones, the standard log rates being given in Scheme 7.8 [75JCS(P2)1614]. The results are mainly consistent with those given in Scheme 7.7 and show that OMe activates more strongly than Me as expected, and show the effects of steric hindrance. The apparently anomalous lower reactivity of the methylated quaternary ions compared to the nonmethylated compounds is probably due to hydrogen bonding of the acidic hydrogen to nucleophilic

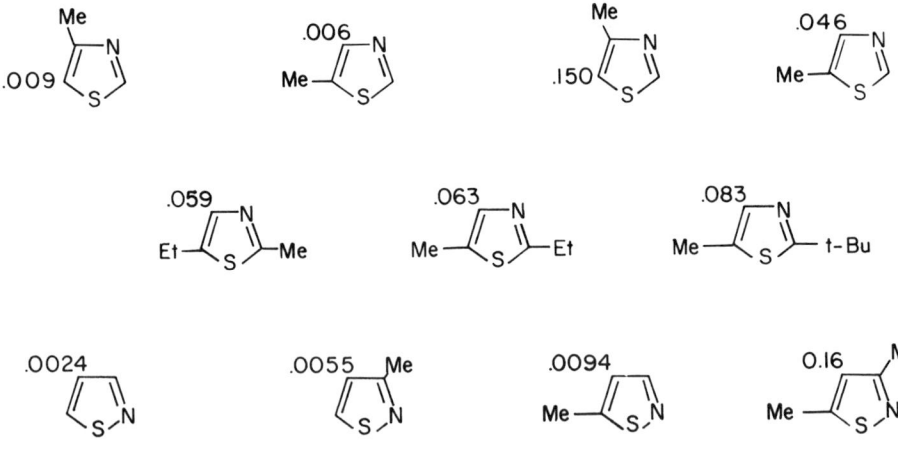

SCHEME 7.7. Rates of nitration relative to a single position of benzene (= 1) for alkylthiazoles and -isothiazoles.

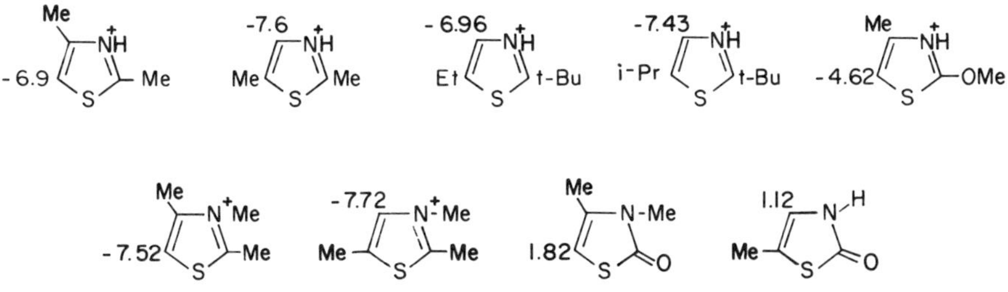

SCHEME 7.8. Standard log nitration rate [75JCS(P2)1614].

species in the medium in the NH cation, making the NH^+ group less electron-withdrawing than NMe^+. The 2-thiazolones are nitrated as the free bases (at least at the lower acidities); this corresponds to the substitution of O^- into the thiazolium ring, which, as expected (cf. phenoxide), causes a very large rate increase. 2-Alkylthiazoles are nitrated ~75% in the 5- and 25% in the 4-position (69JHC575).

c. *Selenazoles*

No quantitative data are available for the nitration of selenazole, though it is said to be more reactive than thiazole; 4-methylselenazole is nitrated at the expected 5-position (48YZ195,197).

d. *Imidazoles*

Tautomerism in imidazole means that the 4- and 5-positions are equivalent on a time average; in the imidazolium ion they are absolutely equivalent. Nitration with mixed acids (65JCS1051; 70CHE465) involves the conjugate acid [72JCS(P2)1654] and gives 4(5)-substitution, with a partial rate factor of 3.0×10^{-9}. 1-Methyl-2-isopropylimidazole gives 57% 4- and 24% 5-nitration (74GEP2354786). The effect of an electron-withdrawing group at the 4-position across the high-order 4,5-bond results in substantial 2-nitration (70CHE614). 2-Nitration can be achieved in 1-substituted imidazoles by using the 2-lithio derivative (of the 1-Me or the 1-Ph compound), which has high charge density at the 2-carbon, hence nitrodelithiation occurs in preference to nitration at the other positions (74URP437763). 1-Phenylimidazole is nitrated (as the conjugate acid) at the para-position of the phenyl ring [72JCS(P2)1654]. 2-(*p*-Fluorophenyl)-imidazole is nitrated 80% at the 4(5)-position, a second nitration occurs either at the 5(4)- or at the meta-position (79JHC1153). 4-*p*-Anisylimidazole undergoes some nitration ortho to the methoxy group, but mainly at the 5-position (77CHE1110). Nitration of 1,4,5-trimethylimidazole 3-oxide (**7.45**) takes place on the free base and at the 2-position, but no rate data are available [77JCS(P1)672].

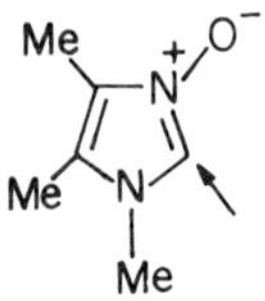

7.45

B. ISOXAZOLES, ISOTHIAZOLES, PYRAZOLES, AND DITHIOLIUM IONS

a. *Isoxazoles*

Nitration of isoxazoles with mixed acids gives only a 3.5% yield of the 4-derivative (59ZOB535), though with nitronium tetrafluoroborate substantially higher yields can be obtained (81USP4288445). Nitration of 5,5'-diisoxazolyl (**7.46**) occurs at the 4- and 4'-positions, whereas the 3,3'-

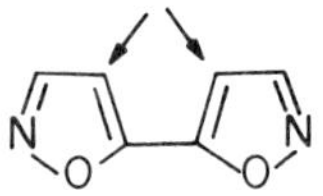

7.46

isomer gives the 4,4'-dinitro isomer directly (60CA17368i). Both 3-and 5-methylisoxazole are nitrated in the expected 4-position (41G327) and 5-methylisoxazole is more reactive because of the bond-fixation effect; it gives a 68% yield (45G131).

In general, the usual pattern of substituent effects is observed for the nitration of isoxazoles (37JA933; 38JA1198; 39JA104). Nitration of 3- and 5-phenylisoxazoles occurs preferentially at the para- and meta-positions, respectively, of the phenyl rings [42G537; 67G1604; 76CI(M)880; 74CHE516]. 5-Phenylisoxazole also gives some 4-nitration (58ZOB359). The possibility that the para substitution observed for 5-phenylisoxazole might be due to 3-phenylisoxazole present as a major impurity in the 5-isomer (65CJC2117) has apparently been excluded (68JOU2057). Some meta substitution of the 5-phenyl ring also occurs in 5-phenylisoxazole and the amount of 4-substitution in 5-arylisoxazoles increases as expected if the phenyl ring contains electron-withdrawing groups. 3,5-Diphenylisoxazole is nitrated preferentially at the 4-position, the para position of the 5-phenyl ring, or the meta-position of the 3-phenyl ring, depending on the conditions used (74CHE516).

Standard log nitration rates (Scheme 7.9) have been determined for some alkyl- and phenylisoxazoles [71JCS(B)2365; 75JCS(P2)1600, 75JCS(P2)1627]. 3-Methyl-5-phenylisoxazole is sufficiently reactive to nitrate (in the para-position of the phenyl ring) as the conjugate acid, and this can be rationalized because this position in the phenyl ring is conjugated with the oxygen atom. By contrast, the para-position in the phenyl ring of 5-methyl-3-phenylisoxazole is conjugated with the electron-withdrawing nitrogen atom, and now nitration occurs on the free base and not on the conjugate acid. This constraint does not apply to the meta-position of the same compound, which therefore undergoes nitration as the conjugate acid. 3,5-Dimethylisoxazole can be nitrated only in the isoxazole ring, and therefore undergoes nitration as the free base. As in the case of the results in Scheme 7.8, methyl substitution at the nitrogen atoms of isoxazole conjugate acids decreases the reactivity, probably again because hydrogen bonding is excluded. 4-Halogeno-3,5-dimethylisoxazoles undergo nitrodehalogenation, the ease of which follows the usual order I > Br > Cl (73CHE1202).

b. *Isothiazoles*

Nitration of isothiazoles in nitric/sulfuric acids gives >90% yield of the 4-nitro derivative and the position of the substitution is unaffected by 3- or 5-alkyl, -halogeno, or -acylamino substituents [59JCS3061; 63AG(E)714]. With the 3-phenyl group conjugated with the electron-withdrawing nitrogen atom, 3-phenylisothiazole is nitrated mainly in the meta position (68CPB160; Table 7.4), whereas 4- and 5-phenylisothiazoles are nitrated mainly at the phenyl ortho- and para-positions (Table 7.4). The

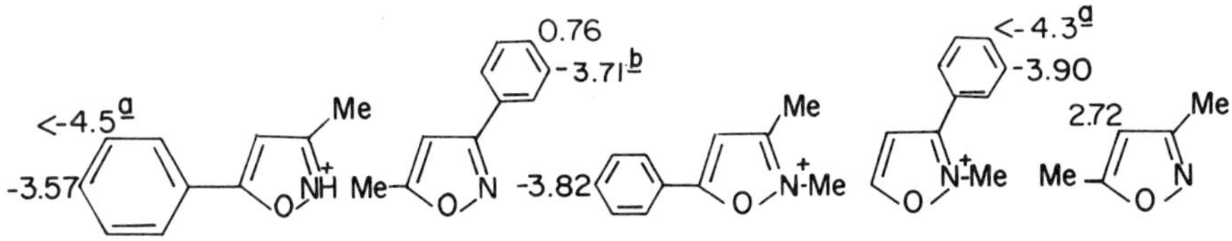

SCHEME 7.9. Standard log nitration rates for isoxazoles [71JCS(B)2365; 75JCS(P2)1600; 75JCS(P2)1627]. *a*Estimated values. *b*For *N*-protonated species.

5-phenyl group is of course also conjugated (though less strongly) with the nitrogen atom, but this is partially outweighed by conjugative electron release from the sulfur. Consequently, a substantial amount (19%) of meta substitution is obtained (Table 7.4). This conjugative electron release from sulfur to the α-position also accounts for the much lower proportion of meta product in the nitration of 2-phenylthiazole compared to 3-phenylisothiazoles (Table 7.4), even though both phenyl groups are conjugated with nitrogen.

Standard log nitration rates have been determined for isothiazole and various methyl derivatives (Scheme 7.10) [75JCS(P2)1620]. All these compounds undergo nitration as the free base, but the dimethyl compound can in addition be nitrated as the conjugate acid at higher acidity, again disclosing the usual pattern of this alternative pathway being available for more reactive compounds. Surprisingly, the 5-methyl compound is slightly *less* reactive than the 3-methyl compound; a possible reason could be that the 3-methyl group reduces rate-decreasing hydrogen bonding at the nitrogen atom by a steric effect. By contrast, another report gives the k_{rel} values for the 3-Me, 5-Me, and 3,5-Me$_2$ derivatives as 2.3, 3.9, and 6.7 (72BSF162). Both sets of data agree in that the effects of the methyl groups are not additive, the dimethyl compound being *more* reactive than expected; no reason for this is apparent. The deactivation of the ring by protonation is (at the 4-position) ~7–8 log units.

c. *Pyrazoles*

Pyrazole is nitrated in the 4-position as the free base in H$_2$SO$_4$ (70–80 wt%) [82CI(L)57] but predominantly as the conjugate acid in 90–99 wt% H$_2$SO$_4$ (65JCS1051). For the latter the partial rate factor is 2.1×10^{-10}, making the pyrazolium ion 14.3 times less reactive than the imidazolium ion. The low 3,4-bond order results in 3-carboxypyrazole also undergoing nitration in the 4-position (59G1539). However, whereas 3-methyl-1-phenylpyrazole is nitrated at the 4-position, for 3-carboxy-1-phenylpyrazole it occurs at the para-position of the phenyl ring (59G1539).

Nitration of 1-phenylpyrazole itself has been subjected to considerable attention. The low basicity of the compound (the nitrogen lone pair is

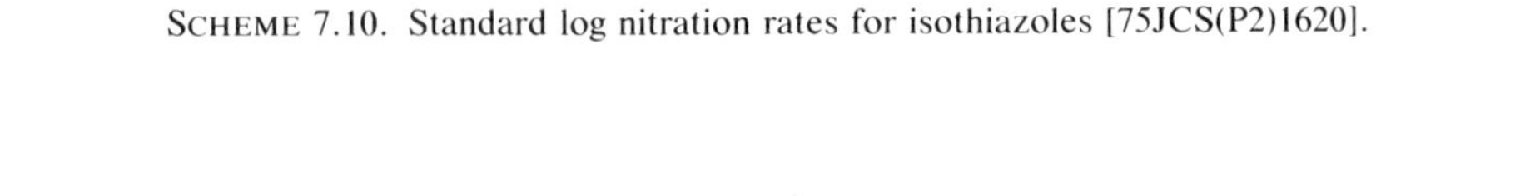

SCHEME 7.10. Standard log nitration rates for isothiazoles [75JCS(P2)1620].

delocalized into the phenyl ring) allows nitration on the free base; nitration in acetic anhydride (probably via the free base) goes into the 4-position (56G797; 63CJC1540) (as it also does for 1,5-diphenylpyrazole), but in sulfuric acid the *p*-nitro derivative is obtained (56G797; 57JCS3024) indicating that the N-2-protonated species is involved. A similar pattern has been found in studies of nitration of other 1-arylpyrazoles. Methyl groups at N-2 or C-5 reduced the nitration rate of 1-phenylpyrazoles; this was attributed to steric hindrance to attainment of coplanarity between the phenyl and pyrazole rings [72JCS(P2)1654], though it should be noted that the electron release should be from N-1 into the phenyl ring rather than vice versa. An alternative explanation is that the methyl groups reduced the conjugation between N-1 and the phenyl groups so that increased protonation (or hydrogen bonding) occurs at N-1.

Nitration of 3- or 5-phenylpyrazole by nitric acid/acetic anhydride gives a mixture of 1-acetyl-3-(5)*p*-nitrophenylpyrazole and 3-(5-)phenyl-1-nitropyrazole [58AC(R)783]. Nitration of 4-phenylpyrazole with the same reagent give substitution in both rings, and in the phenyl ring this is predominantly ortho, with an ortho/para ratio of 4 (72RTC1185) [cf. nitration of biphenyl, [66JCS(B)727]]. For 4-formyl-1-phenylpyrazole, the ortho–para-product ratio is 1.2, presumably because of the greater inductive electron withdrawal of the formyl group.

The high reactivity of the 4-position is such that 4-bromopyrazole undergoes nitrodebromination, but in 1- or 3-phenyl-4-bromopyrazole, nitration in the phenyl rings also occurs (79AJC1727). However, in 4-chloro-5-carboxyl-1-methylpyrazole, nitrodecarboxylation is preferred (82JGU2291); the carboxyl group deactivates the 4-position more than the chloro deactivates the 5-position.

Kinetic studies of the nitration of a range of pyrazoles, hydroxypyrazoles, and pyrazolones have produced the standard log nitration rates given in Scheme 7.11 [74JCS(P2)382, 74JCS(P2)389; 75JCS(P2)1600, 75JCS(P2)1609, 75JCS(P2)1632].

The data show very clearly the rate-lowering effect of methyl groups at the 2- and 5-positions on nitration at the para-position of a 1-phenyl group; this rate reduction has been attributed to steric hindrance to coplanarity of the two aryl rings as noted above. However, entries in Scheme 7.11 dealing with nitration at the pyrazole ring 4-position show that the net effect of electron flow is from the pyrazole ring to the phenyl ring. Thus, the introduction of methyl groups in the ortho-position of the phenyl ring, which also reduces conjugation between the aryl rings through steric hindrance, produces (in the case of the dimethyl compound) a substantial *increase* (30-fold) in the nitration rate at the 4-position of the pyrazole ring; this may be attributed to greater availability of the N-1 lone pair in the pyrazole ring. The difference in rate between 5-methoxy-3-

methyl-4-nitro-1-phenylpyrazole and pyrazolone structural isomers has been attributed to reaction occurring on the neutral (and thus more reactive) oxo form of the latter.

Nitration of 1-methyl-4-(2,4-dinitrophenyl)pyrazole (**7.47**) occurs 25% at the 3-position and 15% at the 5-position of the pyrazole ring, whereas the 2,4,6-trinitro homologue (**7.48**) nitrates only in the 3-position (70JHC707; 71JHC293), probably because of the increased electron withdrawal in **7.48** across the high-order 4,5-bond (cf. the explanation given in the paper that the 5-nitro derivative was formed, but that its rate of decomposition was greater than its rate of formation!)

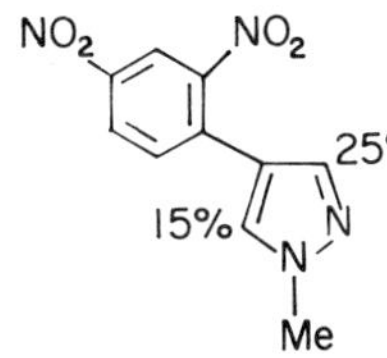

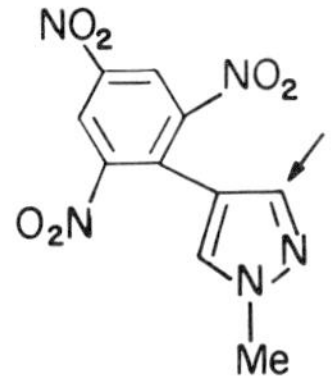

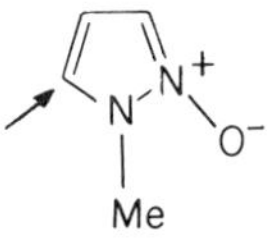

In contrast to the results for pyrazole, 1-methylpyrazole N-oxide (**7.49**) nitrates at the 5-position (72TL2771), which may be attributed to conjugative ($+M$) electron release from oxygen to the 5-position. Some 3-nitration could therefore also be expected, but none was reported.

d. *Dithiolium Ions*

Nitration of phenyl derivatives of 1,2-dithiolium ions **7.50** and **7.51** occurs at positions by arrows where (61JA2934), showing that in the dithiolium ring the positive charge is greater at the 3-(5)-position than at the 4-position.

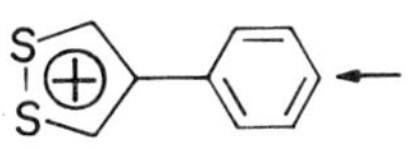

SCHEME 7.11. Standard log rates for nitration of pyrazole and derivatives [74JCS(P2)389; 75JCS(P2)1600, 75JCS(P2)1609, 75JCS(P2)1632]. Positions that have underlined values react as the free bases. Where two values are given, the underlined value is for reaction on the free base (at low acidity); the other value refers to reaction on the conjugate acid.

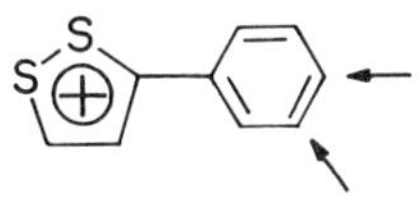

C. Oxadiazoles, Thiadiazoles, Triazoles, and Derivatives

a. *Oxadiazoles*

2,5-Diphenyloxa-1,3,4-diazole (**7.52**) gives the percentage yields of mononitration products shown (61ZOB1919), so the heterocycle has no strong directive effect. More recent work [80JCS(P2)773] has shown that all six possible dinitro derivatives may be obtained. Surprisingly, nitration with nitric acid gives mainly para derivatives, whereas with nitronium tetrafluoroborate mainly meta derivatives are obtained.

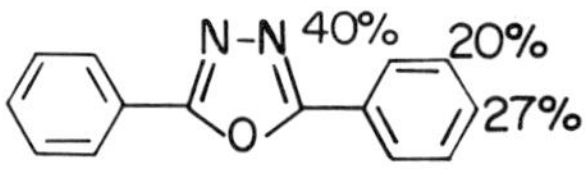

7.52 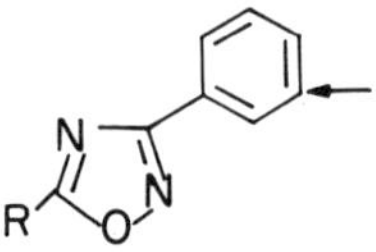   **7.53** 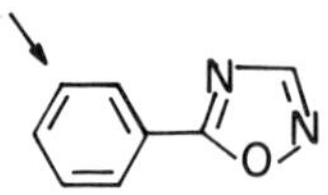   **7.54**

5-Alkyl-3-phenyloxa-1,2,4-diazole (**7.53**), by contrast, nitrates mainly at the meta-position (98% for R = H) (63G1196); the electron density at the 3-position of oxa-1,2,4-diazole in **7.53** is expected to be lower than at the 2-position of oxa-1,3,4-diazole in **7.52**. Similar reasoning accounts for the meta substitution in 5-phenyloxa-1,2,4-diazole (**7.54**) (63G1196). Nitration of 3-arylsydnones (**7.55**) occurs in the sydnone ring rather than in the phenyl ring (69TL579). The sydnone ring is a mesomeric betaine and can be considered as an oxadiazolium analogue of the phenol anion and therefore very reactive toward electrophilic substitution. Note that bond fixation in **7.55** means that the positive pole does not deactivate strongly the adjacent 4-position.

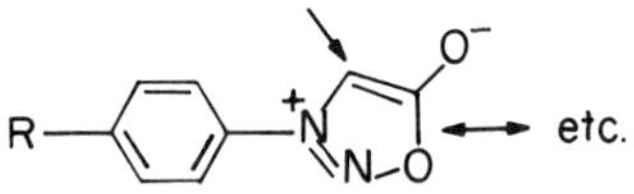

7.55 (R = H, Me, OMe)

b. *Thiadiazoles*

For 1,3,4-thiadiazole, the lower electron supply from the sulfur compared to the oxygen equivalent accounts for the ortho : meta : para rate ratio in the phenyl ring of 1 : 3 : 2 in the 2-phenyl derivative (**7.56**); no substitution took place at the 5-position (53JPJ701).

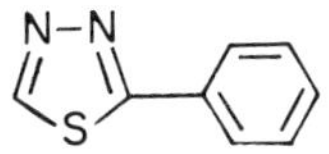

7.56

c. Triazoles

2-(2',4',6'-Trinitrophenyl)-1,2,3-triazole (**7.57**) undergoes nitration at the 4-position (71JHC51); the lone pair on N-2 is evidently still sufficiently electron supplying to compensate for the deactivation by the other nitrogens. However, the 1-isomer (**7.58**), which might be expected to be more reactive, does not undergo nitration (71JHC51). The triazole ring is less reactive than benzene, but not greatly so, as indicated by the nitration of 2-phenyl-1,2,3-triazole, which occurs first at the para-position of the phenyl ring, and then at the 4-position (63CJC274).

Nitration of 1,2,4-triazol-5-ones **7.59** and **7.60** took place on the free bases and gave the standard log rates indicated. The enormously higher (9000-fold) reactivity of the dimethyl compound has been attributed to the electronic effect of the *o*- × *m*-methyl groups (82MI1), but this appears unlikely. Perhaps **7.59** easily undergoes covalent hydration.

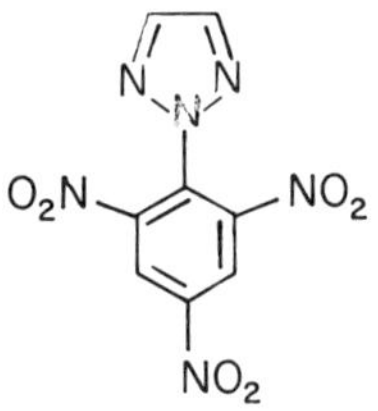

7.57 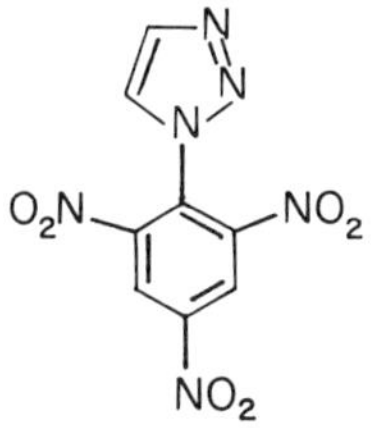   **7.58** 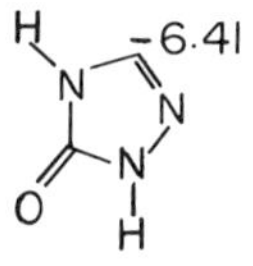   **7.59** 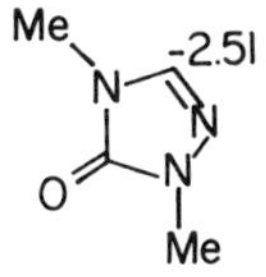   **7.60**

5. Halogenation

A. Oxazole, Thiazole, and Imidazole

The positional reactivity order for bromination of oxazole is 5 > 4 > 2 [59LA(626)83,92], which contrasts with an earlier report that thiazole brominates in the 2-position (34RTC77). 4,5-Diphenyloxazole is reported not to be brominated in either of the phenyl rings (63JCS1363), presumably due to steric hindrance forcing the phenyl rings to be orthogonal to the oxazole ring. 2-Phenyloxazole and 2-phenyltriazole brominate as the free bases in the 5-position [85CS(25)295].

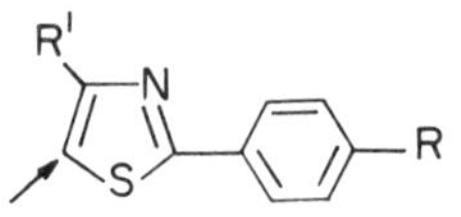

7.61 (R = Me, H, Br, NO$_2$)

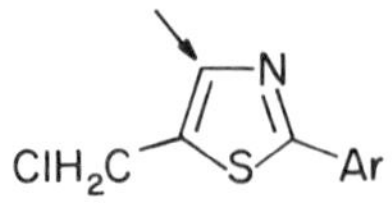

7.62

Bromination rates for compounds (**7.61**) give a Hammett correlation with ρ = −0.69 for R′ = CH$_2$Cl, and −0.75 for R′ = CH$_2$OH [72MI3]. The activating effects of 2-substituents upon the 5-position was shown to be OH, NH$_2$ > Me (39CB1470) and the strong activation of both bromination and chlorination of the 5-position by 2-NH$_2$(NHAc) has been confirmed (47JCS431; 54RTC325). Substituents in the aryl ring of compounds **7.62** activate bromination in the usual electrophilic order [72CA(76)12683]. Compound **7.63** is brominated in both positions indicated, suggesting that both rings are of comparable reactivity [73ACH107]. Bromination of **7.64** (X = H, Br) takes place in the thiazole

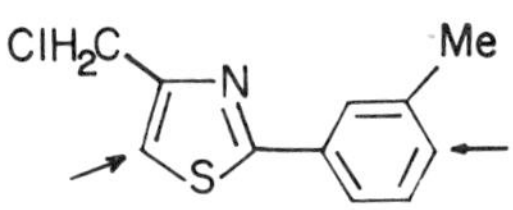

7.63

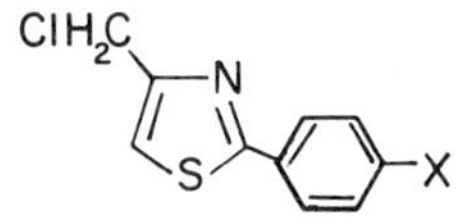

7.64

ring, but in the presence of a silver salt only the phenyl ring of **7.64** (X = Me) is brominated (74ACH381). This is an interesting result because there are very few quantitative (or qualitative) data in aromatic substitution in general relating to the effect of forming silver–aromatic complexes (which lowers the reactivity). The implication here is that thiazole complexes better than benzene, consistent with the higher π density in the former.

The positional reactivity order becomes 2 >> 5 > 4 in the electrophilic bromodechlorination of 2,4,5-trichlorothiazole (76JHC1297). This finding indicates that the 5- and 4-positions are mutually strongly deactivated by the −I effect of the adjacent chlorine, which is enhanced by the short C-4—C-5 bond length.

Imidazole is brominated preferentially in the 4(5)-position, though iodination apparently goes at the 2-position [28JPR(118)33]. The latter anomaly may arise through iodination (and diazonium coupling) taking place upon the conjugate base. Iodination gives a substantial kinetic isotope effect ($k_H : k_D$ = 4.4) showing that C—H bond breakage is partially rate-determining; diazonium coupling on the other hand gives no kinetic isotope effect (58MT1). Bromination without catalysts or in HOAc–

NaOAc gives the tribromo derivative (22JCS947; 73ACS2179) emphasizing the high intrinsic reactivity of imidazole, which is also shown by the ready bromination at the 5-position of 2-methyl-5-nitroimidazole, and the 2-position (85%) of 4-carboxy-5-nitroimidazole (Br_2 in NaOH) (73GER2243015; 82CHE539). This is also evident from the close positional reactivities in bromination of imidazole: ~0.45 (4,5), 0.2 (2); and 1-methylimidazole: 2.3 (5), 1.7 (4), and 0.8 (2). The relative reactivity of the 5-position of 2-methylimidazole was 90 (i.e., the 2-methyl substituent activates the 5-position ~180-fold) (74AJC2331). More recent partial rate factors for imidazole are $f_2 = 1.44 \times 10^{10}$; $f_4 = 2.88 \times 10^{10}$; $f_5 = 3.6 \times 10^{10}$ [78JCS(P2)865]. The greater reactivity of the 1-methyl derivative is also shown by the yield (in bromination by tetrabromocyclohexa-2,5-dienone) being higher (66%) than for imidazole (41%)[72JCS(P2)2567].

Iodination of 1-methylimidazole shows the same substitution pattern as bromination, giving 14% of a mixture of 4- and 5-iodo-, 33% of 4,5-diiodo-, and 2% of triiodo-1-methylimidazole. The 2-position in the latter is the most reactive toward protiodeiodination (79BAU1446), which parallels the observation in bromodechlorination of thiazole (above). Methylimidazoles give the expected reactivity order in iodination, namely 4-Me > 2-Me > (imidazole), and reaction of each compound is base catalyzed. Imidazole and 2-methylimidazole will also undergo uncatalyzed iodination, whereas the 4-methyl isomer apparently does not (80JOC3108). Iododelithiation has been used to iodinate an iodoimidazole in the 2-position (the initial lithiation was carried out on the 1-benzenesulfonyl derivative) (77JHC517); this uses the same principle noted above under nitrodelithiation (Section 4.A.d).

Iodination of nickel(II)-coordinated imidazole at pH 6.5 takes place 19 times slower than for imidazole. This contrasts with the iodination of nickel(II)pyrazole which at pH 6.5 reacts ~4 times *faster* than does pyrazole. The reason for this difference is behavior was not evident, but it was noted that nickel complexes of imidazole are much stronger than nickel complexes of pyrazole. Alteration in acid dissociation constants of the substrate may affect the rates of individual steps of the reaction (66JA5537;72JA2460).

B. ISOXAZOLE, ISOTHIAZOLE, AND PYRAZOLE

Isoxazole is brominated in the expected 4-position in 42% yield (60G356), whereas the 3- and 4-phenyl derivatives are brominated almost exclusively in the phenyl rings (98% para in the latter case) [67G1604; 76CI(M)880]. The result for the 3-phenyl compound is surprising and suggests steric hindrance as the cause. The relative rates of bromination in 85% acetic acid at 50°C of 5-phenyl-, 3,5-diphenyl-, 3-methyl-5-phenyl-, and 5-methyl-3-phenylisoxazole are ~1 : 3 : 30 : 5, the compounds overall being 100–1000 times more reactive than benzene (75CHE643). Bromination of 3,5-diphenylisoxazole in sulfuric acid takes place at the 4-position, whereas the corresponding quaternary ion brominates in the phenyl rings, showing that the free base is involved under the former conditions (71JOU1835). Isoxazoles with other activating groups at the 3- and/or 5-positions are readily brominated and iodinated at the 4-position (68YZ1289).

High yields of 4-derivatives are obtained on bromination of isothiazoles with activating substituents at the 3- and/or 5-positions [63AG(E)714; 64CI(M)207; 72AHC(14)17]; 3-methyl-5-carboxyisothiazole has been brominated despite the strong deactivation across the 4,5-bond by the carboxyl group (63JCS2032). Bromination of 4-phenyisothiazole goes para in the phenyl ring and in the 5-position of the heterocyclic ring (69JHC841).

Bromination of 3-2H-isothiazolones **7.65** give the 4-bromo derivatives, which are further brominated to the 4,5-dibromo derivatives only with difficulty. By contrast, chlorination readily affords the 4,5-dichloro derivative, and stopping the reaction at the 4-chloro stage is difficult (77JHC627, 77JHC725). The anomaly was not explained by the original authors but strongly suggests that the steric hindrance slows the second bromination; such hindrance would be enhanced here by the particularly short 4,5-bond length. This would also explain why chlorination of the 4-methyl compound (**7.66**) was easy while bromination was difficult (77JHC725). Halogenation of **7.65** became increasingly difficult along the series R = alkyl, aralkyl, cyclohexyl, *p*-chlorophenyl.

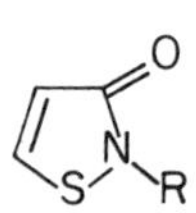

7.65 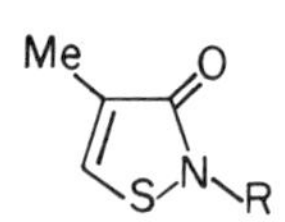   7.66

Pyrazole is chlorinated, brominated, and iodinated in the 4-position [66AHC(6)391] and the partial rate factor for bromination of the 4-position of pyrazole is 7.2×10^6 (i.e., slightly lower than that for the 4-position of imidazole) [78JCS(P2)865]. 3-Methylpyrazole is dichlorinated at the 4- and 5-positions [56LA(598)186] and pyrazole is readily tribrominated [55LA(593)179], a 96% yield of 3,4,5-tribromopyrazole being recorded under alkaline conditions (reaction via the anion) (74FRP

2193823). 1-Phenylpyrazole is always brominated initially in the phenyl ring (66AHC(6)391) (cf. nitration), and further bromination gives a tribromo derivative, probably 1-(p-bromophenyl)-4,5-dibromopyrazole (61JCS2769). 3-Methyl-1-phenylpyrazole is brominated at the para- and 4-positions (61JCS2769). As in the case of imidazole, 1-methylpyrazole is more reactive than pyrazole itself, by a factor of ~2.0, (cf. ~4.0 for imidazole); such N-methylation also facilitates iodination [55LA(593)200]. The combined activation by 3,5-dimethyl substituents in bromination is 3.7×10^3 over pyrazole itself (71AJC1413); the individual 3- and 5-Me effects are estimated as ~15- and 300-fold, respectively, since the 4,5-interaction will be approximately double the 2,5-interaction in imidazole because of the bond-order effect. A deactivating carboxy substituent in the 3-position does not prevent halogenation at the 4-position (82JGU2291), again because of the low C-3—C-4 bond order (cf. nitration). Halogenation of 1-hydroxypyrazole and its 2-oxide also goes into the 4-position (80JOC76). The 2-oxide function should raise the reactivity of the 3- and 5-positions relative to that of the 4-position, but evidently this is insufficient to outweigh the intrinsic higher reactivity of the latter.

A study of the iodination of pyrazole and 1-alkylpyrazoles has led to the conclusion that the reactivity of the anion (conjugate base) is enhanced relative to the neutral species by $10^{9.5-12.8}$ (67JA6218). The reactivity of the 4(5)-positions of imidazole (statistically corrected) to that of the 4-position of pyrazole has been determined as 1.3 (64JA2857), which agrees with the localization energy calculated for these positions of -2.10β and -2.13β, respectively (55AJC100).

Bromination of thienylpyrazoles and their N-methyl derivatives (**7.67** and **7.68**) took place as expected at the 5-position of the thiophene rings. Compound **7.67** (R = Me) is brominated 5.4 times faster than **7.69**, which

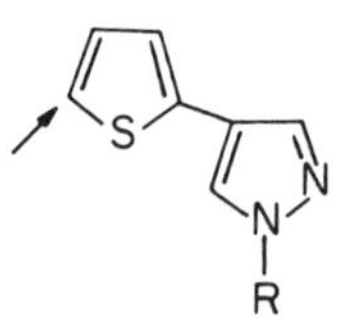

7.67 (R=H,Me) 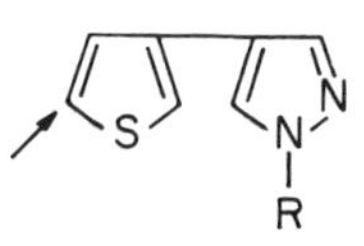   7.68 (R=H,Me) 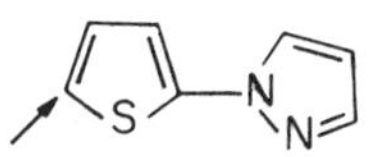   7.69

is also brominated at the 5-position of the thiophene ring [80CS(15)102]. This latter finding is again exactly as expected in view of the well-documented results for 1-phenylpyrazole. Similar results were obtained in nitration.

C. THIADIAZOLES AND TRIAZOLES

Both 1,2,5- and 1,3,4-thiadiazoles can be monohalogenated if the strongly activating NH_2 group is present (64GER1175683; 65ACS2434). The more reactive 1,2,4-triazole can itself be dibrominated in 29% yield [65AG(E)434]. The triazoles are, however, considerably less reactive than benzene, as indicated by bromination of the para-position of the phenyl ring in 2-phenyl-1,2,3-triazole (63CJC2380). Bromination and chlorination of 1,2,4-triazole can also be achieved through rearrangement of the 1-halogeno derivative (69CHE844) (cf. the rearrangement of N-haloamides in benzenoid chemistry).

6. Alkylation, Chloro(hydroxy)alkylation, and Acylation

Because of the ease of N-alkylation, there is but one report of C-alkylation, namely the benzylation of N-alkylpyrazoles in the 4-position at high yield (60ZOB2942).

Chloro(hydroxy)methylation has been more studied. The reactivity order for chloromethylation of isoxazole derivatives at the 4-position for various substituent groups is 5-Ph > 3,5-Me$_2$ > 5-Me > 3-Me > (isoxazole) (57ZOB3210; 58ZOB2376); the effect of bond fixation is again evident in the results for the monomethyl compounds. 1-Benzyl-2-isopropylimidazole is hydroxymethylated first at the 5- and then at the 4-positions (72RTC1383). 2,4-Dialkylimidazoles give 5-hydroxymethyl derivatives in considerable yield in weakly acidic (but not weakly basic) media (74CPB2359). 1,2,5-Thiadiazole is sufficiently reactive to be chloromethylated to the 3,4-bis(chloromethyl) derivative (74CC585), since the first substituent activates the ring toward entry of the second.

There are no reports of acylation of the oxygen- and sulfur-containing heterocycles, presumably because the catalytic conditions that would be necessary (more drastic than for furan or thiophene) result in coordination of the ring nitrogen with the catalyst. N-Substituted pyrazoles and imidazoles can be C-acylated only if special conditions are used. Both 1-methyl- and 1-phenylpyrazole, and 2- and 5-methyl derivatives of the latter, are formylated in the 4-position (57JCS3314; 59JCS1819; 61JCS2733), as are 1,3- and 1,5-dimethylpyrazoles, each in 57% yield (73JOU840). N-Arylpyrazoles are easier to acylate because they are less likely to be protonated under the reaction conditions; benzoylation occurs in high yield in the 4-position even in the absence of catalysts (1889G128; 26CB611; 60ZOB203). Acylation of imidazole in the *2-position* can be achieved in the presence of triethylamine (77LA145), which suggests that the conjugate base is the reacting species. Use of chloroformates in this reaction gave 2-carboxylates, and benzoylation at the 2-position can be achieved in 59–81% yield (78S675). The reaction also works with N-acyl- and N-methoxymethylimidazoles, the N-substituent being subsequently removed readily with dilute acid (77LA159).

7. Sulfonation, Sulfenylation, and Diazonium Coupling

Phenyloxazoles are sulfonated and chlorosulfonated only at the para-position of the phenyl ring (53GEP869490; 55GEP926249; 63JCS1363). If 20% oleum is used, isoxazole can be sulfonated in the 4-position (17%)(59ZOB535). 3,5-Dialkylisoxazoles are more reactive, as expected, and sulfonation can be achieved with chlorosulfonic acid (40G1,11). 3-Methylisothiazole is sulfonated 97% in the 4-position (65JCS7283), 4-methylisothiazole is reactive enough to be sulfonated in low yield in the 5-position [63AG(E)714], and 2,4-dimethylselenazole can be sulfonated in the 5-position (48YZ195).

The tendency toward protonation is particularly marked for imidazole and pyrazole, which, probably because of this, are sulfonated only under fairly drastic conditions. For example, sulfonation of 3-methylpyrazole (at the 4-position) requires 100°C and 20% oleum [1894LA(279)217]. Sulfonation of 3-methyl-1,5-diphenylpyrazole does *not* go into the para-position of the 1-phenyl ring, but rather into the 4-position, and the para-position of the 5-phenyl ring (61JCS3851). Steric interaction between the 1- and 5-phenyl rings and the resulting noncoplanarity with the pyrazole may reduce conjugation between the 1-phenyl ring and the nitrogen lone pair. Sulfonation of 2-methylimidazole requires strong oleum and takes place at the 4-position (24JCS919), whereas sulfonation of 4-methylimidazole with oleum (27JCS2711), and 4-bromoimidazole with chlorosulfonic acid (57JA2188), both occur at the 5-positions.

Chlorosulfonation of 4-(p-X-phenyl)-2-methylthiazoles (X = H, Me, Cl, Br) occurs at the meta-position of the aryl ring in 75–81% yield [77IJC(B)1063]. For the unsubstituted phenyl compound this is very surprising and contrasts markedly with results for nitration (Table 7.4). Sulfonation of imidazole by arylsulfuryl chlorides takes place 35 times faster in acetonitrile than in methanol at 25°C. The ρ factor for the effects of substituents in the aryl ring was ~1.5, and slightly greater for reaction in acetonitrile [73JCS(P2)823].

Sulfenylation of N-methyl-2-thiomethylimidazole with trifluoromethylsulfenyl chloride in THF gives a 43% yield of the 4-trifluoromethylthio derivative (75JHC597).

The reaction of aryldiazonium ions with 1-(4'-sulfophenyl)-3-methylpyrazolium-5-one (**7.70**) takes place at the 4-position, and the ρ factor for the effects of substituents in the aryl ring of the diazo ion is 3.29, similar to that for coupling at the 1-position of 2-hydroxynaphthalene-6-sulfuric acid (69CCC3895). For coupling at the 4-position of 1-phenyl-2,3-dimethylpyrazolium-5-one (**7.71**) the corresponding ρ factor was 3.60 (69CC3905).

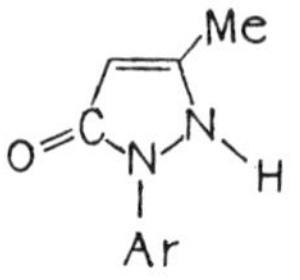

7.70 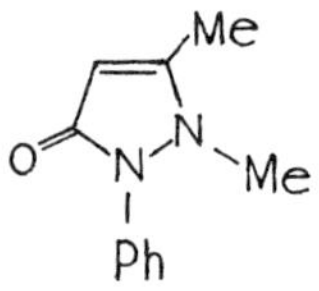 7.71

8. Metallation

A. MERCURIATION

Mercuriation of diphenyloxazoles occurs only in the heterocyclic rings. The order of reactivity, namely 2,4-diphenyl- > 2,5-diphenyl- > 4,5-diphenyloxazole, suggests a positional reactivity order of 5 > 4 >> 2 (66CHE14), though this conclusion is not rigorous because of the differential activating effects of the phenyl rings.

Isoxazole decomposes under mercuriating conditions, though 3,5-disubstituted derivatives may be mercuriated in the 4-position in 90–100% yield (58ZOB359; 60ZOB1269).

From the ease of mercuriation of mono-, di-, and trimethylthiazoles, the positional reactivity order is indicated to be 5 > 4 > 2 (60CA24661c). Surprisingly, 2-acetamido-4-methylthiazole is said not to mercuriate (59JIC434), though the 5-position should be strongly activated; it may be sterically hindered. Steric hindrance must be partly responsible for the reactivity order for substituted thiazoles: 2-Ph- > 4-Me-2-Ph- > 2-Methiazole; in this work the 4,5-dimethyl compound was unreactive [72CA(76)72617].

1-Arylpyrazoles mercuriate in the 4-position (54JCS2293; 55JCS1205; 60ZOB2931), and 3-phenyl-1,2,4-oxadiazole mercuriates in the 5-position (64HCA838), the only electrophilic substitution reported in this heterocylic ring.

B. LITHIATION

The mechanism for this reaction parallels that in base-catalyzed hydrogen exchange, so the positional reactivity order should be 5 > 3 > 4 for 1,2-azoles and 2 > 5 ≥ 4 for 1,3-azoles. Thus, isothiazole is lithiated in the 5-position (64JCS446, 64JCS3114), 1-phenylpyrazole gives a 4 : 1 mixture of the 5-/ortho-derivatives (in 80% yield) (58JA6271), and 1-substituted imidazoles are lithiated at the 2-position (71BAU1429; 73JOC3762; 77JHC517). Because the 4-position in azoles with 1,2-heteroatoms is unreactive, 3,5-dimethylisothiazole is cleaved by *n*-butyllithium, whereas 4-methylisothiazole is lithiated at the 5-position (69%) (70CJC2006). 4-Aryl-2-methylthiazoles can be lithiated in (mainly) the 5-position, but 2,4-dimethylthiazole is lithiated in the side chain (74JOC1192).

1-Methyltetrazole is lithiated in the 5-position, but 1-phenyltetrazole is prone to decompose under the same conditions (71CJC2139). 3,4-Disubstituted-2-3*H*-thiazolthiones (**7.72**) can be lithiated in the 5-position (80S800). 2-(2-Thienyl)thiazole (**7.73**) is lithiated at the 5-position of the thiazole ring (74BSF2099).

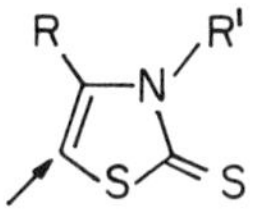

7.72 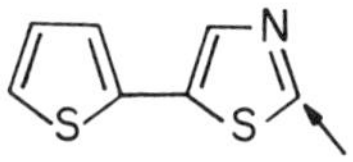   7.73

9. Side-Chain Reactions

A. DETERMINATION OF POSITIONAL REACTIVITIES

Side-chain reactions are able to give accurate quantitative values for the electrophilic reactivities of the free bases, but there have as yet been far fewer such studies than of the conventional electrophilic substitutions already discussed. Reactions can be carried out in solution or in the gas phase. Solution work has been due to Noyce and co-workers using the solvolysis of either 1-arylethyl chlorides, 2-arylprop-2-yl chlorides, or the corresponding *p*-nitrobenzoates. Results for thiazole, isothiazole, and *N*-methylimidazole (in terms of σ^+ values) are given in Scheme 7.12 (73JOC3316, 73JOC3762; 75JOC3381). These demonstrate a number of points.

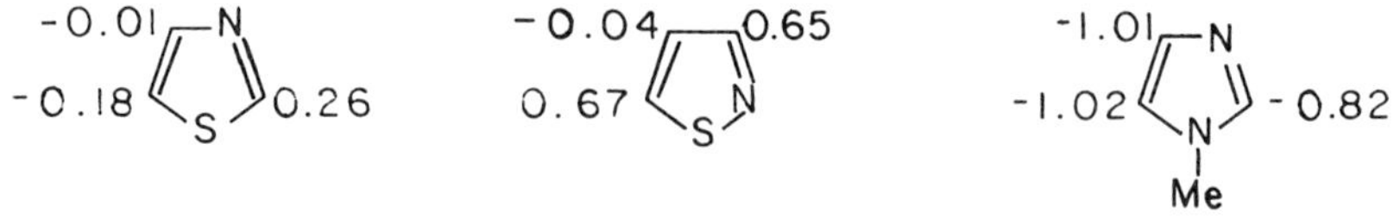

SCHEME 7.12. Values of σ^+ determined via solvolysis (73JOC3316; 73JOC3762; 75JOC3381).

(1) Each compound is less reactive than the corresponding heterocycle which does not contain multiply bonded nitrogen (Chapter 6), as theory requires.

(2) N-Methylimidazole is much more reactive than thiazole, again as required.

(3) The 1,3-azole (thiazole) is more reactive than the 1,2-azole (isothiazole), also as required.

(4) The positional reactivity order in the 1,3-azoles is 5 > 4 >> 2.

(5) The positional reactivity order in the 1,2-azole (isothiazole) is 4 > 3 > 5. The 3- versus 5-order is anomalous, the reactivity of the 5-position appearing to be too low, and possible causes for this are described below.

(6) The α- : β-reactivity ratio for N-methylimidazole is close to 1.0 as it is also in N-methylpyrrole (Chapter 6).

Values of σ^+ have also been determined in the gas phase, from pyrolysis of 1-arylethyl acetates [86J(P2)1265]; results for isothiazole are unpublished. They are shown in Scheme 7.13, and are free from any solvation, protonation, or hydrogen-bonding effects. Noteworthy points here include the following.

(1) The positional reactivity order in thiazole is again the expected one. However, the magnitudes of the σ^+ values are very different from those in Scheme 7.12. The most probable cause for this is that the azoles are *very* susceptible to demands for resonance stabilization of the transition state for a particular reaction. This is not unexpected, because the reactivity of the azoles is the product of two opposing electronic effects from the heteroatoms, which are each large. A small alteration in the demand for resonance in a particular reaction may dramatically upset this balance.

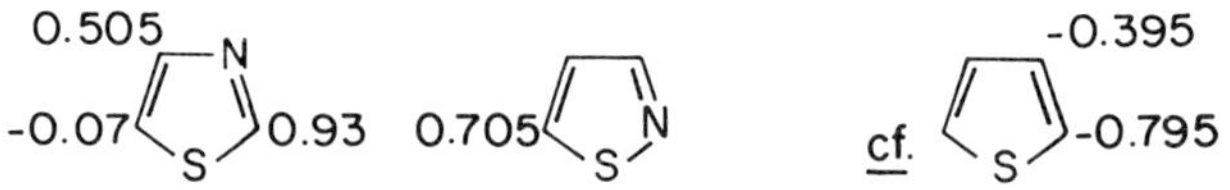

SCHEME 7.13. Values of σ^+ determined from pyrolysis of 1-arylethyl acetates [86JCS(P2)1265).

Further evidence to support this view comes from hydrogen exchange (Scheme 7.1), in which log f for the 4-position of isothiazole is 3.6, and since ρ is -7.5, σ^+ is -0.48, again very different from that given in Scheme 7.12. Available results suggest that the demand for resonance stabilization of the transition state decreases along the series hydrogen exchange > solvolysis > pyrolysis. As a result, the 4-position of thiazole (and possibly also for isothiazole) may be more, or less, reactive than a position in benzene, depending on the electrophile and on the conditions. The para-position in fluorobenzene provides a direct analogue in benzenoid chemistry.

(2) In view of the above, the reactivity of the 5-position of isothiazole in pyrolysis could be expected to be substantially less than in solvolysis, whereas this is not observed. The discrepancy appears to arise because, as already noted above, the reactivity of the 5-position of isothiazole in solvolysis seems too low. This view is also supported by the following argument. The α-position of thiophene is ~0.4 σ units more reactive than the β-position. Because of bond fixation, the nitrogen in azoles deactivates more strongly across the 2,3-positions than across the 2,5-positions. Hence it is impossible, unless secondary factors intervene, for the 3-position of isothiazole to be more reactive than the 5-position, so such a factor must be involved in producing the anomalous result in solvolysis.

Two secondary factors may be considered. One is that hydrogen bonding (which would lower the reactivity) is differentially affecting the positional reactivities. The probe substituent in the solvolysis is quite large, and when adjacent to nitrogen it may partially prevent hydrogen bonding, thus causing the reactivity of the probe site to be enhanced. This would affect the reactivities of the 2- and 4-positions of thiazole and the 3-position of isothiazole, each of which should be substantially more reactive than in pyrolysis, compared to other sites. The data in Table 7.5 support this view, there being a large difference in $\delta\sigma^+$ between the "shielded" positions on the one hand, and the "unshielded" positions on the other. Furthermore, such shielding would be greatest across a bond of high or-

TABLE 7.5
σ^+ Values from Solvolysis and Pyrolysis

Compound	Position	σ^+-Solvolysis	σ^+-Pyrolysis	$\delta\sigma^+$
Thiazole	2	0.26	0.93	0.67
Thiazole	4	-0.01	0.505	0.515
Thiazole	5	-0.18	-0.07	0.11
Isothiazole	5	0.67	0.705	0.035

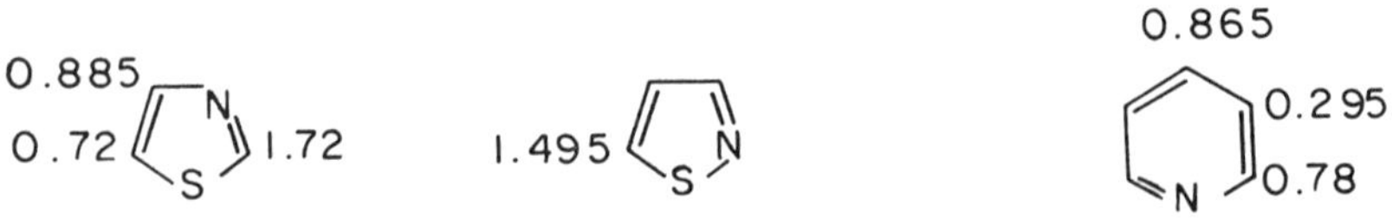

SCHEME 7.14. Deactivating effects of the aza substituent, in terms of σ^+ values, in thiazole, isothiazole and pyrdine.

der, it being shortest, and the difference between the first two entries in Table 7.5 demonstrates this.

The second possibility requires the assumption that the lone pair on nitrogen (but not the sulfur d orbitals) has a smaller space requirement than a C—H bond. Consequently, when the probe is adjacent to nitrogen, the empty p orbitals in the carbocation in the transition state are better able to overlap with the p orbitals in the azole ring, and the reactivity will be enhanced. One must also assume that this is relatively unimportant in pyrolysis, which is reasonable since less charge is developed in the transition state. A parallel discrepancy in reactivity adjacent to nitrogen is evident in pyridine chemistry, in which the results tend to support the former interpretation; this is discussed further in Chapter 9.

(3) Comparison of the reactivity of thiazole, isothiazole, and thiophene (see also Chapter 6, Section 9.A) determined under the same gas-phase conditions permits determination of the effect of the aza substituent at each position in thiazole (assuming that sulfur has the same effect in both thiophene and thiazole) and likewise at the 5-position of isothiazole. This effect (in terms of σ^+ values) is shown in Scheme 7.14, together with, for comparison, the values obtained for pyridine under the same conditions (see also Chapter 9).

Three features are evident from these results. (a) Nitrogen deactivates much more strongly in the azoles than in pyridine. This is the first indication of the phenomenon, and the reasons for it are not yet apparent. (b) Nitrogen deactivates much more strongly across the 2,3-bond compared to the 3,4-bond. This is expected because of bond fixation, and these results provide the first direct quantitative evidence. (c) Nitrogen deactivates more strongly between the 2,3-position compared to the 2,5-positions. This has been indicated by analyses given earlier in this chapter, and the present results provide the first direct quantitative confirmation.

B. Transmission of Substituent Effects

Transmission of substituent effects has been measured in thiazoles only, using compounds **7.74–7.77** (Table 7.6, 73JO3318,3321). Data for compounds **7.74** correlate well with σ^+ values, with $\rho = -6.2$, a similar value applying for compounds **7.75** (excluding the result for the 2-phenyl substituent). This implies that thiazole is a better transmitter of substituent effects than benzene (for which ρ is -5.7), but other factors may contribute to enhancing the ρ factor (see Chapter 6, Sect. 10.C). In compounds **7.75**, the phenyl substituent activates ~20 times more than it should according to its σ^+ value, which has been ascribed to greater coplanarity between the aryl rings than in biphenyl, arising from the more favorable five-membered ring geometry, and the lack of C—H bonds in the heterocycle. The results indicate that the two aryl rings are almost coplanar (73JOC2433). Compounds **7.76** and **7.77** give a very poor Hammett correlation, for not only are the methyl groups exceptionally activating, but the *m*-SMe and *m*-Ph groups *activate* and *m*-bromo should be much more deactivating.

TABLE 7.6
RELATIVE RATES OF SOLVOLYSIS OF SUBSTITUTED THIAZOLES

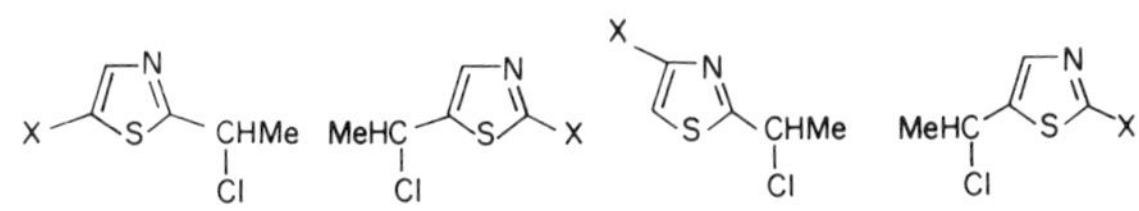

X	(7.74)	(7.75)	(7.76)[a]	(7.77)
OMe	68,000	—	—	—
SMe	9,800	6100	—	5.85
Me	75	74	11.1	6.29
Ph	—	—	3.3	2.43
H	1	1	1	1
Cl	0.29	—	—	—
Br	—	—	—	0.11

[a] In this series the combined effects of 4,5-Me$_2$ substituents was 1275.

Although the results suggest that conjugative effects are transmitted exceptionally strongly between the 2- and 4-positions, and may be enhanced by increased opportunities for coplanarity of substituents and the heterocyclic ring, an alternative possibility is that when nitrogen is flanked by two substitutents, hydrogen bonding is reduced even further and the reaction rates become abnormally high (75MI2).

10. Theoretical Calculations of Reactivity

There have been many molecular orbital calculations performed on the azoles, the majority of them directed toward determining factors other than their electrophilic reactivities. In view of the wide range of σ^+ values for a given position, with activation or deactivation depending upon the reaction studied, it is clear that calculations are unlikely to be able to provide more than the positional order of reactivities since they are unable to take account of the structure of a given transition state. The positional orders can, however, be much more simply obtained by application of a few elementary principles, as follows.

(1) The order of azole reactivity will parallel that found for the analogous five-membered heterocycles containing one heteroatom (Chapter 6) [i.e., NH(R) > O > Se > S].

(2) The azoles will be less reactive than these analogues, and this effect will increase the more multiply bonded nitrogen atoms there are within the ring (i.e., the order of reactivity is five-membered rings with one heteroatom > azoles > diazoles > triazoles).

(3) Nitrogen in positions α to the principal heteroatom will lower the reactivity more than nitrogen in β-positions. [The same rule applies to benzo-substituted six-membered heterocycles (Chapter 11)]. Hence, 1,3-azoles are more reactive than 1,2-azoles. This will be more true for compounds containing O and S rather than NH, because the differences in the α- and β-positional reactivities are greater for the former compounds.

(4) Nitrogen will deactivate most strongly across bonds of high order.

Application of these principles permits prediction of the relative positional reactivities of the azoles as shown in Scheme 7.15. For X = NH(R) the orders 8 > 9 and 11 > 12 may be reversed.

The positional reactivity order in thiazole has been calculated to be 5

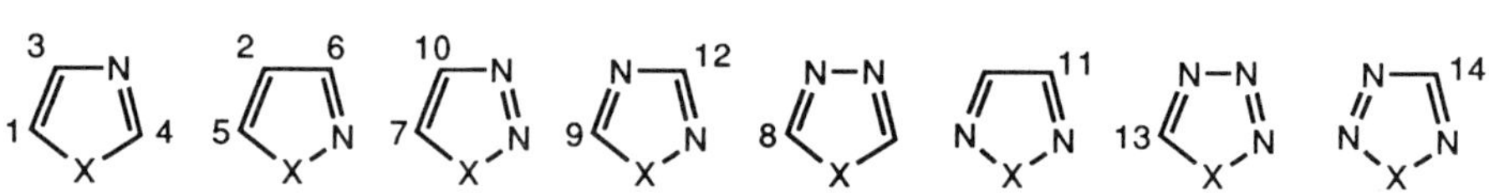

SCHEME 7.15. Predicted positional reactivity order for azoles (1 = most reactive).

TABLE 7.7
π DENSITIES FOR AZOLES CONTAINING A NITROGEN (OR OXYGEN) AT THE 1-POSITION

Position of heteroatoms	Position			
	2	3	4	5
1,3	1.012 (.944)	—	1.006 (.994)	1.142 (1.106)
1,2	—	.917 (.907)	1.090 (1.063)	1.041 (.986)
1,2,3	—	—	.987 (.982)	1.055 (1.003)
1,2,4	—	.856 (.828)	—	.936 (.856)
1,2,5	—	.973 (.924)	.973 (.924)	—
1,3,5	.995 (.939)	—	—	.995 (.939)

> 4 > 2 (69BSF1149) and a plot of the π densities against the log rates of pyrolysis of 1-(thiazolyl)ethyl acetates is linear. The reactivity order for 1,2,5-diazoles from π-density measurements is oxadiazole > selenadiazole > thiadiazole (73CHE1331); the effects of substituents in 1,2,4-triazole have also been calculated (71CHE377).

π-Density measurements for imidazole (I) and pyrazole (P) have indicated the reactivity order 4-P > 4-I > 5-I > 3-P > 5-P > 2-I > 5-(1,2,4-triazole) >3-(1,2,4-triazole); these have been partially confirmed by localization energies (55AJC100; 66CHE413, 66T835). The order appears to be wrong, however, and more reasonable are the π densities calculated by the Pariser-Parr-Pople–self-consistent-field (PPP–SCF) method (Table 7.7, which also includes values for the oxygen analogues). The order for pyrazole and imidazole here is 5-I > 4-P > 5-P > 2-I > 4-I > 3-P > 5-(1,2,4-triazole) > 3-(1,2,4-triazole). The order for oxygen compounds is closely similar, except that the 4-position of oxazole is predicted to be relatively more reactive (70BCJ3344).

Finally, π densities have been calculated for pyrazole by the CNDO/2 method [70JCS(B)1692], the values being different again from those given above, but the correct order, namely. 4 > 5 > 3, is predicted. Methyl at the 3-position raised the density at the 4-position less than did methyl at position 5, consistent with the expected effect of bond fixation, However, methyl at positions 3 or 5 was predicted to *lower* the density at the corresponding meta-position, which is anomalous. For the pyrazole cation, the π densities predicted a lower reactivity, the positional order being 4 > 5 = 3.

CHAPTER 8

Polycyclic Heteroaromatics Containing a Five-Membered Ring

1. General Introduction

This group constitutes a virtually infinite class of heteroaromatic ring systems. However, relatively few of the parent compounds have yet been made, and fewer still have had their quantitative (or indeed qualitative) reactivities measured. The compounds described in this chapter are subdivided as follows: Section 2, compounds with one five- and one six-membered ring; Section 3, compounds with one five and two six-membered rings; Section 4, compounds with two five- and one six-membered rings; Section 5, compounds with two five-membered rings; and Section 6, compounds with three or more five-membered rings.

Higher homologues in each of these categories are possible, and in a few cases such compounds have been made, but there do not appear to be any quantitative reactivity studies as yet. Each category is subdivided according to the number of heteroatoms present.

2. Compounds with One Five- and One Six-Membered Ring

This is the class of compounds in this chapter with the most data available, and consists of benzo[*b*] compounds (**8.1**), the benzo[*c*] isomers (**8.2**), and a unique nitrogen-containing compound, indolizine (**8.3**). The

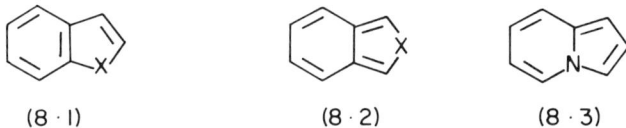

(8 · 1)   (8 · 2)   (8 · 3)

benzo[*c*] compounds are nonbenzenoid in the six-membered ring, and are therefore much less stable than the benzo[*b*] isomers; indeed, there are virtually no quantitative reactivity data. Like indolizine, they possess a very unusual property of *developing* benzenoid character on going to the transition state, which makes them highly reactive, and this is discussed below.

A. Molecules Containing One Heteroatom: Benzo[*b*]furan, Benzo[*b*]thiophene, Benzo[*b*]selenophene, Benzo[*b*]tellurophene, and Indole

a. *Positional Reactivity Order*

Fusion of a benzene ring onto a five-membered heterocycle produces a change in the positional reactivity order associated with the latter. In the single-ring, five-membered heterocycles, 2-substitution is favored over 3-substitution. This can be explained in valence-bond terminology as due to there being three canonicals for the transition state for the former, and only two for the latter (see Chapter 6, Section 10.A).

For the benzo homologues, the corresponding canonicals are shown in Scheme 8.1. Although here 2-substitution is also favored numerically, there is an additional and more important factor: Both canonicals for 3-substitution are benzenoid whereas only one of those for 2-substitution is benzenoid. In contrast, therefore, to the situation in the parent heterocycle, the 3-position is usually (but not always) the most reactive site. The actual result depends upon the inductive effect of X and its importance in a particular reaction. Thus, both benzo[*b*]furan and benzo[*b*]thiophene give the reactivity order 3 > 2 in reactions with transition states near to the ground state, for which conjugative effects are relatively unimportant, and the inductive effect preferentially deactivates the 2-position; this is particularly noticeable for benzo[*b*]furan due to the greater $-I$ effect of oxygen. For reactions with transition states nearer to the Wheland intermediate, conjugative effects become more important, and the 2-position becomes more reactive for benzo[*b*]furan, and about equally reactive

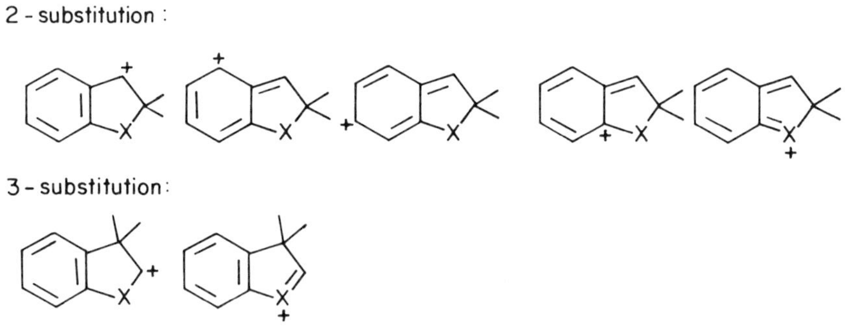

SCHEME 8.1. Canonical forms for the transition state for 2- and 3-substitution of benzo[*b*] compounds.

Sec. 2.A] ONE FIVE- AND ONE SIX-MEMBERED RING COMPOUNDS 183

with the 3-position for benzo[*b*]thiophene [82JCS(P2)1489]. For benzo-[*b*]furan, π-electron densities, which are appropriate for reactions with transition states near to the ground state, predict the order 3 > 2, whereas localization energies, appropriate for Wheland intermediate-like transition states, predict the order 2 > 3. This was originally considered anomalous (70BSF1483) but can now be seen to be normal and expected. A similar trend in reactivity order with nature of the transition state is found for benzo[*b*]thiophene [82JCS(P2)1489]. Other calculations (extended Hückel theory, and CNDO/2) give the order 3 > 2 for indole and 2 > 3 for benzo[*b*]furan, but again suffer from the inability to consider the transition-state structure (68IJQ165).

Additional information may be deduced from Scheme 8.1. The canonical for 2-substitution with the positive charge located on the heteroatom involves disruption of the aromaticity of the benzenoid ring. Consequently, the reactivity of the 2-position of the benzo homologues should be *less* than that of the 2-position of the parent heterocycle and this is true in every case. By contrast, the canonical for 3-substitution with the charge delocalized to the heteroatom does not disrupt the aromaticity of the benzenoid ring, and less resonance energy is lost than in the monocyclic series. Hence the 3-position of the benzo homologues should be *more* reactive than that of the 3-position of the parent heterocycle and this is also true.

The heteroatom can also conjugate with each position in the benzene ring. However, direct conjugation (to the 5- and 7-positions, as in Wheland intermediates **8.4** and **8.5**) should be more favourable than indirect

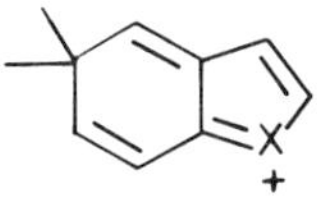

(8·4) 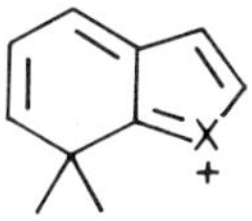   (8·5)

conjugation (to the 4- and 6-position, as in **8.6** and **8.7**) because there is less interruption of bond fixation in the five-membered ring. An additional factor is that structures **8.4** and **8.7** are *p*-quinonoid and therefore more stable than **8.5** and **8.6**. The overall balance of these factors is that the

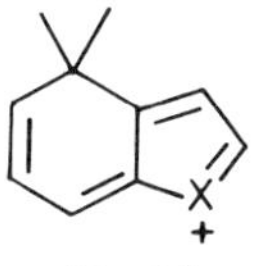

(8·6) 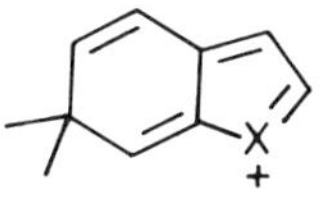   (8·7)

positional reactivity order in benzo[b]thiophene (the only molecule for which it has been determined) is 6 > 5 > 4 > 7, and positions may be activated or deactivated [78JCS(P2)1053; 82JCS(P2)1489]. The reactivity order is correctly predicted by π-electron densities (Hückel) but not by localization energies, which give the order 4 > 7 > 6 > 5 [82JCS(P2)1489]; this is discussed further in Section 2.A.c.

For benzo[b]furan and indole no such precise data are available, but it is possible to adduce some information from the various reactions described below. The positional reactivity orders for these molecules and also for benzo[b]thiophene, which have been calculated by various methods, are given in Table 8.1. In principle the *ab initio* calculations should be the more reliable, but neither the π nor the (σ + π) order is correct for benzo[b]thiophene, suggesting that these are incorrect for the other molecules also. The calculations using the STO-3G basis set certainly wrongly predict the site of most rapid protonation. Notably, only the Hückel calculations give the correct order for benzo[b]thiophene and indeed they are usually the most reliable indicators for electrophilic aromatic substitution.

There are no quantitative data on the transmission of substituent effects between the two rings, though there are some data, described below, on the directing effects of substituents. Activating substituents at the 2-position direct to the 3-position and vice versa, and likewise there is mutual activation between the 4- and 7-positions. Similarly, strongly activating substituents at the 5- or 6-positions activate the corresponding ortho 4- and 7-positions, respectively; the effects across the 5,6-positions are expected to be weaker because of bond fixation. All these effects are straightforward and predictable, and only a few data are given. *Ab initio* (STO-3G) calculations predict that the positional order of protonation (and hence of electrophilic substitution) of 2-methylindole should be 3 > 4 > 5 > 6 > 7 > 2 and for 3-methylindole 3 > 4 > 5 > 6 > 2 > 7 (82T3693); the low reactivity of the 2-position in the latter seems rather unlikely.

Lastly, the 2- versus 3-positional reactivity order for benzo[b]thiophene appears to be affected by steric hindrance from the sulfur d orbitals [82JCS(P2)1489], and this is discussed in detail in Section 2.A.c. For indoles it has been proposed that 2-substitution may take place by initial attachment of the electrophile to the 3-position, followed by migration (69T227).

b. *Reactions*

i. *Acid-Catalyzed Hydrogen Exchange.* Partial rate factors for detritiation of benzo[b]thiophene in trifluoroacetic acid at 70°C (cor-

TABLE 8.1
CALCULATED POSITIONAL REACTIVITY ORDERS FOR BENZO[*b*]FURAN, BENZO[*b*]THIOPHENE, AND INDOLE

Heteroatom	Positional Order	Method	Reference
O	3 > 7 > 2 > 5 > 6 > 4	π-SCF	70BSF1483
O	2 > 4 > 6 > 7 > 5 > 3	LE-SCF	70BSF1483
O	3 > 7 > 5 > 4 > 6 > 2	π-CNDO/2	70BSF1483
O	3 > 7 > 5 > 2 > 4 > 6	$(\sigma + \pi)$-CNDO/2	70BSF1483
O	2 > 6 ≥ 3 > 4 > 5 > 7	$(\sigma + \pi)$-CNDO/2	75JCS(P2)366
O	3 > 7 > 5 > 6 > 4 > 2	π-Hückel	70BSF1483
O	2 > 3 > 6 > 4 > 7 > 5	LE-Hückel	70BSF1483
O	2 > 3 > 7 > 5 > 4 > 6	π-ab $initio$ (LCGO)	74JCS(P2)1893
O	3 > 5 > 7 > 6 > 4 > 2	$(\sigma + \pi)$-ab $initio$ (LCGO)	74JCS(P2)1893
NH	N > 3 > 7 ≥ 2 > 5 > 6 > 4	π-ab $initio$ (LCGO)	74JCS(P2)1893
NH	N > 3 > 7 > 5 > 6 > 4 > 2	$(\sigma + \pi)$-ab $initio$ (LCGO)	74JCS(P2)1893
NH	N > 3 > 4 > 5 > 6 > 7 > 2	YSP charge densities ab $initio$ (STO-3G)	82T3693
NH	N > 3 > 2 > 5 > 6 > 7 > 4	π-Hückel	66PMH142
NH	N > 3 > 2 > 7 > 5 > 6 > 4	SCF	64CA(60)15718d
NH	N > 3 > 2 > 7 > 5 > 6 > 4	VESCF	59AJC152
S	2 > 3 > 7 > 5 > 6 > 4	π-ab $initio$ (LCGO)	74JCS(P2)1893
S	2 > 3 > 5 > 6 > 7 > 4	$(\sigma + \pi)$-ab $initio$ (LCGO)	74JCS(P2)1893
S	3 > 2 > 6 > 5 > 4 > 7	π-Hückel	82JCS(P2)1489
S	3 > 2 > 4 > 7 > 6 > 5	LE-Hückel	82JCS(P2)1489

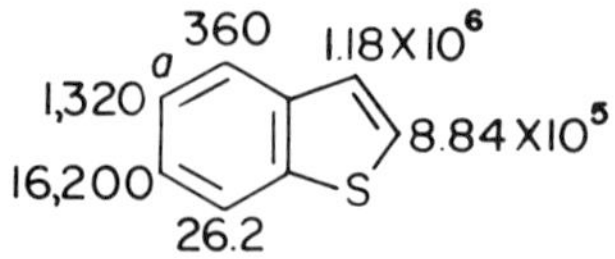

SCHEME 8.2. Partial rate factors for detritiation of benzo[b]thiophene in trifluoroacetic acid at 70°C. *Due to a numerical error, a larger value is given in the original paper.

rected for hydrogen bonding, which sulfur-containing heterocycles undergo in this and probably all other strong acids) have been determined as shown in Scheme 8.2; the derived σ^+ values are given in Scheme 8.3 [72JCS(P2)97; 82JCS(P2)1489]. Under the same conditions, f_2 for thiophene is 9.75×10^7 with $\sigma^+ = -0.913$. The positional reactivity order is correctly predicted by π densities, but not by localization energies and this contrasts with thiophene, thienothiophene, dithienothiophenes, and dithienobenzenes for which the latter reactivity indices are more satisfactory. The significance of the σ^+ values is discussed in Section 2.A.c.

For benzo[b]furan, detritiation in trifluoroacetic acid at 70°C gave $f_2 = 17,000$, hence $\sigma^+ = -0.48$ [62JCS2382; 72JCS(P2)97]. This value was not corrected for hydrogen bonding of the heterocyclic oxygen with the solvent and it is probable that f_2 could be 20% higher, with $\sigma^+ \cong -0.50$. Nevertheless, the 2-position is quite clearly less reactive than the 2-position of benzo[b]thiophene.

For indole there have been a number of studies. Rate coefficients for deuteriation of the 3-position in D_2O/dioxan at 50°C are given in Table 8.2 (71T4171). Of interest here is the very high reactivity; however, the reaction cannot be taking place on the anion as 1-substituted compounds are included.

Rates of dedeuteriation and detritiation of 5-substituted, 3-labeled (deuteriated and/or tritiated) indoles have been determined in both aqueous mineral acid and in acetic acid [72JCS(P2)1618]. Substituent effects in both series were similar [slightly larger for dedeuteriation, which is a

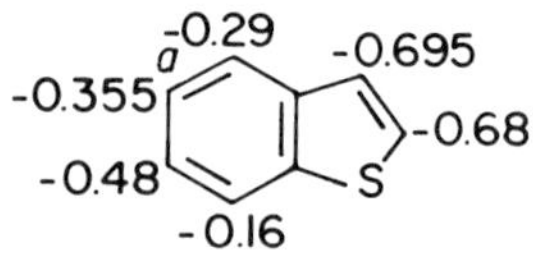

SCHEME 8.3. Values of σ^+ for detritiation of benzo[b]thiophene in trifluoroacetic acid at 70°C. *Due to a numerical error, a larger value is given in the original paper [72JCS(P2)97; 82JCS(P2)1489].

TABLE 8.2

RATE COEFFICIENTS FOR DEUTERATION OF THE 3-POSITION
OF INDOLES[a]

	Substituent			
	H	1-Me	2-Me	1,2-Me$_2$
$10^7 k$ (sec^{-1})	2.0	2.5	29	50
f	1	1.25	14.5	25

[a] In D$_2$O/dioxan at 50°C (71T4171).

more selective reaction (70AK89)]; the substituent effects for detritiation at 25°C are given in Table 8.3. Exchange rates were 10^3 slower in acetic acid and the spread of partial rate factors is larger (larger ρ factors). Values for f_o^{Me} are much higher for both media than that in Table 8.2, which suggests that the exchange mechanism is different. Under both conditions, o-methyl activates less than the 220-fold activation which it produces in benzene (61JCS2388).

The Brönsted coefficients were 0.67 and 0.75 for exchange with H$_3$O$^+$ and HOAc, respectively, and the isotopic rate ratios were consistent with the A-S$_E$2 mechanism, in which the proton is half-transferred in the transition state [72JCS(P2)1116]. However, the ratios of the second-order rate coefficients for exchange (k_2^H/k_2^D) appeared to vary systematically with neither the reactivity of the acid nor that of the aromatic, but this conclusion must be tempered with the fact that the ratios are accurate to only ~10%. More importantly, it was argued that ΔpK (pK$_{substrate}$ − pK$_{catalyst}$) showed no maximum when plotted against k_2^H/k_2^D, in contrast to previous findings [72MI2(217)], whereas one would expect that this kinetic isotope effect should be a function of the position in the transition state of the hydrogen

TABLE 8.3

PARTIAL RATE FACTORS[a] FOR DETRITIATION OF 3-
TRITIOINDOLES IN MINERAL ACID OR ACETIC ACID

	Substituent			
	H	2-Me	5-OMe	5-CN
f(H$_3$O$^+$)	1	81.6	2.28	0.011
f(HOAc)	1[b]	159	2.81	0.09

[a] Relative to indole.
[b] $k = 4340 \times 10^{-7}$ sec^{-1}.

atom being transferred (and therefore a maximum when $\alpha = 0.5$) (61JA2154; 64PAC217).

A possible cause of the discrepancy may lie in the assumption that the most (kinetically) basic site in indole is the 3-position and not nitrogen. Indeed, it had been shown earlier that either site may be the more basic, depending upon the proton-donating acid (62JA2534). It is, of course, a requirement for correlation of isotope effect with ΔpK that exchange take place at the most basic site. For the 5-substituted indoles, the basicity of nitrogen is substantially affected by these substituents, which are directly conjugated with it. The same is unlikely to be true for the 3-position (which is not conjugated with the 5-position), and indeed early work has shown that exchange at nitrogen was faster than at C-3 (38BCJ307, 38BCJ643; 39BCJ453). In D_2O/dioxan, 3-methylindole underwent deuteriation at nitrogen with $k = 1330 \times 10^{-7}$ sec^{-1} at 22°C, which is equivalent to ~10,000 × 10^{-7} sec^{-1} at 50°C (62JA2534). Comparison with the data in Table 8.2 shows that NH exchange is very much faster than CH exchange, and this has been confirmed for exchange in trichloroacetic acid at 25°C. The rates relative to exchange at C-3 in pyrrole (= 1) are pyrrole C-2 (2.7), pyrrole N (15), indole C-3 (12), indole N (1000) [76JCS(P2)388]. These results show that the C-2:C-3 rate ratio for pyrrole is higher (2.7) than in indole (1.6) for exchange under similar conditions (71CC421). They also indicate that nitrogen is the most basic site in both molecules, and that the *ab initio* calculations (Table 8.1) are unsatisfactory. In aqueous TFA, tryptophan [2-amino-3-(3-indolyl)propanoic acid] (protonated at both nitrogens) underwent exchange more rapidly at the 2- and 6- than at the 4-, 5-, and 7-positions (67ACS1674).

Exchange has also been studied under conditions that are basic but do not involve the usual base-catalyzed mechanism of deprotonation. In indole, a proton is removed from nitrogen in a *fast* step, and the anion (with the charge delocalized to the 3-position) undergoes rapid, but rate-determining, proton addition by abstraction from the weak acid water (62JA2534; 63JA2524). This mechanism was indicated by the observation that detritiation of 3-T-2-methylindole is faster than 3-T-1,2-dimethylindole (which cannot form the reactive anion) (63JA2524), and is confirmed by substituents acting in the normal manner for an S_E2 reaction (Table 8.4) [72J(P2)1111, 72JCS(P2)1625]. The very high reactivity of the anion is revealed by the small activating effect of the *o*-methyl group (cf. the much larger values in Table 8.3 for the neutral molecule). 5-Nitro is notably more deactivating than 6-nitro: Neither can conjugate directly with the 3-position, but the 5-nitro group can conjugate with nitrogen, thus affecting the reactivity of the 3-position by a secondary effect.

TABLE 8.4

EXCHANGE RATE COEFFICIENTS AND
PARTIAL RATE FACTORS FOR
DETRITIATION OF INDOLES UNDER BASIC
CONDITIONS AT 25°C

Substituent	$10^5 \, k \, (\text{sec}^{-1})$	f^a
H	1.78	1
2-Me	45.4	25.5
5-NO$_2$	0.785	0.44
5-CN	1.25	0.70
5-Br	2.34	1.31
6-NO$_2$	1.32	0.74
2-Me-5-NO$_2$	27.0	15.2

^{a} Relative to indole.

The canonicals for these interactions are given in structures **8.8** and **8.9**, respectively. The former is of lower energy because there is less bond reorganization required between **8.8** and the ground state compared to **8.9**. The substitutent activation (or deactivation factors A_f [79JCS(P2)381] for **8.8** and **8.9** are 30 and 20, respectively. The observed partial rate factors then follow since a larger value of A_f predicts a larger substituent effect.

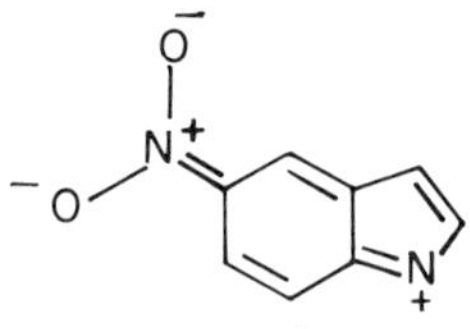

(8·8) 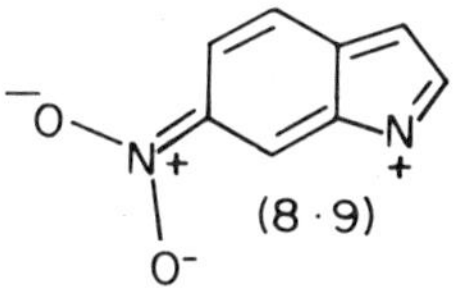   (8·9)

ii. *Base-Catalyzed Hydrogen Exchange.* Rates of base-catalyzed exchange are governed by inductive effects, rather than by stabilities of structures such as those shown in Scheme 8.1. Consequently, as in the five-membered heterocycles, the benzo analogues are also more reactive at the 2- than at the 3-position. The exchange rates have been claimed to be governed by the equation $\log k(140 \, °C) = -9.5 + 8.2\sigma_I$ (69MI2). However, this seems to be inconsistent with the observation that benzo-[*b*]thiophene is *more* reactive than benzo[*b*]furan by a factor of 4 (69MI2; 77JHC95) (cf. 500 for the sulfur/oxygen ratio in the five-membered heterocycles). The higher reactivity of the sulfur-containing compounds has in each case been attributed to stabilization of the intermediate carbanion by ($p \to d$) π-electron withdrawal by sulfur. Any such an effect must operate indirectly, because the carbanion is generated in a σ orbital orthogonal to the p orbital of the aromatic ring [72MI2(271)].

The 2-position of benzo[*b*]thiophene is four times as reactive in dedeuteriation as the 2-position of thiophene, and the corresponding reactivity difference for the 3-positions (under different conditions) is 65-fold (71JGU1945). Since the reactivity difference between the 2- and 3-positions of thiophene is 2.5×10^5 (Chapter 6, Section 2.A), the 2-position of benzo[*b*]thiophene is more reactive than the 3-position by a factor of 1.54×10^4 (which is ~15-fold less than the factor in thiophene).

The above relative rates of deuterium exchange at the 2-positions of benzo[*b*]furan and benzo[*b*]thiophene were confirmed by Attanasi *et al.* [79PS(5)305], who also found benzo[*b*]selenophene to be less reactive than benzo]*b*]thiophene, the relative exchange rates for different heteroatoms being Se (1.0), O (2), S (7). However, in the corresponding benzoazoles, the relative reactivities at the 2-positions for different heteroatoms were S (1.0), Se (4), O (20) [79PS(5)305]. Possible reasons for these latter differences are considered in Section 2.C.

iii. *Nitration and Nitrosation.* Nitration is not very sterically hindered, so the positions of substitution in this reaction ought, in principle, to give a fairly accurate indication of the most reactive sites toward electrophilic substitution. However, in some cases the position of substitution is dependent upon the reaction conditions for reasons that are not entirely clear, though the incursion of nitrosation is a strong possibility.

Benzo[*b*]furan nitrates in the 2-position (mainly) and in the 5-position [1900LA(312)237; 02CB1633; 12CB1596], a positional reactivity order that calculations (Table 8.1) comprehensively fail to predict. Evidence from substituted benzo[*b*]furans (**8.10–8.16**) suggests that the reactivity of the 6-position may be very close to that of the 5-position [64CA(60)493d; 65AC(R)1028, 65BSF1466; 70BSF1029; 73CA(79)105183; 83CJC2287]. In these compounds the 6-position in the benzenoid ring is the most reactive (though **8.13** chloromethylates in the 5-position), but this may be due to activation by the electron-supplying groups in the 2-position that are conjugated with it. By contrast, electron-supplying groups in the 3-position can conjugate with neither the 5- nor the 6-position. That the preference for 5-nitration of 2-acetylbenzo[*b*]furan (**8.17**) is only slight (70BSF1029; 83CJC2287) seems anomalous because here the 6-position should be conjugatively deactivated.

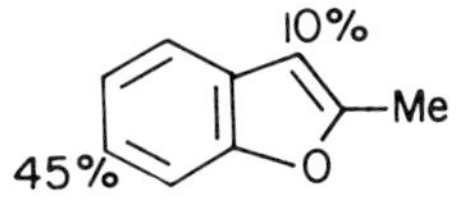

Sec. 2.A] ONE FIVE- AND ONE SIX-MEMBERED RING COMPOUNDS 191

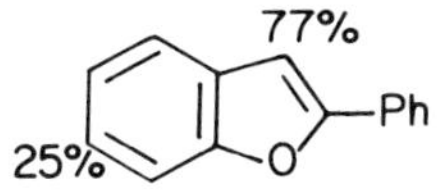

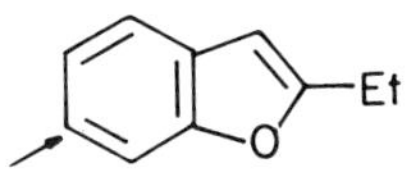

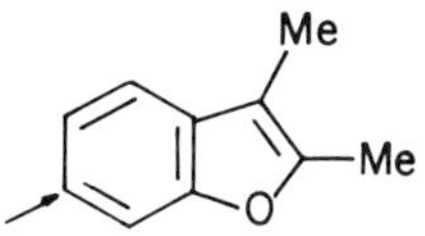

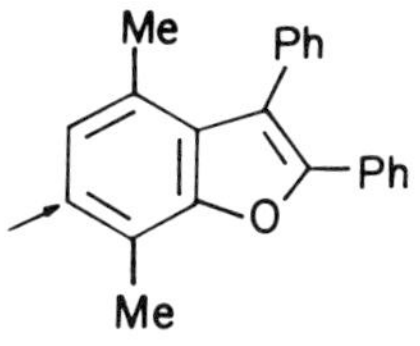

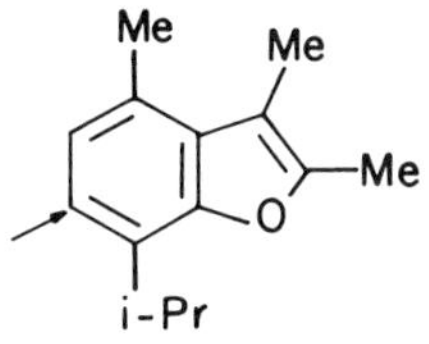

(8·10) benzofuran with 2-Me; 10% (3-position), 45% (benzene ring position)

(8·11) benzofuran with 2-Ph; 77% (3-position), 25% (benzene ring position)

(8·12) benzofuran with 2-Et

(8·13) benzofuran with 3-Me, 2-Me

(8·14) benzofuran with 4-Me, 3-Ph, 2-Ph, 7-Me

(8·15) benzofuran with 4-Me, 3-Me, 2-Me, 7-i-Pr

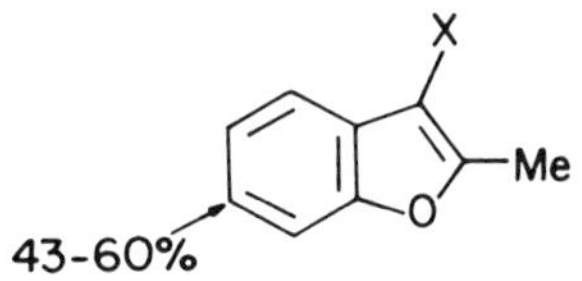

(8·16 X = CHO, COMe, COEt, COPh, CN) 43–60%

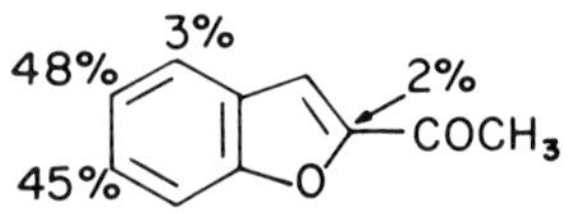

(8·17) benzofuran with 2-COCH$_3$; 3%, 2%, 48%, 45%

Orientations and yields in nitration.

Nitration of benzo[b]thiophene occurs mainly at the 3-position, further nitration going into the benzene ring giving the positional reactivity order (for nitration in acetic acid/acetic anhydride) 4 > 7 > 6 > 5 (with none of the 3,5-dinitro isomer being formed) [69JCS(C)1766]; with nitric acid in acetic acid, the positional order for mononitration is 3 > 4 > 6 (68JHC69). The reactivity order in the benzene ring is at variance with the known positional order 6 > 5 > 4 > 7 established both via hydrogen-exchange (Section 2.A.b.i) and side-chain carbocation reactions (Section 2.A.c). The probability that nitration through nitrosation is partially responsible is reinforced by results for indole described below, and it is notable that some isomer distributions in nitration of benzo[b]thiophenes vary widely with conditions. There appears to be a need to repeat some of this work in the presence of effective nitrous acid traps.

A substantial amount of work provides information on interactions between a substituent in one ring and reaction in the other. These interactions are different for benzo[b]thiophene compared to benzo[b]furan or indole because the availability of canonicals involving tetravalent sulfur (**8.18, 8.19**) means that conjugative effects may be relayed between the 3- and 5- or 7-positions, the 3,5-interaction being the more effective.

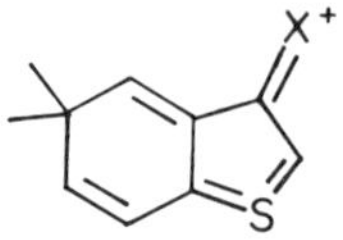

(8·18) 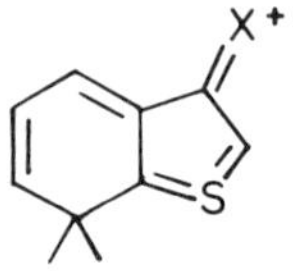   (8·19)

Strongly activating substituents in the benzenoid ring (as in **8.20**) can cause exclusive nitration (and bromination and acylation) in that ring [73J(P1)623, 73J(P1)1196]]. This result may be contrasted with that for **8.21**, which gives both 7- *and* 2-substitution [78JCR(S)10]. Both the 4- and 6-positions are conjugated with the 2-position but the substituent interaction factor A_f [79JCS(P2)381] is higher (12) for the 6,2-interaction (**8.21**) than that (6) for the 4,2-interaction that would apply in (**8.20**). Hence, 2-substitution is observed in **8.21** and not in **8.20**. Conjugation between the 6- and 2-positions also accounts for 50% of the overall nitro product from **8.22** being the 2-derivative [78JCR(S)10]. Nitration (and also bromination and acylation) of **8.23** gave only 5- and 7-derivatives [73JCS(P1)1196], whereas some 2-substitution could have been expected, at least for the less hindered nitration, since the 4- and 2-positions are conjugated. It is possible that reaction occurs here via the phenate anion, which would very strongly activate the benzenoid ring to the exclusion of substitution elsewhere. This suggestions seems to be supported by the relative positional yields for **8.24** [73JCS(P1)623].

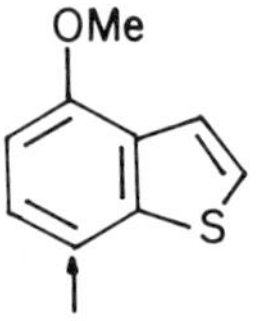

(8·20) 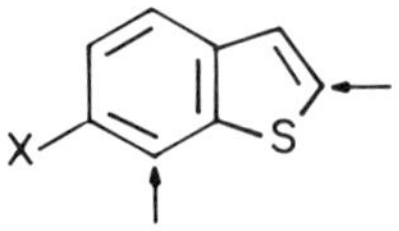   (8·21, X=NHAc,OH) 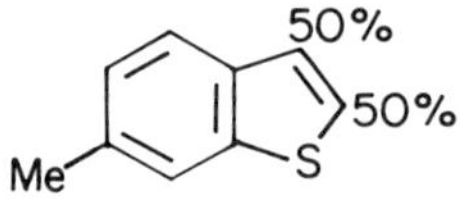   (8·22) 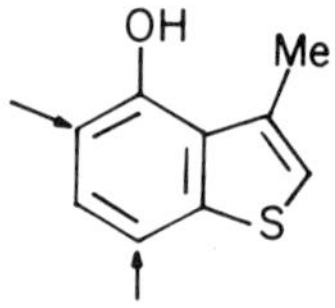   (8·23)

Orientations and yields in nitration.

Nitration of 5-acetylamino- or 5-hydroxybenzo[*b*]thiophene provides an interesting comparison to **8.21** in that 70 and 40% yields, respectively, of the 4-nitro derivatives are obtained (55JA5939; 60JCS938). This shows clearly the effect of bond fixation, so that, of the positions ortho to the 5-substituent, the 4-position is activated more strongly because of the higher order of the 4,5-bond compared to the 5,6-bond. An interesting comparison to **8.22** is the nitration only in the 3-position of 7-methylbenzo[*b*]thiophene (65NKZ853); here the substituent cannot conjugate with the 2-position. One would certainly have expected some 6-nitro derivative as well, and its absence is probably an analytical omission.

Structures such as **8.18** and **8.19** must have real significance, as shown by the positional yields in nitration of 3-nitrobenzo[*b*]thiophene (**8.25**) (68MI5) and of 3-acetylbenzo[*b*]thiophene (**8.26**) [70JCS(C)933], though it

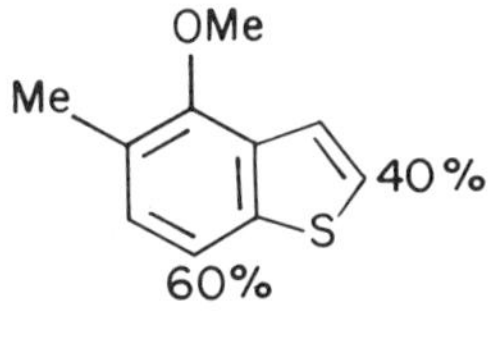

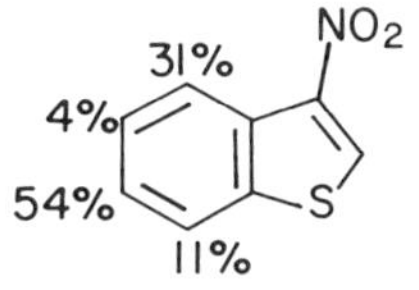

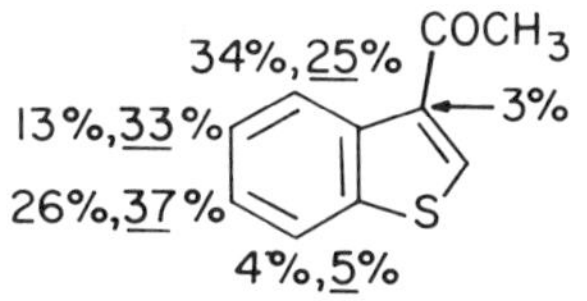

Orientations and yields in nitration.

is less certain in the latter case because of the change in yield with conditions. Such differences, also evident in **8.27–8.29** [69JCS(C)2755; 71J(C)3405], are most probably due at least in part to the incursion of nitrosation, which may also account for the positional order in nitration of 3-nitrobenzo[*b*]thiophene being given as 6 > 5 > 4 > 7 [69JCS(C)1766].

An interesting result is the high yield of the 2-nitro derivative in nitration of **8.30** (68JHC69). Evidently the high order of the 2,3-bond facilitates the $+M$ effect of bromine rather than its $-I$ effect, so that there is little deactivation of the 2-position by the 3-bromo substituent.

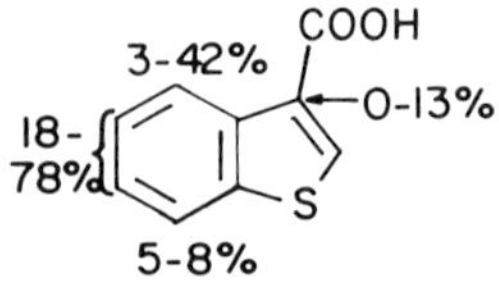

(8·27), 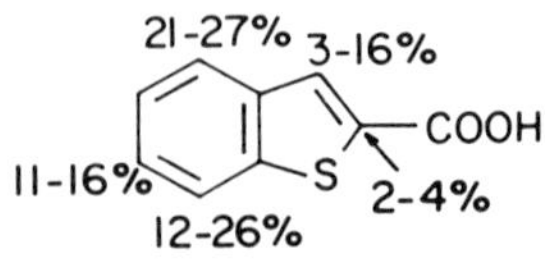 (8·28), 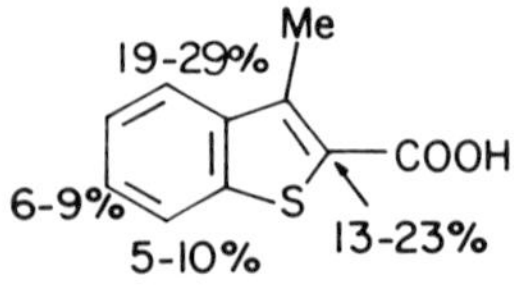 (8·29), 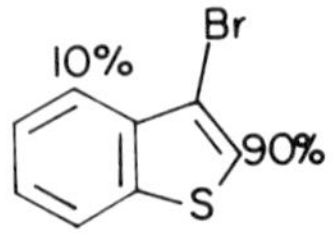 (8·30)

Orientations and yields in nitration.

Ipso substitution of bromine accompanied nitration of **8.31**, **8.32**, and **8.33** [70JCS(C)1949; 71JCS(C)3052; 72JCS(P1)265]; in the latter two compounds, the displaced bromine substituted into starting material to give **8.34** and **8.35**, respectively. The yields of 6- versus 4-product in **8.33** compared to **8.32**, and **8.35** compared to **8.34**, again demonstrate the higher order of the 4,5- versus the 5,6-bond. *Ipso* substitution also occurred in nitration of **8.36**, and side-chain substitution and oxidation of the 2-methyl group (15%) accompanied nitration of **8.37** [72JCS(P1)265]. Lastly, protiodebutylation (10%) accompanied nitration and nitrodebutylation of **8.38** (yield varied with conditions), and the 2-carboxylic acid then nitrated in the 4- (0.5–3%), 6- (4–9%), and 7-position (the yield of the 7-isomer could not be determined because it could not be resolved from the 3-butyl-4-nitro-2-carboxylic acid). Nitrodecarboxylation *and* protiodebutylation also produced some 2-nitrobenzo[*b*]thiophene [72JCS(P1)414]. Isomer yields for other benzo[*b*]thiophenes, but which do not lead to useful reactivity order information, are given in an excellent review by Iddon and Scrowston [70AHC(11)177].

Nitration of indole by benzoyl nitrate gives 3-nitroindole (35%) [68JCS(C)2145]. Nitration of some alkyl-substituted indoles by nitric acid/sulfuric acid (63JOC2262; 79JOU528) takes place upon the protonated species, and in such cases the rates pass through a maximum at 90% sulfuric acid in the usual way. 2-Methyl-, 1,2-dimethyl-, 2-*t*-butyl-, and 2,3-dimethylindole are each nitrated at the 5-position, the former two compounds in 84 and 82% yield, respectively, the latter two compounds at similar rates. This rate similarity suggests that in the protonated species the 5-position is much the most reactive of the benzenoid ring positions. In the protonated species it is probably the 3-position that has become

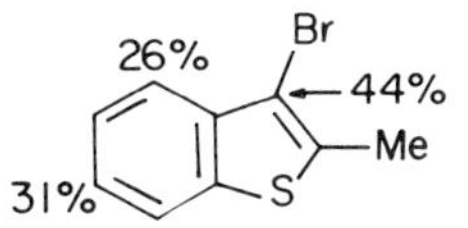

Sec. 2.A] ONE FIVE- AND ONE SIX-MEMBERED RING COMPOUNDS 195

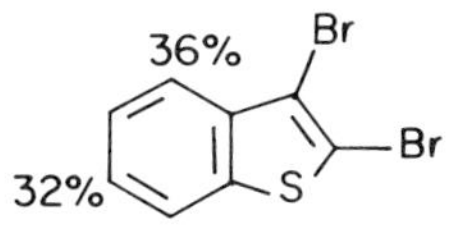

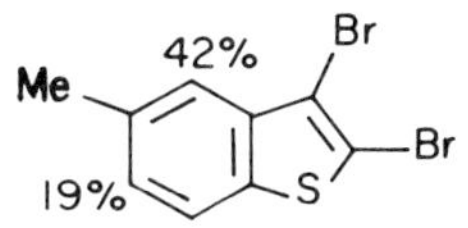

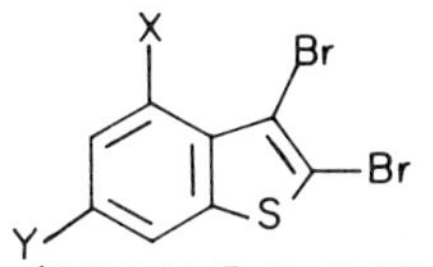

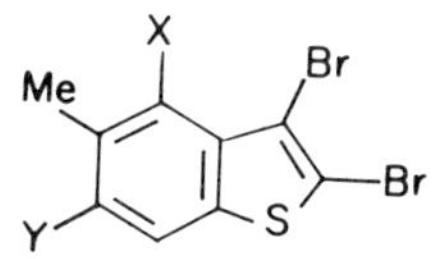

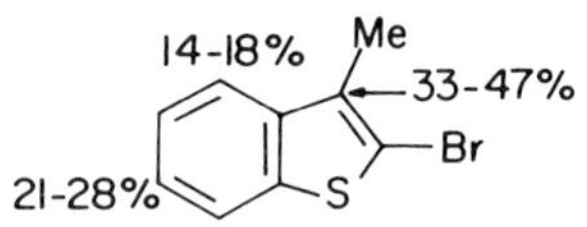

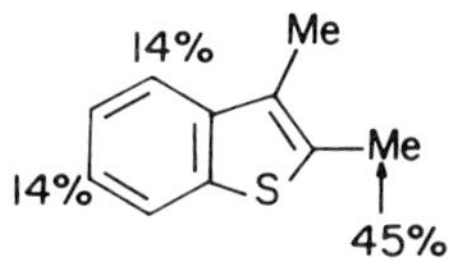

Orientations and yields in nitration.

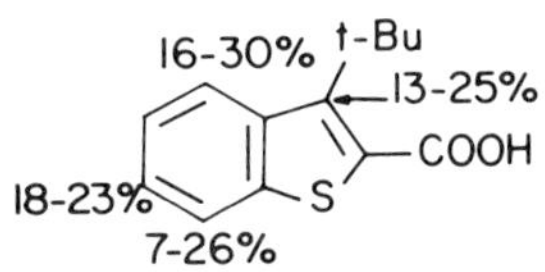

protonated, and hence there is no 3-substitution. Moreover, 2-alkyl groups cannot conjugate with the 5-position, which agrees with the similarity of rates for 2-*t*-butyl- and 2,3-dimethylindole. 5-Methoxy-1,2-dimethylindole is nitrated 10^3 times faster than 1,2-dimethylindole, with substitution occurring at the 6-position. Since the 4-position would (as a result of bond fixation) be more activated by 5-OMe than the 6-position, this suggests a positional reactivity order of 5 > 6 > 4. Nitration of 7-methoxy-2,3-dimethylindole (and the *N*-methyl derivative) also involved the protonated species and took place at the 6-position (74CHE930). The high 6,7-bond order should make activation of the 6-position comparable to that of the 4-position, so the preferred 6-substitution again suggests the reactivity order proposed above.

The species involved in nitration of 2-methylindole with excess concentrated nitric acid is not known, but is probably the free base after the entry of the first nitro group. The final product is 2-methyl-3,4,6-trinitroindole and a variety of intermediate mono- and dinitro compounds can be obtained; nitrodeacylation can also be used (65JOC3457) (Scheme 8.4). These results again show that electron-supplying substituents in the 2-position preferentially activate the 6-position. Evidence that these results relate to free base nitration comes from nitration of compounds **8.39–8.42** in nitric acid/sulfuric acid (65T1923). The conjugate acid is most probably

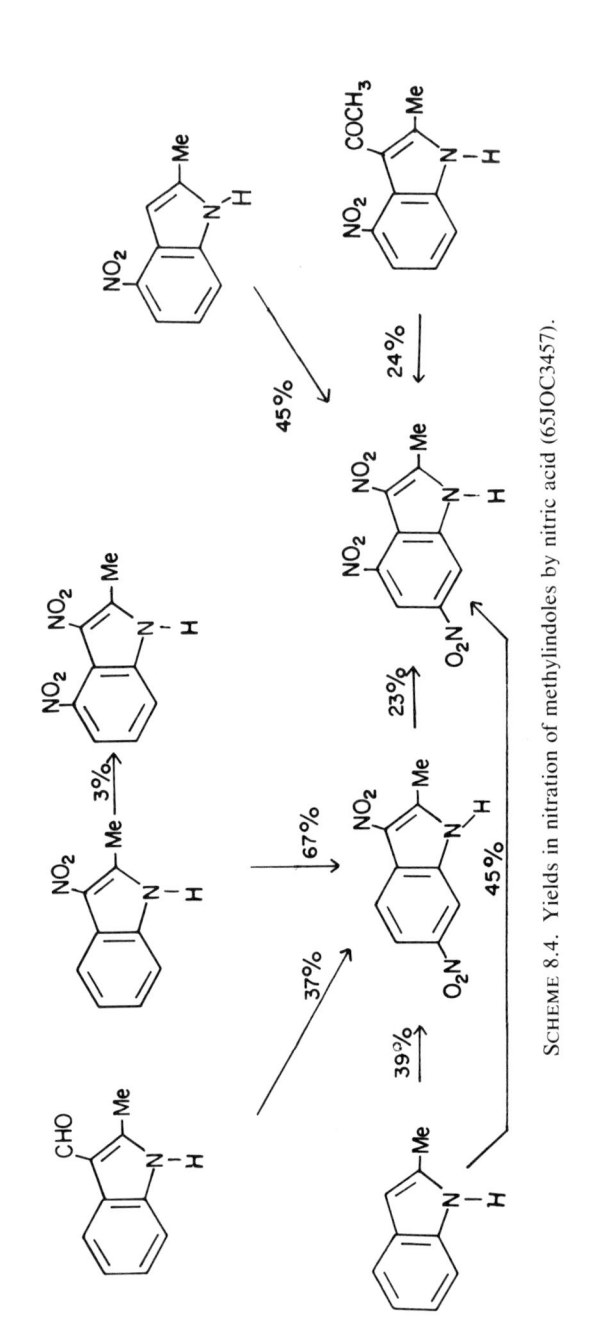

SCHEME 8.4. Yields in nitration of methylindoles by nitric acid (65JOC3457).

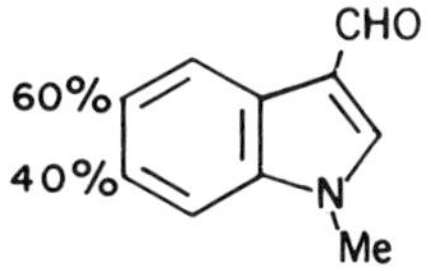

Isomer yields in nitration.

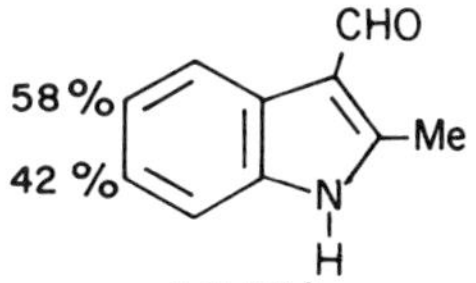

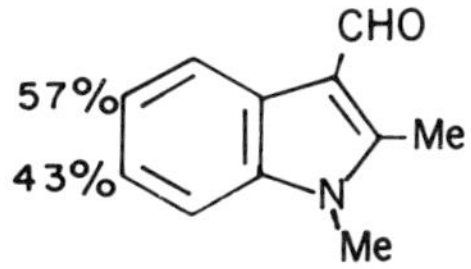

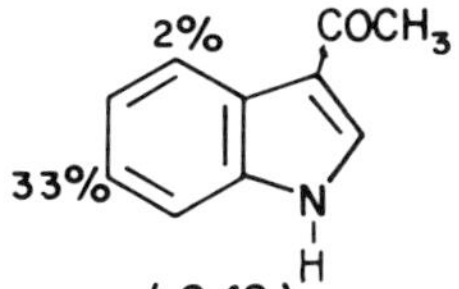

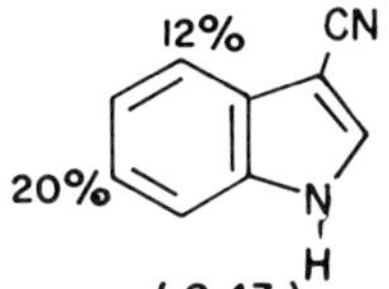

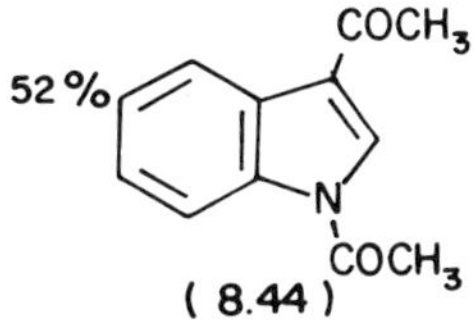

involved here, which would explain the difference in the nitration pattern for 3-formyl-2-methylindole (**8.40**) (cf. Scheme 8.4). The 6-position is the preferred site of nitration by nitric acid of indoles with electron-withdrawing groups in the 3-position (**8.42, 8.43**). The 5-position reacts in **8.44**, and this may involve the conjugate acid as the reaction was carried out in concentrated sulfuric acid (66JOC70).

Clear evidence that indoles may undergo nitration as the free base or conjugate acid comes from nitration of 2,3-diphenylindole (65T823). Nitration by nitric acid in acetic acid involves the free base and gives the 6-nitro derivative. By contrast, nitric acid in sulfuric acid gave initial substitution at the para position of the 3-phenyl ring (this position is conjugated with nitrogen), and further substitution went into the 5-position (as noted for other protonated species implicated or postulated above). These results indicate that the benzenoid ring in the 2,3-diphenylindole free base is more reactive than the 3-phenyl substituent, whereas in the protonated species the reverse applies. Similar results were obtained with 2-methyl-3-phenylindole. However, 3-methyl-2-phenylindole is nitrated (in sulfuric acid) in the 5-position (80%); the 2-phenyl ring is deactivated by conjugation in the conjugate acid so no substitution occurs in this.

Nitration of 2-phenyl-3-substituted indoles at the 3-position indicated a possible role that nitrosation may play in governing the isomer distribution [81JCS(P2)628]. The 3-substituents were mostly electron withdrawing, so reaction probably occurred on the free base. Their ease of removal (71CHE1401) followed the order $CH_2OH > N_2Ph > COCH_3 > CHO > COOEt$ [71CHE1401; 81JCS(P2)628], and in addition to 3-nitro-2-phenylindole, 3,4- and 3,6-dinitroindoles were obtained; this follows from the fact that both 2-phenyl and the nitrogen activate the 4- and 6-positions. Nitration of 3-nitro-2-phenylindole gave 81% of 3,6-dinitroindole. Nitration of 3-nitroso-2-phenylindole gave the 3,5-dinitro derivative, believed to be formed through nitration at the 5-position of the Wheland intermediate (i.e., the protonated species **8.45**). The dominant tautomeric forms of the conjugate acids of substituted indole are protonated at the 3-position and so there seems to be general consistency in that the conjugate acid nitrates in the 5- and the free base in the 6-position. Nitrosation at the 3-position of 2-phenyl-3-substituted indoles also gave 3-nitro derivatives with, in reaction with 3-benzyl-2-phenylindole, nitro(so) substitution in the benzenoid ring [84JCS(P2)165].

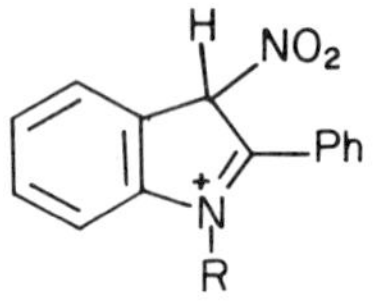

(8·45, R = H, alkyl)

Second-order rate coefficients $10^2 \times k_2$ have been determined for nitrosation at the 3-position at 0°C (by rate-limiting attack of nitrousacidium ion) of the following substituted indoles: 2-Ph (390), 1,2-Me$_2$ (320), 2-Me (290), 2-Me-5-NO$_2$ (150), 5-CN (12.4), 1-Me-5-NO$_2$ (7.0), 5-NO$_2$ (2.7) [73JCS(P2)918]. These data show the normal substituent effect features; for example, 1-Me activates ~twofold and 2-Me activates ~fivefold. The compounds are much more reactive than amines toward nitrosation, so the mechanism is believed to involve initial attack at nitrogen followed by rearrangement. Nitrosation can also take place under the same conditions via nitrous anhydride and nitrosyl chloride as electrophiles, and the rate-limiting step also depends upon the substituents. For the first four of substituted indoles listed in this paragraph, formation of the reagent or diffusion of the reagents is rate-limiting, whereas for the latter three, proton loss is rate-limiting under certain conditions.

iv. *Halogenation.* The ease with which these compounds undergo halogen addition is reflected in the halogenation of benzo[*b*]furan and benzo[*b*]thiophene, and especially of the former, which being the less aromatic has the higher 2,3-bond order. Thus, bromination of benzo[*b*]furan and its 3-Me, 2-Me, and 2,3-Me$_2$ derivatives proceeds via trans-adduct formation, decomposition of which leads to 2-substitution for the former two compounds, 3-substitution for 2-methylbenzo[*b*]furan, and side-chain substitution (in either of the methyl groups) for 2,3-dimethylbenzo[*b*]furan. A similar adduct is formed in chlorination of 2,3-dimethylbenzo[*b*]thiophene [76JCS(P2)266].

In the chlorination of benzo[*b*]furan, both trans and cis adducts are formed, presumably because there is less steric hindrance here to the formation of the cis adduct. Both adducts eliminated HCl to give 3-chlorobenzo[b]furan, the cis adduct decomposing more rapidly (77JHC359), which follows from the required trans disposition of hydrogen and chlorine. Bromination (and acetylation) of 2-phenylbenzo[b]furan gave the 3- and 3,6-derivatives [64CA(60)15808e] (the 6-position is conjugated with the phenyl group) and likewise 2,3-diphenylbenzo[b]furan is brominated in the 6-position [66CA(65)5429e]. Second- and third-order rate coefficients for the 2-bromination of benzo[b]furans are given in Table 8.5 (74BCJ1267), and adduct formation occurred here also. The results for the methyl derivatives demonstrate very nicely the greater 2,6- versus 2,5-conjugative interaction. The ρ factor (for the effects of 5-substituents) was -2.21, very close to that (-2.24) for bromination of styrenes, indicating that both reactions involve similar adducts.

Since the 2- and 3-positions are the most reactive sites in benzo[b]thiophene, it follows that chlorination gives 3-chloro- (69%), 2,3-dichloro- (28%), and 2-chlorobenzo[b]thiophene (3%) (relative yields) [51JA2614, 51USP2571742; 55JCS1565; 66CJC2283; 68JCS(B)397]. Bromination gives a higher yield (92%) of the 3-derivative (67AJC313), with 3- : 2-product ratios of 77–91 : 1, [71JCS(B)79], which might be expected for a more selective reaction. Values of σ^+ for the 2- and 3-positions have been calculated as -0.61 and -0.77, respectively (molecular chlorination and bromination), and as -0.39 and -0.69, respectively (positive bromination) [73JCS(P2)1250]; these values are discussed further in Section 2.A.c.

The high selectivity of bromination may also account for the relative insensitivity of the site of reaction to existing substituents. Thus, 5-bromo- (65NKZ1067), 2-fluoro- (63JOC1420), 2-phenyl- (66JMC551), 2-methyl- (53JA3278; 66AJC1909), and 7-methylbenzo[b]thiophene (65NKZ853; 66BSF3618) all undergo bromination at the 2-position. Likewise 3-methyl- (52JA2185), 3-bromo- (67NKZ755), and 7-chloro-3-methylbenzo[b]thi-

TABLE 8.5
RATE COEFFICIENTS FOR BROMINATION OF
BENZO[b]FURANS AT THE 2-POSITION

Substituent	$10^2 \, k_2 \, (M^{-1} \, \text{sec}^{-1})$	$k_3 \, (M^{-2} \, \text{sec}^{-1})$
H	0.64	3.6
5-Me	2.96	35.9
6-Me	8.28	616
5-MeO	3.48	261
5-Cl	0.053	0.17

ophene [72JCS(P1)1404] are each brominated in the 3-position, the latter in 79% yield compared to 36% in nitration, which may again reflect the selectivity differences. Chlorination of 3-bromobenzo[b]thiophene also gave 2-substitution, and also *ipso* substitution, leading to all possible product combinations arising from bromination by the displaced bromine (73IJS233); the lower selectivity of chlorination compared to bromination is reflected in the 3-:2-product ratio of 44 [71JCS(B)79]. 2,3-Dimethylbenzo[b]thiophene is brominated at the 6-position, as expected (66BSF3055), as is also 2,3-dibromo-5-methyl- (70%) and 2,3-dibromobenzo[b]thiophene (80%) [70JCS(C)1949]. The last results further confirm that the 6-position is the most reactive in the benzenoid ring, because the 6-position should be more deactivated by the 2-bromine than the 5-position is by the 3-bromine. In the bromination of 2,3-dibromobenzo[b]thiophene in silver sulfate/sulfuric acid (Br^+ electrophile), a small amount of the 4-bromo derivative was obtained though the expected 5-isomer was not reported [70JCS(C)1949]. Chlorination of 2,3-dimethylbenzo[b]thiophene occurred in the 2-methyl group [68JCS(B)397]. In the side-chain substitution of the 2,3-dimethyl derivatives of the benzo[b] compounds, the electrophile enters the methyl group attached to the *least* reactive of the 2- or 3-position. Bromination of **8.21** went into the 2- and 7-positions, as in nitration [78JCR(S)10]. Iodination of benzo[b]thiophene goes only in the 3-position (52JA4950; 66CJC2283).

Chlorination of indole with sodium hypochlorite can take place through an N-chloro intermediate, which rearranges with base to the 3-chloro product (78JOC2639). However, the generality of this mechanism for other halogenations has not been established. Indole is brominated 60% in the 3-position (54ZOB1265) and this can be raised to 88% using 2,4,4,6-tetrabromocyclohexa-2,5-dienone [72JCS(P1)2567] and to 96% with bromine in DMF. Similar yields were obtained with 1-Me-, 2-Me-, 2-Ph-, 5-Cl-, and 5-$COCH_3$-substituted indoles (82S1096), and also in iodinations (82S1096). Bromination of 5-ethoxycarbonylindole also occurs exclusively at the 3-position, and as with benzo[b]thiophene, the high selectivity of bromination tends to override substituent effects. If the 3-position is blocked, then bromination takes place at the 2-position. Yields for NBS bromination of 3-phenylindole were 98% (74T2123) and for indole with 3-CH_2X substituents (X = electron-withdrawing group) yields were 85% (75CPB2990). Bromination of 2-methylindole with excess bromine gives the 3,6-dibromo product [76CI(M)220], substitution in the 6-position being expected as a result of conjugation with the methyl group.

Evidence from bromination suggests that the 5- and 6-positions are close in reactivity. For example, if both 2- and 3-positions are blocked as in **8.46**, in which X and Y are electron-supplying groups, then substitution goes into the 6-position. However, if X is a donor and Y is an acceptor, then substitution goes into the 5-position [81CA(94)156679], the 6-position being conjugatively deactivated; this switch in order suggests that the 5- and 6-positional reactivities cannot be too dissimilar in the first place. Moreover, if X and Y are both only alkyl groups, then bromination occurs in the 5-position (71CHE1406), as does nitration. Again, a small change in the relative electron-supplying abilities of X and Y is sufficient to bring about this switch in orientation. This is also shown by the bromination of **8.47**, in which the 7-alkyl group is sufficient to switch the preferred substitution site back to the 6-position (71CHE1406).

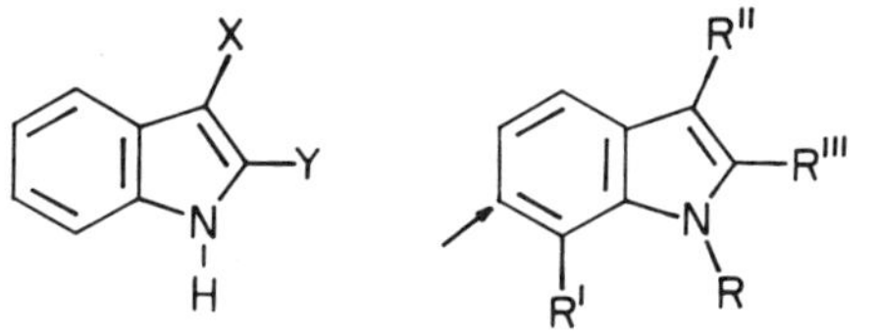

(8·46)　　　　(8·47, R-R''' = alkyl)

Iodination of indole, 2-methylindole, and the corresponding 1,3-D_2 derivatives have shown the kinetic isotope effects k_H/k_D to be 2 and 3, respectively [68AC(R)1435]. Substituent effects for the iodination reaction (Table 8.6) [69AC(R)799] are particularly interesting because they demonstrate very clearly that conjugative effects from the benzenoid ring are relayed very poorly to the 3-position, so much so that 5-OMe is less activating than 5-Me.

Bromination of benzo[*b*]selenophene occurs at both the 2- and the 3-positions (72CHE13).

v. *Alkylation and Acylation.* *t*-Butylation of benzo[*b*]furan with 2-methylpropene–H_3PO_4/Kieselguhr gives the 2-, 3-, and other isomers in the ratio 8:10:3 (69BAU2446), but with *t*-butyl chloride/$ZnCl_2$ the 2:3 ratio becomes 1:2 (71CHE144). Under the latter conditions, the activation energies for both positions are similar (~8.6 kcal mol^{-1}), and the reaction is zeroth order in heteroaromatic, so that the rate-determining step is formation of the polarized complex (71CHE953). The 3-isomer is

TABLE 8.6
SUBSTITUENT EFFECTS ON THE RATE OF 3-IODINATION OF SUBSTITUTED INDOLES

Substituent	H	1-Me	2-Me	5-Me	5-OMe	6-OMe	5-Br	5-Cl
$k_{rel.}$	1	1.47	358	3.07	2.25	2.18	0.071	0.069

thermodynamically less stable than the 2-isomer (presumably through steric interaction with the 4-hydrogen), and this explains why t-butylation with t-butanol–$ZnCl_2$/Al_2O_3 at 200°C gives an overall yield of 60%, with 52–68% of 2-, and 5–9% of 3-t-butylbenzo[b]furan (together with 25–29% of phenolic products) (71CHE956). Chloromethylation occurs at the 2-position (54% yield) [53CA10519c; 63CA(58)5606h], whereas 2-phenylbenzo[b]furan chloromethylates in the 5-position (cf. nitration, which goes into the 6-position) (65BSF1466). A possible reason for the 5-orientation could be coordination of the Friedel–Crafts catalyst with oxygen.

Isopropylation of benzo[b]thiophene with propene, 2-propanol, or 2-chloropropane, catalyzed by various acids, occurs mainly at the 2-position (78–92%) together with a little 3-substitution (62JOC2026). This is somewhat anomalous, being the only electrophilic substitution of benzo-[b]thiophene that gives a preponderance of 2-substitution. By contrast, t-butylation with 2-methylpropene, catalyzed by sulfuric or phosphoric acid, gave 67 and 75%, respectively, of the 3-derivative (53USP2652405; 56JOC584). A more recent report showed that with polyphosphoric acid as catalyst, the 3-isomer (71%) is accompanied by 22% of the 2-isomer; these yields became 89 and 6%, respectively, with t-butanol-H_2SO_4 as the reagent [72JCS(P1)414]. Alkylation with t-amyl alcohol, pent-1-ene, or hex-1-ene also gives the 3-derivative (53USP2652405; 66JOC3093). Because there will be steric hindrance from the sulfur d orbitals as well as the 4-hydrogen, it is unlikely that there is a substantial difference in thermodynamic stability between the 2- and 3-t-butyl derivatives, in contrast to the t-butylbenzo[b]furans.

Indole is alkylated at the N- or at the 3-position, depending upon the conditions. For example, alkylation of the anion in liquid ammonia gives 100% N-alkylation (54CB127) and with benzyl bromide in DMSO the anion yields 92–97% of N-benzylindole (73OS1758). However, with aziridinium tetrafluoroborate, the product was mainly 3-(2-aminoethyl)indole together with a little of the 2-isomer [67AG(E)178]. The proposal that the 2-substitution of indole takes place through 3-substitution followed by rearrangement (69T227) has been shown not to be a significant pathway for 2-alkylation of 3-alkylindoles (72TL5277). Alkylation with xanthydrol (**8.48**) in the presence of acid gave a 78% yield of the 3-isomer. 2-Methyl- and 3-methylindole gave the expected 3- and 2-substitution, respectively, while 2,3-dimethylindole gave 5- (45%) and 6-substitution (11%), indicating fairly similar reactivity of the 5- and 6-positions (84JHC1485). Hydroxymethylation of indole with paraformaldehyde goes 82% into the 3-position (51G613). In reaction with the alkylating reagent **8.49**, the relative reactivities of π-excessive heterocycles were thiophene (1), furan (10^3), pyrrole (10^5), indole (5×10^5), N-methylindole (3×10^6), 2-methyl-

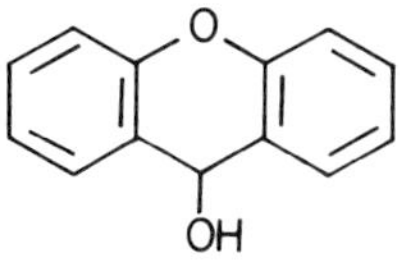

Sec. 2.A] ONE FIVE- AND ONE SIX-MEMBERED RING COMPOUNDS 203

(8·48) 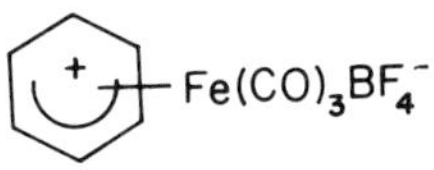 (8·49)

indole (3.5×10^6) (73CC540). Cyanoethylation of indole and 2-methylindole with vinyl cyanide gave 80 and 81%, yields, respectively, of the 3-(2-cyanoethyl) derivatives (56ZOB557). The reaction between methyl chloroacetate and photoexcited indole gives all seven monosubstitution products (72CPB2163).

Isomer ratios for the acylation of benzo[b] furan under various conditions are given in Table 8.7 [71JCS(B)79; 73JCS(P2)1250]; from these data, values of σ_2^+ and σ_3^+ were determined as -0.575 and -0.48, respectively [78JCS(P2)1053].

Phosphoric acid-catalyzed acylation by anhydrides also gives mainly the 2-isomer (e.g., acetic anhydride gives 55% of 2-acetylbenzo[b] furan) (64JCS173). The FeCl$_3$-catalyzed reaction of substituted benzoyl chlorides gives 35–40% yields of 2-aroylbenzo[b]furans [80CA(92)215176]. The variation in the isomer ratio with temperature (Table 8.7) is unlikely to be a thermodynamic effect arising from steric hindrance from the 4-hydrogen, since the change is in the wrong direction. More probably it is due to the increase in the reactivity of the electrophile with increasing temperature, which produces a diminishing 2- : 3-rate ratio (see Section 2.A.c).

2-Cyanobenzo[b]furan is acetylated in the 4-, 5-, and 7-positions (84JHC177). The 6-position, like the 4-position, is conjugatively deactivated by the 2-cyano group, but the canonical for the 2,6-interaction (**8.50**) is of lower energy than that (**8.51**) for the 2,4- [i.e., the former has

TABLE 8.7
ISOMER RATIOS (2 : 3) FOR ACYLATION OF BENZO[b]FURAN AND BENZO[b]THIOPHENE

Reaction	Reagent	T (°C)	Benzo[b]furan	Benzo[b]thiophene
Acetylation	Ac$_2$O, SnCl$_4$	25	7.2	0.15
Acetylation	Ac$_2$O, SnCl$_4$	75	3.5	0.28
Acetylation	Ac$_2$O, SnCl$_4$	130	1.7	0.56
Acetylation	Ac$_2$O, SnCl$_4$	200	—	0.71
Acetylation	CF$_3$COOCOMe	75	8.0	0.27
Benzoylation	Bz$_2$O, SnCl$_4$	75	8.4	0.53

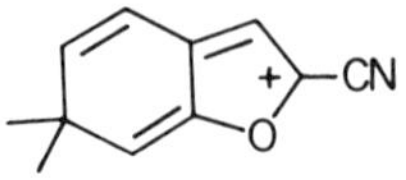

(8·50) 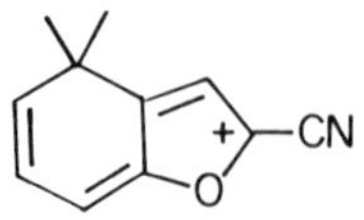   (8·51)

the higher substituent interaction factor A_f (79JCS(P2)381)]. The cyano group therefore destabilizes more effectively the transition state of **8.50** compared to that of **8.51**. Benzo[*b*]furan-2-carboxamide also acetylated at the 5-position (84JHC177). The greater-electron-supplying ability of a methoxy group compared to the heterocyclic oxygen is shown by the position of acetylation in **8.52** and **8.53** (65BSF1473; 82JHC279); in addition, the orientation in **8.54** shows the effect of the high 4,5- versus 5,6-bond order. The 2-nitro derivative of **8.53** is also acetylated in the 7-position (82JHC279). The 6- and 4-substitutions of 2,3-dimethylbenzo[*b*]furans containing either a bromine or a chlorine in any position of the benzenoid ring (70BCJ3496, 70BSF3601) may also be interpreted in terms of the bond fixation in that ring. With a methyl group in the five-membered ring, the latter becomes almost equally reactive to that of the methoxy-substituted benzenoid ring in **8.55**, and more reactive in **8.56** (65BSF1473).

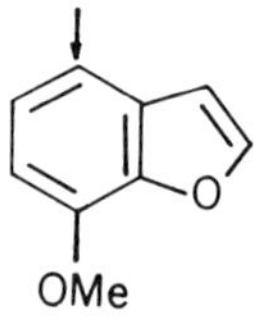

(8·52) 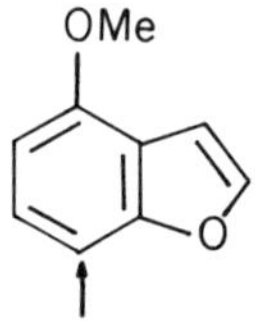   (8·53) 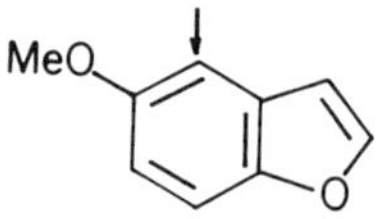   (8·54)

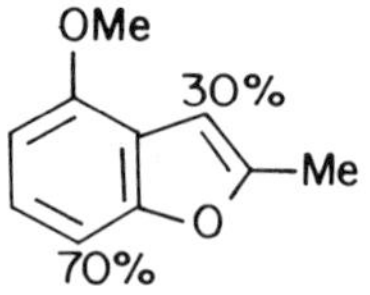

(8·55) 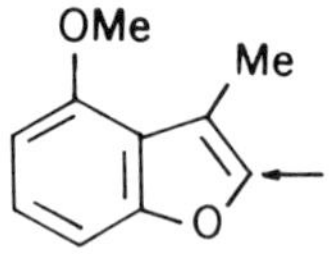   (8·56)

Acylation of benzo[*b*]thiophene goes preferentially into the 3-position, with the isomer ratio given in Table 8.7 [73JCS(P2)1250]. These results lead to σ_2^+ and σ_3^+ values of -0.49 and -0.58, respectively. In contrast to benzo[*b*]furan, the proportion of the 2-isomer formed increases with increasing temperature, but the interplay of many factors here (steric hindrance from the 4-hydrogen *and* the sulfur *d* orbitals, as well as the change in reactivity of the electrophile) obscures the precise reasons for this variation.

A wide range of substituent effects has been determined for the acetylation of benzo[*b*]thiophene, and the more important results are given in Scheme 8.5 in terms of isomer yields or the main site of substitution [52JA766,2285; 60T(10)215; 61BSF1534, 61JOC359,363; 64BSF1525; 65N KZ99,637,647, 65NKZ1067; 67BRP1058468, 67JCS(C)2084, 67M-2039, 67NKZ751; 78JCR(S)10; 79JHC1029]. In general, these show that electron-supplying substituents at the 2- or 3-positions activate the adjacent position. If the 3-position is blocked, electron-supplying substituents at the 2- or 3-positions activate the adjacent position. If the 3-position is blocked, electron-supplying substituents at the 2-position activate the 6-position making it even more reactive than the 5-position, the next most reactive site in the benzenoid ring (see Section 2.A.c). Hence, 5-substitution may or may not accompany 6-substitution here. Electron supply from the 2- to the 4-position is less effective and coupled with the substantially lower reactivity of the 4-position (Section 2.A.c) means that 4-substitution is often not observed. Conversely, electron-donor substituents at the 6-position, and to a lesser extent at the 4-position, preferentially activate the 2-position, such that 2-substitution becomes preferred over 3-substitution, which is not difficult because of the fairly close reactivity of the 2- and 3-positions (Section 2.A.c). Likewise, if both 5- and 6-positions contain electron-supplying substituents, then since the 6,2-conjugative interaction is greater than the 5,3-conjugative interaction, 2-substitution is again the *expected* result [cf. 70AHC(11)177]. Electron-supplying substituents at the 5- and (less effectively) at the 7-position activate (but less strongly) the already most reactive 3-position so substitution occurs there. Electron-withdrawing substituents at the 6- and (less effectively) the 4-position deactivate the 2-position, and the effect from the 5- and 7-position upon the 3-position will be correspondingly smaller. If the 2- and 3-positions are blocked with electron-withdrawing substituents, then the 6-position should be more effectively deactivated by the 2-substituent than will the 5-position be deactivated by the 3-substituent, so both 6- and 5-substitution should be observed.

Results of additional note are that 4-chlorobenzo[*b*]thiophene gives 2-substitution (84%) and the 6-isomer gives mainly 2- (64%) together with some 3-substitution (26%) [78JCR(S)10]. This suggests that the $+M$ (mesomeric donor) effect of chlorine is more effectively transmitted between the two rings than the $-I$ (inductive acceptor) effect, which is not unreasonable. This view is supported by the preferential 3-acetylation of 5-

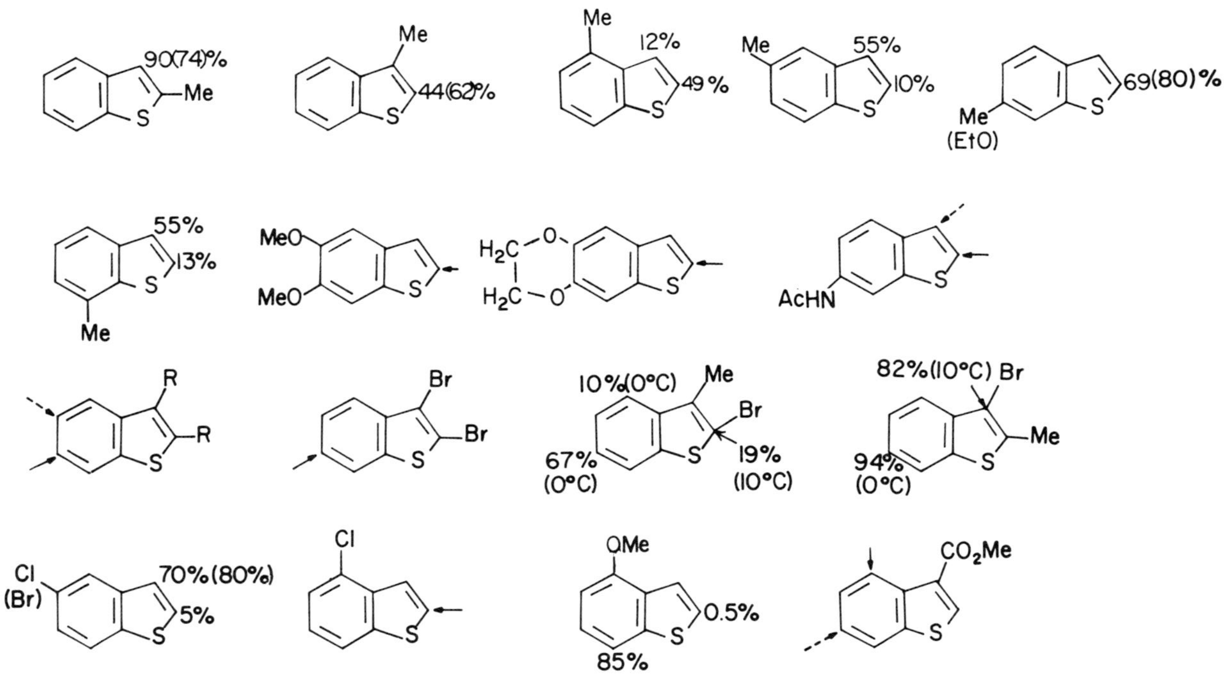

SCHEME 8.5. Isomer yields and positions of substitution in acetylation of benzo[b]thiophenes.

chloro- and 5-bromobenzo[b]thiophene [67JCS(C)2084]. Less explicable is the 4-acetylation said to accompany 6-acetylation of 2-bromo-3-methylbenzo[b]thiophene (67NKZ751), whereas 6- and 5-acetylation would be expected. Moreover, the change to *ipso* substitution in this work (almost total for the 3-bromo-2-methyl isomer) for a mere 10°C increase in temperature suggests an experimental artifact. The 3-methoxycarbonyl substituent deactivates the 6- and 4-positions (the former more effectively as expected), giving a 3 : 1 ratio for acetylation (79JHC1029). For benzoylation [which is *less* sterically hindered, 72MI2(181)] the ratio of 6- : 4-substitution is substantially lower, as expected. The high steric hindrance to acetylation also shows up in reaction of 6-ethoxybenzo[b]thiophene, no 7-acetylation being found despite strong activation by the *o*-ethoxy group. The same is found for acetylation of 6-acetylaminobenzo[b]thiophene, and this may be attributed to steric hindrance arising from the adjacent 6-substituent and sulfur *d* orbitals. By contrast, 4-methoxybenzo[b]thiophene, in which the 7-position is activated similarly to the previous case, is almost exclusively 7-substituted, and no 5-substitution occurs because of steric hindrance. The small amount (0.5%) of 2-substitution here reflects the poor 4,2-conjugative interaction. Lastly, the high steric hindrance to acetylation is confirmed by nitration and bromination of 6-acetylaminobenzo[b]thiophene, which gives both 2- *and* 7-substitution [78JCR(S)10]. Additional quantitative data which may be interpreted in the above manner are given in the literature [70AHC(11)327].

As in alkylation, acylation of indole can take place either at nitrogen or at C-3. Thus, trifluoroacetylation gave 20, 50, and 15%, respectively, of the C-3-mono-, N-mono-, or C-3,N-diacylated products (76T2595). By contrast, pyrrole gives exclusive substitution at carbon, and carbazole exclusive substitution at nitrogen, the differences being attributed to the relative loss of ground-state resonance in each molecule on going to the transition state for N-substitution. This is confirmed by a comprehensive set of Vilsmeier–Haack acetylation data given in Table 8.8 [77JCS(P2)1284]. Rates here are calculated relative to the 2-position of indole (= 1), and data for pyrroles and carbazoles are included for comparison.

A less comprehensive set of data for Vilsmeier–Haack formylation (72CC427) (which gives 90% 3-substitution of indole) [77CA(86)171800] confirmed the main features summarized below.

(1) C-Alkylation produces only a small increase in the rate of N-acylation of indole.

(2) 3-Alkylation produces only a small increase in reactivity of the 2-position, as does 2-alkylation upon the reactivity of the 3-position (~fourfold each), which reflects both steric hindrance, and the high reactivity of indole.

TABLE 8.8
Relative Rates for Acetylation with Dimethylacetamide–COCl$_2$ at 25°C

Compound	Substitution position			
	N	C-2	C-3	Benzenoid ring (position)
Pyrrole		5000		
1-Methylpyrrole		5785		
3-Methylpyrrole		2.1 × 10^5		
Indole	59	1	2620	
1-Methylindole			4250	
2-Methylindole			1 × 10^4	
3-Methylindole	68	3.8		
4-Methylindole	102		2050	
3-t-Butylindole	54			
1,2-Dimethylindole			9040	
1,3-Dimethylindole		3.95		
2,3-Dimethylindole	59			1.9 (6)
1,2,3-Trimethylindole				2.0 (6)
Carbazole	434			16 (3)
N-methylcarbazole				25 (3)
1,2,3,4-Tetrahydrocarbazole	32			1.8 (7)

The 1- and 4-methyl groups are each the same distance from the 3-position and can only act inductively. The former produces a small activation, and the latter a small deactivation resulting from superimposition of steric hindrance upon a comparable activation. As in the case of benzo-[b]thiophene, the 6-position appears to be the most reactive in the benzenoid ring. Acetylation of 1-acetyl-2,3-dimethylindole also occurred in the 6-position (73% yield) (47JCS1631), but for 2-methyl-5-nitroindole, 2% was acetylated at the 3-position and a very surprising 64% at nitrogen, which ought to be conjugatively deactivated (63JOC2262). From Table 8.8, the positional reactivity order ($k_{rel.}$) in indole is 3 (2620) > 1 (59) > 2 (1) >> 6 (?). Results for carbazole are described in Section 3.

The 3-position of indole is acetylated by acetic anhydride/formic acid to give 3-(1-acetyl-1-hydroxyethyl)indole (70–77%) (72CC77). Cyanoacetylation of indole occurs in 84% yield; 2-methylindole gives 81% yield but 2-phenylindole does not react, probably due to steric hindrance (80CB3675). Acetylation of 2-ethoxycarbonylindole with acid chlorides of strong acids (e.g., chloroacetic, p-nitrobenzoic) occurs at the 5-position (84H241), which follows from conjugative deactivation of the 6-position and steric hindrance at the 3-position. The ratio of N to C-3 α-bromopropionylation [by N-(2-haloacyl)pyridinium salts] appears to be extraordinarily temperature sensitive, being 100 (20°C), 1.25 (40°C), and 0.12 (60°C) (73T971), probably as a result of N to C-3 rearrangement. α-Bromobutyrylation at 60°C gave only a minor amount of the 3-derivative, but logically this does not seem to be a steric effect (cf. 73T971) since N and C-3 are equidistant from *peri*-hydrogens.

Benzoylation of benzo[*b*]selenophene gives mainly 2- together with a little 3-substitution, as does acetylation (72BSF3955), though another report gives the 3 : 2 ratio as 1.5 in acetylation (72CHE13).

vi. *Other Substitution Reactions.*

a. Metallation. Lithiation of benzo[*b*]furan, benzo[*b*]thiophene, benzo[*b*]selenophene, and indole occurs at the 2-positions [72CHE13; 77JCS(P1)887], the carbanions generated being stabilized by inductive electron withdrawal by the heteroatom. This parallels base-catalyzed hydrogen exchange in view of the similarity of the mechanism. For benzo[*b*]thiophene and benzo[*b*]furan, the next most reactive site is the 7-position and again this follows from the inductive effect of the heteroatom. If an electron-withdrawing group is at the the 2-position of benzo[*b*]furan then lithiation occurs in the 3-position, as expected, and vice versa [84JCS(P1)2839]. Yields of up to 85% of 2-lithiobenzo[*b*]thiophene have been reported (73CHE953), and with *n*-BuLi in *N,N, N', N'*-tetramethylenediamine (TMEDA), lithiation gives 55% 2-substitution and 32% of 2,7-disubstitution [77JCS(PI)887].

Mercuriation of indole by mercuric acetate in acetic anhydride is accompanied by N-acetylation, giving 61% of 1-acetyl-3-acetoxymercuriindole. The 2- and 3-methyl derivatives of indole and of benzo[*b*]selenophene are substituted in the 3- and 2-positions, respectively [70CHE254; 71CHE1401; 72CHE18].

b. Diazonium coupling. Reaction between *p*-nitrophenyldiazonium chloride and indole gives 3-*p*-nitrophenylazoindole. Between pH 4 and 6 at 0°C, the reaction is first-order in both reagents, shows no kinetic isotope effect, and has a rate independent of pH. The rate-determining step is thus attack of the ion upon the neutral indole molecule (57JCS2398).

c. Thiocyanation and selenacyanation. Reaction of thiocyanogen with indole gave an 8% yield of 3-thiocyanoindole. Surprisingly, indole-2-carboxylic acid also gave the 3-thiocyano derivative in 92% yield (60JA2742). The reaction of selenocyanogen with indole gave 70% yield of 3-selenocyanoindole (63ACS268).

d. Sulfonation. Sulfonation of benzo[*b*]furan occurs at the 3-position (53CA10519) [and this is probably true also for benzo[b]thiophene (61M677)], whereas chlorosulfonation of 3-methyl- and 3,5- and 3,7-dimethylbenzo[*b*]thiophene takes place at the 2- and 6-positions [74CA(81)105154; 76CA(84)89931], which follows from a combination of intrinsic reactivity, steric hindrance, and bond-fixation effects. An early report of the sulfonation of indole (49JGU763) is largely incorrect. Scheme 8.6 shows product orientations from more recent determinations (51JGU1415; 73T669) and these may each be rationalized in terms of the usual substituent effects.

e. Phosphonylation. Reaction of indole with phosphoramidates gives mainly N-substitution [e.g., P(NEt$_2$)$_3$ gave 85% of (**8.57**)], but with phosphonoamidates, mainly 3-substitution [e.g., MeP(O)(NBu$_2$)$_2$ gave 79% of **8.58**] (80JGU618). The latter reagent is more electrophilic than the

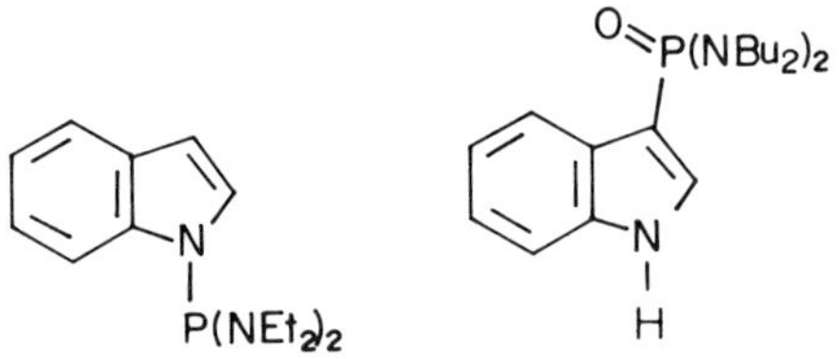

former, and this may account for the difference in positional selectivity. Another possible explanation is that the N-substituted compound in the latter case simply rearranges to the 3-position more quickly.

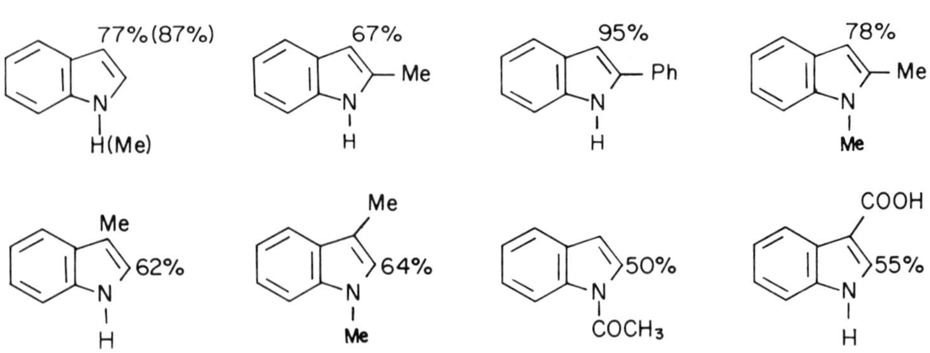

SCHEME 8.6. Isomer yields in sulfonation of substituted indoles.

f. Acyldesilylation. Acetylation of 4-trimethylsilylindole (**8.59**) went in the 3-position, whereas the *N*-acetyl derivative underwent acetyldesilylation at the 4-position. By contrast, the reaction of the N-acetyl compound with chloroacetyl chloride gave 37% of 6-chloroacetylation (together with 46% of protiodesilylation). The lack of acyldesilylation here was thought to be a steric effect (84JOC4409), but more probably arises from the greater reactivity of the electrophile in chloroacetylation. The balance between *ipso* substitution and deprotonation elsewhere in the ring has been found to vary according to the reactivity of the electrophile in other reactions (see, e.g., 76MI2).

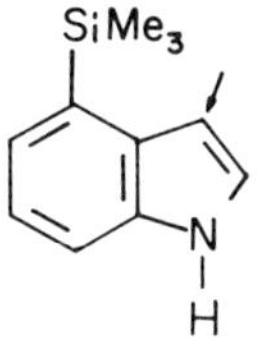

(8·59)

g. Protiodemetallation. Protiodesilylation of 2- and 3-trimethylsilylbenzo[*b*]thiophene gave partial rate factors of 39.6 and 40.7, leading to σ^+ values of -0.33 and -0.34, respectively (61JCS4921). This reaction shows greater similarity in the reactivity of the 2- and 3-positions than does any other (a possible reason is given in Section 2.A.c).

A comprehensive kinetic study has been made of the rates of base-catalyzed cleavage of compounds $ArXR_3$ (X = Si, Ge, Sn; Ar = 2-furyl, 2-thienyl, 2-benzo[*b*]furyl, and 2-benzo[*b*]thienyl). Rates increased with increasing base concentration; relative rates are given in Table 8.9 [76JCS(P2)925]. Product and kinetic isotope effects were determined and analysis of these and the reactivity pattern confirmed that, for the silicon- and germanium-containing compounds, the rate-determining step is effectively formation of the free carbanion [Eq. (8.1)]; subsequent reaction of the carbanion with a proton from the solvent is fast. For the tin compounds, electrophilic attack by the incoming proton has made significant progress in the transition state [Eq. (8.2)] [72MI2(348)].

$$ArXR_3 + OMe^- \rightleftharpoons Ar^- + XR_3OMe \quad (8.1)$$

$$ArSnR_3 + OMe^- + MeOH \rightleftharpoons Ar\cdots^-SnR_3OMe \quad (8.2)$$
$$ H\text{- -}OMe$$

TABLE 8.9

RATES OF CLEAVAGE[a] OF ArXR$_3$ COMPOUNDS RELATIVE TO THOSE OF PhCH$_2$XMe$_3$

Ar	R	X	
		Sn	Si
PhCH$_2$	Me	1	1
2-Thienyl	Me	43,000	43
	Et	1,360	0.64
	i-Pr	2.6	—
2-Furyl	Me	68,000	15.8
	Et	2,100	—
2-Benzo[b]thienyl	Me	90,000	690[b]
	Et	3,580	5.5
	i-Pr	11.3	—
2-Benzo[b]furyl	Me	155,000	560
	Et	7,400	3.3

[a] By NaOMe in MeOH at 50°C.
[b] The value for the corresponding germanium compound was ~0.5.

The negative charge is partially delocalized onto the aromatic ring. The importance of this electrophilic attack accounts for the different reactivity order for the various heterocycles. For both mechanisms, base attack is sterically hindered and increasingly so the larger is R, as the results show. The variation in rate with the size of R is smaller for the tin compounds as tin is larger to begin with.

vii. *Side-Chain Reactions.* Studies of side-chain reactivity have given a precise indication of the total positional reactivity order in benzo[b]thiophene, showing that the positional reactivities are very reaction dependent, considerably more so than is the case for the monocyclic five-membered heterocycles. Three reactions have been studied: solvolysis of 1-arylethyl acetates (69JA7381) and 1-arylethyl chlorides in solution ($\rho = -4.8$ at 25°C) (74JOC2828), and pyrolysis of 1-arylethyl acetates in the gas phase ($\rho = -0.66$ at 600 K) [78JCS(P2)1053]. The two last reactions (for which all positions were measured) gave the same positional reactivity order: $3 > 2 > 6 > 5 > 4 > 7$ (as does acid-catalyzed hydrogen exchange; Section 2.A.b). The σ^+ values differ considerably, however, being, respectively, $-0.56, -0.49, -0.42, -0.34, -0.25$, and -0.11 in solvolysis, and $-0.54, -0.39, -0.32, -0.29, -0.155$, and $+0.10$ in pyrolysis. The difference is due to the differing polarizabilities of each position and is discussed fully in Section 2.A.c. The reactivity of the 2- and

3-position in solvolysis of 1-arylethyl acetates yields corresponding σ^+ values of -0.43 and -0.52, which are significantly different from those obtained in solvolysis of the chlorides, for the reason mentioned above.

Two of these reactions have been used to measure the reactivity of the 2- and 3-positions in benzo[b]furan, the corresponding σ^+ values being -0.47 and -0.46 (from solvolysis of 1-arylethyl acetates, 69JA7381), and -0.225 and -0.495 [from pyrolysis of 1-arylethyl acetates, 78JCS(P2)1053]. Solvolysis of 1-arylethyl acetates has been used to determine the relative reactivity of the 2-positions of benzo[b]selenophene and benzo[b]tellurophene compared to the corresponding position in benzo[b]furan and benzo-[b]thiophene (77G339), the $k_{rel.}$ values being 1.7, 5.2, 1.5, and 1, respectively. From these data the corresponding σ^+ values for the 2-positions of benzo-[b]sellenophene and benzo[b]tellurophene may be calculated as -0.475 and -0.58, respectively. All of these results are discussed in Section 2.A.c.

c. *Quantitative Aspects of the Reactivity Data*

The rate data collected in the foregoing sections are gathered, in terms of σ^+ values, in Table 8.10. The main features of these results are summarized in the following subsections.

i. *Benzo[b]furan.* Although the σ^+ values for the 3-position are essentially constant, those for the 2-position show a wide variation, with the reactivity increasing as the demand for resonance stabilization of the reaction transition state increases. In the reaction with the lowest demand, pyrolysis of 1-arylethyl acetates, the $-I$ effect of oxygen at the 2-position is only partially outweighed by the $+M$ effect, resulting in the 2-position being less reactive than the 3-position. By contrast, in acetylation, in which the demand is high, the 2-position is more reactive than the 3-position, and in the reaction of intermediate demand, solvolysis of 1-arylethyl acetates, both positions are about equally reactive. This variation in reactivity with nature of the transition state is predicted by calculations [78JCS(P2)1053].

ii. *Benzo[b]thiophene.* Here again the σ^+ values are not constant, and generally increase with increasing demand for resonance stabilization of the transition state. Similarly, the effect is greatest at the 2-position, but since the $-I$ effect of sulfur is less than that of oxygen, the difference between the reactivities of the 2- and 3-positions is never so great as with benzo[b]furan. Under no conditions does the 2-position become more reactive than the 3-position. This may stem from a lower $+M$

TABLE 8.10
σ^+ Values from Various Reactions[a]

Reaction	Benzo[b]furan		Benzo[b]thiophene						Benzo[b]-selenophene	Benzo[b]-tellurophene
	2	3	2	3	4	5	6	7	2	2
Pyrolysis, 1-arylethyl acetates	−0.225	−0.495	−0.39	−0.54	−0.155	−0.29	−0.32	+0.10		
Solvolysis, 1-arylethyl acetates	−0.47	−0.46	−0.43	−0.52					−0.475	−0.58
Solvolysis, 1-arylethyl chlorides			−0.49	−0.56	−0.25	−0.34	−0.42	−0.11		
Protiodetritiation[b]			−0.68	−0.695	−0.29	−0.365	−0.48	−0.16		
Protiodesilylation			−0.33	−0.34						
Positive bromination			−0.39	−0.69						
Acetylation	−0.575	−0.48	−0.49	−0.58						
Molecular chlorination			−0.61	−0.77						
Molecular bromination			−0.61	−0.77						

[a] For references, see the appropriate preceding section.
[b] For hydrogen exchange of indole, σ^+ for the 3-position is calculated to be −1.55 (Section 2.B.b).

effect of sulfur compared to oxygen but may also be partially a steric effect, as described below. The increase in σ^+ values with demand for resonance stabilization of the transition state is also shown very nicely by the data for each position in benzo[*b*]thiophene, which show a *regular* increase in value throughout. It is probably significant that the positions most affected by the $-I$ effect of sulfur, namely the 2- and 7-positions, show the greatest variation in σ^+ value—indeed the 7-position is actually deactivated in the pyrolysis. Despite the variations in σ^+, it is noteworthy that in all three reactions in which all the positional reactivities have been determined, the positional order is the same: $3 > 2 > 6 > 5 > 4 > 7$.

Given the clear indication of the variation in σ^+ with transition-state structure, some anomalies are evident. First, in reactions with high demand for resonance stabilization of the transition state (viz. acetylation and molecular halogenation), one could reasonably expect the 2-position to be *more* reactive than the 3-position. Similarly, in a reaction with a low demand for resonance stabilization (e.g., protiodesilyation), the 2-position ought to be much *less* reactive than the 3-position. Taylor *et al.* have suggested that the discrepancy arises through steric hindrance at the 2-position arising from the sulfur d orbitals [82JCS(P2)1489]. This would reduce 2-substitution for acetylation and molecular halogenation, reactions which are sterically hindered, but would enhance 2-substitution for protiodesilylation, since it is sterically accelerated. Confirmation that this is the likely course of the discrepancy comes from benzoylation of benzo-[*b*]thiophene. Benzoylation is well known to be less sterically hindered than acetylation, and gives twice as much 2-substitution relative to 3-substitution than does acetylation under the same conditions (Table 8.7) [82JCS(P2)1489].

It should be stressed that there is nothing exceptional about the benzo[*b*] compounds in not providing constant σ^+ values. Proper analysis of the literature data for practically every aromatic shows the σ^+ values to be variable. For some molecules that are not very polarizable, the variation in σ^+ is quite small, leading to the erroneous conclusion of their constancy. The presently described molecules are particularly polarizable and show that satisfactory treatment of reactivity data can generally be achieved only through the use of a multiparameter equation, notably the Yukawa–Tsuno equation (59BCJ971).

iii. *Benzo[b]selenophene and Benzo[b]tellurophene.* The reactivity of the 2-positions of these molecules and the lower heterologues follows the order (for the heteroatom) Te > Se > O > S. This is similar to that found for the five-membered heterocycles (Chapter 6); clearly the order may vary according to the nature of the reaction.

B. Molecules Containing One Heteroatom: Benzo[c]selenophene, Isoindole, and Indolizine

The benzo[c] derivatives (**8.2**) of the five-membered heterocycles do not contain a true Kekulé benzenoid ring and so tend to undergo other reactions under the conditions for electrophilic substitution. Like indolizine, however, they are very reactive toward electrophilic substitution for a reason that does not appear to have been noted previously, namely that in passing to the transition state, a true benzenoid ring is generated (e.g., **8.60** and **8.61** for benzo[c]selenophene and indolizine, respectively) [87JCS(P2)591].

(8·60)   (8·61)   (8·62)

a. Reactions

Because the nitrogen in indolizine (pyroccoline, pyrrolo[1,2-a]pyridine) is sp^2 hybridized, it cannot be protonated: Such protonation would make the nitrogen sp^3 hybridized and the molecule nonplanar and nonconjugated. By contrast, protonation can readily occur at the 1- and 3-positions (giving, e.g., **8.61**, E = H); 3-protonation is favored thermodynamically (maximum delocalization of the nitrogen lone pair). Electrophilic substitution usually takes place at the 1-position, possibly because the 3-position is subject to $-I$ inductive electron withdrawal by the adjacent nitrogen. Acid-catalyzed hydrogen exchange (D$_2$O, 200°C) gave the rate coefficients in Table 8.11; the positional reactivity order, 1, 3 >> 2 > 7 > 5 > 6 > 8 is close to that, 3 > 1 >> 2 > 5 [or 5 > 2] ≈ 7 > 6 > 8 predicted by π-electron densities (47TFS87; 71T851). In contrast, local-

TABLE 8.11
Rate Coefficients for Deuteration of Indolizine[a]

Position	1,3	2	5	6	7	8
$10^7 k$ (sec^{-1})	>500,000	1300	4.6	0.67	6	<0.1

[a] In D$_2$SO$_4$ at 200°C.

ization energies predict the order as 1 > 3 (47TFS87) (which is incorrect, see Table 8.12); this failure may stem from the very high reactivity of indolizine, such that the reaction transition state will be far from the Wheland intermediate. Partial rate factors for exchange under standard conditions (TFA, 70°C) and the positional reactivity orders are discussed further in Section 2.B.b.

The effect of the high reactivity of indolizine also shows up in substituent effects that are very small, as they should be according to the reactivity–selectivity principle. The exchange-rate coefficients for deuteration in D_2O/dioxan at 50°C are given in Table 8.12 (71T4171), and from these data the methyl substituent effects may be calculated. Notable features are [87JCS(P2)591] summarized below.

(1) Comparison with the data for exchange of indole under the same conditions (Table 8.2) shows the 3-position of indolizine to be 1800 times more reactive than the 3-position of indole. This is a measure of the effect of forming the benzenoid transition state for indolizine exchange.

(2) Substituent effects are very small (~10 times smaller than in other aromatics) [e.g., naphthalene [68JCS(B)1112] or thiophene (Chapter 6, Section 1.A)].

(3) The 1,3-interactions (meta) are similar in each direction ($f^{Me} \simeq 3.9$) and are smaller than the 2,1- or 2,3-interactions (ortho), as they should be.

(4) Because of bond fixation, the 2,3-interaction ($f^{Me} = 8.9$) is smaller than the 2,1-interaction ($f^{Me} = 7.9$).

(5) The combined 1,2-Me_2 and 2,3-Me_2 substituent effects are approximately additive.

TABLE 8.12

RATE COEFFICIENTS AND METHYL SUBSTITUENT EFFECTS FOR DEUTERATION OF INDOLIZINE[a]

Substituent	Position	$10^4 k$ (sec^{-1})	f^{Me}
H	1	0.7	—
H	3	3.6	—
1-Me	3	14	3.9
2-Me	1	5.5	7.9
2-Me	3	32	8.9
3-Me	1	2.7	3.85
1,2-Me_2	3	200	55
2,3-Me_2	1	28	40

[a] In D_2O/dioxan at 50°C.

Nitration of indolizine and 2-methylindolizine in weak acid occurs at the 3-positions, but under more strongly acidic conditions reaction occurs on the 3-protonated conjugate acid in the 1-position [72JCS(P1)2954]. Nitration of 1-methyl-2-phenylindolizine in aqueous sulfuric acid at 25°C involves the conjugate acid, and thus takes place at the 4'-position of the phenyl ring; because the "substituent" on the phenyl ring is a cation, the ring is deactivated giving $f_4 = 0.10$ [79JCS(P2)312]. The high reactivity of the 3-position is such that if the 1- and 2-positions are blocked by activating groups, then 3-substituents such as alkyl, acyl, and diazonium groups may be replaced by nitronium ion. The same will also occur at the similarly very reactive 1-position if the 2- and 3-positions are blocked (84JCS(P2)165).

Indolizine is formylated almost exclusively at the 1-position (75JHC379), and with ethyl benzoylacetate ($PhCOCH_2CO_2Et$), 2-methylindolizine gives a 70% yield of the 3-ketone (62JGU1515). The higher reactivity of the 3-position is here aided by the greater activation by 2-methyl across the 2,3- compared to the 2,1-bond.

2-Methyl- and 2-phenylindolizine are each sulfenylated by CF_3SCl at both the 1- and the 3-positions giving the SCF_3 derivatives. With an acetyl substituent at the 3-position, substitution still takes place at the 1-position (81JFC67), so acetyl does not deactivate appreciably, which is consistent with the small substituent effects in this very reactive system.

The 1- and 3-positions in benzo[c]selenophene are, as evident from **8.60**, identical and very reactive. Hence reaction with mercuric chloride gives the 1,3-bis (chloromercuri) derivative [79CA(91)73755].

Under the conditions given in Table 8.12, N-methylisoindole (**8.62**) exchanges at the 1-position with $10^4 k = 24$ sec^{-1} (71T4171) and so is significantly more reactive than indolizine. The significance of these results is discussed in Section 2.B.b.

b. *Quantitative Aspects of the Reactivity Data*

From the rate of detritiation of the 1-position of azulene in acetic acid and trifluoroacetic acid at 70°C, and the relative rates of exchange of azulene, indolizine, N-methylisoindole, and indole under the conditions given in Tables 8.2 and 8.12, partial rate factors and σ^+ values have been determined, as given in Table 8.13 [87JCS(P2)591].

The most noticeable feature of these results is the very high reactivity of the compounds. Indeed, N-methylisoindole is the most reactive aromatic yet known. The high reactivity derives from the fact that, on going to the transition state, aromaticity is *created* whereas for a normal elec-

TABLE 8.13
PARTIAL RATE FACTORS AND σ^+ VALUES FOR DETRITIATION[a]

Compound	Position	f	σ^+
Azulene	1	9.93×10^{13}	-1.60
Indolizine	1	1.93×10^{16}	-1.86
	3	9.93×10^{16}	-2.01
	2	$<5 \times 10^{13}$	<-1.57
	5,6,7,8	$<2.3 \times 10^{11}$	<-1.30
N-Methylisoindole	1	6.62×10^{17}	-2.04
Indole	3	5.52×10^{13}	-1.57

[a] In trifluoroacetic acid at 70°C [87JCS(P2)591].

trophilic substitution it is destroyed [87JCS(P2)591]. Moreover the greater reactivity of N-methylisoindole compared to indolizine probably occurs because the transition state for the exchange of the former is benzenoid (**8.63**) and therefore more stable than that for the latter, which is pyridinoid (**8.61**).

Thus, valence bond theory once again provides an easily visualized explanation of the aromatic reactivities (cf. Sections 2.A.c and 3.A.c). It also provides a clear explanation of the positional reactivity order in indolizine: $3 > 1 \gg 2 > 7 > 5 > 6 > 8$ (Tables 8.11 and 8.12). The 1- and 3-positions are much more reactive than the others because reaction at these sites produces a pyridinoid transition state; the 3-position is the most reactive because it involves maximum delocalization of electrons. The 2-position is the next most reactive because of secondary relay of electron release from the 1- and 3-positions. In the six-membered ring, the 6- and 8-positions are conjugated with nitrogen, but this produces an antiaromatic five-membered ring (e.g., **8.64**). Consequently, these are the *least* reactive sites in the six-membered ring; the 8-position is here *less* reactive than the 6-position because maximum conjugation is involved for reaction at the former. Secondary relay effects in this ring make the 5- and 7-positions relatively unreactive, but not so much as the 6- and 8-positions [87JCS(P2)591]. Similar application of valence bond theory accounts for the positional reactivities in cycl[3,2,2]azine, and also its reactivity relative to indolizine (Section 4.A)

Pyrrolo[1,2-*a*]quinoline is a benzo derivative of indolizine and its 2,7-dimethyl derivative (**8.65**) is nitrosated, acetylated, diazo coupled, and formylated in the expected 1-position in 90, 40, 90, and 70% yields, respectively (79JHC393). Nitration also goes in the 1-position if the conditions are not strongly acidic, in which case it goes in the 6-position. This is undoubtedly due to N-protonation, and it should be noted that this is easier than for indolizine because the loss of aromaticity is less. Bromination gave addition across the high-order 4,5-bond (cf. halogenation of **8.142**).

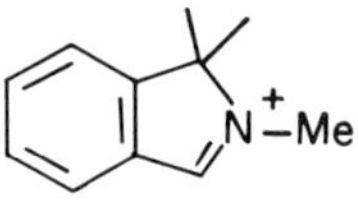

(8·63) 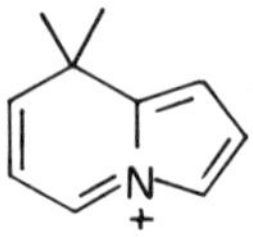   (8·64) 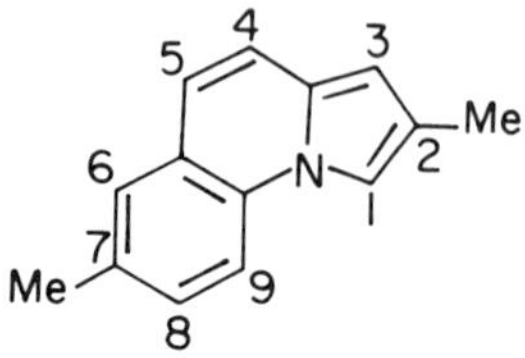   (8·65)

C. MOLECULES WITH MORE THAN ONE HETEROATOM IN THE FIVE-MEMBERED RING

Molecules described in this section are compounds **8.66–8.70**.

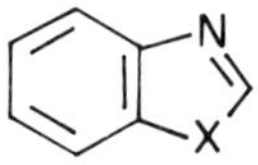

(8·66)   (8·67)   (8·68)

X = NH, Benzimidazole
    O, Benzoxazole
    S, Benzothiazole
    Se, Benzoselenazole

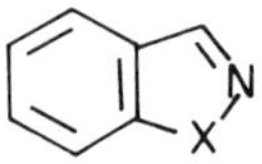

X = NH, 1H-Indazole
    O, 1,2-Benzisoxazole
    S, 1,2-Benzisothiazole
    Se, 1,2-Benzisoselenazole

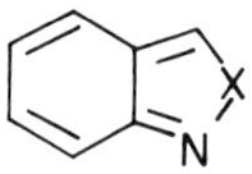

X = NH, 2H-Indazole
    O, 2,1-Benzisoxazole
    S, 2,1-Benzisothiazole
    Se, 2,1-Benzisoselenazole

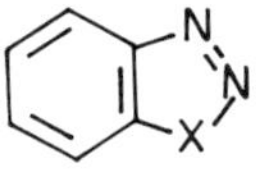

(8·69)   (8·70)

X = NH, 1H-Benzotriazole
    O, 1,2,3-Benzoxadiazole*
    S, 1,2,3-Benzothiadiazole
    Se, 1,2,3-Benzoselenadiazole

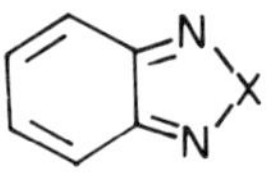

X = NH, 2H-Benzotriazole
    O, 2,1,3-Benzoxadiazole*
    S, 2,1,3-Benzothiadiazole
    Se, 2,1,3-Benzoselenadiazole

*These molecules are also known as benzofurazanes.

a. *Positional Reactivity Order*

The positional reactivities have been calculated for some of those molecules by MO methods, though the most reactive sites can be more easily ascertained by a few moments work with a pencil. Thus, for **8.66–8.68**, conjugative release by the heteroatom X and withdrawal by the nitrogen atom, respectively, distribute charges as shown in **8.71–8.73**. Electrophilic substitution should therefore take place at the 5- and 7-positions in each molecule, with a possibility of some 3-substitution in **8.72** and **8.73**; some 4- and 6-substitution could be observed under favorable conditions in **8.71**, in which conjugative electron withdrawal from the benzenoid ring by nitrogen is not possible. Compounds **8.73** will be the most reactive because there is no loss of benzenoid resonance in the transition state; indeed, it is *gained* on 3-substitution (cf. bromination of 2*H*-indazole, below). Likewise, **8.69** and **8.70** give **8.74** and **8.75**. In **8.75**, the $-I$ effect of the nitrogens could be expected to lower somewhat the reactivities of the 4- and 7-positions relative to the 5- and 6-positions, but this is likely to be outweighed by the greater interruption of bond fixation required to place a positive charge on the 4(7)-positions compared to the 5(6)-positions. Overall, compounds **8.75** will be much more reactive than **8.74**, since aromaticity is not lost in the transition state. The reactivity order should therefore be 5,7 (**8.73**) > 5,7 (**8.71**) > 5,7 (**8.72**) > 5,7 (**8.74**) ⩾ 4,7 (**8.75**) > 5,6 (**8.75**) > 4,6 (**8.74**), and the reactivities of compounds **8.71–8.73** should be close to that of benzene because the effects of the heteroatoms approximately cancel.

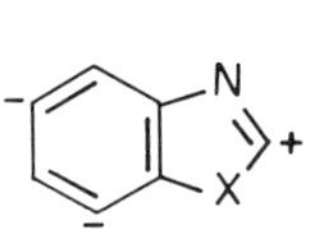

(8·71) 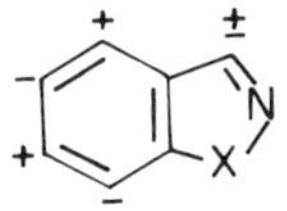   (8·72) 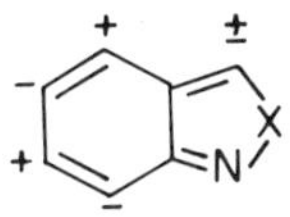   (8·73) 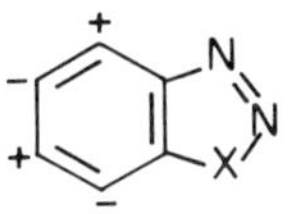   (8·74) 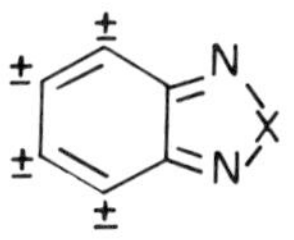   (8·75)

π Densities have been calculated for both the benzisoxazoles (**8.76**, **8.77**) [60T(10)81] and for benzoxazole (**8.78**) (74CHE166). These will underestimate the extent of substitution at the site (5-position in each case) furthest from oxygen, but indicate that, in general, 5- and 7-substitution

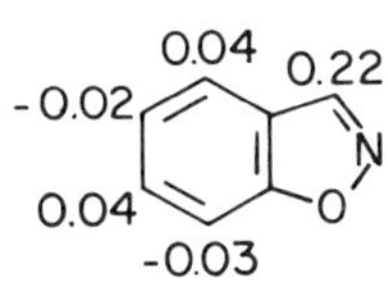

(8·76) 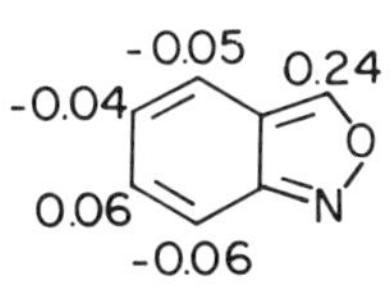   (8·77) 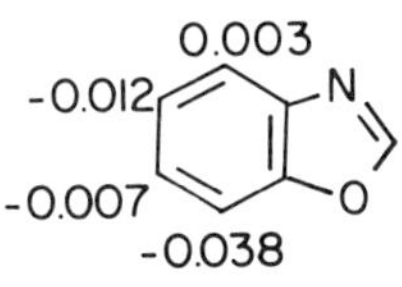   (8·78)

is expected (as indicated by **8.71–8.73**). However, for benzothiazole, π densities appear to indicate that the 7-position will be deactivated while the 5-position will be activated [67TCA(9)181]. Similar anomalies are encountered with benzimidazole, two sets of calculations which use the same method (SCF LCAO–MO) having produced entirely different results. One set (71CHE1443) gives the positional reactivity order as 7 > 5 > 4 > 6 (all activated) with the 3-position deactivated, whereas the other gives 5 > 6 > 7 > 4 > 3 with only the 5-position activated [67TCA(9)181]. Similar lack of agreement is encountered in calculations for compounds **8.70**, π densities for one set indicating the 5(6)-position to be more reactive than the 4(7)-position (all deactivated for X = O, S, Se) the reactivity order for X being O > S > Se (73CHE1331). By contrast, another set of π densities and localization energies both show each position to be activated in 1,2,3-benzothiazole (**8.70**, X = S) but again give the 4(7)-position as the most reactive [66TCA(5)401]. Lastly, CNDO/2 calculations give the positional reactivity order as 5 > 7 > 4 > 6 in indazole and 5,7 > 6 > 4 in benzimidazole [78JCS(P2)865].

b. *Reactions*

i. *Acid-Catalyzed Hydrogen Exchange.* As yet there have been no studies of this reaction for these compounds.

ii. *Base-Catalyzed Hydrogen Exchange.* Detritiation of benzimidazole and its 1-Me and 1-Et derivatives at various pH values showed the exchange rate to be independent of pH at pH 5–11, but to decrease rapidly at higher or lower values. This seems to be consistent only with rate-determining attack of hydroxide ion upon the protonated substrate, giving an ylide intermediate [73JCS(P2)432]. The effects of substituents (Table 8.14) appear then to derive from alteration of the concentration of the protonated species. The alternative A-S_E2 mechanism (which would give similar substituent effects) is ruled out by the decreased rate at low pH, and also because benzimidazole is too unreactive to undergo acid-catalyzed exchange under these conditions. From the data in Table 8.14, one may calculate the activating effects of a 4- and a

TABLE 8.14
RATE COEFFICIENTS FOR DETRITIATION OF BENZIMIDAZOLES[a]

Substituent	H	1-M	1-Et	1-*i*-Pr	5,6-Cl$_2$	4,5,6-Cl$_3$	4,5,6,7-Cl$_4$
$10^5 k$ (sec^{-1})	78.7	246	215	163	19.1	7.5	2.93

[a] In neutral solution at 85°C.

5-chloro substituent as 2.55 and 2.02, respectively, which accords with inductive effects being predominant in the base-catalyzed mechanism. This mechanism is also consistent with the decreasing rates with increasing size of the 1-alkyl groups, the reaction being well established to be slightly sterically hindered [72MI2(266)]. At 50°C, log k for deuteriation of the 2-position of 1-methylbenzimidazole is -4.1 (77CHE1235).

Benzothiazole is less reactive toward exchange than benzoxazole (64TL845). This has been more recently confirmed, the relative reactivities at the 2-positions of benzoazoles (**8.66**) being for X = S (1.0), Se (4), O (20), with benzimidazole (X = NH) being more reactive than any of these [68CA(69)85848; 77G359]. Reactivity of the sulfur-containing compound is considerably lower relative to the O and Se analogues than is the case for benzo[b]thiophene, etc; the reason for this is not yet understood [79PS(5)305]. Thus, the activating effect of nitrogen on going from **8.1** to **8.66** for different heteroatoms X is 520 (S), 14,850 (Se), and 41,400 (O).

The effects of substituents in the benzenoid ring on the rate of exchange at the 2-position of benzothiazole have also been measured (Table 8.15) (76JHC1021). These parallel those for benzimidazole (Table 8.14) and notable is the small difference between the effects at the 5- and the 6-positions. This may be attributed to the fact that only inductive effects are important in base-catalyzed exchange, and they may be transmitted through either heteroatom with comparable facility.

iii. *Nitration*. Nitration of benzothiazole occurs mainly in the 6-position (52JPJ1263) and the most comprehensive investigation gives the positional yields as 20% (4), 9% (5), 50% (6), and 20% (7) (61JCS2825); the yield of the 6-isomer can be as high as 63% using metal nitrates in trifluoroacetic anhydride (81JOC3056). 6-Substitution is also predominant in nitration of 2-substituted derivatives, the yields being 75% (2-Cl) and 72% (2-OMe) (this latter is accompanied by 10% of 2-hydroxy-6-nitrobenzothiazole); however 2-nitrobenzothiazole did not undergo nitration (47CA754b; 52JPJ1263). No combination of normal electronic effects can account for the 6-substitution, and it may be that the mechanism is not straightforward. For example, nitration of benzoxazole, which gives

TABLE 8.15
RATE COEFFICIENTS FOR DEUTERIATION OF BENZOTHIAZOLES[a]

Substituent	H	5-Me	6-Me	5-Cl	6-Cl	5-NO$_2$	6-NO$_2$	6-NH$_2$
$10^4 k$ (sec^{-1})	1.21	0.70	0.65	5.00	6.19	66.3	64.3	0.177

[a] With NaOD at 25°C.

6- and 5-nitro derivatives in the ratio 3 : 1, has been shown to take place via the intermediates shown in Scheme 8.7 (79CJC937), and thus 6-substitution results from the high para-directing effect of the NHCHO substituent.

Nitration of 6-methoxy- and 6-chlorobenzotriazole takes place mainly in the 7-position (together with a small amount of 5-substitution) (55CA8257h), again demonstrating bond-fixation effects (i.e., the 6,7-bond is of higher order than the 5,6-bond).

Benzimidazole is nitrated at the expected 5/6-position (through tautomerism the 5- and 6-positions are equivalent here) (48RTC45), a 90% yield being obtained from the 2-methyl derivative (52JGU1069); the 2-CF$_3$ derivative is also nitrated in the 5/6-position (73AJC2725) as are 1-alkylbenzimidazoles (66CCC113) and 2-phenylimidazole (66CCC1093). The conjugate acid is involved in the last two cases and probably in the others as well. 4-Acetylaminobenzimidazole is nitrated in the 5- and 7-positions in 14 and 34% yields, respectively (53ZOB951), and this illustrates a feature of which there are many reports (as there are in general for benzo derivatives of five-membered heterocycles) of the ortho- and para-directing effects of electron-supplying substituents at the 4- and 7-positions; further details are inappropriate here. Bond fixation (emphasized in one early paper, 52JGU1069) is clearly responsible for the 57 : 43 product ratio for the 4- and 6-isomers in nitration of 5-methylbenzimidazole (58ZOB62) and, through enhancing deactivation, also accounts for the lack of (*ortho*) 4-nitration of 5-nitrobenzimidazole, the products being 5,6-dinitro- (54%) and 5,7-dinitro- (21%) -benzimidazole (63JCS736).

1,2-Benzisoxazole is nitrated in the expected 5-position [26LA(449)63] as are the 3-methoxycarbonyl (78CPB3498) and 3-methyl derivatives. The latter is nitrated as the free base in 80–90wt% H$_2$SO$_4$, and as the conjugate acid at higher acidity [77JCS(P2)47]. The 5-methyl and 5-chloro derivatives are nitrated in the 4-position (77IJC1058,1061), which is the *expected* result arising from bond fixation [cf. 81AHC(29)10], since 5-methyl activates the 4-position most, and 5-chloro deactivates it least. 2,1-Benzisoxazole is also nitrated in the expected 5-position [65CC408; 66T(Suppl 7)49] and again bond fixation causes the 5-chloro derivative to be nitrated in the 4-position (81JHC1081) and the 6-chloro (and 6-methoxy) derivative in the 7-position (70JOC1662; 77JOC897). 5-Chloro-3-

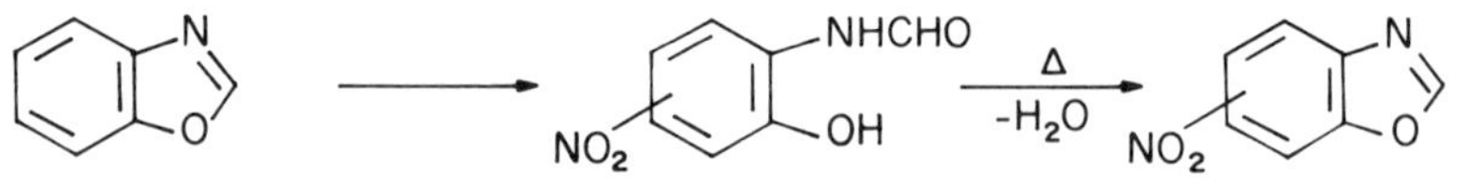

SCHEME 8.7. Mechanism of nitration of benzoxazole.

phenyl-2,1-benzisoxazole is nitrated in the 7- and para-positions (81JHC1081), because the oxygen conjugates with the para-position of the phenyl ring (via nitrogen), while the phenyl ring activates the 7-position and deactivates the 4-position.

1,2-Benzisothiazole is nitrated in the expected 5- and 7-positions [63AC(R)1860], reported yields being 44 and 39%, respectively. In this latter work, the 5-OH, 5-OMe, 5-NHAc, and 5-Br substituents gave 4-substitution in 72, 93, 60, and 81% yields, respectively [80JCR(S)197], again due to bond fixation, and the 4-Br substituent gave 5- (41%) and 7-substitution (42%), as expected. 2,1-Benzoisothiazole gives mainly 5- together with some 7- and 4-nitration [69JCS(C)2189]. 1,2-Benzoisoselenazole is nitrated in the 5- and 7-positions in a 55:45 ratio (75JHC1091).

In indazole, electron release by the NH group (to the 5-, 7-, and 3-positions) is considerably greater than that in the above compounds. Indazole is also quite basic and it is expected that nitration should involve the conjugate acid; this has been confirmed for nitration at acidity below 90 wt% H_2SO_4, which gives the 5-nitro and then the 5,7-dinitro derivatives [78JCS(P2)632]. Nitration in the less acidic nitric acid/acetic anhydride gives the 3-nitro and 3,5-dinitro compounds (71JOC3084) and may involve the free base. The high reactivity is also shown by the formation of 3,5,7-trinitroindazole from 3-nitro-, 3,5-dinitro-, and 5,7-dinitroindazole, of 3,5,6-trinitroindazole from 3,6- and 5,6-dinitroindazole, and 3,4,6-trinitroindazole from 4,6-dinitroindazole (77JOU1192); 3-chloroindazole also gives the 5-nitro derivative (78PHA419).

1,2,3-Benzothiadiazole is nitrated in the 4- and 7-positions (48JCS-1006). The former result is difficult to explain without involving **8.79** as a significant contributor to the overall resonance hybrid, and thereby deactivating the 5-position.

Nitration data for 2,1,3-compounds (**8.70**) point to them having high degrees of bond fixation. Thus, although conjugative electron withdrawal by nitrogen places a positive charge on the 5- and 7-positions (hence by symmetry upon them all), structure **8.80** involves less interruption of the ground-state structure than does **8.81** (i.e., **8.80** has a low substituent interaction factor) [79JCS(P2)381]. Structure **8.80** is also *p*-quinonoid and therefore more stable, and these explanations are of course interrelated. As a consequence of this, 2,1,3-benzofurazane (83AJC1227), 2,1,3-benzo-

(8·79)   (8·80)   (8·81)

selenadiazole, and 2,1,3-benzothiadiazole (57JOC507) are each nitrated in the 4-position, the latter in 91% yield (53ZOB1552)[1]. The bond fixation effect accounts for the fact that 5-methyl-2,1,3-benzoselenadiazole [58JOC(22)610] and 5-chloro-2,1,3-benzothiadiazole (57ZOB2599) are each nitrated in the 4-position, and may contribute to the observation (83AJC1227) that 4-nitrobenzofurazane could not be dinitrated whereas 5-nitrobenzofurazane could. In the former, bond-order effects would cause strong deactivation of the 5-position, whereas in the latter the 6-position would only be weakly deactivated for the same reason. This accounts for the fact that 4,7-dibromo-2,1,3-benzothiadiazole could be dinitrated (80CPB1909) and also shows the generally higher reactivity of the 2,1,3-compounds compared to the 1,2,3-isomers for both the S and Se series (other examples may be gleaned from the literature) because of the poorer ground-state stability of the former.

iv. *Halogenation.* A comprehensive kinetic study of the bromination of benzimidazole and indazole [78JCS(P2)865] permits calculation of the partial rate factors given in Scheme 8.8. These show that the benzo derivatives are less reactive than the parent five-membered heterocycles (Chapter 7, Section 5.A), as is the general trend. 1*H*-Indazole is more reactive than benzimidazole and this may be due to the relative distance of the aza substituent, which can only deactivate the sites inductively. The zero deactivating effect of *o*-bromine is consistent with the direction of substitution into the 4-position by 5-halo substituents, as noted above. This again probably derives from the high order of the 6,7-bond, such that the conjugative effect can outweigh its $-I$ effect; this behaviour for halogens was predicted by Norman and Taylor [65MI1(296)] and has been observed (for *p*-bromo, and -chloro) in the NMR shifts of the cations produced for substituted benzyl alcohols in superacids (81JOC1646). Substit-

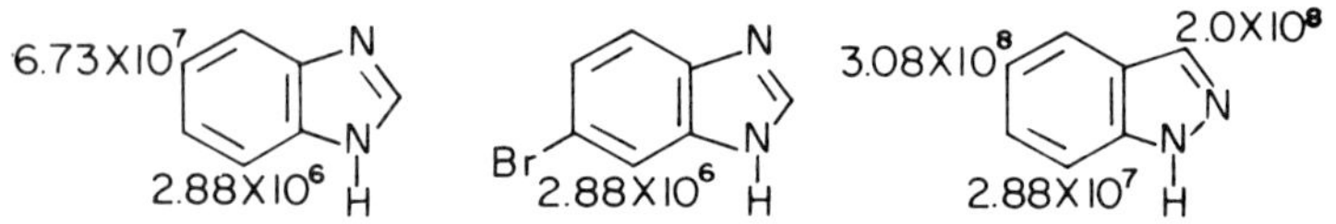

SCHEME 8.8. Partial rate factors for bromination benzimidazole, 5-bromobenimidazole, and indazole.

---

[1]The argument that 2,1,3-benzothiadiazole nitrates in the 4-position because protonation of nitrogen occurs, the resultant positive charge being cancelled by conjugative electron withdrawal (57JOC507), is misleading and involves a frequently cited error. The positive charge is created in a σ orbital and cannot therefore be delocalized by π-electron withdrawal (though, of course, the nitrogen does become more electropositive through protonation).

uents (2-CF$_3$, 2-Me) in the five-membered ring of benzimidazole do not affect the predominant 5-substitution [73AJC2725, 73CA(78)58306].

Bromination of 2-phenyl-2H-indazole occurs mainly in the 3-position (89% yield) and subsequently in the 5- and 7-positions (84JOC3401). The high reactivity of the 3-position here may be attributed to the gain of benzenoid aromaticity on forming the transition state **8.82**.

(8·82)

Bromination of 1,2-benzoisothiazole gives the 5-Br (32%), 7-Br (37%), 4,7-Br$_2$ (3%), and 5,7-Br$_2$ (2%) derivatives. Electron-donor 5-OMe, 5-OH, and 5-NH$_2$ substituents direct bromination into the 4-position in 87–95% yield, as does the 5-Br substituent (24%), which also gives the 5,7-Br$_2$ (10%) and 4,5,7-Br$_3$ (34%) derivatives [80JCR(S)197].

Bromination of 5-methyl- and 5-bromofurazane occurs in the 4-position (74JHC813), but on the whole compounds of type **8.70** tend to become disubstituted because of their high reactivity. Thus 2,1,3-benzothiadiazole is brominated in the 4,7- as well as the 4-position and chlorination of the 5-amino compound also occurred in the 4-position (63JGU223). Polyhalogenation was reported for both the benzothia- and benzoselenadiazoles (64JGU3063). Benzotriazole gave 90% of the 4,7-dichloro derivative and 10% of the 4,5-derivative (75JOU889,902).

v. *Metallation.* Benzoxazole (83JOM159), benzothiazole (85H-295), and benzimidazole (76CHE1399) are each lithiated in the expected 2-position. In magnesiation with EtMgBr, the reactivity order at the 2-positions was benzothiazole > 4- and 5-chloro-1-methylimidazole > 1-methylbenzimidazole >> 1-methylimidazole (69JGU1816).

vi. *Other Reactions.* Benzimidazoles can be acylated at the 2-position in 58–88% yield in the presence of triethylamine (77LA145). 2-Alkylbenzimidazoles are sulfonated in the 5-position (75ZPK2241) and 1H-indazole is sulfonated (with 20% oleum) in the 7-position (50BSF466, 1278). 2,1,3-Benzothiadiazole is sulfonated in the 4-position (90%) as is the 5-amino derivative (91%) (64JGU1265). The 5- and 4-methyl derivatives are sulfonated in the 4- and 7-positions, respectively [64CA(60)10670]. 2-Hydroxybenzoxazole (benzoxalone) sulfonates in the 6-position (41JA879). 2,1,3-Benzothiadiazole is chloromethylated to give the 4,7-bis(chloromethyl) derivative, and the 5- and 4-Cl compounds give the 4- and 7-derivatives, respectively (64JGU2491).

D. MOLECULES WITH HETEROATOMS IN EACH RING

Compounds described in this section include compounds of general types **8.84–8.88**, though studies have not yet been carried out on all of them. In addition, there are compounds with nitrogen at bridgeheads, with structures **A–D** being typical examples.

(8·83)

X = O, Furopyridine
S, Thienopyridine
NH, Pyrrolopyridine

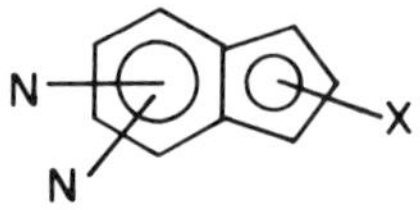

(8·84)

X = O, Furopyrazine -pyridazine, -pyrimidine
S, Thienopyrazine, -pyridazine, -pyrimidine
NH, Pyrrolopyrazine, -pyridazine, -pyrimidine

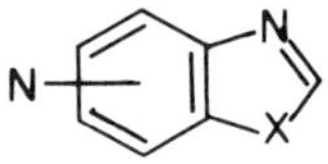

(8·85)

X = O, Oxazolopyridine
S, Thiazolopyridine
NH, Imidazolopyridine

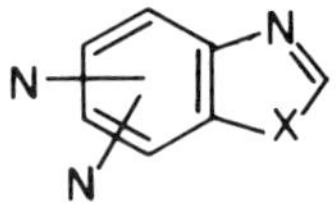

(8·86)

X = O, Oxazolopyrazine, -pyridazine, -pyrimidine
S, Thiazolopyrazine, -pyridazine, -pyrimidine
NH, Imidazolopyrazine, -pyridazine, -pyrimidine

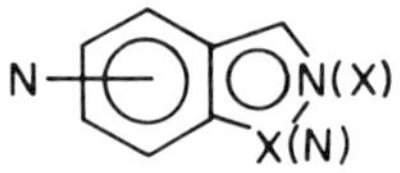

(8·87)

X = O, Isoxazolopyridine
S, Isothiazolopyridine
NH, Pyrazolopyridine

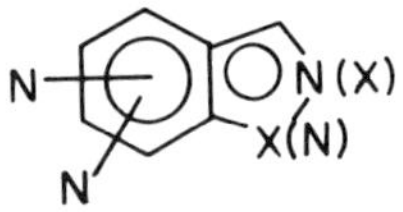

(8·88)

X = O, Isoxazolopyrazine, -pyridazine, -pyrimidine
S, Isothiazolopyrazine, -pyridazine, -pyrimidine
NH, Pyrazolopyrazine, -pyridazine, -pyrimidine

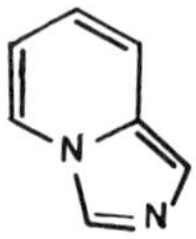

A

Imidazo[1,5-*a*]pyridine

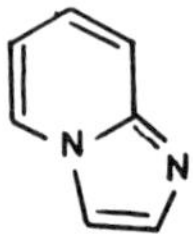

B

Imidazo[1,2-*a*]pyridine

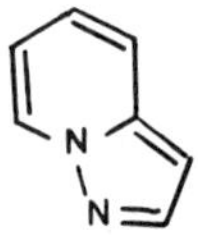

C

Pyrazolo[1,5-*a*]pyridine

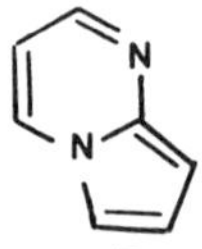

D

Pyrrolo[1,2-*a*]pyrimidine

a. Positional Reactivity Order

There have been a number of calculations of these, and the results are generally those which may be deduced from a simple combination of conjugative and bond-order effects. Data have been given in the form of electron densities (σ, π, or both) or frontier electron densities. These latter and π densities on the whole agree on the most reactive site within a molecule, but tend to disagree on the subsequent order. Frontier electron densities for thienopyridines, thienopyridazines, and thienopyrazines are given in Scheme 8.9 (72ACS2601; 76JHC581), and again the very high reactivity predicted for the 1-positions of the thieno[c] compounds, due to the formation of a 6π aromatic transition state, is notable.

This same factor accounts for the 3-substitution predicted for imidazo[1,2-a]pyridine, imidazopyrimidines, and imidazol[1,2-a]pyrazine by both frontier electron densities (Scheme 8.10) and π-electron densities (not shown) (74JHC1013; 82AJC1761). The former indicate the 5- or 7-positions to be the next most reactive, whereas the latter indicate it to be the 2-position which is more probably correct (cf. indolizine); in general, frontier electron densities are poorer indices of aromatic reactivity.

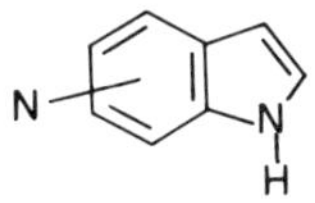

(8·89)

For imidazo[1,2-a]pyridine, two sets of data are shown (from different authors) and the values disagree both in magnitude and positional order (the same is true of the π densities); this reflects the lack of reliability of current MO methods. For pyrrolopyridines (azaindoles) (**8.89**), the 3-position is predicted to be the most reactive site in each case (83T2851).

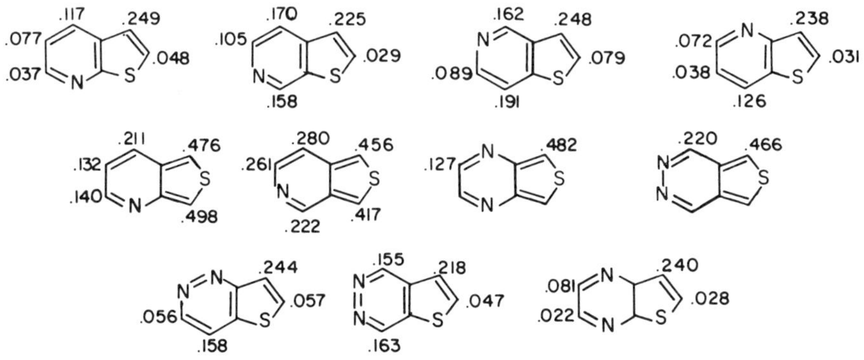

SCHEME 8.9. Frontier electron densities for thienopyridines, thienopyridazines, and thienopyrazines.

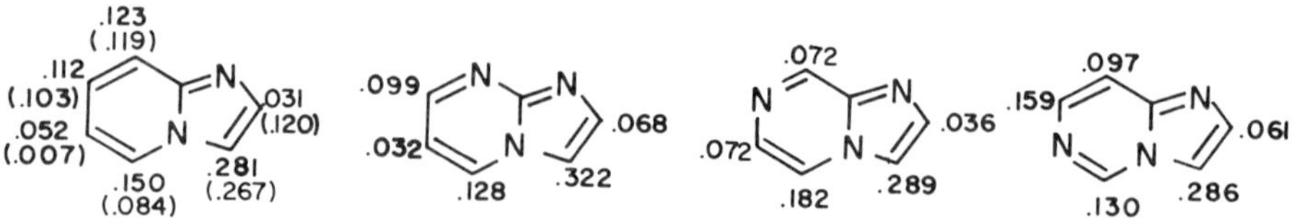

SCHEME 8.10. Frontier electron densities for imidazo[1,2-a]pyridine, imidazopyrimidines, and imidazo[1,2-a]pyrazine.

b. Reactions

i. *Acid-Catalyzed Hydrogen Exchange.* Standard deuteration rate coefficients (100°C, pH = 0) have been determined for deuteration of furopyridines, thienopyridines, pyrrolopyridines (azaindoles), selenopyridines, and borazathienopyridines in D_2SO_4 [73JCS(P2)1072; 77JHC893; 78JCS(P2)861]. Rate–acidity profiles showed that most compounds underwent exchange as the conjugate acids; the logarithms of the derived partial rate factors are given in Scheme 8.11. The 3-positions are

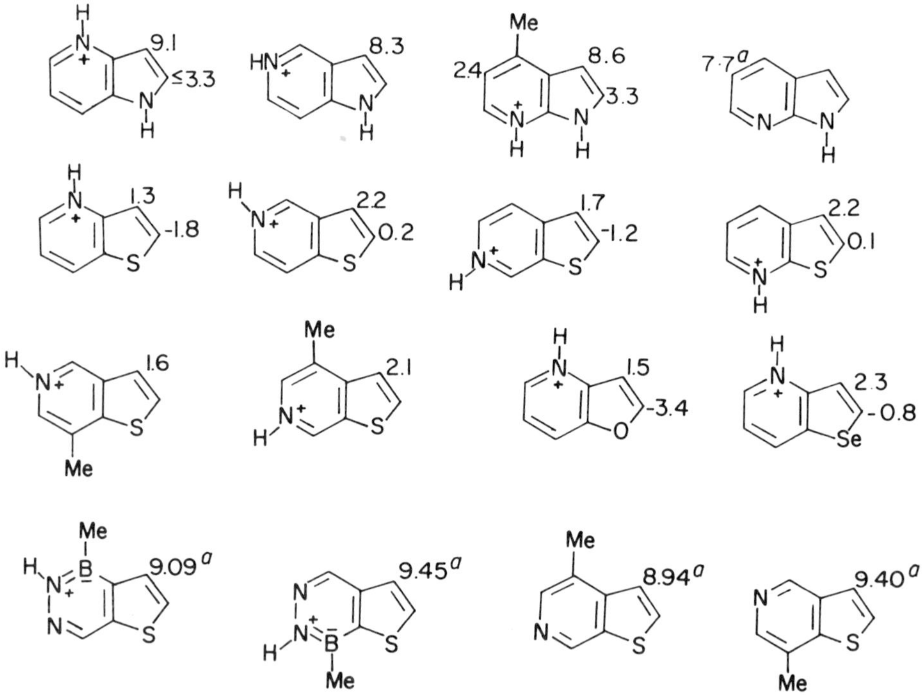

SCHEME 8.11. Logarithms of partial rate factors for deuteriation (100°C, pH = 0). [a]Data from exchange of the neutral species.

the most reactive, with reactivity decreasing according to the nature of the heteroatom in the order NH > Se > S > O. Deactivation by the positive pole is insufficient to outweigh the activation of the 3-position by the heteroatom, and this is due to poor cross-conjugation (the benzenoid character of the six-membered ring becomes disrupted), which was noted in Section 2.A. In general, positive poles at the 4- and 6-positions have a larger effect upon the reactivities of the 3-positions than those at the 2-positions, whereas positive poles at the 5- and 7-positions have the greatest effect upon the 2-positions, and this follows the expected ability to conjugatively withdraw electrons in each case. Methyl substituent effects are small for the same reason noted above. In one case, deactivation is observed, though this may be an indication of the extent of experimental error arising from the extrapolations needed to obtain these data. The free bases are ~10^7 times more reactive than the conjugate acids. The order of reactivity of the thienopyridines is incorrectly predicted by frontier electron densities (Scheme 8.9).

Data have also been obtained for deuteration of B-hydroxy derivatives of borazathienopyridines (which react as the anhydrides) (**8.90** and **8.91**) and their 5- and 7-methyl, 4- and 6-methyl derivatives, respectively (77JHC893). Rate coefficients were obtained at 57, 65.2, and 72.9°C for exchange at the 2- and 3-position; the data for the 3-position at 65.2°C and the average k_3/k_2 rate ratios at 65.2 and 72.9°C are given in Table 8.16. The 3-positions are in each case more reactive than the 2-positions, but by only a relatively small factor. Both **8.90** and **8.91** have similar reactivities at the 3-positions, but the 2-positions are more reactive in compounds **8.91** (R = H, Me). This is difficult to understand, because electron withdrawal from the 2-position in **8.91** as in **8.92** should make this position relatively *less* reactive. Methyl-substituent effects are small and rather

TABLE 8.16
RATE COEFFICIENTS AND k_3/k_2 RATE RATIOS FOR DEUTERATION OF BORAZATHIENOPYRIDINES **8.90** AND **8.91**

Compound	$10^5 k_3$ (sec^{-1}) (65.2°C)	$k_3/k_2{}^a$
4,5-Borazathieno[2,3-c]pyridine (**8.90**)	7.67	6.0
5-Methyl-4,5-borazathieno[2,3-c]pyridine	10.1	4.5
7-Methyl-4,5-borazathieno[2,3-c]pyridine	5.70	4.9
5,7-Dimethyl-4,5-borazathieno[2,3-c]pyridine	8.83	3.4
7,6-Borazathieno[3,2-c]pyridine (**8.91**)	8.83	2.3
6-Methyl-7,6-borazathieno[3,2-c]pyridine	13.0	3.7
4-Methyl-7,6-borazathieno[3,2-c]pyridine	9.83	1.8
4,6-Dimethyl-7,6-borazathieno[3,2-c]pyridine	13.8	2.7

^{a} Average of values obtained at 65.2 and 72.9°C.

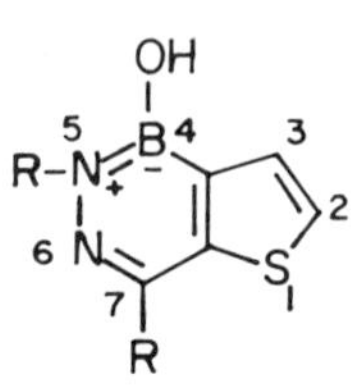

(8·90) 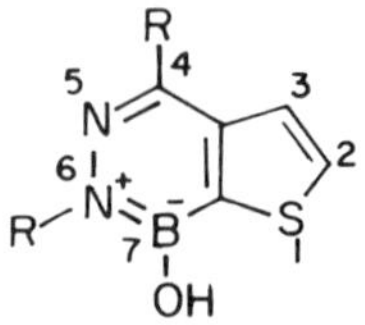   (8.91) 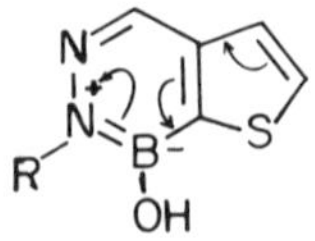   (8.92)

random, which may reflect the experimental difficulties encountered in this work (e.g., differences in the extent of equilibrium between the hydroxy compounds and the anhydrides) (77JHC893).

The azaindolizines **8.93–8.95** protonate at the non-bridgehead nitrogen in the six-membered ring. However, protonation at C-3 competes, and for pyrrolo[1,2-*b*]pyridazine (**8.96**) protonation at C-3 is stated to be exclusive even though calculations (π densities, localization energies) predict protonation at N-5 (71JOC3087; 85JOC1324). The pyridine-like nitrogen in **8.96** will be expected to be the least basic of the isomers because of the adjacent electron-withdrawing bridgehead nitrogen.

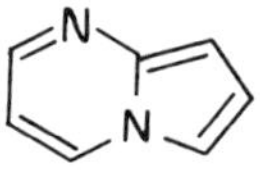

(8.93) 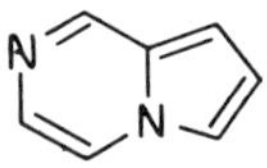   (8·94) 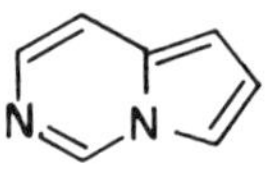   (8·95) 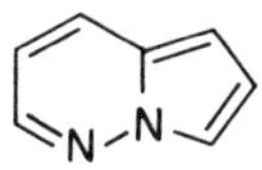   (8·96)

Half-lives for deuteriation in D_2SO_4 of imidazo[1,2-*b*]pyridazine (**8.97**) and 1,2,4-triazolo[4,3-*b*]pyridazine (**8.98**) have been reported as 24 hr and 2 min, respectively, with the 7- and 8-methyl derivatives of **8.98** also giving half-lives of ~2 min (71M837). These results could be rationalized by the assumption that **8.97** reacts as the conjugate acid, and **8.98** as the free base. Indeed, the N-methyl cations of **8.98** did not exchange, but the same was true also of the 1-methyl cation of **8.97**, so the picture is somewhat confused.

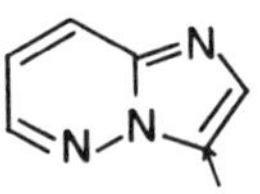

(8·97) 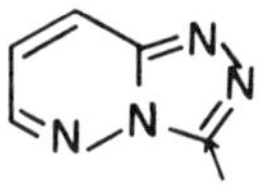   (8·98)

Half-lives have also been reported for deuteriation in 3 *M* D_2SO_4 at 100°C of imidazo[1,2-*a*]pyridine (**8.99**), imidazo[1,2-*a*]pyrimidine (**8.100**),

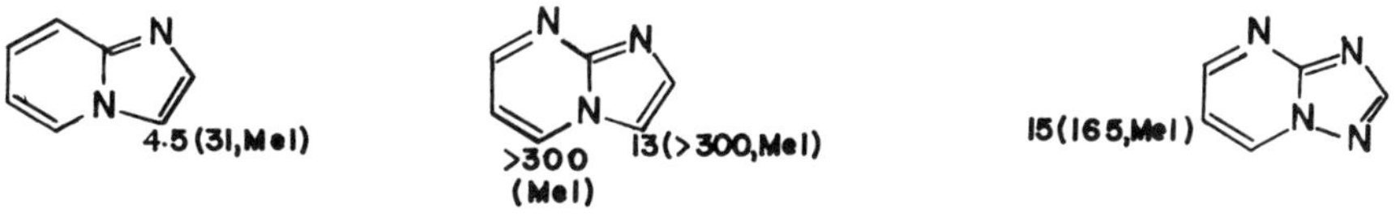

Sec. 2.D] ONE FIVE- AND ONE SIX-MEMBERED RING COMPOUNDS 233

4·5 (31,Me1)   >300 13(>300,Me1)   15(165,Me1)
                (Me1)

(8.99)   (8.100)   (8.101)

SCHEME 8.12. Half-lives (hr) for deuteriation of free bases (and methodides).

and 1,2,4-triazolo[1,5-*a*]pyrimidine (**8.101**), and some of their quaternary salts; the values (in hr) are given Scheme 8.12 (67CC377). Exchange takes place preferentially at the 3-positions of **8.99** and **8.100** and at the 5-position of **8.101**, in which the 2-position is strongly deactivated by electron withdrawal by the 3-nitrogen acting across the high-order 2,3-bond. The free bases were the dominant exchanging species as shown by the lower reactivities of the quarternary ions.

ii. *Base-Catalyzed Exchange.* Rates of detritiation of 2-tritioimidazo[5,4-*d*]pyrimidine (purine) (**8.102**), of its 1-*i*-propyl and 1-*t*-butyl derivatives, and of 2-tritioimidazo[5,4-*b*]pyridine[2] (**8.103**) have been measured between pH 2.08 and 11.5 [73JCS(P2)1889]. For **8.102** and **8.103** a bell-shaped rate–pH profile was obtained (as in the case of benzimidazole, Section 2.C.b.i) and likewise interpreted in terms of rate-determining attack of hydroxide ion upon the conjugate acid to give an ylide intermediate. At pH 6.25, the value of 10^6 k(obs) (sec^{-1}) at 85°C for **8.102**, its 1-*i*-propyl and 1-*t*-butyl derivatives, and **8.103**, were 32.0 104, 55.3, and 162, respectively. Thus, **8.103** is more reactive than **8.102**, as expected on the basis of the former giving a higher concentration of conjugate acid. The differences in the values of the alkyl derivatives reflect the usual steric hindrance to base-catalyzed exchange. At higher pH, hydroxide ion-catalyzed exchange on the neutral species became increasingly important for the alkyl derivatives of **8.102**, which then become much more reactive than the other compounds, and also the rate differences between the 1-propyl and *t*-butyl compounds increased to ~fourfold. Detritiation of adenine (**8.104**, R = H) in water at 85°C is 2000 times slower at the 6- than at the 2-position (71CC394; note that different numbering is used in this paper). At the 2-positions, purine (**8.102**, R = H) and adenine (**8.104**, R = H) exchange at similar rates, and adenosine (**8.104**, R = β-D-ribofura-

---

[2] Described incorrectly in the original paper as the [4,5-*b*] isomer.

(8·102 R=H, i-Pr, t-Bu)   (8·103)

(8·104 R=H, β-D-ribofuranosyl)

nosyl) is twice as reactive. At the 2-position purine is 25 times less reactive than benzimidazole (**8.66**, X = NH), so the regular increase in reactivity along the series **8.101** (R = H) < **8.104** < **8.66** (X = NH) indicates a common mechanism, being that noted above.

The half-lives have been determined for deuteriation (in NaOD) of imidazo[1,2-*b*]pyridazine, 1,2,4-triazolo[4,3-*b*]pyridazine, and tetrazolo-[1,5-*b*]pyridazine and some of their derivatives (Scheme 8.13) (71M837). The results are much as expected: The greater the number of nitrogens, the faster the rate; rates are larger for quarternary ions than for the corresponding neutral species; and the rate is reduced (in some cases quite markedly) by methyl substituents, presumably due to steric effects. Steric effects may also account for the *unexpected* and large deactivation by *o*-chlorine.

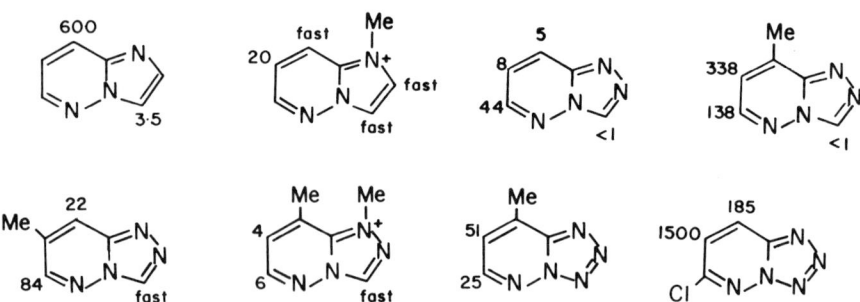

SCHEME 8.13. Half-lives (min) for base-catalyzed deuteriation at 95°C.

Sec. 2.D] ONE FIVE- AND ONE SIX-MEMBERED RING COMPOUNDS 235

Exchange in these compounds is rapid at the 3-positions because of the highly favorable resonance (**8.105**) producing a 6π-pyridazinoid ring and generating a positive charge at the bridgehead. This strongly inductively stabilizes the negative charge created in the transition state for 3-exchange. The importance of such stabilization by positive poles is also demonstrated by the rapid exchange at the 2- and 8-positions in 1-methylimidazolo[3,2-*b*]pyridazinium ion.

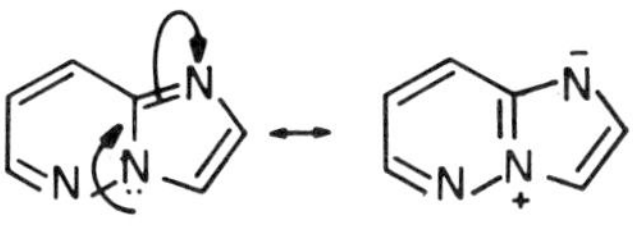

(8·105)

Similar factors can be seen to be responsible for the relative magnitudes of the second-order rate coefficients for deuteration of imidazo[1,2-*a*]pyridines and imidazo[1,2-*a*]pyrimidines at 35°C (in some cases 65°C), given in Scheme 8.14 (68JOC1087). Exchange is rapid at the 2-position when there is a positive pole at the 1-position. When the 3-position is blocked by nitrogen but the 5-position is free, then exchange at 5-positions is rapid due to favorable resonance (**8.106**). Likewise exchange is retarded by the presence of C-methyl groups and greatly accelerated by the presence of positive poles. Similar reasons account for the orientation of the base-

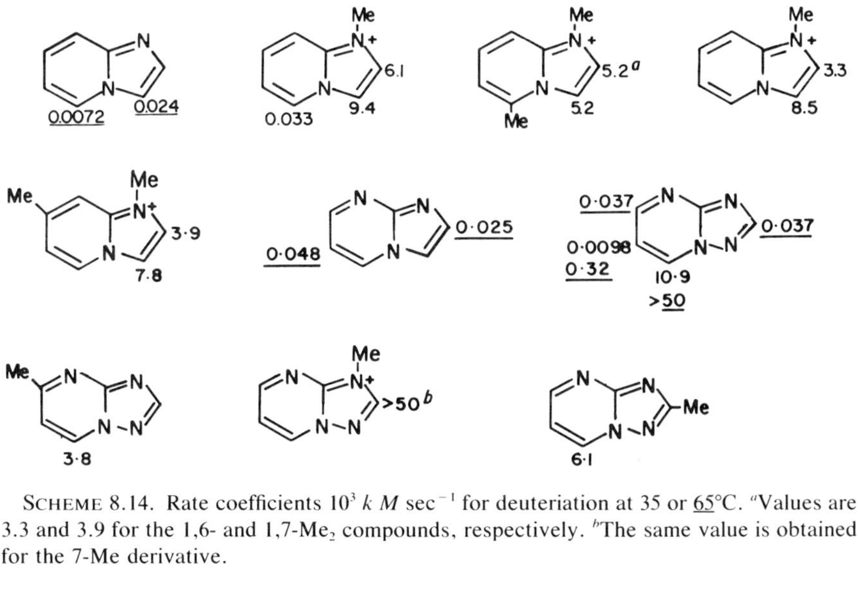

SCHEME 8.14. Rate coefficients $10^3 \, k \, M \, \text{sec}^{-1}$ for deuteration at 35 or 65°C. *^a*Values are 3.3 and 3.9 for the 1,6- and 1,7-Me$_2$ compounds, respectively. *^b*The same value is obtained for the 7-Me derivative.

catalyzed deuteriation at the 3- and 5-positions of imidazo[1,2-*a*]pyrazine (**8.107**) (75JHC861).

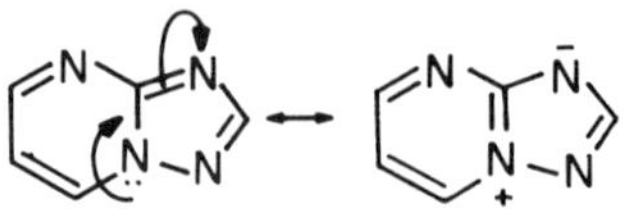

(8·106) 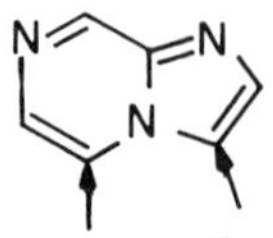   (8·107)

Under basic conditions, hypoxanthin (**8.108**), its 1- and 3-methyl derivatives, and 5-chloro-1-methylhypoxanthin,[3] all underwent exchange at the 2-positions (conjugatively activated by the carbonyl group as well as the 3-nitrogen). If, however, the 2-position is blocked (as in the 2,6-dimethyl derivative) or there is a potential positive pole at the 4-nitrogen (as in the 4-methyl and 2,4-dimethyl derivatives) then exchange occurs at C-5 [73JCS(P1)789]. The exchange rate coefficient of $68 \times 10^{-1} \sec^{-1}$ for the 2-position of thiazolo[4,5-*c*]pyridine (**8.109**) makes this compound 56 times more reactive than benzothiazole (Section 2.C.b.ii), due to activation by the additional nitrogen [76T399].

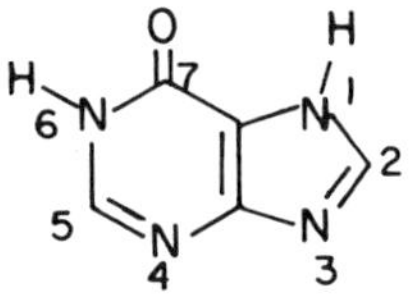

(8.108 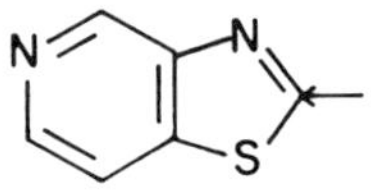)   (8.109)

iii. *Metallation.* In discussing compounds in this section and in Section 2.D.b.iv, the numbering used is as shown in **8.110**, except for compounds with bridgehead nitrogen, for which the numbering is as in **8.111**; in some of the original papers different numbering systems have been used.

Lithiation of compounds **8.110** generally takes place at the 2-position [as, e.g., in thieno[3,2-*b*]pyridine (84JHC785), furo[3,2-*c*]pyridine (83T1777), and thieno[2,3-*b*]pyridine (74JHC355)]. However, in thieno[2,3-*b*]pyrazine lithiation occurs at the 5- and 6-positions in a 17:83 ratio (80JHC1019). Compounds of type **8.111** undergo lithiation as follows: imidazo[1,2-*a*]pyridine, 3-position (or 5- if the 2- and 3-positions are

---

[3] Note that in the original paper a different numbering system was used.

(8.110, X = O, S, NH)   (8.111)

blocked) (72JHC1157; 83S987); pyrrolo[1,2-*a*]pyrazine, 8-position; imidazo[1,2-*a*]pyrimidine, 7-position; 1,2,4-triazolo[1,2-*a*]pyrimidine, 5- and 7-positions (72JHC1157).

iv. *Other Reactions.* Few studies have produced relative rate data. Available nonquantitative work has shown that for compounds of type **8.110** (X = O) substitution occurs at the 2-position, and for **8.110** (X = S, NH) it takes place at the 3-position [e.g., 84AHC(36)394]. Halogens tend to add across the 2,3-bond, especially for **8.110** (X = O). Substitution in compounds of type **8.111** occurs at the 1- or 3-positions (both if they are unoccupied).

A kinetic study of the bromination of imidazo[1,2-*a*]pyridine showed reaction to take place on the neutral species, at the 3-position, and 2000 times faster than at the 5-position in imidazole (74AJC2349), the formation of a stable 6π-pyridinoid ring in the transition state for the former accounting for this reactivity difference.

Rate coefficients for nitration at the 3-position of pyrrolopyridines (azaindoles) with nitrogen in the 4-, 5-, 6-, and 7-positions of indole (the 7-aza compound also contained a 4-methyl substituent) were reported as 2.28, 5.91, 2.11, and 3.91 sec^{-1}, respectively (79CHE1195). These results are curious since one would expect the 5-aza compound to be less reactive because of the ability to conjugatively lower the lone-pair density on the pyrrole nitrogen (cf. results for hydrogen exchange, Scheme 8.11); the results appear to parallel those for hydrogen exchange of thienopyridines. Nitration of 2-(2-furyl)imidazo[1,2,-*a*]pyridine **8.112**, X = O) occurs first in the furan ring and then in the azaindolizine ring [80CA(92)215319], indicating that the former is more reactive. However, it is probable that under the nitration conditions the azaindoline ring is protonated, because in the brominations of **8.112** (both X = O and X = S), bromine first enters the azaindolizine ring (65BAU1391; 72CHE627).

Substitution (usually nitration or halogenation, but in a few cases other reactions) has been reported at the 3-position of compounds **8.110**; that is, for furo[2,3-*b*]pyridine (and for the corresponding 3,2-*b*, 2,3-*c*, and 3,2-*c* isomers) (75JHC705; 84JHC725), 2-substituted furo- or thieno[2,3-*b*]-pyrazines (78RTC151), thieno[2,3-*b*]pyridine (and the 3,2-*b*, 2,3-*c*, and 3,2-*c* isomers) (70AK249, 70JHC373; 74JHC205; 84JOC785), thieno[2,3-*b*]pyrazine, thieno[2,3-*d*]pyridazine, thieno[2,3-*d*]- and thieno[3,2-*d*]pyrimidine [68CR(C)(267)697; 80JHC1019], pyrrolo[2,3-*b*]pyridine (and the 3,2-*b*, 2,3-*c*, and 3,2-*c* isomers) [69TL1909; 77CA(82)57518; 79CHE695; 82JHC665], pyrrolo[2,3-*d*]pyrimidine (69JHC207), imidazo[1,2-*a*]pyrazine (75JHC861), imidazo[1,2-*b*]pyrazine (68T239), and imidazo[1,2-*a*]pyrimidine (77H929). Imidazo[4,5-*c*]pyridin-2-one (**8.113**) is substituted in high yield at the 7-position [79CA(91)175261] as is 4-methyl-4,5-borazathieno[2,3-*c*]pyridine (**8.114**) (66TL2967).

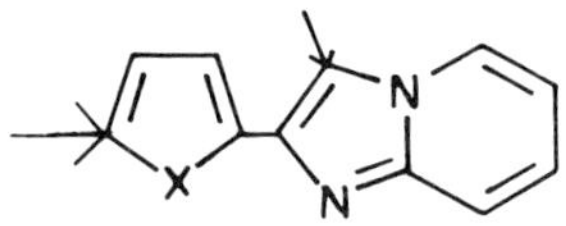

(8.112, X = O, S 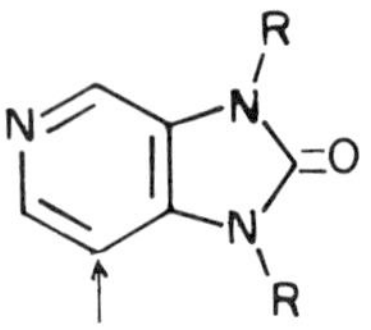)   (8.113 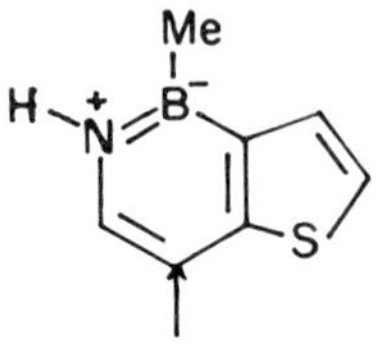)   (8.114)

Among compounds of type **8.111**, substitution has been reported to occur at the 3-position of imidazo[1,2-*a*]pyridine, or at the 5-position if the 3-position is blocked by methyl. 5-Ethoxyimidazo[1,2-*a*]pyridine is nitrated to give approximately equal amounts of the 3,6- and 3,8-dinitro products (65JOC4085; 76JOC3549; 82AJC1761); this result suggests the reactivity of the five-membered ring is comparable to that of ethoxybenzene. Pyrrolo[1,2-*c*]- and pyrrolo[2,1-*b*]pyrimidine are formylated in the 1- and 3-positions, and in the 3-position, respectively (76JOC351; 77JOC2448). Imidazo[1,5-*a*]pyridine (**8.115**) is substituted preferentially at the 1- but also at the 3-position; further substitution occurs at the 5-position if the 3-position is blocked [72BSF2481; 75JHC379; 75JOC3373; 80JCS(P1)959]. Pyrazolo[1,5-*a*]pyrimidine is substituted in the 3- and 6-positions (corresponding to the 1- and 6-positions in **8.111** (75CJC119)).

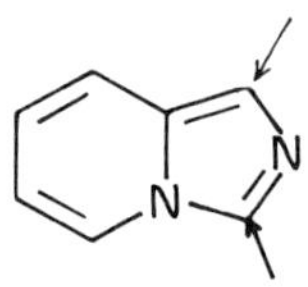

(8.115 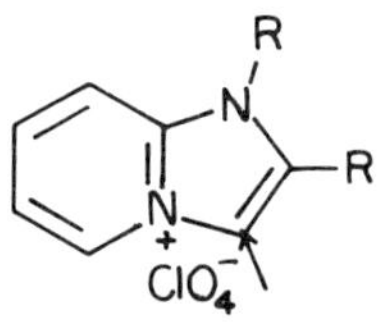)   (8.116, R = H, Me)

The preferential 1-substitution in some of the compounds just discussed compared to preferential 3-formylation in indolizine (**8.61**) led Fuentes and Paudler to suggest that the electrophile coordinates with the bridgehead nitrogen of indolizine (75JHC379). However, this postulate is both highly unlikely and unnecessary. Indolizine substitutes at the 3-position because this is much the most reactive site, and the preferential 1-substitution in imidazo[1,5-*a*]pyridine arises from the high 2,3-bond order so that the 2-nitrogen deactivates the 3-position more strongly than the 1-position [87JCS(P2)591]. 5-Methylimidazo[1,5-*a*]pyridine formylates exclusively at the 1-position (75JHC379) and this may be attributed to steric hindrance, which strongly affects this reaction [87JCS(P2)591]. Imidazo[1,2-*a*]pyrazine is halogenated in the 3- and 5-positions (77JOC4197), and imidazo[1,2-*a*]pyrimidinium salts (**8.116**) are substituted quite readily at the 3-position (74JOU2489), presumably because the low 3,4-bond order reduces the ability of the positive pole to deactivate.

3. Compounds with One Five- and Two Six-Membered Rings

A. Molecules Containing One Heteroatom

In discussing these compounds (dibenzofuran, dibenzothiophene, dibenzoselenophene, and carbozole), it should be noted that the numbering system differs between dibenzofuran and dibenzothiophene on the one hand (**8.117**) and carbazole on the other (**8.118**). An added complication is that in early literature the numbering, especially for dibenzofuran, followed the system used for carbazole, so that the numbering given in this section may not always correspond to that given in the original papers. To avoid complications in general discussion both here and in Section 3.A.c, the positions will be referred to as shown in **8.119**.

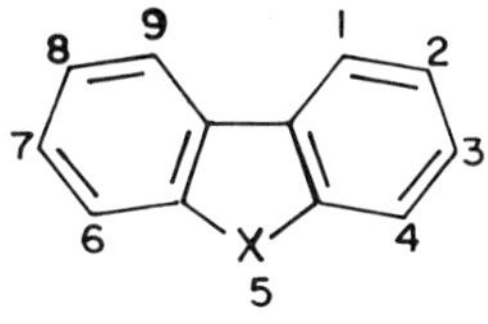

(8.117, X = O, S 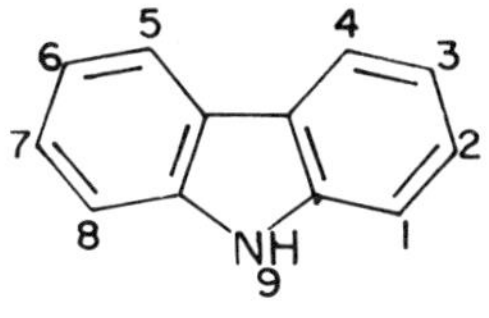)   (8.118 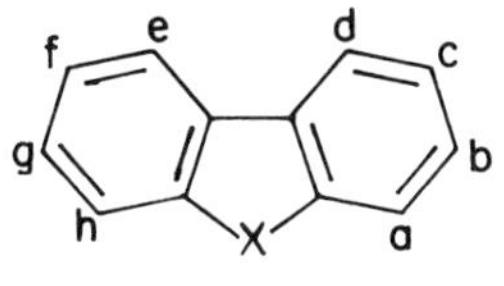)   (8.119)

a. *Positional Reactivity Order*

In these compounds the two benzenoid rings tend to behave independently, so that substitution in one ring has little effect upon substitution in the other. Hence symmetrical disubstitution is very easy, and careful control of conditions is needed to avoid this. Conjugation is possible between positions b (or d) and e (or g) only through the intermediacy of structures such as **8.120**, which not only have lost the benzenoid character of two rings, but also have unfavorable bond fixation in the five-membered ring. Inter-ring conjugation is therefore very weak. Direct conjugation between positions a (or c) and f (or h) is impossible for the oxygen- and nitrogen-containing compounds, and is only possible for dibenzothiophene through the intermediacy of **8.121**, which is also of high energy and unfavorable.

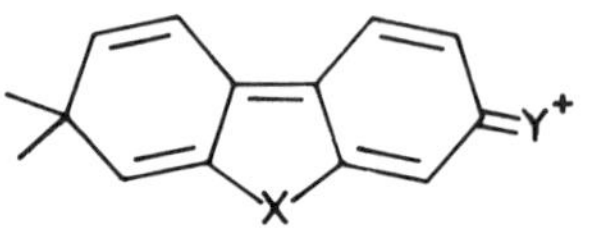

(8.120, X = O, S, NH 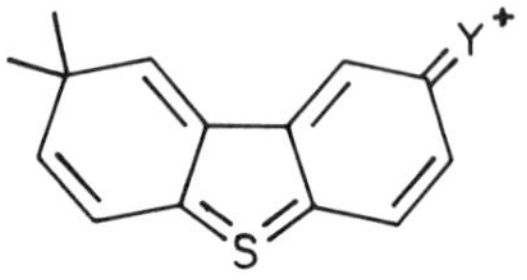)   (8.121)

These compounds are the benzo homologues of those described in Section 2.A. Valence bond theory demonstrates nicely how the site of highest reactivity in the benzo[*b*] compounds, the 6-position, becomes the c(f)-position in the corresponding dibenzo compounds. As noted in Section 2.A.a, the 5- and 6-positions of the benzo[*b*] compounds are more reactive than the 4- and 7-positions, because the principal resonance structures (**8.4** and **8.7**) are *p*-quinonoid. Overall, however, the 6-position is the most reactive because there are five canonical forms for the transition state (Scheme 8.16) compared to four canonicals for the transition state for 5-substitution (Scheme 8.15). However, for the dibenzo compounds, each of the four canonicals for substitution at position c(f) has an intact benzenoid ring, whereas for substitution at position b(g) two of the canonicals (including the most important one, which has the positive charge delocalized onto the heteroatom) have lost the benzenoid character present in the ground state. This is sufficient to make position c(f) more reactive than position b(g) in the dibenzo compounds.

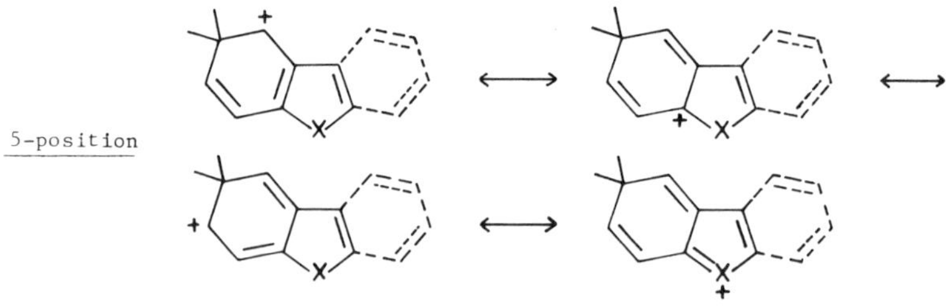

5-position

SCHEME 8.15. Transition-state canonicals for substitution at the 5-position of benzo[*b*] compounds and the corresponding position (c) in the dibenzo homologue.

Sec. 3.A] ONE FIVE- AND TWO SIX-MEMBERED RING COMPOUNDS 241

6-position

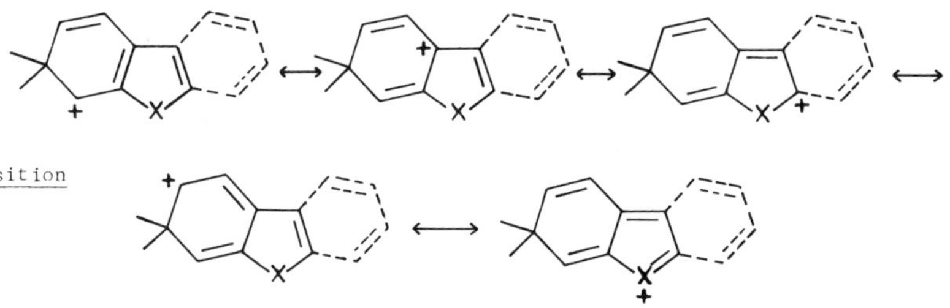

SCHEME 8.16. Transition-state canonicals for substitution at the 6-position (b) of benzo[b] compounds and the corresponding position in the dibenzo homologue.

Attempts to predict the positional reactivity orders in these compounds by MO calculations have had only limited success. The problem is that the molecules can be considered as substituted biphenyls (**8.122**), in which case positions b and d would be the most reactive (as they are in fluorene). Alternatively, they can be considered as derivatives of diphenyl ether, etc., as in **8.123**, in which case positions a and c, which are,

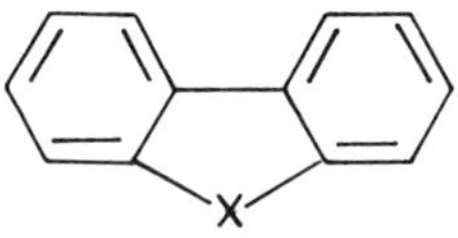

(8.122 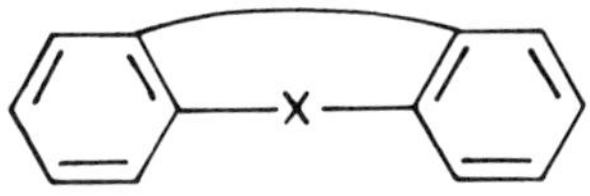)   (8.123)

respectively, ortho and para to the heteroatom, would be the most reactive. The importance of this latter aspect should decrease for different heteroatoms X along the series NH > O > S, so that for carbazole it is possible that (in the absence of steric hindrance) the positional reactivity order could be c > a > b > d, as predicted by π-density calculations (Table 8.17). However, no data are available from acid-catalyzed hydrogen exchange, the only reaction cabable of providing unambiguous infor-

TABLE 8.17
CALCULATED POSITIONAL REACTIVITY ORDER FOR DIBENZOFURAN, DIBENZOTHIOPHENE, AND CARBAZOLE

Compound	Positional order	Method	Reference
Dibenzofuran	a > c > b > d	π densities, HMO	69IJQ33
Dibenzofuran	c > b > a > d	CNDO/2	77NKK1518
Dibenzothiophene	c > a > b > d	π densities, HMO	58CA14317i
Dibenzothiophene	a > c > b > d	SCF (PPP)	68M16
Carbazole	c > b > a > d	π densities	47TFS87

mation because of the freedom from steric effects. Of the other calculations, only the CNDO/2 calculations for diphenyl ether predict the correct positional order.

b. *Reactions*

i. *Acid-Catalyzed Hydrogen Exchange.* Exchange rates for detritiation of dibenzofuran in trifluoroacetic acid have led to the partial rate factors shown in Scheme 8.17 (61JCS5077), from which may be derived the σ^+ values shown in Scheme 8.18 [72JCS(P2)97]. It is probable that the positional reactivities are reduced somewhat (of the order of 10%) by hydrogen bonding, which was subsequently realized to occur in trifluoroacetic acid. However, the positional reactivity orders (viz. 2 > 3 > 4 > 1) should not be affected by such bonding.

For dibenzothiophene, partial rate factors and σ^+ values have also been determined (Schemes 8.17 and 8.18) [61JCS5077; 72JCS(P2)97]. These show the same reactivity order, 2 > 3 > 4 > 1, as for dibenzofuran; again the true values may be a little higher after correction for hydrogen bonding. The results show that the 2-position of dibenzofuran is more reactive than the 2-position of dibenzothiophene, as expected in view of the greater $+M$ electron release by *p*-oxygen. The unexpectedly low reactivity of the 4-position of dibenzofuran relative to that in dibenzothiophene (and not reflected in the reactivities of the ortho positions of diphenyl ether and diphenyl sulfide, 61JCS5077) has been explained in terms of strain effects [68JCS(B)1559], described fully in Section 3.A.c. It should be noted here that one treatment of the partial rate factors for detritiation of dibenzothiophene given in a review [74AHC(16)181] is mathematically incorrect, and leads to erroneous impressions of the reactivity.

Rates of exchange of *N*-methylcarbazole with H_2SO_4 in EtOD were faster at 80°C than for *N*-phenylcarbazole at 125°C (41JA358). This is expected, providing exchange takes place at the 3-position (this was not ascertained). Exchange probably occurs on the free base, because if this were not the case the observed rates should have been somewhat closer in view of the greater ease of protonation of N-methyl compounds. More-

SCHEME 8.17. Partial rate factors for detritiation of dibenzofuran and dibenzothiophene in trifluoroacetic acid at 70°C (61JCS5077).

Sec. 3.A] ONE FIVE- AND TWO SIX-MEMBERED RING COMPOUNDS 243

SCHEME 8.18. Values of σ^+ for dibenzofuran and dibenzothiophene.

over, the basicity of carbazole is low because if nitrogen becomes sp^3-hybridized, the molecule suffers a substantial loss of resonance energy.

ii. *Nitration.* Interpretation of nitration data for dibenzofuran is complicated by widely differing descriptions of the melting points of the various isomers (discrepancies up to 20°C), suggesting that isolated fractions were in some cases far from pure. It is also anomalous in giving mainly 3-substitution under most conditions (e.g., 80% yield, together with some 4-isomer) with fuming nitric acid in acetic acid (30JCS2267), though another report gave the main by-product as the 2-isomer (23CB2498). Nitric acid in acetic acid gave 52% of the 3-isomer [82CA(96)68737] and this becomes 80% with nitric acid in trifluoroacetic acid (82BCJ629); fuming nitric acid gave 3,8-dinitration (74USP3792017). Only with ethyl nitrate/AlCl$_3$/MeNO$_2$ is the 2-isomer predominant, though the yield was only 28% (82BCJ629). Nitration by nitric acid in acetic anhydride gives 1-, 2-, and 3-isomers in the ratio of 1:2:2 (57J345; 58JCS3079), from which partial rate factors of $f_1 = 47$, $f_2 = 94$, and $f_3 = 94$ were calculated, and from the analytical method it may be calculated that f_4 is not greater than 12 [68JCS(B)1559]. It is probable that the greater tendency for 3-substitution may arise from hydrogen bonding to the oxygen since the highest yield of this isomer is obtained under the most protic conditions.

Nitration of dibenzothiophene by HNO$_3$ gives oxidation to the 5,5-dioxide, but with HNO$_3$/Ac$_2$O, 40% of the 3-nitro derivative together with 40% of the dioxide is obtained (36JCS1435); the dioxide is also nitrated 96% in the 3-position (74GEP2457082). 3-Acetylaminodibenzothiophenes nitrated in the 4-position (77%) (53JOC1492) rather than the 2-position, which might *a priori* have been expected. This result may be attributed to the lower energy of structure **8.124**, in which conjugation in the central ring is uninterrupted, compared to structure **8.125**.

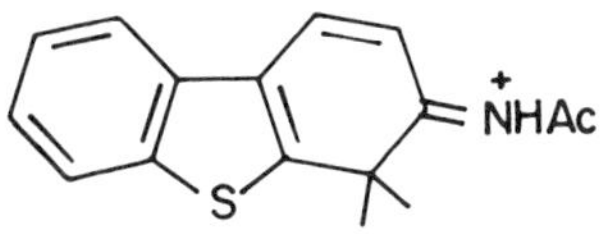

(8.124) 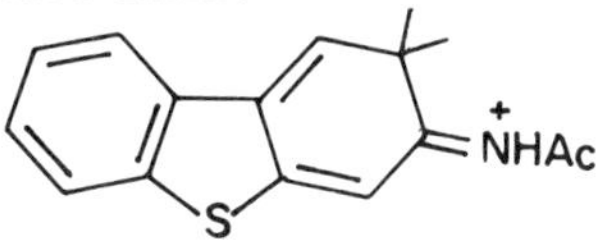   (8.125)

Dibenzoselenophene is nitrated (nitric acid/acetic acid) 48% in the expected 2-position, but also 25% in the 4-position (55JA1061), implying that the positional reactivity order may be different from that in the oxygen and sulfur analogues.

Nitration of carbazole is straightforward and gives mainly the 3-nitro derivative together with some 1-isomer (24CB555; 31JCS3283). A 3-:1-isomer ratio of 7:3 has been reported (84T1857). The reaction can be accompanied by N-nitrosation (24CB555). It has been suggested that nitration (and of the 1-position in particular) may occur via *N*-nitration, followed by rearrangement, since rearrangement of *N*-nitrosocarbazole in the presence of an oxidizing agent (the N-nitro compound could not be prepared) gave the same 7:3 product ratio for the 3- and 1-isomers (84T1857). If the 9-position is blocked, higher yields of the 3-isomer are obtained (84T1857) (e.g., 85% from 9-ethylcarbazole) (74BSF183), though this could of course arise from the extra electron release to the (para) 3-position.

Nitration in acetic anhydride of 9-(*p*-tosyl)carbazole gave the 1-, 2-, and 3-isomers in 28, 19, and 53% yields, respectively, and for 9-acetyl- and 9-nitrosocarbazole these became, correspondingly, 10, 48, and 42%, and 24, 0, and 66% (84T1857). Nitration of carbazole itself under those conditions gave partial rate factors of $f_1 = 32{,}100$; $f_2 = 1100$; and $f_3 = 77{,}600$ (58JCS3079).

The activation energies for nitration of carbazole, 9-methylcarbazole, and 3-nitrocarbazole have been determined as 102, 75, and 117 kJ mol^{-1}, respectively (73JPU1237), and the strong activation by the 9-methyl group is apparent.

iii. *Halogenation.* Chlorination of dibenzofuran (Cl$_2$/Fe) gives the expected 2-chlorodibenzofuran (36%) together with 4% of the 2,8-dichloro derivative (55JOC657). Bromination of dibenzothiophene (Br$_2$/CS$_2$ gives 40% of the 2,8-dibromo derivative (47JA1920). Bromination of carbazole with 1,3-dibromo-5,5-dimethylhydantoin gives 37% of 3-bromocarbazole, and with an excess of the reagent, 68% of the 3,6-dibromo compound (51M11). Bromine in pyridine is reported (67KKZ63) to give a truly remarkable 100% of 3-substitution *and* 7% of 1-substitution! Under the same conditions chlorination and iodination went mainly in the 3-position (67KKZ63). Partial rate factors for the chlorination of *N*-acetylcarbazole have been determined as f_1, 4300; f_2, 8600; f_3, 122,000; f_4, 8,600; and for carbazole itself, $f_1 > 10^7$; $f_3 > 10^8$ [66JCS(B)521]. These results show the strong activation at positions ortho and para to nitrogen and are discussed further in Section 3.A.c.

iv. *Alkylation and Acylation.* Ethylation of dibenzofuran occurs in the expected 2-position (5OUSP2500732), while chloromethylation gave 97% of 2,8-bis(chloromethyl)dibenzofuran [73CA(78)84153]. However, chloromethylation of dibenzothiophene gives 50% of 2,7-bis(chloromethyl)dibenzothiophene (76BAU2609) a result difficult to explain. The hydrogen-exchange data (Section 3.A.b.i) show that the 2- and 3(7)-positions are more nearly equal in dibenzothiophene than in dibenzofuran. Possibly in reactions with very reactive electrophiles where the demand for resonance stabilization by the heteroatom is not so important, dibenzothiophene behaves more like a substituted biphenyl (**8.122**) so that the reactivity of the 3(7)-positions becomes relatively more important.

Benzoylation of dibenzofuran with benzoyloxypyridine gives 98% of 2-benzoyldibenzofuran (80S139). Benzoylation with benzoyl chloride in nitrobenzene at 15°C gives the partial rate factors shown in Scheme 8.19 (77NKK1518). Benzoylation involves a slightly less reactive electrophile than hydrogen exchange; although the ρ factor for benzoylation is not known it will be very close to that, -9.1, for acetylation. Using this value gives σ^+ values for the 2- and 3-positions identical to those found from hydrogen exchange (Section 3.A.b.i). However the 1- and 4-positions are less reactive in benzoylation than in hydrogen exchange, as a result of steric hindrance. Other workers also report mainly 2-benzoylation with, under some conditions, 2,8-dibenzoylation of dibenzofuran (54JA6407; 73NKK1505).

Benzoylation of various 2-substituted dibenzofurans (NHAc, Cl, Br, NO$_2$ substituents) gave 84–95% of the 8-benzoyl derivatives (72NKK387), showing the poor inter-ring conjugation. With 3-substituents, 8-derivatives were obtained in 56–67% yield, accompanied by some 7-derivatives, in yields said to be 0.6% (NHAc), 31% (Br), and 3% (Cl). This wide variation indicates experimental or typographical error, which may be the cause of the curious results for benzoylation of methyldibenzofurans (Scheme 8.20) (74NKK1708). Dibenzofuran is acetylated in the 2-position (5OUSP2500734) and with acetyl chloride/AlCl$_3$/EtNO$_2$ at 25°C, the relative rates of acetylation of toluene, dibenzofuran, and dibenzothiophene

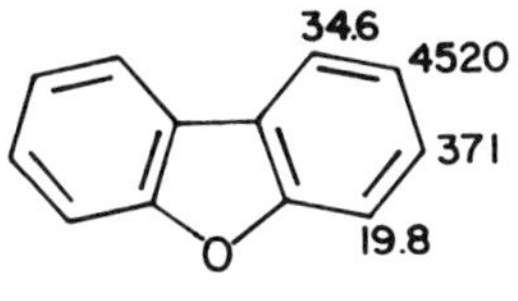

SCHEME 8.19. Partial rate factors for benzoylation of dibenzofuran with benzoyl chloride at 15°C (77NKK1518).

SCHEME 8.20. Reported isomer yields for benzoylation of methyldibenzofurans (74NKK1708).

are 1 : 5.94 : 5.34 [69AC(R)787]. The closeness of the second and third values implies that the dibenzofuran oxygen is partially coordinated to the aluminum chloride.

Monoacetylation of dibenzothiophene occurs mainly (50%) at the 2-position, together with some 4-substitution, (38JA2628). Under suitable conditions, 2- and 4-formylation can be obtained in a 3 : 2 ratio (69AJC1963). Surprisingly, diacetylation gives the 2,6-disubstituted product (72JOC-3355) though diformylation goes in the expected 2,8-positions (77JIC-1151).

Carbazole tends to be acylated under kinetic control in the 9-position [63CA(58)2422c; 76T2595], though 3,6-diacetylation has been reported from both carbazole (35JCS741) and 9-methylcarbazole (68%) (63CA2422d). 9-Benzoylcarbazole is further acylated in the 2-position (35JCS741), apparently due to electron withdrawal from nitrogen by benzoyl. The most recent study of the acetylation of 9-methylcarbazole with acetyl chloride/AlCl$_3$/MeNO$_2$ gives 3- (79.9%), 1- (3.2%), 9- (4.3%), and 3,6-substitution (10%) [75CA(82)16654]. The activation energies for acetylation with acetic anhydride of 9-methylcarbazole are 20.1, 22.0, and 23.1 kcal mol^{-1} for catalysis by HClO$_4$, Mg(ClO$_4$)$_2$, and H$_2$SO$_4$, respectively [74CA(81)168829].

v. *Protiodesilylation.* This reaction has been used to provide the only complete set of reactivity data (other than those from hydrogen exchange) for dibenzofuran and dibenzothiophene (61JCS4921). The partial rate factors together with that for the 3-position of 9-ethylcarbazole are given in Scheme 8.21. These show that, as expected, the positions para to the heteroatom decrease in reactivity along the series carbazole > dibenzofuran > dibenzothiophene. The results for all positions of the latter two compounds parallel those for hydrogen exchange (Scheme 8.16) except that the 1-position of dibenzothiophene appears to be too reactive in desilylation. This could be due to steric acceleration known to affect desilylation (for example, biphenyl gives an $f_o : f_p$ ratio of 2.2 (61JCS4921).

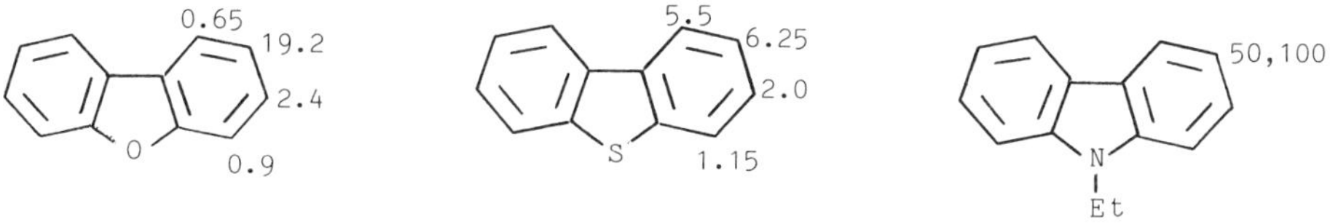

Sec. 3.A] ONE FIVE- AND TWO SIX-MEMBERED RING COMPOUNDS 247

SCHEME 8.21. Partial rate factors for protiodesilylation of dibenzofuran, dibenzothiophene, and 9-ethylcarbazole (MeOH–HClO$_4$).

If this is the cause, then it should also have enhanced the reactivity of the 1-position of dibenzofuran, so it seems possible that one or the other of these values is in error.

vi. *Other Reactions.* Sulfonation of dibenzofuran occurs at the 2-position (49JA1593) and this takes place regardless of the presence of an amino or nitro substituent at the 8-position (51MI3), which emphasizes the lack of effective inter-ring conjugation. Dibenzothiophene is sulfonated at the 2- and 8-positions [34CR(198)2260] and the 5,5-dioxide is substituted at the 3- and 7-positions (i.e., meta to sulfur) [79JCS(P2)224]. Carbazole is sulfonated in the 1-, 3-, and 6-positions (49CA6205f), though under controlled conditions 83–87% of 3-, and 85–98% of 3,6-substitution can be achieved (50MI1).

Thiocyanation of carbazole occurs at the 1- and 3-positions [73CA-(78)111049], 9-methylcarbazole is thiocyanated at the 3- and 6-positions [77CA(87)151948], and 3-nitrocarbazole is thiocyanated at the 6-position [78CA(88)152349]. The last result again shows the weakness of inter-ring conjugation.

Mercuriation of 9-ethylcarbazole occurs at the expected 3-position and more readily than in dibenzofuran (at the 4-position) (36JOC146). Metallation with alkyllithium goes into the position nearest to the heteroatom [i.e., the one most activated by the $-I$ (inductive withdrawal) effect] (38JOC120). Lithiation at the 4-position (ortho to the heteroatom) is more rapid for dibenzofuran than for dibenzothiophene, which in turn is more reactive than 9-phenylcarbazole (39JA951; 40JA2606; 41JA2479; 45JA877). Reported yields of the 4-lithio derivative are 62% for dibenzothiophene (41JA2479) and 73% for dibenzoselenophene (54JA5775). 9-Ethylcarbazole is lithiated in the 1-position as expected. However, 9-phenylcarbazole is dilithiated not in the expected 1,8-positions but rather in the 2'- and 6'-positions of the phenyl ring (25%) (43JA1729). This is probably because steric hindrance will force the phenyl ring orthogonal to the plane of carbazole so that there will be little conjugative electron release from nitrogen to the phenyl ring to outweigh the (conformation-independent) $-I$ effect of nitrogen; by contrast electrons may still be released conjugatively to the 1- and 8-positions of the carbazole ring.

c. Quantitative Aspects of the Reactivity Data

i. Comparison of the Reactivities of Benzo[b] and Dibenzo[b,d] Compounds. At present this comparison is possible only for the sulfur compounds, using the hydrogen exchange data from Schemes 8.3 and 8.18 (Scheme 8.22). The reactivity of positions c and d (see **8.119**) is effectively the same in both molecules. Position b is less reactive in dibenzothiophene for reasons described in Section 3.A.a. The higher reactivity of position a in dibenzothiophene may be due to the fact that this is the site most susceptible to the $-I$ (inductive withdrawal) effect of sulfur, and in the dibenzo compound sulfur can satisfy its demand for electrons from two rings rather than one.

ii. The Effect of Strain on the Positional Reactivities. The α-position of the benzene ring in indane has long been known to be substantially less reactive than the α-position of tetralin (the Mills–Nixon effect). This has been explained in terms of strain since two-thirds of the canonicals representing the transition state for α-substitution have a double bond in the five-membered ring as opposed to one-third of those for β-substitution (and one-half of those representing the ground state) (see Scheme 8.23) (65T1665). Taylor showed that this argument could be extended to strained aromatics in general [e.g., benzocyclobutene, biphenylene, triptycene [67JCS(B)780; 68C1, 68JCS(B)1402; 71JCS(B)536], benzo-1,3-dioxole [72MI2(243)], and fluoranthene [77JCS(P2)866]], and also to the dibenzo derivatives of the five-membered heterocycles [68JCS(B)1559].

The partial rate factors given in the preceding sections are gathered in Table 8.18, along with, for comparison, those for fluorene (**8.126**, X = CH$_2$) and the open-chain analogues (**8.127**) in the same reactions [68JCS(B)1559]. Position a in **8.126** is α to the five-membered ring and

SCHEME 8.22. Values of σ$^+$ for comparable positions in benzo[b]thiophene and dibenzothiophene.

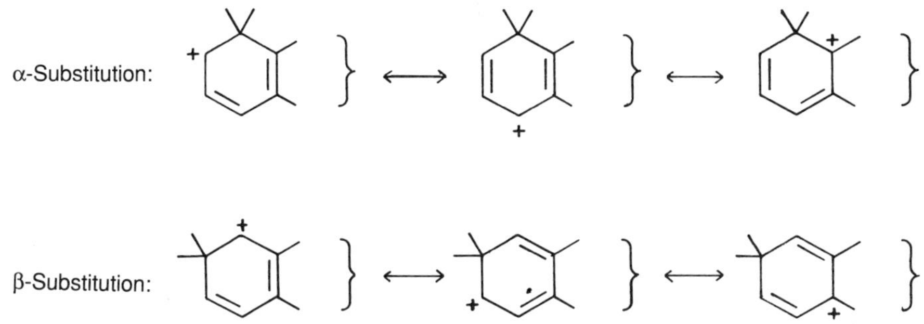

Sec. 3.A] ONE FIVE- AND TWO SIX-MEMBERED RING COMPOUNDS 249

α-Substitution:

β-Substitution:

SCHEME 8.23. Canonicals representing the transition state for α- and β-substitution in a benzene ring adjacent to a cyclic substituent.

position c is β to it. They each correspond to positions a and c in **8.127**, differing only in that a and c in **8.126** are each meta to the second bridge, which should therefore produce a constant effect at each. The effect of closing the ring may therefore be quantitatively assessed by comparing the ratios $\log f_a : \log f_c$ in **8.126** and **8.127**. Reference to Table 8.18 shows the following results.

(1) In *every* case the ratios are smaller in **8.126** compared to **8.127**. This indicates that position a in **8.126** has attenuated reactivity because of the increase in strain accompanying formation of the transition state for reaction at this site.

(2) Where the ratios have been determined for the same reaction for compounds with different heteroatoms, the decrease in the ratio on forming the ring is smaller for the sulfur compounds (in which the five-membered ring is less strained) than for the corresponding compounds where X is O or CH_2. Thus, in detritiation the ratio decreases by 28% (CH_2), 22% (O), and 11% (S), and in desilylation by 104% (O) and 31% (S).

(3) Similar comparisons of the reactivities of positions b and d should reveal a similar reduction of the $\log f_d : \log f_b$ ratio in **8.126** compared to biphenyl **8.128**. The comparison is less unequivocal here because (1) there is a very large reactivity difference between the two sets of positions and (2) differences in coplanarity between biphenyl and the dibenzo com-

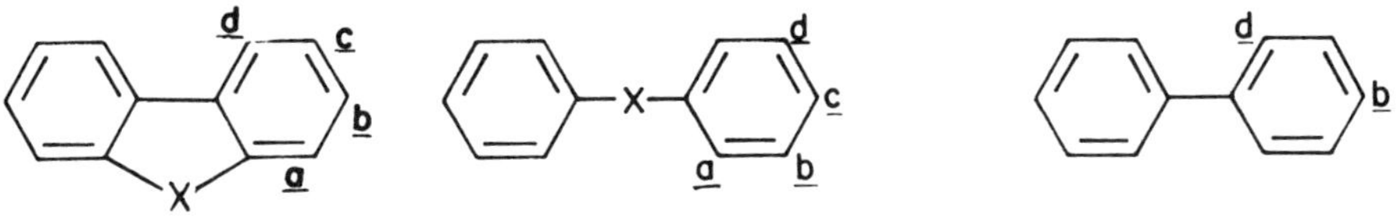

(8.126)   (8.127, X = CH_2, O, S, NH)   (8.128)

TABLE 8.18
PARTIAL RATE FACTORS FOR ELECTROPHILIC SUBSTITUTIONS

Reaction	X	Compounds of structure **8.126**					Compounds of structure **8.127**			
		Partial rate factors					Partial rate factors			
		a	b	c	d	$\log f_a : \log f_c$	a	b	c	$\log f_a : \log f_c$
Detritiation	CH_2	21	17,000	125	5,500	0.63	47	3.8	117	0.81
	O	160	313	3,670	135	0.62	6,930	—	30,000	0.86
	S	362	430	1,830	272	0.785	3,300	—	9,830	0.88
Desilylation	O	0.9	2.4	19.2	0.65	−0.02	8.7	—	88.5	0.48
	S	1.15	2.0	6.25	5.5	0.077	1.30	—	10.7	0.111
Nitration	CH_2	*	2,040	60	940	—	13	—	32	0.74
	O	<12†	94	94	47	0.65	117	—	234	0.88
	NH	32,100	1,100	77,600	—	0.92	837,000	—	575,000	1.03
Molecular	CH_2	—	300,000	—	39,000	—	—	—	—	—
chlorination	NAc	4,300	8,600	122,000	8,600	0.715	6,200	—	24,200	0.865

*The correlations in this table indicate a value of 10–15 for f_a for fluorine in nitration.
†No substitution was detected at this position. The value given is the *maximum* that could apply without the isomer being detected.

pounds may affect the relative conjugative electron release to the ortho and para positions, as well as steric acceleration in desilylation. Nevertheless, the majority of results give lower values of $\log f_d : \log f_b$ in **8.126** compared to biphenyl [68JCS(B)1559].

B. MOLECULES CONTAINING TWO OR MORE HETEROATOMS

There have been few studies of the reactivity of molecules in this class. Calculations (π density) indicate that the positional order in naphtho-[2,3-c]-1,2,5-thiadiazole (**8.129** X = S), is 4 > 6 > 5, whereas localization energies and frontier electron densities predict 4 > 5 > 6 [66TCA(5)401]; the high reactivity of the 4-position follows from the benzenoid character generated in the transition state (**8.130**). CNDO/2 calculations for naphtho[2,3-c]-1,2,5-oxa-, -thia-, and -selena-diazoles (**8.129**, X = O, S, Se) predict the order 6 > 4 ⩾ 5 (73CHE1331). For the corresponding [1,2-c] compounds (**8.131**) the order is predicted by the same method to be in each case 8 > 4 > 6. For the oxadiazole (**8.131**, X = O) the positions of substitution are 4 and 8 (chlorination and sulfonation) and 4 and 6 (nitration) (73CHE1331). Reports for the thiadiazole (**8.131**, X = S) disagree, bromination and sulfonation apparently giving 8-substitution (73CHE1331), whereas nitration gives 7 > 4 [72CA(77)101477], and halogenation gives polysubstitution [72CA(77)126512]. It is possible that the extent of protonation at nitrogen under the various conditions will affect the isomer distribution.

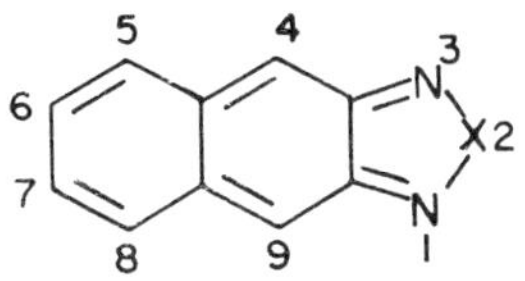

(**8.129**, X = O, S, Se 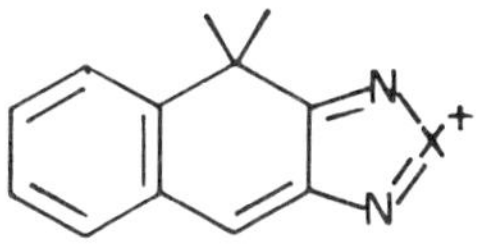)   (**8.130** 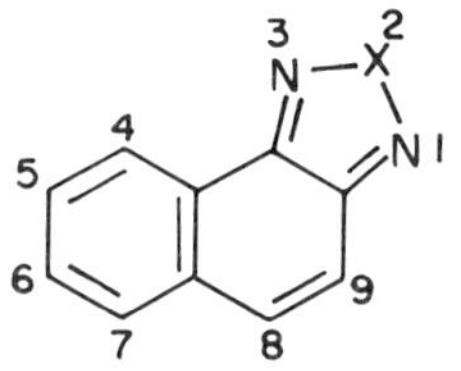)   (**8.131**, X = O, S, Se)

Calculations predict the most reactive sites in thieno[2,3-b]quinoline (**8.132**) to be the expected 2- or 3-positions [77ZN(B)1331], but there is little agreement on the positional reactivity order in 6-methylindolo-[2,3-b]quinoxaline (**8.133**) (84CHE687). π Densities predict the order 7 > 9 > 1 (Hückel) or 7 > 9 > 3,4 (CNDO/2), localization energies (Hückel) predict 4 > 7 > 1, superdelocalizabilities predict 4 > 1 > 7, and frontier electron densities predict 1 > 4 > 3 (Hückel) or 4 > 1 > 2 (CNDO/2). The observed nitration and bromination at the 9-position serve to underline the inadequacy of all these theoretical methods. Localization energies predict that pyrrolo[1,2-a]quinoxaline (**8.134**) should have a positional reactivity order 1 > 3 > 6 > 2 [71JCS(C)2018]; the high reactivity of the 1- and 3-positions again follows from the increase in aromaticity on going to the transition state. The 1-position is the most reactive in sulfonation, bromination, and chlorination, but the 3-position is slightly more reactive than the 1-position in iodination and nitration [67JCS(C)-1164; 68J(C)2848].

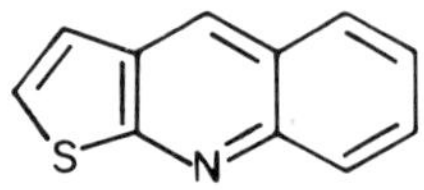

(8.132) 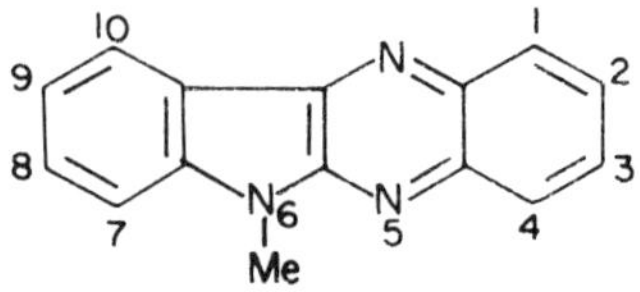   (8.133) 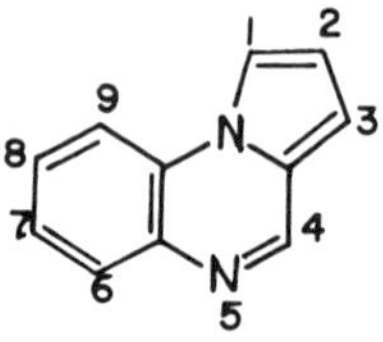   (8.134)

The log rates of base-catalyzed deuteration at 50°C of *N*-methylnaphthoimidazole derivatives **8.135–8.137** are as shown (77CHE1235). Naphtho[2,3-*d*]-1-methylimidazole (**8.135**) is therefore more reactive than 1-methylbenzimidazole (for which log k is -4.1), whereas naphtho-[2,1-*d*]- and -[1,2-*d*]-1-methylbenzimidazoles (**8.136, 8.137**) are less reactive. This probably derives from the C-2—N bond order in the five-membered rings being greater in **8.135** (since the ring is not fully aromatic) than in **8.136** and **8.137**. Electron withdrawal by N-3 from C-2 will be greater in the former, thereby increasing stabilization of the carbanionic transition state.

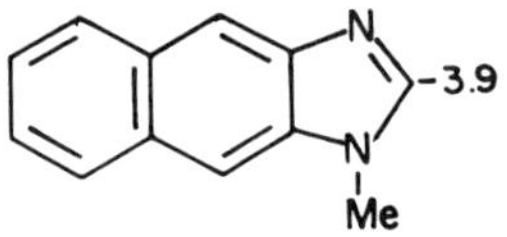

(8.135) 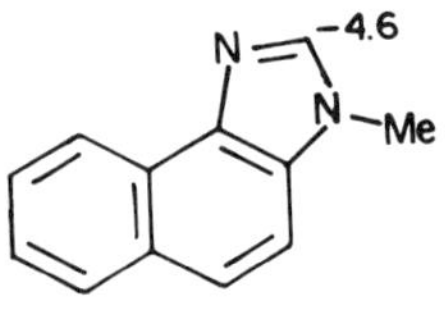   (8.136) 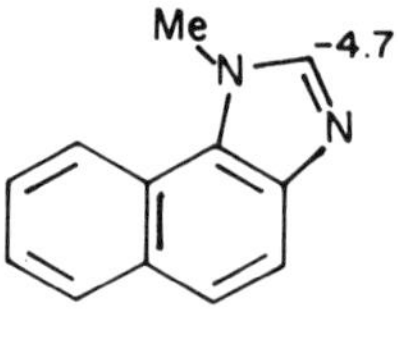   (8.137)

Benzothieno[3,2-*d*]pyrimidine (**8.138**) is said to be brominated and nitrated exclusively at the 8-position in 48 and 80% yield, respectively (80JHC1399), though 6,8-disubstitution had been reported earlier (72T3277). Disubstitution also occurred with the 2- and 4-oxo derivatives. Thieno[2,3-*b*]quinoline (**8.132**) is brominated and iodinated at the 3-position, gives a 2,3-addition product with chlorine, and is nitrated at the 2- and 3-positions [80JCR(S)201], in agreement with the calculations above. Pyrido[1,2-*a*]benzimidazole (**8.139**) is nitrated at the 8-position (55J3275); this position is expected to be strongly activated because the transition state (**8.140**) gains as well as loses benzenoid resonance compared with the ground state (**8.139**).

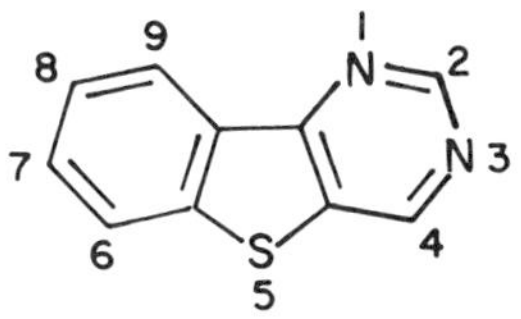

(8.138) 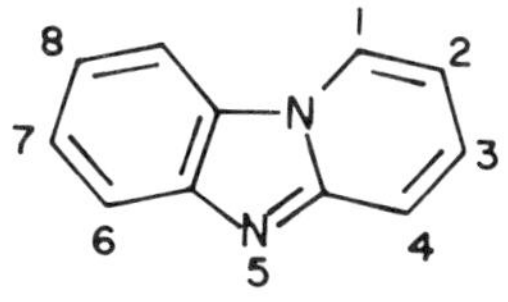 (8.139) 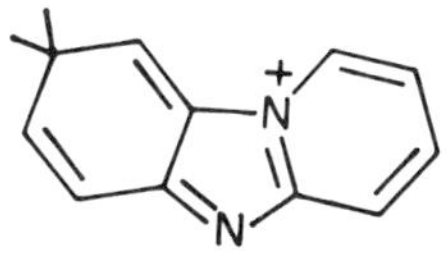 (8.140)

Thieno[3,2-*f*]quinoline (**8.141**) is nitrated, brominated, and acetylated in the 2-position in 80, 57, and 29% yield, respectively [70JCS(C)2334]. The corresponding *N*-oxide is nitrated in the 8- and 9-positions [79IJC(B)342]. The bromination of *s*-triazolo[3,4-*a*]isoquinoline (**8.142**) in the 3-position [74CA(80)3443] is expected because the transition state will have increased benzenoid character. Nitration, however, occurs at the 7- and 9-positions (75CB3762) almost certainly due to N-1 or N-2 protonation, and chlorine adds across the high-order 5,6-bond (71CB3925). Naphtho[1,2-*d*]imidazole (**8.143**) is brominated in the 5-position (i.e., para to NH) (54ZOB488).

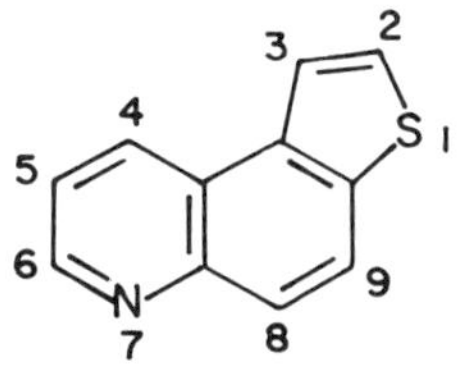

(8.141) 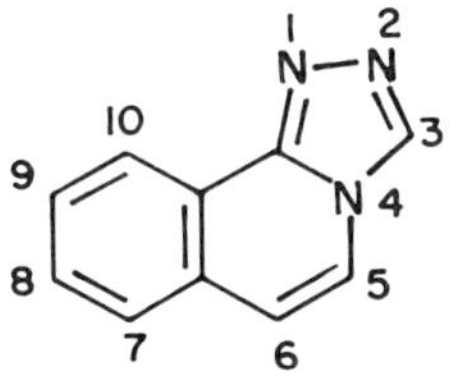 (8.142) 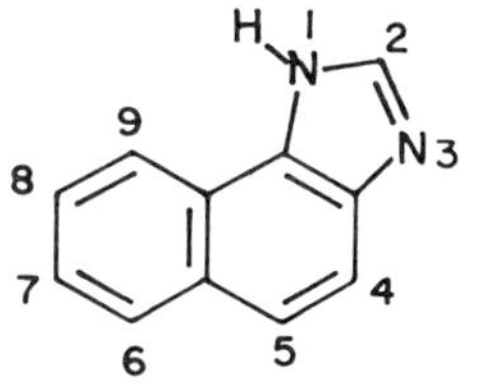 (8.143)

4. Compounds with Two Five- and One Six-Membered Ring

Potentially there are a very large number of molecules in this section; they fall into four main categories.

(1) Linear tricycles, in which a six-membered ring is flanked by two five-membered rings (e.g., **8.144**, **8.145**) to give analogues of anthracene. These compounds are not very stable because of the unfavorable bond fixation in one of the five-membered rings.

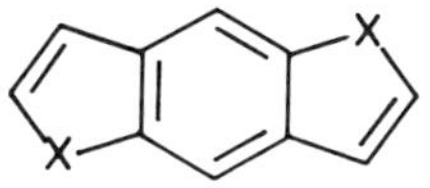

(8.144) 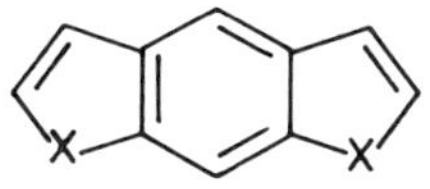 (8.145)

(2) Angular tricycles which are the corresponding analogues of phenanthrene (e.g., **8.146–8.148**); these compounds are more stable than those in group (1) above.

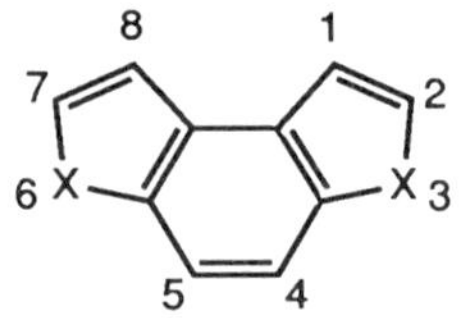

(8.146) 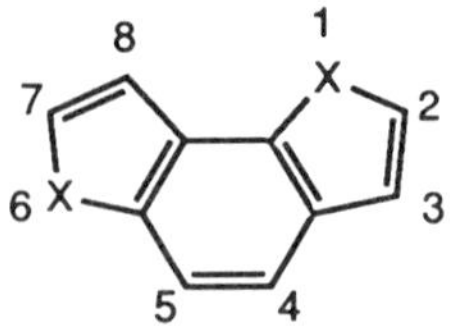   (8.147) 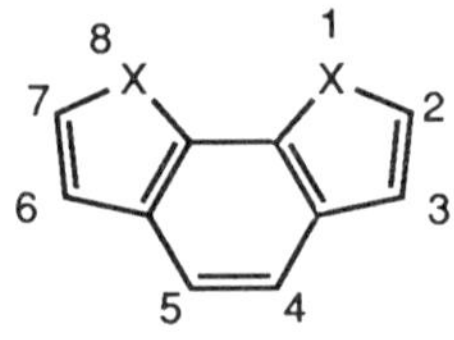   (8.148)

(3) Compounds in which a five-membered ring is flanked by one five- and one six-membered ring (e.g., **8.149, 8.150**).

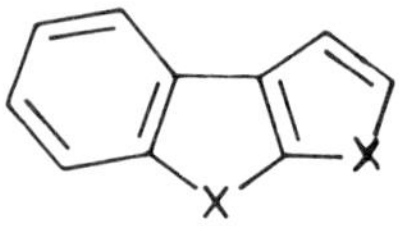

(8.149) 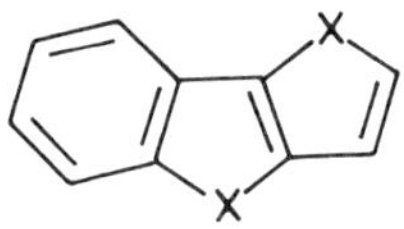   (8.150) 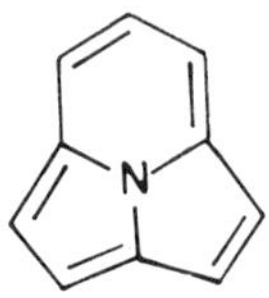   (8.151)

(4) Analogues of cycl[3,2,2]azine (**8.151**), in which all three rings are mutually fused.

For unsubstituted structures **8.144–8.150** alone (with X = O, S, Se, or NH) the number of compounds is 29 if both heteroatoms X are identical, but if the heteroatoms are different it rises to 89. If there are two or more heteroatoms in any one of the three rings then the number of compounds becomes very large indeed. However, there have as yet been very few studies of the reactivities of any of these molecules.

Different nomenclature systems are in use and a given molecule may be described (e.g., in *Chemical Abstracts*) under different names. For example, compound **8.146** (X = S) has been described as benzo[1,2-*b* : 4,3-*b*′]dithiophene, but this name (used in order to match the number–letter sequence in square brackets with the ring fragments) is misleading since it clearly is not a dithiophene, such a name being more appropriate to compounds of type (3) above. On the other hand, the more meaningful name dithieno[1,2-*b* : 4,3-*b*′]benzene (cf. dibenzothiophene, etc.) has the number–letter sequence the wrong way around (though there seems no reason why [*b*-1,2] etc. could not come into general use). A more rigorous name is thieno [2′,3′ : 1,2]benzo[4,3-*b*]thiophene, and thieno[3,2-*e*][1]-benzothiophene could also be used.

A. Acid-Catalyzed Hydrogen Exchange

The only quantitative and comprehensive determination of the reactivity of molecules in this section has been through hydrogen exchange of the dithienobenzenes (**8.146–8.148**, X = S) and of cycl[3,2,2]azine (**8.151**). For detritiation in anhydrous trifluoroacetic acid at 70°C, the partial rate factors (corrected for hydrogen bonding) for the dithienobenzenes are as shown in Scheme 8.24 and the corresponding σ^+ values calculated from these are given in Scheme 8.25 [83JCS(P2)813]. Hydrogen bonding produces attenuation of the rate of exchange in TFA by an amount similar to that found with thienothiophenes and in general there is a parallel between the extent of hydrogen bonding and the number of sulfur atoms present in the heterocycle (see Section 6 for tabulated data).

The data in Schemes 8.24 and 8.25 show the following results.

(1) The positional reactivity order in each molecule is sulfur $\alpha > \beta >$ positions in the benzene ring.

(2) Relative to thiophene and benzo[*b*]thiophene the order of reactivity at the α-position to sulfur is thiophene > dithienobenzenes > benzo[*b*]thiophene, and at the β position is dithienobenzenes > benzo[*b*]thiophene > thiophene.

(3) In compound **8.147** (X = S) the lower reactivity of the 5-position ($\sigma^+ = -0.645$) compared to the 4-position ($\sigma^+ = -0.70$) reflects the lower reactivity of the 7-position in benzo[*b*]thiophene compared to that of the 4-position.

The fact that the $\alpha > \beta$ reactivity order parallels that in thiophene rather than benzo[*b*]thiophene has also been observed in nitration and bromination [77CS(12)97], and was regarded as being explicable only by "sophisticated MO methods." However this is not necessary. Reference to Scheme 8.1 shows that in benzo[*b*]thiophene, 3-substitution is favored over 2-substitution because both canonicals for the former are benzenoid, whereas only one is for the latter. If, however, a second thiophene ring is fused to the benzene ring, then because of the bond fixation in thiophene, the nonbenzenoid canonicals for 2-substitution become much

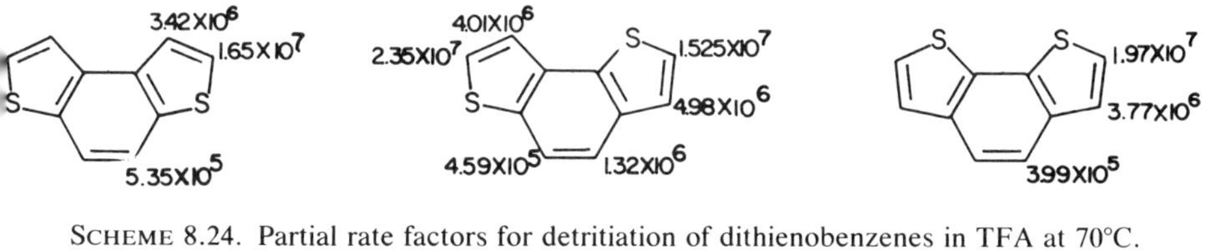

SCHEME 8.24. Partial rate factors for detritiation of dithienobenzenes in TFA at 70°C.

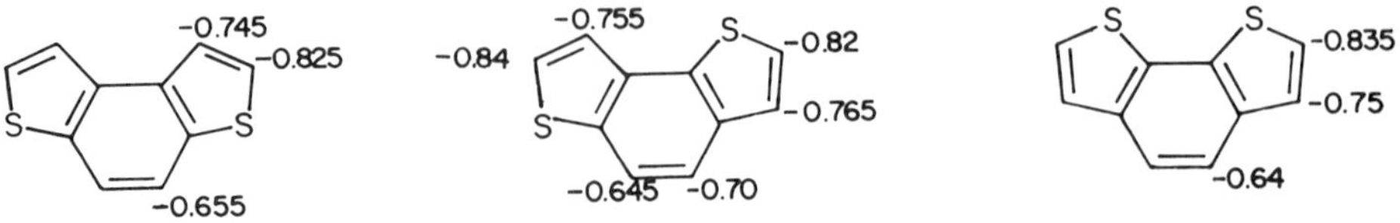

SCHEME 8.25. Values of σ^+ for detritiation of dithienobenzenes in TFA.

more stabilized (e.g., as in **8.152**). Moreover the charge in the benzenoid ring can be delocalized into the second thiophene ring, and onto sulfur in particular (e.g., **8.153**). By contrast, for 3-substitution no delocalization of charge into the second thiophene ring is possible. Both factors therefore combine to produce preferential α-substitution.

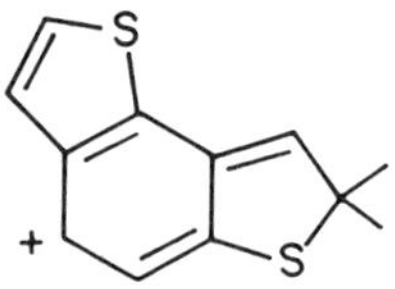

(8.152 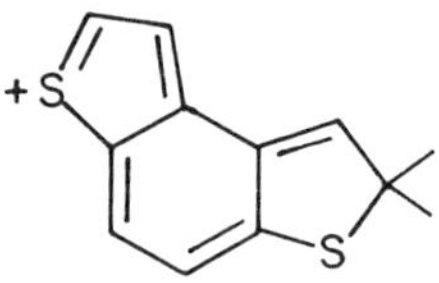)   (8.153)

Both π-densities and localization energies have been calculated for these molecules by the Hückel method with $\beta_{CS} = 0.6$ and $\alpha_S = \alpha_C$ (i.e., $h = 0$) (Table 8.19). Each position is assigned a number, and for the correspondence of these to the positions as given in (**8.146–8.148**), see Table

TABLE 8.19

HÜCKEL LOCALIZATION ENERGIES AND π-ELECTRON DENSITIES

Compound	Position	No.	π Densities	$\Delta L_r^+ / -\beta^a$
8.146, X = S	1	1	1.176	1.877
	2	2	1.119	1.745
	4	3	1.060	2.063
8.147, X = S	2	4	1.120	1.740
	3	5	1.175	1.840
	4	6	1.064	1.996
	5	7	1.060	2.043
	7	8	1.117	1.732
	8	9	1.177	1.848
8.148, X = S	2	10	1.120	1.756
	3	11	1.174	1.862
	4	12	1.063	2.043

[a] For benzene the value is 2.536.

8.19. Comparison of the observed and predicted reactivity orders are as follows.

Observed: 8 > 10 > 2 > 4 > 5 > 9 > 11 > 1 > 6 > 3 > 7 > 12.
Predicted (localization energies): 8 > 4 > 2 > 10 > 5 > 9 > 11 > 1 > 6 > 7 = 12 > 3.
Predicted (π densities): 9 > 1 > 5 > 11 > 10 > 4 > 2 > 8 > 6 > 12 > 3 = 7

It is evident that localization energies are excellent for predicting the results here (and are totally within a given molecule) whereas π densities are very poor. Localization energies also correctly predict (using the same parameters) the reactivity order in thiophene, the thienothiophenes (Section 5) and the dithienothiophenes (Section 6). By contrast for benzo-[*b*]thiophene (Section 2), only π densities correctly predict the overall order whereas neither method works for dibenzothiophene (Section 3). Other π-density calculations for the dithienobenzenes [69ZN(B)12] also fail.

For cycl[3,2,2]azine (**8.151**) the partial rate factor for detritiation of the 1-position in trifluoroacetic acid at 70°C has been determined as 4.34 × 10^{13}, the σ^+ value being -1.56 [87JCS(P2)591]. The high reactivity of cycl[3,2,2]azine is due, as in the case of idolizine, to formation of a pyridinoid ring in the transition state (**8.154**) [87JCS(P2)591]. However, the 1-position is 445 times *less* reactive than the corresponding position in idolizine (cf. **8.63**, and Table 8.13). This may be attributed to the greater ground-state stability of cycl[3,2,2]azine, which has two 10π-electron circuits, whereas indolizine has only one [87JCS(P2)591].

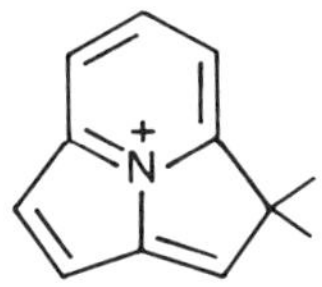

(8.154)

B. Other Reactions

There are insufficient data to permit a general description of the reactivity patterns, which are particularly dependent upon the combination of heteroatoms and their positions. π-Density calculations have been carried out for a few molecules (in addition to those given in Table 8.19). Values

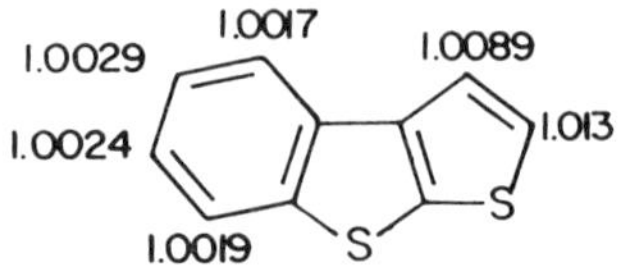

(8.155) 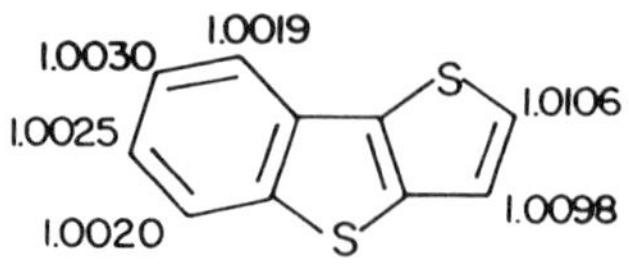   (8.156)

for thieno[2,3-*b*][1]benzothiophene (**8.155**) and thieno[3,2-*b*][1]benzothiophene (**8.156**) were calculated with the parameters $\beta_{CS} = 0.6$, $h = -0.1$, and these calculations predicted 2-substitution, as found in acylation, bromination, nitration, and metallation [though the latter depends upon the $-I$ (inductive withdrawal) effect of sulfur] [71JCS(C)463, 71JCS(C)1308]. The high bond order causes 2-bromination and 2-acylation of the 3-methyl derivatives, and 3-bromination of the 2-methyl derivatives. 3-Acetylation of 3-methylthieno[2,3-*b*][1]benzothiophene was accompanied by some 6-acetylation [71JCS(C)1308], which may be attributed to the high steric hindrance to acetylation. Nitration of 2-formylthieno[2,3-*b*][1]benzothiophene took place at the 2-position (nitrodeacylation) and this is attributable to the high conjugative deactivation of the 3-positions arising from the high 2,3-bond order.

Bromination, acylation, and nitration of selenolo[2,3-*b*][1]benzothiophene (**8.157**) and thieno[2,3-*b*][1]benzoselenophene (**8.158**), and their [3,2-*b*] isomers take place preferentially at the 2-positions [76CA(84)135510]. Acetylation of 1-methyl[1]benzofuro[3,2-*b*]pyrrole, and the thieno and selenolo analogues (**8.159**, X = O, S, or Se) gave the following isomer yields: 100% (2-position), 0% (3-position) (X = O); 32% (2), 68% (3) (X = S); 42% (2), 58% (3) (X = Se). For the [2,3-*b*] isomers (**8.160**, X = S, Se), the yields were 66% (2) and 34% (3) in each case (83JHC61). These results appear to demonstrate that for compounds **8.160**, conjugation between oxygen and the 2-position (**8.162**) outweighs the preferential direction into the 3-position by the NMe group. However, for compounds **8.159**, this resonance disrupts the 6π benzenoid ring in the transition state (**8.161**) so conjugative electron release from S or Se is insufficient to produce 2-substitution. By contrast, in **8.162** there is no loss of the 6π structure so conjugation here will be stronger, hence both S and Se compounds give preferential 2-substitution. Bromination and acylation at the 2-position of 4*H*-furo[3,2-*b*]indole (**8.163**) is aided by *N*-benzoylation. For the *N*-benzoyl derivative, the yields of 2-derivatives were 61% (bromination), 57% (benzoylation), and 43% (acetylation) (78JHC123).

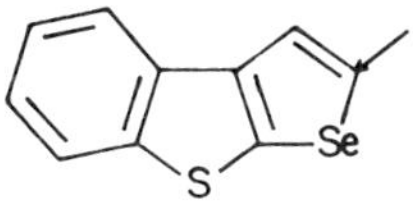

Sec. 4.B] TWO FIVE- AND ONE SIX-MEMBERED RING COMPOUNDS 259

(8.157) 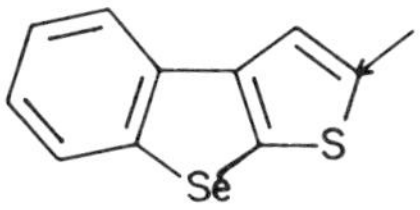   (8.158)

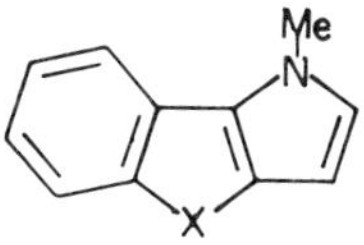

(8.159) 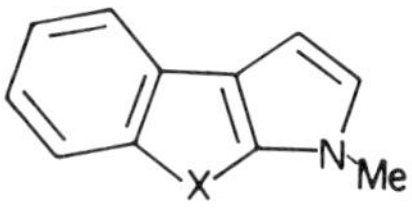   (8.160) 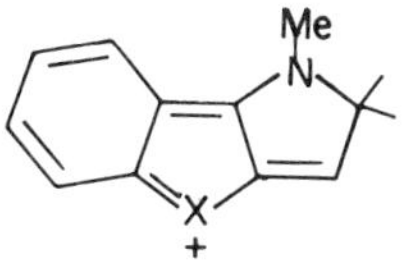   (8.161) 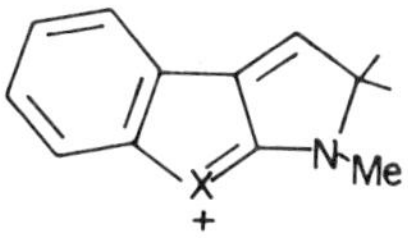   (8.162)

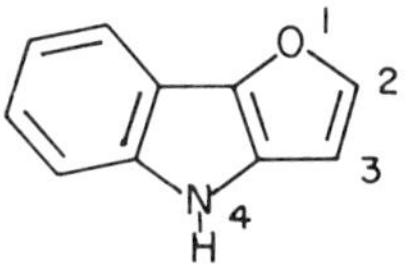

(8.163)

π-Density calculations for 1H,5H-pyrrolo[2,3-f]indole (**8.164**) and 3H,6H-pyrrolo[3,2-c]indole (**8.165**) indicate 3,7- and 1,8-substitution, respectively, as found in acylation and diazonium coupling (83CHE871). MO calculations indicate that the 1- and 3-positions are the most reactive ones in 4H-pyrrolo[1,2-a]benzimidazole (**8.166**) (74CHE230). The same result is more easily deduced and understood as follows. Although the lone pair on N-4 can be delocalized to the 1-, 3-, 5-, or 7-positions, the important factor is delocalization of the lone pair from N-9 to the 1- and 3-positions (e.g., as in **8.167**). Although the 6π pyrrole ring of the ground

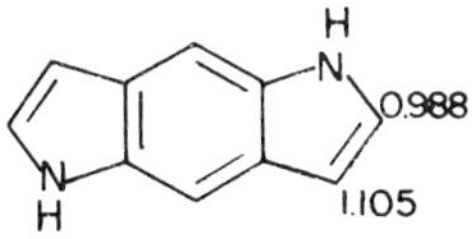

(8.164) 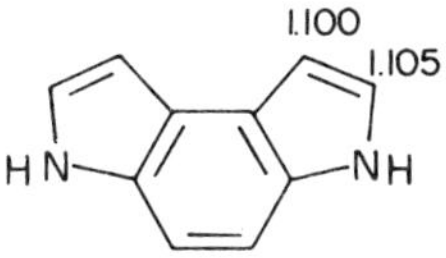   (8.165) 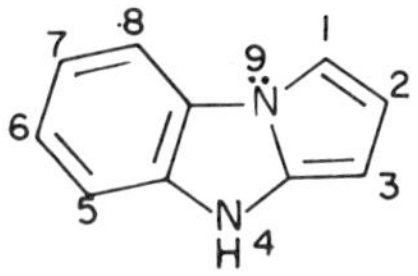   (8.166)

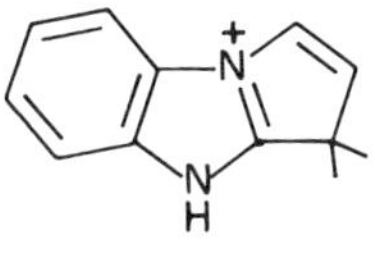

(8.167)

state is lost in the transition state, a 6π imidazole ring is gained, so the transition state will be stabilized, and substitution at the 1- and 3-positions will be very easy. Substitution at the 1-position involves the maximum delocalization of the N-9 lone pair, so this position should be the most reactive, as found in acylation [67CHE723; 69CA(70)68249] and nitrosation [71CA(74)76377]. Deuteriation also takes place at the 1- and 3-positions (76CHE64) and the 1,3-dimethyl derivative is preferentially protonated at the 1-position (though the 1-methyl and 3-phenyl derivatives are preferentially protonated at their 3-positions) (72CHE1023). The selectivity of the protonation position was temperature dependent (76CHE64), so that both kinetic and thermodynamic factors are involved. Nitration of 4H-pyrrolo[1,2-a]benzimidazole has not been achieved, presumably because under nitrating conditions protonation with consequent deactivation occurs. By contrast the 1- and 3-acetyl derivatives are readily nitrodeacylated [71CA(74)76377]; here conjugation between the nitrogen lone pairs and the acetyl groups will substantially reduce the basicity and hence the extent of N-protonation.

Conjugation from the bridgehead nitrogen without loss of aromaticity in the transition state is evidently responsible for the 1-substitution of 2-phenylpyrrolo[2,1-b]benzothiazole (**8.168**) (67G1286), 3-substitution of 2-phenylimidazo[2,1-b]benzothiazole (**8.169**) (67G1286), 1-substitution in 3-phenylimidazo[5,1-b]benzothiazole (**8.170**) (66CHE210), 1-substitution in 3-phenylimidazo[5,1-b]benzimidazole (**8.171**) (72KFZ22), 3-substitution in 1-phenylimidazo[5,1-b]benzimidazole (**8.172**) (73CHE366), 3-substitution of 2-alkylpyrazolo[2,3-a]benzimidazole (**8.173**) (68USP3369897), and 3-substitution of 1,3,4-triazolo[3,4-b]benzothiazole (**8.174**) (70CHE849). Likewise, isoindolo[1,2-b]benzothiazoles (**8.175**) substitute where shown (63JGU1946).

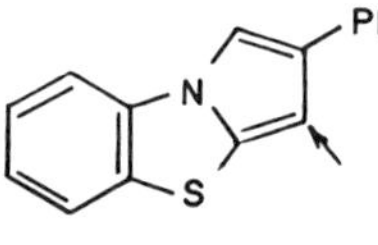

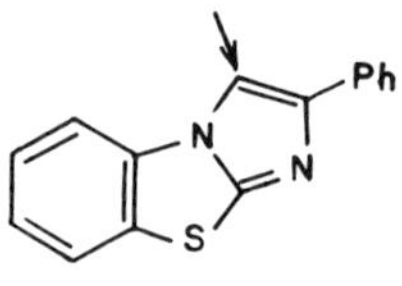

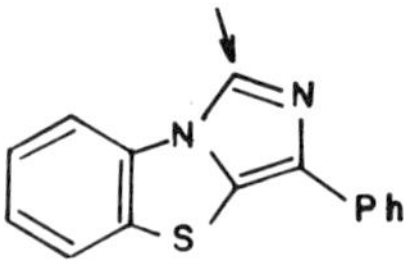

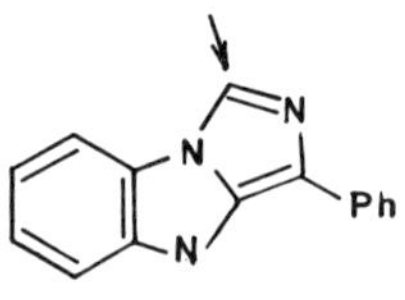

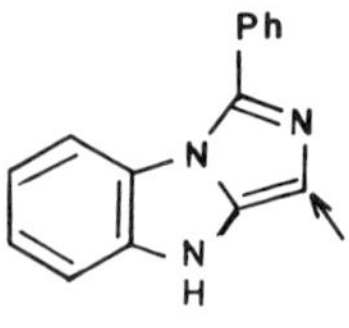

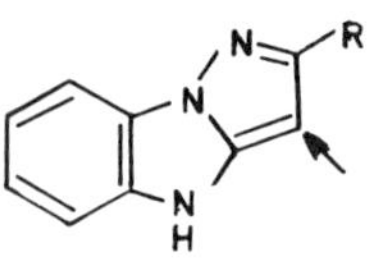

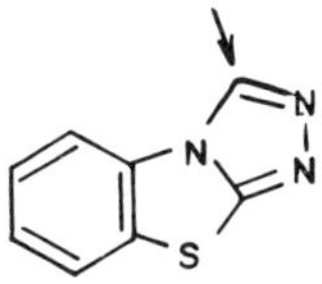

Sec. 4.B] TWO FIVE- AND ONE SIX-MEMBERED RING COMPOUNDS 261

(8.174) 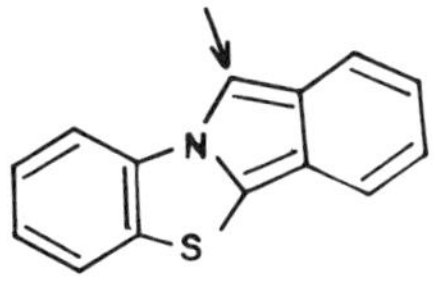   (8.175)

Benzo[1,2-d:4,5-d']diimidazole (**8.176**) is halogenated in the 4- and 8-positions (50JCS1515; 52JGU1069; 58JGU2214) and nitrated in the 4-position (52JGU1069); this may be attributed to both strong deactivation of the 2-position by N-3 due to the high 3,2-bond order, and also to delocalization of the lone pair from N-1 to C-4 to give the transition state **8.177**. This latter has lost a 6π benzenoid ring, but gained a 6π imidazole ring and is therefore particularly stable. Similar reasoning accounts for the 4-nitration and 4,5-dibromination of benzo[1,2-d:3,4-d']diimidazole (**8.178**) (73CHE95).

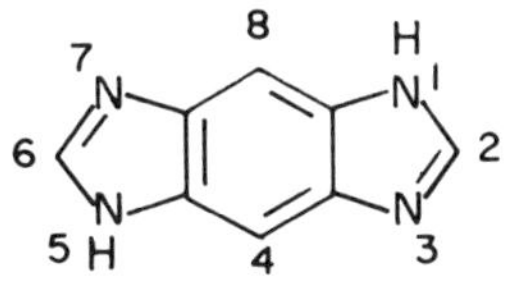

(8.176) 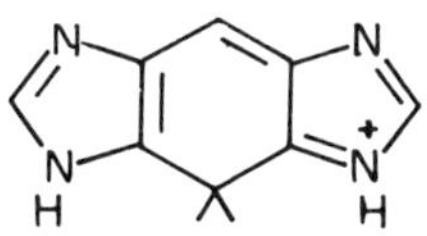   (8.177) 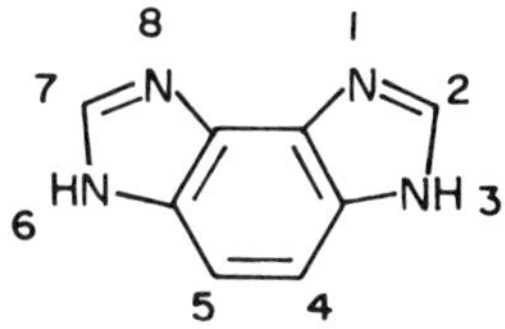   (8.178)

Nitration of the dithienobenzene **8.146** (X = S) gave 59% 4-, 35% 2-, and 6% 1-substitution, and bromination gave the 2- and 1-derivatives in a 5:3 ratio [77CS(12)97]. The bromination result is consistent with the hydrogen-exchange data and σ^+ values given above (Schemes 8.24, 8.25), but the nitration is anomalous, as is often the case. By contrast, both nitration and bromination of **8.148** (X = S) gave mainly the 2-derivative, as required by the σ^+ values. Further bromination went into the 7-, 3-, and 6-positions, in that order.

Acetylation of 2,3-dimethylthienofurobenzene (**8.179**) differs from the above results in going into the 5-position of the benzene ring (65BSF1473); this finding is attributable to the strong para activation by oxygen. Nitration of benzothieno[3,2-b][1]benzothiophene (**8.180**) goes

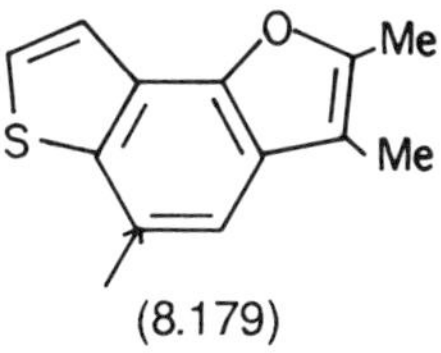

(8.179) 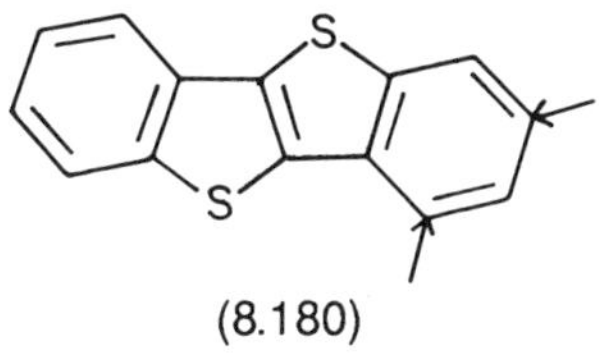   (8.180)

into the 3- and 5-positions (80JOU391) but it is not certain that this shows these to be the sites of highest electrophilic reactivity (cf. nitration of benzo[b]thiophene, Section 2.B).

There have been two studies of molecules containing one seven- and two five-membered rings. Deuteration, acylation, nitration, and halogenation of 2-phenylcyclohepta[4,5]pyrrolo[1,2-a]imidazole (**8.181**) occurred in the 3- and 10-positions, as predicted by π densities. The π values (positions) are 1.22 (10), 1.10 (3), 1.07 (2), and <1.0 (others) [84JCR(S)390]. Hydrogen exchange showed the 10-position to be the most reactive, though the more sterically hindered nitration and acylation occurred preferentially at the 3-position. The high reactivity of the 3- and 10-positions arises because both are conjugated with N-4; in addition, substitution at the 10-position creates a 6π aromatic ring in the transition state (**8.182**). Deuteration of some dithieno- and furothieno-annelated tropylium ions (**8.183, 8.184**) took place at the heterocyclic β-positions, as predicted by π densities (73ACS2257).

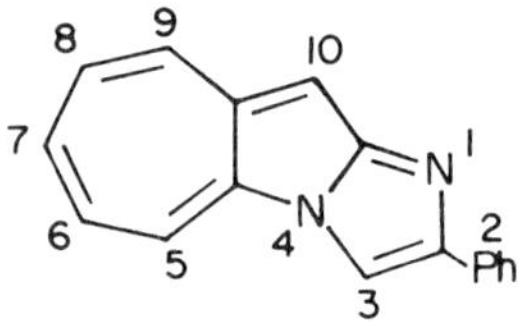

(8.181)

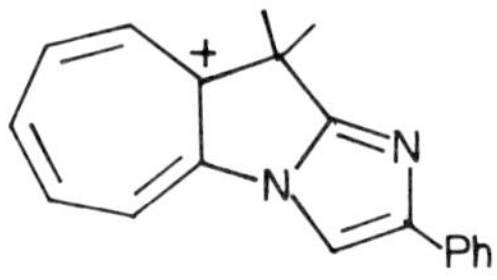

(8.182)

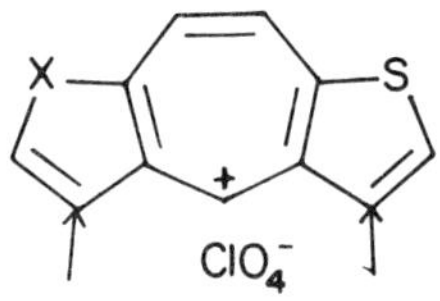

(8.183, X=O, S)

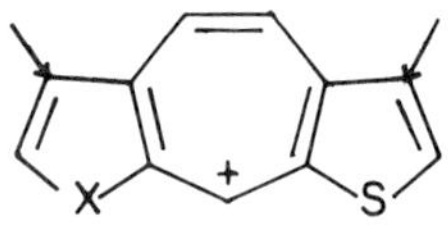

(8.184, X=O, S)

5. Compounds with Two Five-Membered Rings

Most work has been concerned with molecules of types **8.185–8.188**, and in particular the thienothiophenes, selenolothiophenes, and selenoloselenophenes, (**8.185–8.187**). Many MO calculations have been carried out, and all of them are unsatisfactory in some respect. For example, π densities (calculated by the SCF-CNDO/2 method) indicate that, as expected, the 2-positions of **8.185** and **8.186** should be more reactive than the 3-positions; however, the 3-positions are predicted to be deactivated, which is incorrect. Positions α and β to selenium were predicted be be,

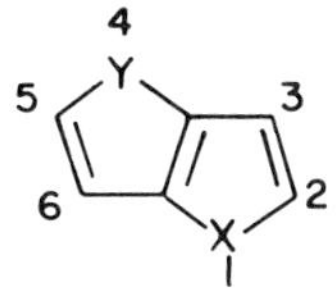

(8.185) 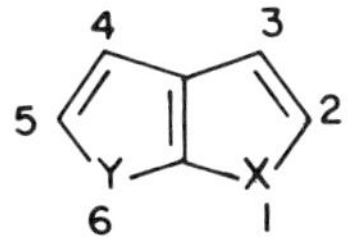   (8.186) 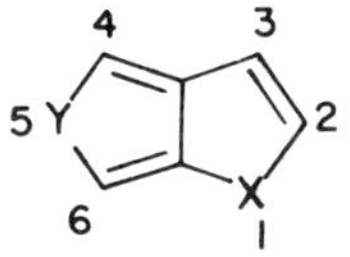   (8.187)

respectively, more activated and more deactivated than those α and β to sulfur. For compounds **8.187**, the positional order is predicted by charge densities to be 4 >> 6 > 2 > 3, and by localization energies to be 4 > 6 > 2 > 3 regardless of the nature of X and Y (80CHE142; 81BAU1089); other workers give the order (for X = Y = S) as 6 > 4 = 2 > 3 (70ACS23). Since reaction at positions 4 or 6 involves no loss of aromaticity on forming the transition state (e.g., **8.189**), these positions should be the most reactive, the order 4 > 6 being expected since the former position is conjugated with the heteroatom at the 1-position; the 2 > 3 positional order is expected because position 2 is conjugated with the heteroatom at the 5-position.

For 4H-furol[3,2-b]pyrrole (**8.185**, X = O, Y = NH) π densities predict the order 6 > 3 > 2 > 5 (the 5-position being deactivated) (80CCC2949). However, a 5-phenyl substituent was predicted to deactivate the 6-position, which is almost certainly wrong. More reasonable seem the π densities (SCF) which predict the orders 5 > 6 > 3 > 2 for 4H-thieno[3,2-b]pyrrole (**8.185**, X = S, Y = NH), 5 > 4 > 3 > 2 for 6H-thieno[2,3-b]pyrrole (**8.186**, X = S, Y = NH), and 6 > 4 > 3 > 2 for 5H-thi-eno]2,3-c]pyrrole (**8.187**, X = S, Y = NH), all positions being activated (71T4045). However, the reliability of these calculations must be considered with the fact that they also predict the 2-position of thiophene to be more reactive than the 2-position of pyrrole. Total electron densities (*ab initio*) predict the reactivity order in 4H-thieno[3,2-b]pyrrole and 6H-thieno[2,3-b]pyrrole to be 2 > 6 > 3 > 5 and 2 > 4 > 3 > 5, respectively. This method predicts the β-positions of both pyrrolopyrroles (**8.185, 8.186**, X = Y = NH) to be the most reactive [85JCS(P2)97].

π Densities (SCF) indicate that the most reactive sites in 1H-pyrrolo[1,2-a]imidazole (**8.190**), 1H-pyrrolo[1,2-b]-*sym*-triazole (**8.191**), 1H-pyrrolo[2,1-c]-*sym*-triazole (**8.192**), and 1H-pyrrolo[1,2-a]imidazole

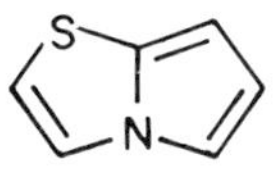

(8.188) 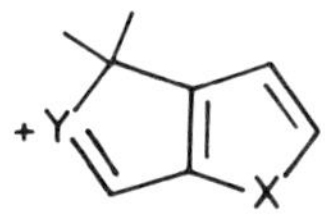   (8.189) 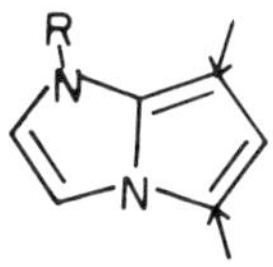   (8.190)

(**8.193**) are as indicated (74CHE230). These positions are comparable to the 1- and 3-positions of indolizine (**8.61**) and are very reactive for similar reasons [i.e., in the transition state for reaction (e.g., **8.194**), there is no loss of aromaticity since both **8.190** and **8.194** each contain one completely conjugated 6π five-membered ring]. The relative resonance stabilization should not be so great as with indolizine, so the reactivity should be lower, but there are as yet no quantitative experimental data with which to confirm this prediction.

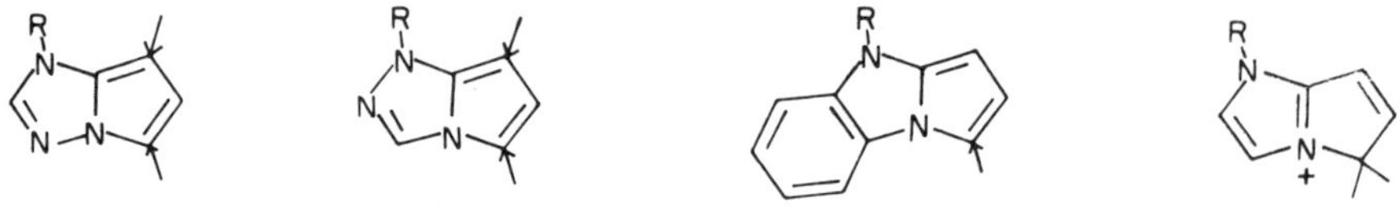

(8.191)   (8.192)   (8.193)   (8.194)

The ability of thienothiophenes and selenoloselenophenes to transmit substituent effects has been determined by ^{1}H and ^{13}C NMR, and also calculated by the CNDO/2 method [82CS(20)208; 83CS(22)22]. As expected, resonance effects are better transmitted between the 2- and 5-positions in **8.185** than in **8.186**, since only in the former is direct conjugation between these positions possible. Both resonance and inductive/field effects were greater in the sulfur- relative to the selenium-containing compounds, but in general, replacement of sulfur by selenium caused only minor perturbations of data, which follows from the fact that the main mode of transmission of substituent effects in five-membered heterocycles must be via the carbon chain (cf. Chapter 6, Section 9.C).

For selenolo[2,3-*b*]selenophene (**8.186**, X, Y = Se) comparisons of the transmissions of substituent effects between the 2-position on the one hand, and the 3-, 4-, and 5-positions on the other, showed the following generalizations.

(1) Field/inductive and resonance effects are both transmitted most effectively between the 2- and 3-positions. This is as expected.

(2) The field/inductive effect is greater between the 2- and 5-positions than between the 2- and 4-positions.

(3) The resonance effect is greater between the 2- and 4-positions than between the 2- and 5-positions. This is explained in terms of the weak conjugative interaction arising from structure **8.195**, there being no comparable interaction between the 2- and 5-positions. Additional MO calculations are described under hydrogen exchange below.

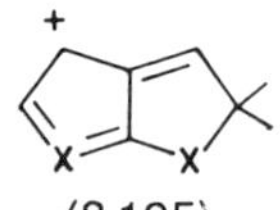

(8.195)

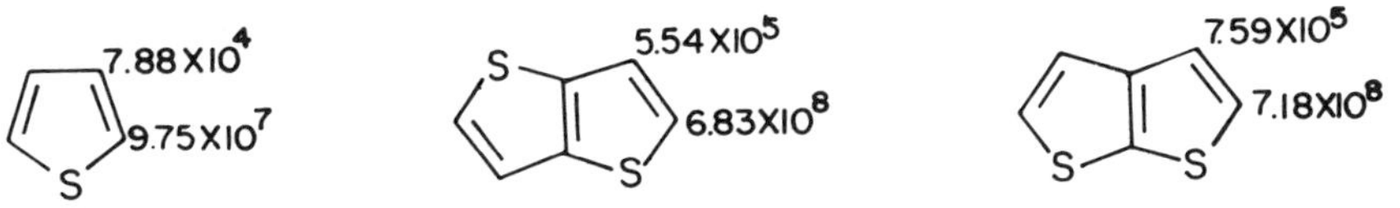

SCHEME 8.26. Partial rate factors for detritiation of thiophene and thienothiophenes in TFA at 70°C.

A. ACID-CATALYZED HYDROGEN EXCHANGE

Detritiation in trifluoroacetic acid at 70°C has provided the only fully quantitative data for two fused five-membered rings; partial rate factors and σ^+ values for thienothiophenes are given in Schemes 8.26 and 8.27, respectively, along with data for thiophene obtained under the same conditions [82JCS(P2)295]. These values were corrected for the effect of hydrogen bonding, which is more severe in the thienothiophenes than in the thiophene. It has been found that, in general, the extent of hydrogen bonding parallels the number of sulfur atoms in the heterocycle. The results demonstrate a number of features.

(1) As in thiophene, the α-positions are in each case more reactive than the β-positions.

(2) At both α- and β-positions the thienothiophenes are more reactive than thiophene.

(3) At both positions the [2,3-*b*] isomer is more reactive than the [3,2-*b*] isomer.

(4) The relative reactivities at the α-position of thieno[2,3-*b*]thiophene (**8.186**, X = Y = S), thieno[3,2-*b*]thiophene (**8.185**, X = Y = S) and thiophene are 7.4 : 7.0 : 1.0; for the β-position the corresponding ratios are 9.6 : 7.0 : 1.0. The former rates are in excellent agreement with those (viz. 7.8 : 7.1 : 1.0) obtained for dedeuteriation in TFA at 25°C (70JGU1609), although this is partly fortuitous since the effect of the higher ρ factor under the latter conditions and the absence of correction for hydrogen bonding cancel each other out. In dedeuteriation, the relative reactivities of the β-position of the thienothiophenes (determined for the α,α'-diethyl derivatives) (70JGU1609) were also in the order found for detritiation.

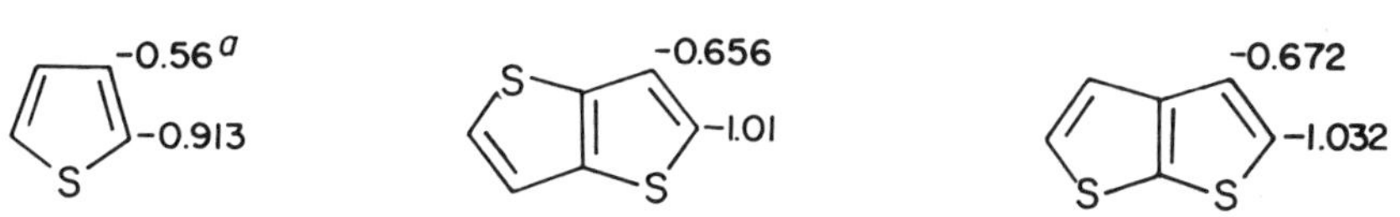

SCHEME 8.27. Values of σ^+ for detritiation of thiophene and thienothiophenes in TFA. [a] Owing to a typographical error this was given as −0.50 in the original paper.

(5) The higher reactivity at the α-position of the [2,3-*b*] isomer can be readily understood because conjugation between sulfur and the 2-position in the transition state involves less disturbance of the ground-state bond structure than is the case for the [3,2-*b*] isomer (cf. **8.196** and **8.197**). Thus the "substituent" interaction factor A_f is higher for **8.196** than for **8.197** [79JCS(P2)381].

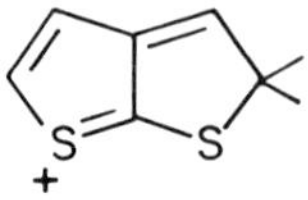

(8.196) 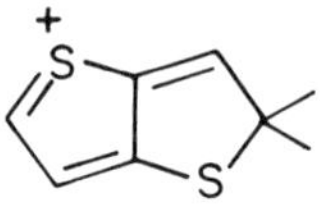   (8.197)

(6) Comparison of the positional reactivities for detritiation with those obtained in other electrophilic substitutions (Table 8.20) [82JCS(P2)295] shows good agreement with the results for acetylation (which has a similar ρ factor, −9.1, to that for the detritiation, −8.75) but disagreement with the chlorination and formylation data [72CS(2)137]. These latter give the reactivities of the α-positions of the thienothiophenes as the reverse of that in detritiation. Since chlorination and formylation have substantially higher ρ factors than acetylation or detritiation (compare the last column in Table 8.20), the transition states for the former will be nearer to the Wheland intermediate. It appears, therefore, that thieno[3,2-*b*]thiophene is more polarizable than thieno[2,3-*b*]thiophene.

(7) Hückel calculations of localization energies ($\beta_{CS} = 0.6$, $\alpha_S = \alpha_C$) give the positional orders as 2 (thiophene) > 2 [3,2-*b*] > 2 [2,3-*b*] > 3 [2,3-*b*] > 3 [3,2-*b*] > 3 (thiophene) [82JCS(P2)295]. The reactivity of the 2-position of thiophene is clearly predicted incorrectly, and the order for the 2-positions of the thienothiophenes reflects that found for chlorination and formylation, which have transition states nearer to the Wheland intermediate. (Localization energy calculations assume, of course, that

TABLE 8.20
RELATIVE REACTIVITY OF THIENOTHIOPHENES IN ELECTROPHILIC SUBSTITUTION

Reaction	Thiophene		T[2,3-*b*]T		T[3,2-*b*]T		2-T[2,3-*b*]T/2-Thiophene
	2−	3−	2−	3−	2−	3−	
Acetylation	200	1	634	6.3	594	4.8	3.17
Detritiation	1235	1	9110	9.6	8670	7.0	7.4
Chlorination	250	1	5975	23.9	8300	<8.3	23.9
Formylation	>1000	1	>34,000	68	>44,400	—	34

the transition state *is* the Wheland intermediate.) CNDO/2 calculations of localization energies (66JCP389; 67JCP158) are better in that the reactivity of the 2-position of thiophene is correctly predicted to be less than the 2-position of the thienothiophenes, the rest of the order remaining the same. π Densities predict the order 2 [3,2-*b*] > 2 [2,3-*b*] > 2 (thiophene) > 3 [3,2-*b*] > 3 (thiophene) > 3 [2,3-*b*] (70ACS23). Total electron densities for the thienothiophenes, calculated by the *ab initio* method [85JCS(P2)97], predict the order for the 2-positions as found in detritiation, though the order for the 3-positions is predicted incorrectly.

(8) Annelation of thiophene by benzene, and by thiophene, produces the changes in σ^+ values shown in Scheme 8.28. Whereas benzene annelation changes the relative α- : β-reactivities, annelation by thiophene (in either configuration) produces much the same effect at both positions. The former result arises because in benzo[*b*]thiophene both canonical forms of the resonance hybrid for β-substitution are benzenoid, whereas only one of those for α-substitution is benzenoid) (i.e., the resonance energy of benzene plays an important role). The resonance energy of thiophene is considerably lower, so that its differential effect on the stabilities of the resonance hybrid for α- and β-substitution is smaller. An additional factor is that two of the canonical forms for the transition state for α-substitution of thienothiophenes involve the sulfur π electrons (cf. only one for benzo[*b*]thiophene). For β-substitution, however, there is only one structure involving the sulfur π electrons in each case [82JCS-(P2)295].

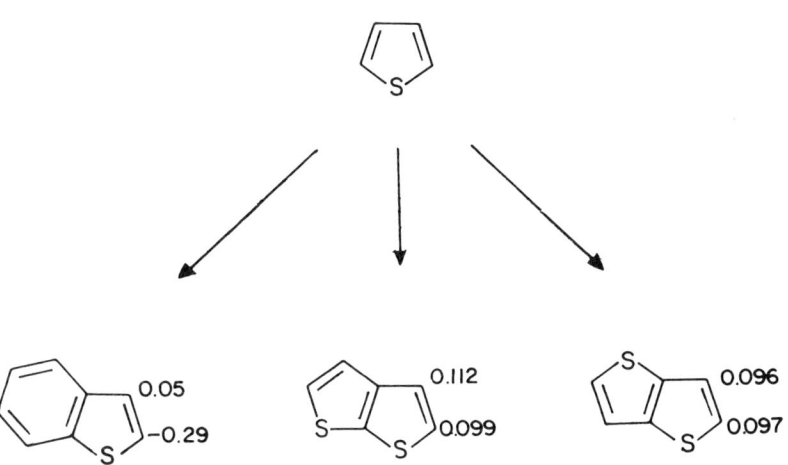

SCHEME 8.28. Change in σ^+ values due to annelation of thiophene.

From the rates of deuteriation of the 5- and 7-positions of 6-methylpyrrolo[2,1-*b*]thiazole (**8.198**) in D_2O/dioxan at 50°C (71T4171) and the reactivity relative to other aromatics, the partial rate factors have been determined under standard conditions (detritiation, TFA, 70°C) as 1.68×10^{18} and 1.19×10^{17}, respectively, the corresponding σ^+ values being -2.08 and -1.95 [87JCS(P2)591]. Exchange at these positions, and the extremely high reactivity, parallels the behavior of pyrroloazoles **8.190**–**8.192** and derives from the same cause, namely that the transition state is of comparable aromaticity to the ground state (cf. **8.194**) [87JCS(P2)591]).

Deuteriation of thieno[3,2-*c*]isoxazole (**8.199**, X = S) and selenolo-[3,2-*c*] isoxazole (**8.199**, X = Se) occurred at the 3-position in each case [76CS(10)165]. This is the expected result because of the following factors.

(1) The X group cannot conjugate directly with the 2-position.
(2) Nitrogen conjugatively deactivates both the 2- and the 6-position. The latter process involves formation of an aromatic 6π five-membered ring (**8.200**)
(3) Oxygen can conjugate directly only with the 6-position. This process is favorable as it also involves formation of a 6π five-membered ring, but this is cancelled out by the effect of the nitrogen noted in (2).

Nitration and bromination of (**8.199**) (X = S) also went into the 3-position, but for the selenium compound, attempted nitration caused decomposition, and bromination produced 2,3-addition [76CS(10)165].

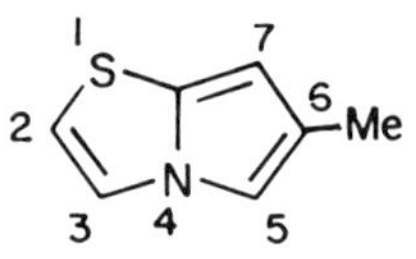

(8.198) 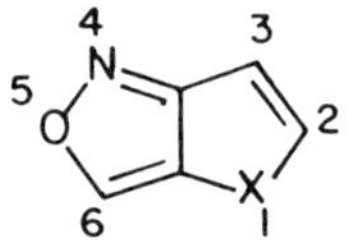   (8.199, X = S, Se) 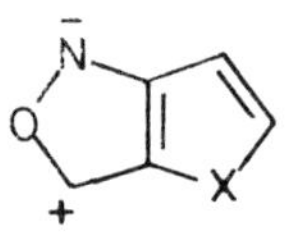   (8.200)

B. Base-Catalyzed Hydrogen Exchange

The relative rates of deuteriation of the α- and β-positions of thiophene and the thienothiophenes under base-catalyzed conditions have been determined as in Scheme 8.29 (71JGU1945). Since the rate of deuteriation of the 2-positions of thiophene has been determined under similar conditions to be 2.5×10^5 that of the 3-position Chapter 6, Section 2), it follows that for each molecule the β-position is less reactive than the α-position. The increased reactivity of the thienothiophenes compared to thiophene follows from the greater inductive electron withdrawal in the former (two

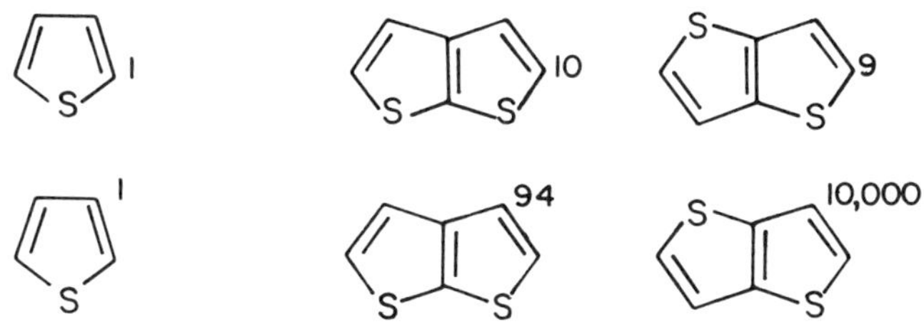

Sec. 5.C] COMPOUNDS WITH TWO FIVE-MEMBERED RINGS 269

α-positions[a]   1   10   9

β-positions[b]   1   94   10,000

SCHEME 8.29. Relative rates for base-catalyzed deuteriation of thiophene and thienothiophenes. [a] t-BuOK in t-BuOH; [b] t-BuOK in DMSO.

S atoms). The exceptionally high reactivity of the 3-position of thieno[3,2-b]thiophene was thought to arise from initial coordination at the adjacent sulfur atom.

Consistent with the results for base-catalyzed exchange, are the more rapid reactions of n-butyllithium with the thienothiophenes compared to thiophene (68ACS63).

C. OTHER REACTIONS

A quantitative study of acetylation, chlorination, and formylation of thieno[3,2-b]thiophene (**8.185**, X, Y = S), selenolo[3,2-b]thiophene (**8.185**, X = S, Y = Se), and selenolo[3,2-b]selenophene (**8.185**, X, Y = Se) gave the rates relative to thiophene given in Table 8.21 [80CS(15)206]. Only α-substitution was observed (due to the high selectivity of the reactions), and the results clearly show that replacement of S by Se increases the reactivity (cf. the reactivity of selenophene and thiophene, Chapter 6).

Acetylation, formylation, trifluoroacetylation, and nitrosation of 6-

TABLE 8.21
RELATIVE RATES OF ELECTROPHILIC SUBSTITUTION

	Acetylation	Chlorination	Formylation
Thiophene	1	1	1
Thieno[3,2-b]thiophene	3	33	44.5
Selenolo[3,2-b]thiophene	5	139	204
Selenolo[3,2-b]selenophene	8.3	347	635

methylpyrrolo[2,1-*b*]thiazole (**8.198**) all took place in the 5-position in 98, 88, 98, and 70% yield, respectively [66JCS(C)1908]. The high reactivity of the 5-position follows both from this being the site of maximal delocalization of the lone pair from nitrogen, and also because methyl will activate strongly across the high-order 5,6-bond. The strong electron release to the 5-position means that it readily undergoes protonation, and this is thought to be the reason for nitration (which occurs under strongly acidic conditions) occurring in the para position of the benzene ring of 6-phenylpyrrolo[2,3-*b*]thiazole (**8.201**).

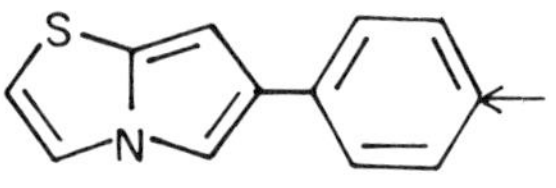

(8.201) 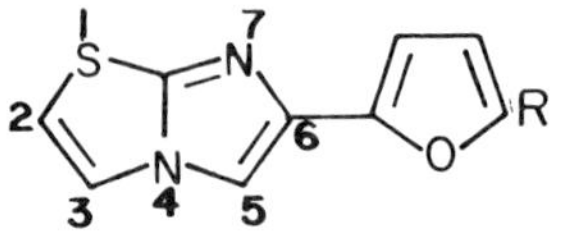   (8.202 R = H, NO$_2$) 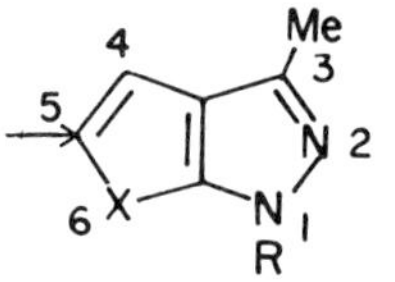   (8.203, X = S, Se)

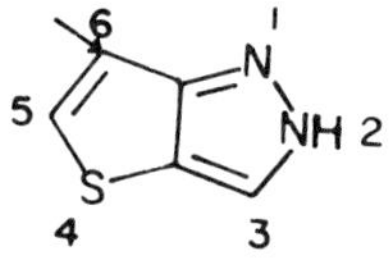

(8.204)

Selenoformylation of 6-methylpyrrolo[2,1-*b*]thiazole (**8.198**) and the 5,6- and 6,7-dimethyl derivatives also goes into the vacant and most reactive 7- and 5-positions, respectively [79JCS(P1)2334]. Nitrosation of 6-(2'-furyl)imidazo[2,1-*b*]thiazole (**8.202**, R = H) takes place at the 5-position (72CHE1223) for the same reasons that 6-methylpyrrolo[2,1-*b*]thiazole reacts at the 5-positions. Bromination and diazo coupling of the 6-phenyl derivative both go into the 5-position. However, nitration goes into the para position of the benzene ring [77HC(30)1], which may again be attributed to protonation under the strongly acidic conditions. Compound **8.202** (R = H) is of course less reactive than the pyrrolo analogue, and this is shown by the fact that the 5-nitro derivative (**8.202**, R = NO$_2$) cannot be nitrosated, though bromination succeeds. The 7-nitrogen in **8.202** is unable to conjugate with any position and so deactivates only by its weaker $-I$ inductive effect. Similarly, neither N-1 nor N-3 is able to conjugatively deactivate the thiazole ring in thiazolo[3,2-*b*]1,2,4-triazoles (**8.205**) or thiazolo[3,2-*d*]tetrazoles (**8.206**), which explains their ready substitution at the 2-position [71CA(75)140864; 74JHC459]. Indeed, **8.206** is nitrated more rapidly than dimethylthiazole (68CHE314), suggesting that those are nitrated as the free base, and protonated species, respectively.

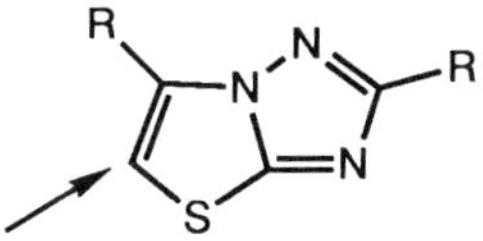

(8.205)

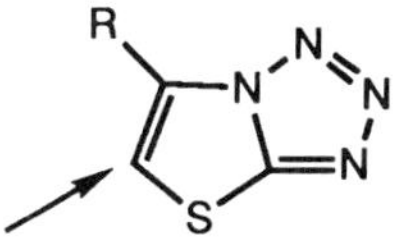

(8.206)

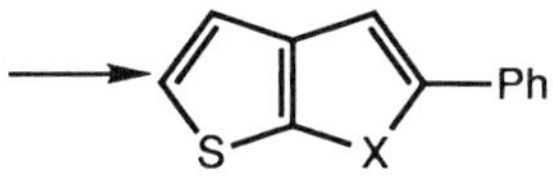

(8.207, X=NMe, O, S)

Formylation and bromination of 1,3-disubstituted 1H-thieno- and -selenolo[3,2-d]pyrazoles (**8.203**) takes place at the 5-position. For the 3-methyl-1-phenyl derivatives, the rates of formylation relative to that of thiophene were 0.61 and 1.77 for the S and Se compounds, respectively (73JOU2216). Although the lone pair on N-1 is conjugated with the 5-position, the 5-position is also conjugated with N-2, and as a result there is little overall activation. Moreover, conjugation between N-1 and C-5 reduces the aromaticity of the N-containing ring and is therefore unfavorable.

Bromination of 2H-thieno[3,2-c]pyrazole (**8.204**) takes place by contrast at the 6-position [77CS(12)1], which is equivalent to the 3-position in thieno[3,2-c]isoxazole (**8.199**, X = S), and the explanations advanced to account for the latter apply to **8.204** also. Rates of acetylation of compounds **8.207** relative to that of thiophene for different heteroatoms X were found to be NMe (9.4), O (5.0), S (1.5); the relative rate was 0.13 for **8.208**. For the selenophenes **8.209**, the rates relative to that of selenophene for different heteroatoms X were NMe (8.1), O (4.3), S (0.96) (76JOU1550). Formylation of selenopheno[2,3-c]thiophene (**8.210**) occurred at the 4- and 6-positions in a 3:2 ratio (81BAU1089). Both these positions are very reactive because the respective transition states create aromaticity in the selenium-containing ring. Lithiation went mainly in the 6-position (6-:4-isomer ratio = 4.6) because the *I* (inductive withdrawal) effects of both heteroatoms are greatest there. Imidazo[1,2-a]imidazole (**8.211**) is also substituted in the 5-position, the yield in nitration under mild conditions being 66% (73JOC1955).

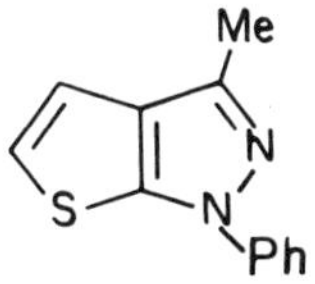

(8.208)

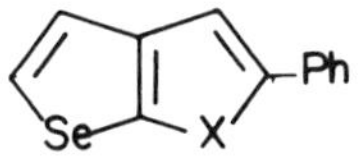

(8.209, X=NMe, O, S)

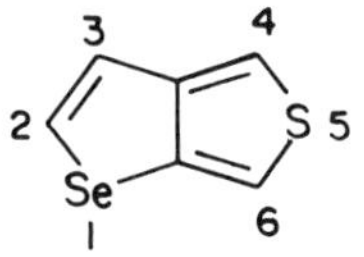

(8.210)

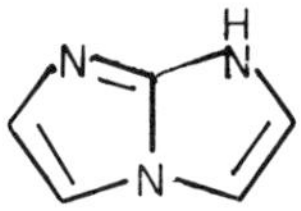

(8.211)

6. Compounds with Three or More Five-membered Rings

ACID-CATALYZED HYDROGEN EXCHANGE OF DITHIENOTHIOPHENES

The only quantitative study of the electrophilic reactivity of molecules in this category concerns the detritiation of all positions of the isomeric dithienothiophenes in trifluoroacetic acid at 70°C [82JCS(P2)301]. The rate data were corrected for hydrogen bonding, which was more severe than for the thienothiophenes (Section 5) because of the greater number of sulfur atoms in each molecule. There is a rough parallel between the extent of hydrogen bonding and the number of sulfur atoms, as the data in Table 8.22 show. The differences in exchange rate between 100% CF$_3$COOH and 35% CF$_3$COOH/65% CH$_3$COOH decrease with the severity of the attenuation in rate in the former medium, due to the effect of hydrogen bonding. Partial rate factors are given in Scheme 8.30 and the σ^+ values in Scheme 8.31.

The main features of these results are summarized below.

(1) The dithienothiophenes show very high reactivity toward electrophilic substitution, comparable to that of phenol.

(2) The α-positions are in each case more reactive than the β-positions.

(3) The order of reactivity for dithieno[2,3-b : 3′,2′-d]thiophene (**8.212**) dithieno[3,2-b : 2′,3′-d]thiophene (**8.213**), and dithieno[2,3-b : 2′,3′-d]thiophene (**8.214**) are for the α-positions **8.212** > **8.214** > **8.213**, and for the β-positions **8.214** > **8.212** > **8.213**. However, the reactivity of the α-positions on one hand, and the β-positions on the other, are almost independent of structure.

(4) The effects of annelation, in terms of the increment in σ^+ values, are shown in Scheme 8.32. In general, the increment in the σ^+ value is

TABLE 8.22
EFFECT OF HYDROGEN BONDING IN DETRITIATION[a]

Compound	k(difference)	Number of S atoms
Mesitylene	5230	0
Thiophene	2420	1
Benzo[b]thiophene	2200	1
Thienothiophenes	950	2
Dithienothiophenes	550	3

[a] As measured by rate differences between 100% CF$_3$CO$_2$H and 35% CF$_3$CO$_2$H–CH$_3$CO$_2$H (35:65).

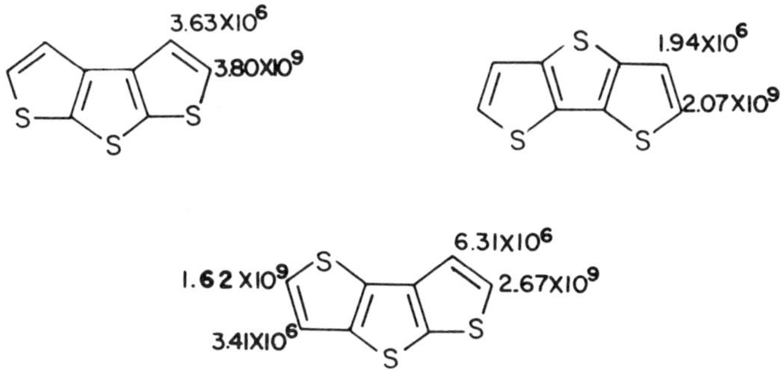

SCHEME 8.30. Partial rate factors for detritiation of dithienothiophenes in TFA at 70°C.

(8.212)

(8.213)

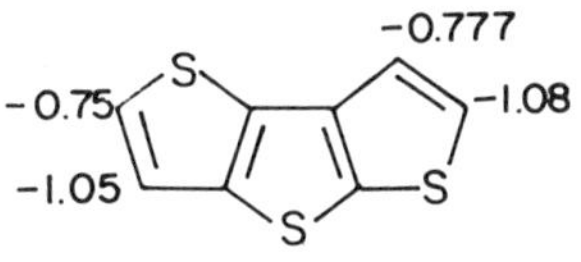

(8.214)

SCHEME 8.31. Values of σ^+ for detritiation of dithienothiophenes in TFA.

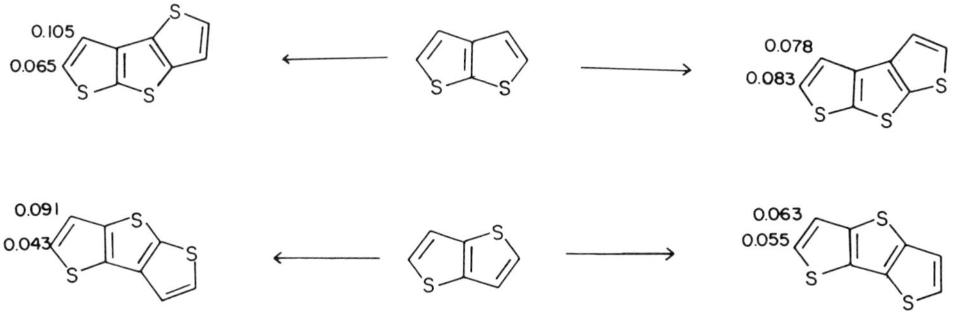

SCHEME 8.32. Increase in σ^+ value produced by annelation in dithienothiophenes.

TABLE 8.23
HÜCKEL LOCALIZATION ENERGIES FOR
DITHIENOTHIOPHENES

Compound	Position	No.	$L_r^+/-\beta$
8.212	2	1	1.6044
	3	2	1.8173
8.213	2	3	1.6445
	3	4	1.8100
8.214	2	5	1.6097
	3	6	1.8030
	5	7	1.6464
	6	8	1.8153

greater when the annelating sulfur atom is on the same side of the molecule as the position undergoing substitution.

(5) The positional reactivity order is quite well predicted by localization energy (Hückel) calculations using $\beta_{CS} = 0.6\beta_{CC}$ and $\alpha_S = \alpha_C$ (Table 8.23). The observed reactivity order is $1 > 5 > 3 > 7 > 6 > 2 > 8 > 4$, and the predicted reactivity order is $1 > 5 > 3 > 7 > 6 > 4 > 8 > 2$.

The agreement is quite remarkable considering the very small differences in reactivity for like positions. Small differences in polarizability of the molecules could cause slight change in the order for other electrophilic substitutions.

Part III

Six-Membered Heterocyclic Rings

CHAPTER 9

Heteroaromatics Containing One Six-Membered Ring

1. Introduction

Comprehensive accounts of electrophilic substitution in pyridines and pyridine N-oxides have been given previously [60HC(14,1-4); 67MI1; 70RCR627; 74HC(14,1-4)]; these contain many data concerning the qualitative aspects of substituent effects. In addition, there have been reviews of heteroaromatic reactivity (71PMH55), base-catalyzed hydrogen exchange [74AHC(16)1; 76CHE1397], and electrophilic substitutions of 3-hydroxypyridines [75RCR823; 76CHE955].

A. Compounds Considered

Compounds covered in this chapter include the following.

(1) Pyridines and substituted pyridines (**9.1**), and their protonated derivatives (**9.2**).

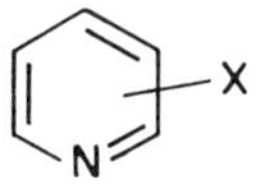

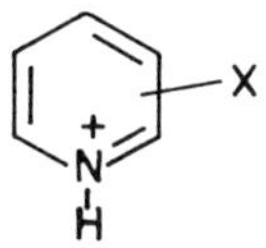

(2) The 2- and 4-pyridones (hydroxypyridines) (**9.3, 9.4**) and their protonated derivatives (e.g., **9.5, 9.6**).

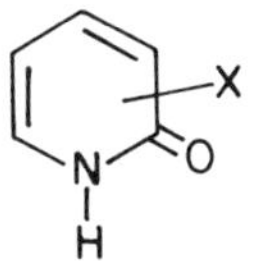

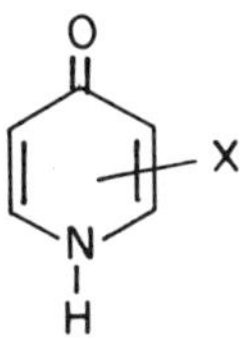

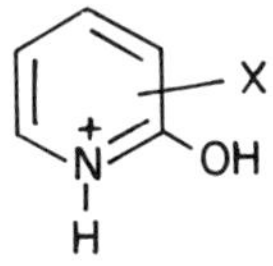

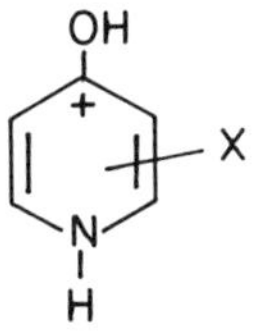

(3) Pyridine *N*-oxides (**9.7**) and protonated derivatives (**9.8**).

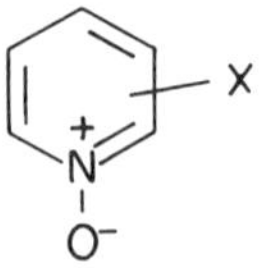

(9.7) 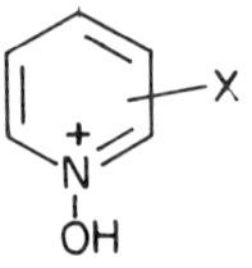   (9.8)

(4) Pyrones (e.g., **9.9**) and thiapyrones (e.g., **9.10**).

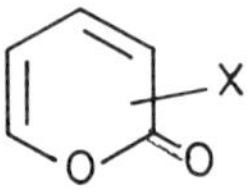

(9.9) 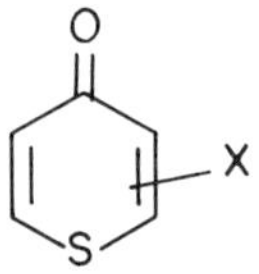   (9.10)

(5) Pyrylium ions (**9.11**).

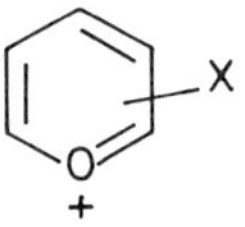

(9.11)

(6) Arsabenzene (**9.12**).

(9.12)

(7) Phosphorins (**9.13**).

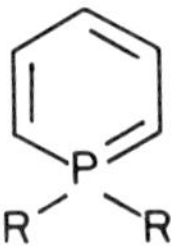

(9.13)

B. REACTIVITY PATTERNS

a. *Pyridines*

Pyridines are less reactive toward electrophilic substitution than are the corresponding benzenoid compounds because of the electron-withdrawing inductive ($-I$) and mesomeric ($-M$) effects of the ring nitrogen. These effects place substantial positive charge on the 2- and 4-carbon atoms (**9.14**), which are therefore considerably less reactive than the car-

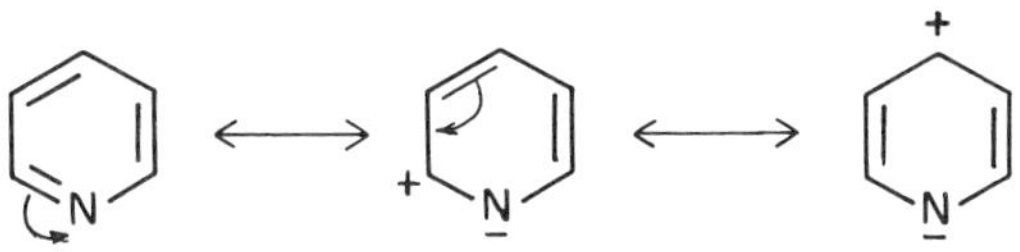

(9.14)

bon atoms in benzene. The 3-position of pyridine is also deactivated, but to a smaller extent. The positive charge is relayed mostly to the 4-position in the highest occupied molecular orbital (HOMO), so that the 4-position is less reactive than the 2-position. Pyridine (pK 5.2) is readily protonated in moderately strong acid, and in the pyridinium cation electron withdrawal from all three positions is substantially increased. The resultant decrease in reactivity corresponds to ~ 1.0 σ unit at each position. Much of the work described in this chapter has concerned the elucidation of the nature of the species undergoing substitution under the various conditions. In general, electron-supplying substituents tend to favor reaction on the conjugate acid, whereas electron-withdrawing substituents favor reaction on the free base. This is partly because the higher the pK_a of the substrate, the lower the concentration of the free base species. Additionally, the activation energy for substitution of the conjugate acid becomes unfavorably high in pyridines with electron-withdrawing substituents.

Moreover, as has been recognized, pyridine is strongly hydrogen bonded in protolytic media (hence, for example, its high water solubility). Such hydrogen bonding substantially reduces the reactivity toward electrophiles, and as a consequence reactivity parameters determined for the free base in solution in H-donor solvents differ markedly from those determined in the gas phase. This difference is ~ 0.3 σ units at each position.

b. *Pyridine N-Oxides*

The positive charge on nitrogen withdraws electrons by an inductive (or field) effect more strongly from all positions than in unprotonated pyridine. However, mesomeric electron release ($+M$) from oxygen returns electrons to the 2- and particularly to the 4-position. The result is that the 2- and 4-positions are more reactive, whereas the 3-position is less reactive than in pyridine. In acidic media, protonation at oxygen reduces the mesomeric electron release and all positions become less reactive. The 2- and 4-positions are especially strongly affected, and 3-substitution then dominates in the electrophilic substitution of the 1-hydroxypyridinium cation.

c. *Pyrones and Thiapyrones*

In both 2- and 4- isomers the 3- and 5-positions (activated by conjugation with the ring heteroatom) are more reactive than the 2-, 4-, and 6-positions (deactivated by conjugation with carbonyl oxygen). Corresponding positions in thiapyrones should be somewhat less reactive.

d. *Pyrylium Ions*

Pyrylium ion should be even less reactive than the corresponding pyridinium cation; in fact, no simple electrophilic substitutions are known.

e. *Arsabenzene*

Arsenic is slightly more electropositive than carbon, so arsabenzene should be a little more reactive than benzene, arising from the resultant $+I$ effect of the heteroatom. However, arsabenzene is reported to be considerably more reactive than benzene, especially at the α- and γ-positions, due to the additional high polarizability of the arsenic *p* orbitals.

f. *Phosphorins*

These compounds undergo electrophilic substitution at the 4-position, but their quantitative reactivities are not yet known.

2. Acid-Catalyzed Hydrogen Exchange

A. METHYLPYRIDINES

Pyridine itself is too unreactive toward acid-catalyzed hydrogen exchange for rates to be measurable, but results have been obtained for exchange at the 3- and 5-positions of 2,6-dimethylpyridine (**9.15**), 2,4,6-trimethylpyridine (**9.16**), and 1,2,4,6-tetramethylpyridinium ion (**9.17**)

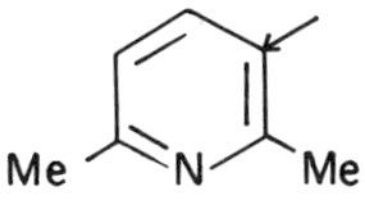

(**9.15**) 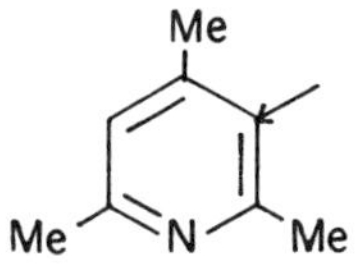 (**9.16**) 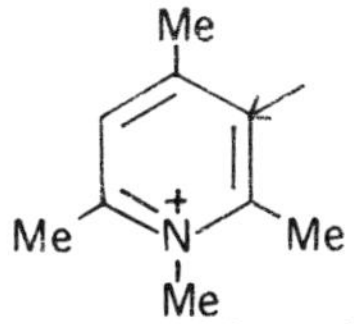 (**9.17**)

(63JCS3753). Rate profiles for the former two compounds indicate reaction upon the conjugate acid (expected in view of the rather high pK_a values), with slopes of the log k vs. $-H_0$ plots at 204°C of 0.69 and 0.56, respectively. Thus, the more reactive compound gave the smaller slope. This conforms to the reactivity–selectivity principle, since at higher acidity the electrophile is more reactive, giving rise to a smaller spread of rates between substrates. Involvement of the protonated species was confirmed by exchange measurements for the 1,2,4,6-tetramethylpyridinium ion (**9.17**), which gave a log k vs. $-H_0$ slope of 0.77 at 182°C (cf. 0.66 for 2,4,6-trimethylpyridine); **9.17** was also about twice as reactive as **9.16**. This rate difference is due to the activating effect of the *m*-methyl; comparison of exchange rates of **9.15** and **9.16** showed *o*-methyl to activate 160-fold at 204°C, in reasonable agreement with the value (250) that applies in detritiation of toluene in sulfuric acid at 25°C (60JCS3301). (Pyridine is less reactive, so a larger factor would be expected, but this is more than compensated by the higher temperature of measurement.) These exchange data indicate that substitution of =NH$^+$— for =CH— in benzene produces a deactivation of the meta position by a factor of $\sim 10^{18}$, approximate because very large extrapolations of data were necessary. The entropy of activation determined in this work became more negative as the acidity of the exchange medium increased. This was attributed to more rapid desolvation with increasing acidity of the singly charged reagents compared to the doubly charged transition state.

Results for methoxy-substituted dimethylpyridines are described in section 2.C.

B. Aminopyridines

Exchange of 4-aminopyridines (**9.18**) takes place at the 3-position on the monoprotonated species [67JCS(B)1219], as expected. In 3-aminopyridine, exchange occurs at the 2-position: This is strongly deactivated by the adjacent nitrogen and at low acidities at 170°C exchange occurs on the free base [or through reaction of the conjugate acid with D_2O via the ylide (**9.19**)].

At higher acidity exchange occurs on the conjugate acid and the rate increases with increasing acidity until the H_0 value corresponding to the pK_a for the second protonation is reached, the rate–acidity profile slope then becoming zero [73JCS(P2)1072].

In 4-amino-2,6-dichloropyridine (**9.20**), electron withdrawal and consequent low pK_a causes exchange to take place on the free base at low acidity, with a changeover to reaction on the conjugate acid at high acidity. Thus, at H_0 −0.4, **9.20** is apparently ~100-fold more reactive than **9.18**, whereas at H_0 −5.5 it is threefold less reactive. Exchange in the 3- and 5-methyl derivatives (**9.21**, **9.22**) of 2-aminopyridine at 158°C and 107°C, respectively, takes place as expected on the conjugate acids, and the increasing rate with increasing acidity again ceased at a point corresponding to the second pK_a of the substrate [67JCS(B)1219; 71JCS-(B)2363]. The exchange at the more reactive 5- and 3-positions of 4- and 6-methyl-2-aminopyridines (**9.23**, **9.24**) is similar, but changes over to reaction on the *free base* at low acidity [73JCS(P2)1072]; a similar reaction on the free base should be observed for exchange in **9.21** and **9.22**, but this was not examined at such low acidity.

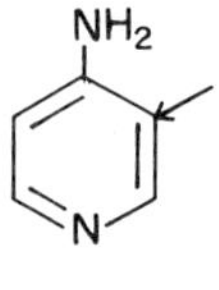

(9.18) 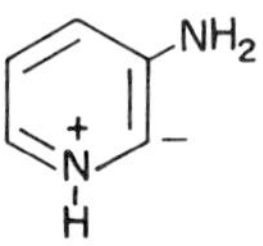   (9.19) 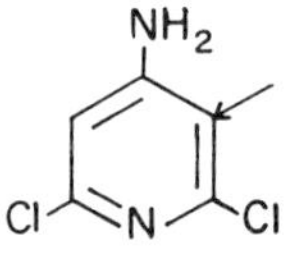   (9.20) 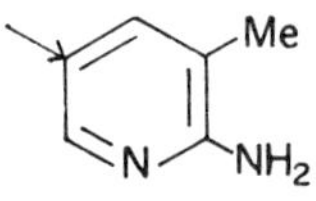   (9.21) 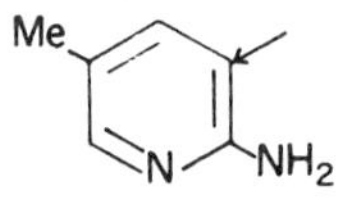   (9.22)

For 5-chloro-4-aminopyridine (**9.25**) and its *N,N*-dimethyl derivative (**9.26**), exchange at 158°C took place predominantly on the free base at lower acidity and on the conjugate acid at higher acidity (the changeover point occurs at a higher acidity for **9.26** than for **9.25**). The rate–acidity profile became horizontal (i.e., zero slope) at very high acidity for **9.25** due to second protonation, but this was not observed for **9.26**]71JCS(B)2363].

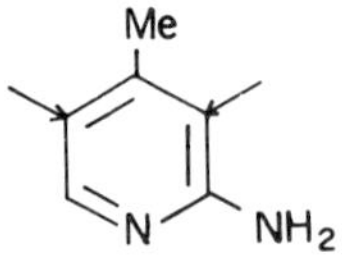

(9.23) 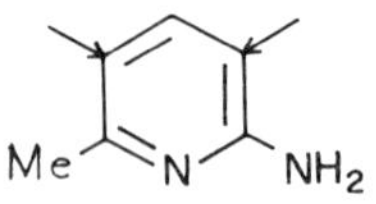   (9.24) 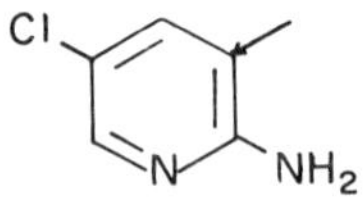   (9.25) 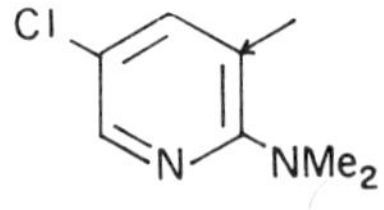   (9.26)

C. Pyridones and Hydroxypyridines

Both 4-pyridone (**9.27**) and 1-methyl-4-pyridone (**9.28**) undergo exchange at the 3-position at 170°C on the free base, as shown by the rapid increase in exchange rate with increasing acidity at pH above the protonation pK_a but a zero slope of the log k vs. $-H_0$–rate profile in more acidic media]65CI(L)1384; 67JCS(B)1226]. The similarity in exchange rate of **9.27** and **9.28** showed that 4-pyridone reacted as **9.27** and not as the 4-hydroxypyridine tautomer; moreover, 4-methoxypyridine (**9.29**) did not undergo exchange under similar conditions. 2-Pyridone (**9.30**), and its 3- and 5-methyl derivatives (**9.31**, **9.32**), all showed similar behavior. The free base is also involved in sulfuric acid-catalyzed exchange at the 3- and 5-positions of 2,6-dimethyl-4-pyridone (**9.33**) and its 1-hydroxy derivative (**9.34**) [67JCS(B)1226].

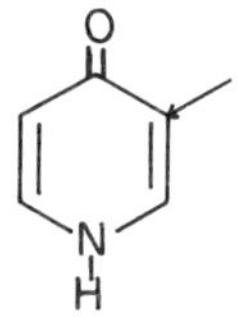

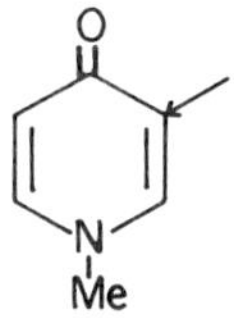

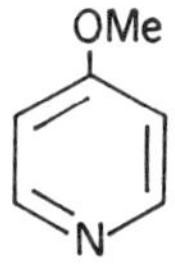

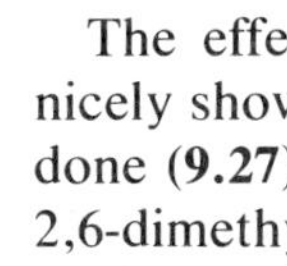

The effects of substituents on the form of the substrate reacting is nicely shown by comparison of the rate profiles for **9.27** and **9.33**. 4-Pyridone (**9.27**) exchanges as the free base even at H_0 −10, whereas for its 2,6-dimethyl derivative (**9.33**), the reaction takes place mainly via the conjugate acid at $H_0 \sim -3.5$. 1,2,6-Trimethyl-4-pyridone (**9.35**) shows the changeover at even lower acidity ($H_0 \sim -2.7$). At high acidity, **9.33**, **9.35**, and 4-methoxy-2,6-dimethylpyridine (**9.36**) all react at similar rates and show similar dependence of rate upon acidity. This indicates that all react as the conjugate acids of type **9.37**, and excludes the unlikely alternative **9.38**. [In [68JCS(B)866], curve C of Fig. 3 refers to 4-methoxy-2,6-dimethylpyridine, and not as stated]. At lower acidity the similarity in rate persists for **9.33** and **9.35**, but **9.36** is much less reactive. Hence, the 4-pyridone **9.33** reacts as such and not as the 4-hydroxypyridine tautomer.

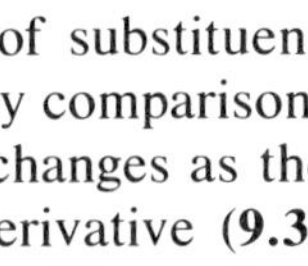

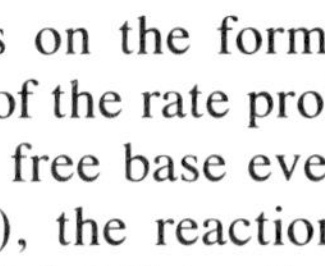

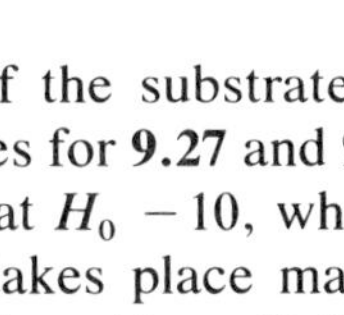

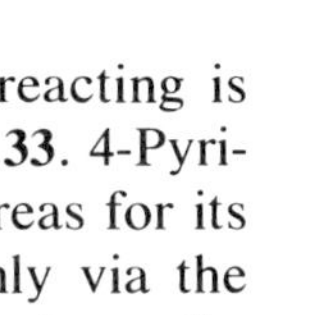

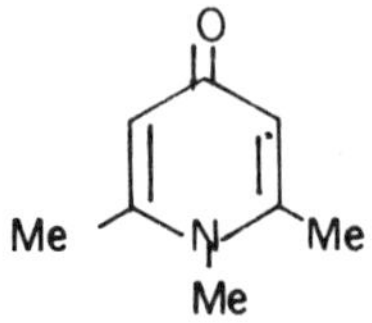

(9.35) 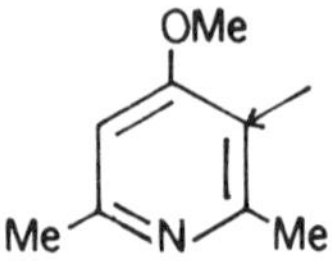 (9.36) 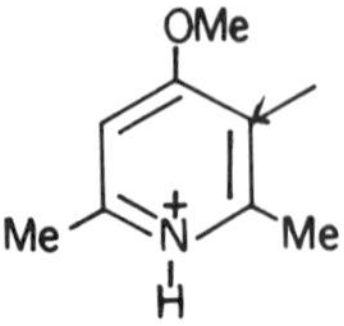 (9.37) 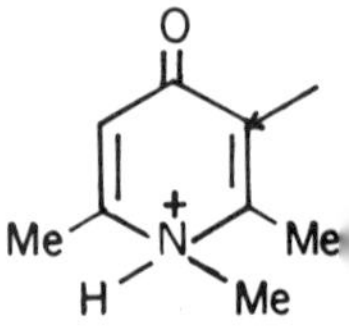 (9.38)

At high acidity, 2,6-dimethyl-1-hydroxy-4-pyridone (**9.34**) reacted as the conjugate acid [68JCS(B)866] with a rate similar to that of 4-methoxy-2,6-dimethylpyridine N-oxide, so both must then react as conjugate acids of the structure type **9.39**. In the intermediate acidity range pD +1 to H_0 −2.5, the rate–acidity profile slope was zero for **9.34**, indicating reaction on a neutral form. Since 4-methoxy-2,6-dimethylpyridine N-oxide (**9.40**) was much less reactive, the neutral form reacting must be **9.34** and not **9.41**.

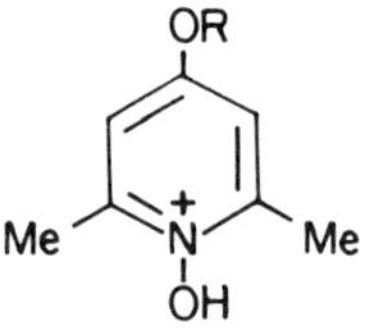

(9.39, R = H, Me) 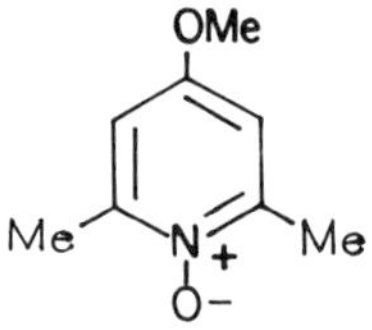 (9.40) 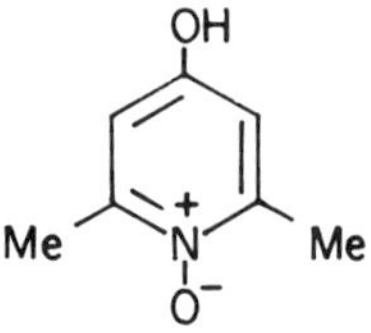 (9.41)

From this hydrogen-exchange work 2-pyridone was found to be slightly more reactive than 4-pyridone; in 2-pyridone a 3- or 5-methyl group activates the corresponding 5- or 3-positions ~twofold, in good agreement with the value obtained in methylpyridines (see Section 2.A). A semiquantitative study of the reactivity in methylpyridones towards exchange also indicated the positional reactivity order as 2(6) > 4 [68CPB715], and a further study indicated that in 2-pyridone the reactivities of the 3- and 5-positions are similar [72BAU1166].

D. Pyridine N-Oxides

Acid-catalyzed exchange has not been established in pyridine N-oxide itself. The exchange of a number of derivatives has been studied [64CI(L)1576]. 3,5-Dimethylpyridine N-oxide (**9.42**) reacts as the free base at the 2(6)- and 4-positions giving a zero slope of the rate–acidity profile [67JCS(B)1222]; the 2- and 6-positions are slightly more reactive than the 4-position, as expected. In 2,4,6-trimethylpyridine N-oxide

(**9.43**), the more reactive positions are blocked so that reaction can occur only at the 3(5)-position [67JCS(B)1222]. Exchange of the conjugate acid is observed: The difference in reactivity of the free base and conjugate acid is much less for the 3,5- than for the 2,4,6-positions, and the higher ratio of protonated to nonprotonated form is decisive. In this work a side reaction prevented a rate versus acidity study of exchange in 2,6-dimethylpyridine N-oxide, but comparison of the rate of exchange at H_0 -9.3 and 258°C with that in **9.43** gave a value of f_4^{Me} of 50. This compares with a partial rate factor of 20 for the 4-methyl group in the pyridine series under the same conditions. The similarity of these values indicates that the exchange in 2,6-dimethylpyridine N-oxide proceeds as a species of the same charge type as that in the 2,4,6-trimethyl analogue. Exchange at the 3(5)-positions of 4-methoxy-2,6-dimethylpyridine N-oxide (**9.40**) also occurs on the conjugate acid [68JCS(B)866].

3-Hydroxypyridine N-oxide (**9.44**) also reacts as the free base in the range H_0 -2.0 to -8.0 [67JCS(B)1222], but it is not clear if **9.44** or **9.45**

(9.42)   (9.43)   (9.44)   (9.45)

is the species undergoing exchange. With 3,5-dimethoxypyridine N-oxide (**9.46**), exchange was observed only in the 2(6)-position [activated by o-OMe × p-OMe]; any exchange at the 4-position [activated by $(o\text{-OMe})^2$] must be slower by a factor of at least 100 [67JCS(B)1222]. At high pD values, rapid base-catalyzed exchange occurred in all ring positions. At low pD values, but above pK_a 0.90, exchange occurred on the free base species and a rapid increase in rate occurred with increasing acidity. At acidity greater than pK_a, the rate decreased with increasing acidity, the negative slope (-0.25) of the rate–acidity profile reflecting here, as when observed elsewhere, the difference between the acidity functions for carbon and oxygen (or nitrogen) protonation.

4-Aminopyridine N-oxide (**9.47**) undergoes exchange only as the conjugate acid [68JCS(B)864], again probably due to the combined effects of high pK_a and hence increased protonation of the oxide function, and the strongly activated adjacent 3-position. The less reactive 2-amino-5-bromopyridine N-oxide (**9.48**) undergoes exchange on the free base at lower acidity, and as the conjugate acid at higher acidity [71JCS(B)2363].

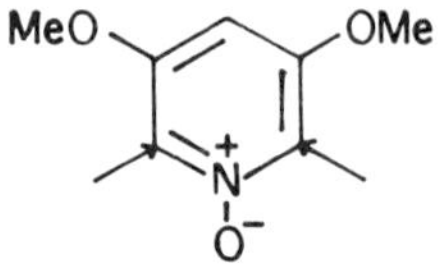

(9.46) 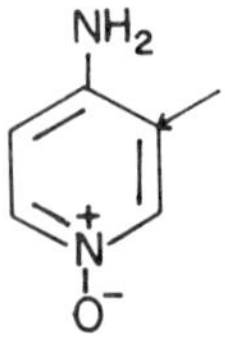   (9.47) 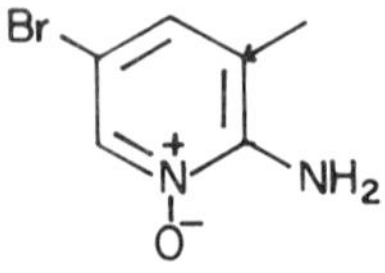   (9.48)

E. Pyrones and Thiapyrones

2.6-Dimethylpyrone (**9.49**, X = O, R = Me) undergoes exchange at the 3- and 5-positions via the free base over a wide range of acidity. The same is true of 2,6-dimethylthiapyrone (**9.49**, X = S, R = Me), which was, however, at pD = 0 and 148.3°C, 16-fold less reactive [68JCS(B)-866]. This sequence parallels the generally higher reactivity of furans compared to thiophenes toward electrophilic substitution.

However, not all such exchanges take place by simple electrophilic substitution. Hydrogen exchange of 4-pyrone (**9.49**, X = O, R = Me) at the 3,5-positions has been shown to occur via ring opening (64JOC2678, 64JOC2682), because although 4-pyrone itself gave 3(5)-exchange under neutral or acid conditions, both 3-methyl- and 2,6-dimethyl substituents inhibited exchange. Moreover, carboxy substituents at the 2- and 6-positions increase the exchange rate.

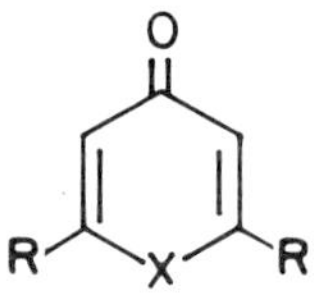

(9.49 X = O,S) 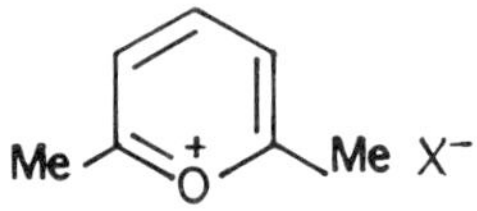   (9.50 ; X = SbCl$_6$,ClO$_4$)

F. Pyrylium Ions

A single study of 2,6-dimethylpyrylium hexachloroantimonate or perchlorate (**9.50**) showed it to undergo exchange at the 3(5)-position in deuterioacetic acid, but not at the 4-position. However, this is too rapid to involve simple electrophilic substitution; exchange also took place at the 2-methyl groups 17-fold faster than the ring 3-hydrogen, and an addition–elimination mechanism has been suggested (71T681).

G. ARSABENZENE

Deuteriation of arsabenzene (**9.12**) in TFA/CH$_2$Cl$_2$ at 100°C has led to partial rate factors under standard conditions (TFA, 70°C) of $f_2 = 2500$, $f_4 = 400$ (81JOC881). The corresponding σ^+ values, ~ -0.39 and -0.30, compared with the 2-, 3-, and 4-σ values (0.3, 0.4, 0.1) indicate very strong mesomeric electron release, though, unusually, this is more effective at the 2-position than at the 4-position. The strong electron release appears to be due to the high polarizability of the p orbitals of arsenic.

H. SUMMARY OF KINETIC DATA

The foregoing data are summarized in Table 9.1, which includes all of the primary kinetic features; additional information is given in Table 3 of [73JCS(P2)1065].

TABLE 9.1
STANDARDIZED REACTIVITY DATA FOR ACID-CATALYZED HYDROGEN EXCHANGE

Substituent	T (°C)	Positions	Species charge	$-H_o$ range	$\dfrac{d[\log k_2(\text{stoich})]}{d(-H_o)}$	$-\log k_2^a$
Substituted pyridines and pyridones						
2-NH$_2$-3-Me	158	5	+	-0.4–5.2	0.47	7.95
2-NH$_2$-5-Me	107	3	+	0.5–4.0	0.67	8.05
2-NH$_2$-4-Me	148	3	+	0.6–2.4	0.93	7.84
2-NH$_2$-4-Me	148	3	0	-0.7–0.3	-0.21	0.81
2-NH$_2$-4-Me	148	5	+	0.6–2.4	0.63	7.62
2-NH$_2$-4-Me	148	5	0	-0.7–0.3	-0.39	0.54
2-NH$_2$-6-Me	158	3	+	0.8–1.6	0.78	7.40
2-NH$_2$-6-Me	158	3	0	-0.7–0.1	-0.14	0.48
2-NH$_2$-6-Me	158	5	+	0.1–1.6	0.66	6.90
2-NH$_2$-6-Me	158	5	0	-0.7–0.1	-0.37	0.28
2-NH$_2$-5-Cl	158	3	+	4.6–6.9	0.56	9.86
2-NH$_2$-5-Cl	179	3	+	4.4–6.2	0.50	9.46
2-NH$_2$-5-Cl	158	3	0	0.8–1.6	0.00	3.87
2-NH$_2$-5-Cl	179	3	0	0.5–1.3	0.00	3.78
2-NMe$_2$-5-Cl	158	3	+	4.0–6.3	0.62	9.22
2-NMe$_2$-5-Cl	158	3	0	3.1–4.0	0.00	2.16
3-NH$_2$	176	2	+(min)	0.8–2.0	0.00	8.95
3-NH$_2$	176	2	+(maj)	0.2–0.8	1.06	8.54
3-NH$_2$	176	2	0	$-(1.1$–0.4)	0.00	3.08
4-NH$_2$	107	3, 5	+	-0.5–4.8	0.66	7.28
4-NH$_2$	146	3, 5	+	0.1–3.0	0.54	7.76

(*continued*)

TABLE 9.1 (continued)

Substituent	T (°C)	Positions	Species charge	$-H_o$ range	$d[\log k_2(\text{stoich})]/d(-H_o)$	$-\log k_2''$
4-NH$_2$	170	3, 5	+	0.1–2.9	0.54	7.66
4-NH$_2$-2,6-Cl$_2$	107	3, 5	+	4.0–4.6	0.98	9.05
4-NH$_2$-2,6-Cl$_2$	122	3, 5	+	4.1–4.6	1.05	9.18
4-NH$_2$-2,6-Cl$_2$	107	3, 5	0	0.8–3.7	−0.23	2.88
4-NH$_2$-2,6-Cl$_2$	122	3, 5	0	−0.7–3.5	−0.12	3.04
2,6-Me$_2$	204	3, 5	+	5.5–6.2	0.99	15.50
2,4,6-Me$_3$	204	3, 5	+	4.1–5.3	0.88	12.40
2,4,6-Me$_3$	180	3, 5	+	4.2–6.0	0.66	11.80
2,4,6-Me$_3$	209	3, 5	+	3.5–5.6	0.63	11.37
4-OMe-2,6,Me$_2$	100	3, 5	+	4.0–8.4	0.69	0.80
4-OMe-2,6,Me$_2$	100	3, 5	+	1.1–2.5	0.00	0.96
2-Pyridone	128	3, 5	0	2.2–7.1	−0.16	4.75
3-Me-2-pyridone	103	5	0	1.7–7.0	−0.15	4.26
5-Me-2-pyridone	120	3	0	1.6–6.7	−0.12	4.73
4-Pyridone	170	3, 5	0	−4.3–7.1	−0.12	4.85
4-Pyridone	187	3, 5	0	−4.3–6.9	−0.13	4.80
1-Me-4-pyridone	170	3, 5	0	−3.0–7.2	−0.08	4.77
2,6-Me$_2$-4-pyridone	108	3, 5	+	4.0–8.6	0.50	8.46
2,6-Me$_2$-4-pyridone	108	3, 5	0	−2.2–2.3	−0.15	2.81
1,2,6-Me$_3$-4-pyridone	100	3, 5	+	3.3–8.1	0.56	8.64
1,2,6-Me$_3$-4-pyridone	100	3, 5	0	1.1–2.4	0.00	3.22
Substituted pyridine N-oxides						
4-OH-2,6-Me$_2$	100	3, 5	+	5.4–7.8	0.52	8.74
4-OH-2,6-Me$_2$	100	3, 5	0	1.3–3.2	0.00	3.96
4-OH-2,6-Me$_2$	100	3, 5	−	(3.2–1.1)	−0.55	−0.87
4-OH-2,6-Me$_2$	100	3, 5	−	(6.3–4.9)	0.00	−0.54
4-NH$_2$	107	3, 5	+	0.5–4.6	0.76	7.92
2-NH$_2$-5-Br	158	3	+	2.7–5.1	0.46	9.81
2-NH$_2$-5-Br	182	3	+	2.6–5.2	0.45	9.71
2-NH$_2$-5-Br	158	3	0	−0.4–0.6	0.00	7.12
2-NH$_2$-5-Br	182	3	0	0.6–1.3	0.00	7.08
3,5-Me$_2$	230	2, 6	0	0.1–4.9	−0.21	8.13
3,5-Me$_2$	230	4	0	0.1–4.9	−0.13	8.38
2,4,6-Me$_3$	185	3, 5	+	5.4–6.7	0.46	11.42
2,4,6-Me$_3$	202	3, 5	+	2.8–5.0	0.75	12.08
2,4,6-Me$_3$	216	3, 5	+	3.8–6.3	0.57	11.87
3,5-(OMe)$_2$	119	2, 6	+	2.2–5.3	0.43	6.48
3,5-(OMe)$_2$	119	2, 6	0	−3.1–2.2	−0.21	4.09
4-OMe-2,6-Me$_2$	100	3, 5	+	4.1–8.8	0.62	9.68

(continued)

TABLE 9.1 (continued)

Substituent	T (°C)	Positions	Species charge	$-H_o$ range	$d[\log k_2(\text{stoich})]/d(-H_o)$	$-\log k_2{}^a$
Substituted pyrones and thiapyrones						
2,6-Me$_2$-4-pyrone	148	3, 5	0	1.1–7.4	−0.21	5.86
2,6-Me$_2$-4-thiapyrone	148	3, 5	0	1.0–8.1	−0.10	6.66
Chromone	180	3	0	1.9–5.4	0.00	6.71
Chromone	180	6	0	1.9–5.4	0.00	7.94
Chromone	180	8	0	2.8–5.4	0.00	8.50
Thiachromone	180	3	0	1.8–5.4	0.00	6.95
Thiachromone	180	6, 8	0	1.8–5.4	0.00	8.89

^{a} Average values.

3. Base-Catalyzed Hydrogen Exchange

The relative positional reactivity of pyridine toward base-catalyzed exchange would *a priori* be expected to be 2 = 4 > 3. However, in the transition state for 2-substitution, a lone pair is generated in a σ orbital adjacent and coplanar to that on nitrogen (**9.51**). This is unfavorable, making 2-substitution comparatively difficult. The effect is particularly marked in a six-membered ring, being much less serious in a five-membered ring (see Chapter 7, Section 3.A). Thus, with sodium in liquid ammonia at −25°C, the relative rates of deuteration of the 4-, 3-, and 2-positions of pyridine were $10^3:10^2:1$ [64CA(60)6721]. However, in aqueous sodium hydroxide at 220°C, exchange was reported to occur exclusively at the 2-position (64CPB1384). A subsequent report confirmed that the positional order is dependent upon the base concentration: In dilute aqueous sodium hydroxide, 2-exchange dominated whereas in 1 *M* sodium hydroxide, the positional rate ratio becomes 3.0:2.3:1 for the 4-, 3-, and 2-positions, respectively (67JA3358).

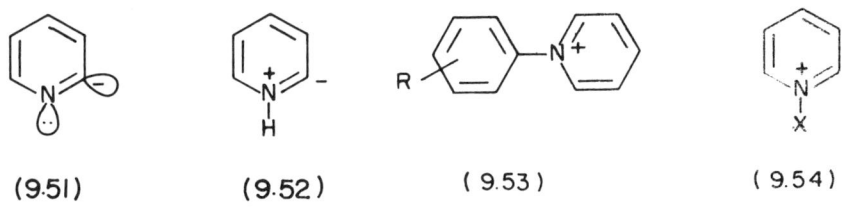

(9.51)   (9.52)   (9.53)   (9.54)

The reason for this dependence is evident from studies of base-catalyzed exchange in the pyridinium cation. Here protonation of nitrogen removes the unfavorable lone pair interaction and also creates a strongly stabilizing positive charge, the transition state for 2-substitution being the ylide (**9.52**). The relative 4- : 3- : 2-position rates for **9.52** are 1 : 3 : 1500 (68JA5939; 69JOC1405). A similar rate spread, such as 1 : 10 : 1500 is obtained for the neutral species of pyridine *N*-oxide (68JA5939). It follows then that under weakly basic or neutral conditions exchange must involve ylides and takes place preferentially at the 2-position, because **9.52** is much more stable than the other isomeric ylides (68JA5939; 81CJC1022). Consistent with this explanation is the fact that the concentration of sodium hydroxide at which exchange occurred principally at the 3- rather than the 2-position in 4-aminopyridine was higher than for pyridine, and higher still for 4-(*N,N*-dimethylamino)pyridine, since the basicity of the nitrogen atom is increased, facilitating ylide formation (68TL421).

Exchange in *N*-arylpyridinium ions (**9.53**) also occurs essentially exclusively at the 2(6)-position and rates correlate with the σ^0 values of the substituents R, with $\sigma = 0.78$ (75CCC1163). Earlier, a correlation had been found with σ_I values for the effects of substituents X in **9.54**, giving $\rho_I = 15$ (though the methyl point misses the correlation line by 2 log units). The relative rates for different substituents X were O^- (1.0), PhCH$_2$ (110), Me (140), Ph (530), ND$_2$ (2700), OMe (1.3 × 10^6). From this work, substitution of =N$^+$H— for =CH— in benzene was estimated to accelerate base-catalyzed exchange by a factor of 10^{14}–10^{16} (70JA7547).

In methanol containing amines, exchange of *N*-methylpyridinium ion was said to be specific-base catalyzed (73JOC829). However, one of the curves shown suggests incursion of general base catalysis. Methyl substituents were said to *increase* the rate of base-catalyzed exchange of 3-hydroxypyridine (72BAU1169). This parallels the effects of 2-alkyl substituents on the rate of base-catalyzed exchange in indole (Chapter 8, Section 2.A.b.ii). A similar explanation may therefore apply [i.e., the negative charge on oxygen (formed by rapid proton abstraction) is delocalized into the ring, the carbanion then carrying out rate-determining abstraction of a proton from the solvent].

Rates of exchange with NaOMe/MeOH at 50°C of the 2-, 4-, and 6-positions of 3-chloropyridine *N*-oxide and of the 2-position of 3,5-dichloropyridine *N*-oxide, relative to the 2-position of pyridine *N*-oxide, were 1840, 0.37, 12.2, and 11,800, respectively. These showed that the normal carbanion mechanism applied and emphasized further the importance of inductive effects in base-catalyzed exchange. The activating effects of substituents (relative to a position in benzene, at 50°C) were calculated as 2-Cl, 1800; 4-Cl, 9; 2-N$^+$O$^-$, 3.8 × 10^9; 4-N$^+$O$^-$, 7.6 × 10^5 (69JOC1405).

TABLE 9.2
RATE DATA FOR BASE-CATALYZED HYDROGEN EXCHANGE OF PYRIDINES UNDER STANDARD CONDITIONS[a]

Substituent	Position of exchange	$-\log k$	E (kcal mol^{-1})	$\log A$ (sec^{-1})
H	2	6.7		
	3	6.3		
	4	6.1	20.3	4.6
2-OMe	3	5.3		
3-NH$_2$	4	5.7	28.6	9.4
3-F	4	2.9	15.3	5.2
3-Br	4	4.4	18.7	5.5
3-OMe	4	4.5	34.0	13.5
3-NO$_2$	2	No exchange		
	4	3.4	23.2	8.9
4-OMe	3,5	5.6	31.6	11.5
4-NMe$_2$	2,6	6.0		
	3,5	5.7		
2-NH$_2$-5-NO$_2$	4	3.9		
3,5-Br$_2$	4	1.7		
3,5-(OMe)$_2$	4	2.9		

[a] Standard conditions, 0.6 M MeOK in MeOD at 140°C.

TABLE 9.3
RATE DATA FOR BASE-CATALYZED EXCHANGE OF PYRIDINE N-OXIDES UNDER STANDARD CONDITIONS[a]

Substituent	Position of exchange	$-\log k$	E (kcal mol^{-1})	$\log A$ (sec^{-1})
H	2	5.3		
	3	7.3	18.6	5.3
	4	8.0	24.0	7.6
2-OMe	6	5.6		
3-F	2	0.3	22.0	14.6
3-Br	2	2.0	23.0	13.6
3-OMe	2	3.9	23.7	12.2
3-NO$_2$	2	-0.3	14.3	8.9
4-OMe	2,6	5.2		
	3,5	5.8		
4-Cl	2,6	4.3		
	3	4.5		
4-NMe$_2$	2,6	6.5	26.3	11.2
	3	8.5		
3,5-Br$_2$	2,6	1.3	19.6	10.9
	4	2.3	26.2	14.4
3,5-(OMe)$_2$	4	5.7	27.9	13.2

[a] Standard conditions, 0.1 M MeOK in MeOD at 50°C.

The N-oxide group is a powerful activator on account of its inductive ($-I$) effect, but is less activating than N^+H, as expected.

Finally, rate coefficients for base-catalyzed exchange of pyridines under "standard" conditions (0.6 M MeOK in MeOD at 140°C) and for pyridine N-oxides (0.1 M MeOK in MeOD at 50°C) are given, in terms of the log values, in Tables 9.2 and 9.3, respectively, along with Arrhenius data (74CHE1397). The wide variations in log A values are difficult to interpret, and similar variations occur with diazines (Chapter 10).

4. Nitration

A. PYRIDINES

Most studies have been carried out using nitric acid/sulfuric acid; fewer have utilized nitric acid/acetic anhydride. Use of nitronium tetrafluoroborate produces a stable N-nitro intermediate, and the corresponding N-nitroso derivative is obtained with nitrosonium tetrafluoroborate (64TL2117; 65JOC3373). One study with molten nitrates and pyrosulfate (which gives a small concentration of nitronium ion) reported 3-nitration of pyridine though no yields were quoted (67CC966). In general, nitration of pyridine is very difficult because of the easy protonation. With potassium nitrate in oleum at ~300°C only 15% of 3-nitropyridine is obtained (12CB428), and indeed other workers have found considerably lower yields than this (30RTC552). As expected, methyl, hydroxy, and amino substituents increase the yield (through activation) and, less obviously, so too can certain electron-withdrawing groups (by promoting reaction on the free base).

The first estimate of the reactivity of pyridine toward nitration indicated the 3-position to be 10^{12} times less reactive than a position in benzene [62CI(L)1057], suggesting that the conjugate acid was the reacting species. As in acid-catalyzed hydrogen exchange, whether the free base or conjugate acid is involved depends largely on the substituents present, electron-supplying groups causing reaction on the latter species. In general, pyridine derivatives with pK_a > 1 will nitrate as the conjugate acid, whereas those with pK_a < −2.5 will nitrate as the free bases [67AG(E)608]. Thus, the 2,4,6-trimethylpyridine and 1,2,4,6-tetramethylpyridinium ion gave similar rate profiles, that for the former passing through the usual maxiumum in ~90 wt% H_2SO_4 and paralleling that for 4-chlorotrimethylanilinium ion. This shows that the trimethyl compound reacts as the conjugate acid [67JCS(B)1204]. Calculation of the theoretical rate of reaction via the 2,4,6-trimethylpyridine free base showed that it would be nearly two orders of magnitude greater than the encounter rate, further confirming that the free base cannot be involved [67JCS(B)1204]. The 1,2,4,6-tetramethylpyridinium ion was slightly less reactive, suggesting a steric effect, perhaps arising from buttressing or hindrance to solvation. Nitration of 2,4-dimethylpyridine at the 3- and 5-positions occurred in 12 and 13% yield, respectively (64RC1321); both positions evidently possess similar reactivity.

Both 2,4- and 2,6-dimethoxypyridines are nitrated as their conjugate acids at the β-positions (5- and 3-positions, respectively); the second nitration of the latter compound involves a less reactive and less basic substrate, which, like 2,6-dichloropyridine, nitrates as the free base [67JCS(B)1204]. From the rates of nitration of 2,6-dimethoxypyridine and m-dimethoxybenzene it was deduced that pyridinium ion is 10^7 times less reactive than benzene; in contrast, if data for 2,4,6-trimethylpyridine are used, the value becomes 10^{20}. 2-Nitration of 3,5-dimethoxypyridine also takes place on the conjugate acid, but 6-nitration of 2-nitro-3,5-dimethoxypyridine takes place on the free base. This indicated a method of obtaining the maximum yield of the mononitro derivative from 3,5-dimethoxypyridine (viz. to nitrate in very strong sulfuric acid to ensure reaction on the conjugate acid) [67JCS(B)1211]. From the nitrations of 2,6-dichloropyridine (3-position, as the free base) and 2-methoxypyridine (5-position, as the conjugate acid), values of log k_2^o were determined as -9.01 and -8.68, respectively [83CHE514]. Activation parameters for dichloro- and dimethoxypyridines indicate that compounds with $\Delta S^\ddagger$ (e.u.) values of 12–16 react as the free bases, whereas those with values of -18 to -20 react as the conjugate acids [67JCS1204, 67JCS1211].

Nitration of 2-phenylpyridine takes place on the conjugate acid, yielding the partial rate factors shown in structure **9.55** [68JCS(B)862]. These are comparable to the partial rate factors for the $CH_2NMe_3^+$ substituent, and are interesting because they indicate ortho and para orientation with deactivation as do halogens; similar behavior has been observed in hydrogen exchange of pentafluorobiphenyl [73JCS(P2)253]. Another determination gives partial rate factors for **9.56** that differ in absolute rather than relative values. The values for the 4-pyridinium substituent (**9.57**) show more clearly the ortho and para orientation, and in particular the reactivity of the ortho position in the phenyl ring becomes proportionally greater (compare with **9.55** and **9.56**), as a result of increasing distance from the positive pole. Deactivation in the 4-benzylpyridinium ion (**9.58**) is considerably less than that in the 4-phenylpyridinium cation since the pole is further away, and countering activation is provided by the methylene

294   9. HETEROAROMATICS WITH A SIX-MEMBERED RING   [Sec. 4.A

(9.55) partial rate factors: 14×10^{-5} (para), 6.2×10^{-5}, 1.3×10^{-5} on 2-phenylpyridinium

(9.56) partial rate factors: 2.2×10^{-5} (para), 8.5×10^{-6}, 1.9×10^{-6} on 2-phenylpyridinium

(9.57) partial rate factors: 5.9×10^{-4} (para), 2.1×10^{-4}, 1.3×10^{-4} on 4-phenylpyridinium

(9.58) partial rate factors: 0.7 (para), 0.026, 0.001 on 4-benzylpyridinium

group. This work also indicates that the deactivating effect at the meta position of the phenyl ring of protonated phenylpyridines is most easily understood as the field effect of a pole [71JCS(B)712].

Nitration of 2-anilinopyridine by nitric acid/sulfuric acid gave a mixture of mono substitutions at the 2′- and 4′-positions in the phenyl ring in poor overall yield. However, with excess nitric acid, the 2′,4′-dinitro derivative was obtained in near quantitative yield. By contrast, nitration in nitric acid/acetic anhydride gave products of substitution in both the phenyl and the pyridyl rings, the latter at the 3- and 5-positions. This suggests that the conjugate acid was involved under the former conditions and the free base under the latter (72CPB2678). Calculations supported this conclusion and also indicated that the free base was 2-anilinopyridine, not the 2-pyridone anil tautomer (**9.59**) (72CPB2686). Nitration of 2-picrylaminopyridine (**9.60**) by nitric acid/sulfuric acid gave 7% of 3- and 93% of 5-substitution; the rate–acidity profile indicated that the cation was involved; E_A was 12.9 kcal mol^{-1} (80CHE272).

Nitration of 2-, 3-, and 4-N,N-dimethylaminopyridine afforded the yields shown (**9.61–9.63**); the low yields from the 3-isomer (accompanied by nitroso and other by-products) reflect the fact that the substituent is meta to the most reactive site in pyridine [72JCS(P2)1940]. Moreover, in **9.61** and **9.63**, electron withdrawal by the ring nitrogen reduces the second pK_a for formation of a dication by protonation at the dimethylamino group, whereas such dication formation occurs to a considerable extent in **9.62**. Hence, **9.61** and **9.63** nitrate as the monocation, which is the majority species. For 2-N,N-dimethylamino-3- and 5-nitropyridines, the slope of the Moodie–Schofield plots indicated that in the low-acidity region, nitration takes place on the free base (at the encounter rate). For the 5-nitro isomer, the free base is involved even in the high-acidity region. For 4-N,N-dimethylamino-3-nitropyridine no definite conclusion regarding the reacting species could be drawn [72JCS(P2)1940].

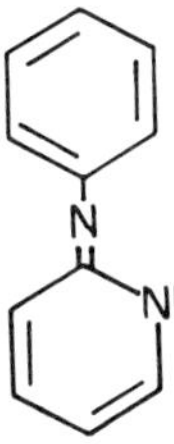

(9.59) 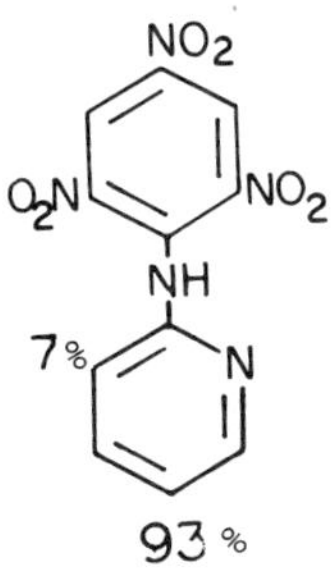   (9.60) 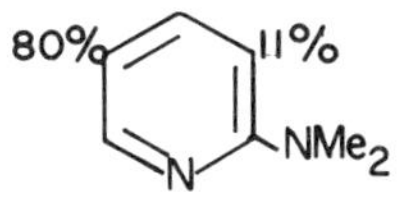   (9.61) 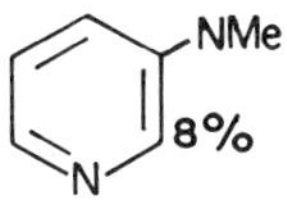   (9.62) 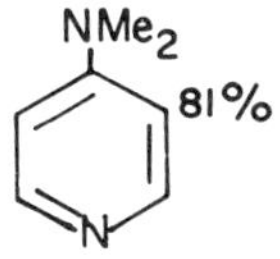   (9.63)

Nitration of the 1-methyl cations of 2- and 4-*N*-methylaminopyridine and 2-*N*-methylamino-5-nitropyridine (**9.64–9.66**) has been studied [72JCS(P2)1950]. 1-Methyl-2-*N*-methylaminopyridinium cation (**9.64**) gave a similar rate profile to that for 2-*N*,*N*-dimethylaminopyridine, confirming that the latter underwent nitration as the conjugate acid. Curiously, **9.64** was fivefold (high acidity) to 4000-fold (low acidity) times less reactive than 2-*N*,*N*-dimethylaminopyridine. This was attributed to steric interactions in the *N*-methyl compound. Cation **9.66** underwent nitration immeasurably slowly under conditions sufficient to nitrate 2-*N*,*N*-dimethylamino-5-nitropyridine, confirming (see above) that the latter reacts as the free base. Cation **9.65** undergoes nitration at rates closely similar to those for 4-*N*,*N*-dimethylaminopyridine, confirming that the latter is nitrated as the conjugate acid [72JCS(P2)1950].

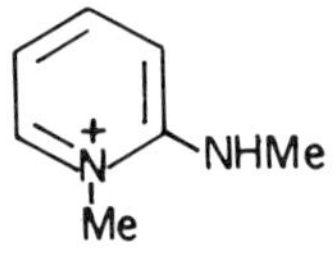

(9.64) 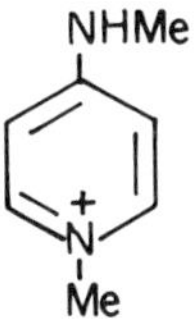   (9.65) 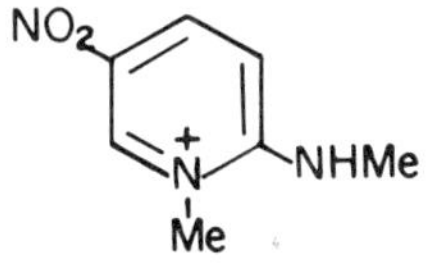   (9.66)

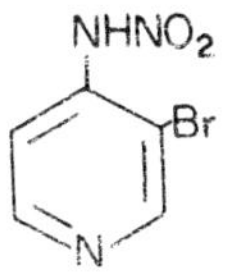

(9.67) 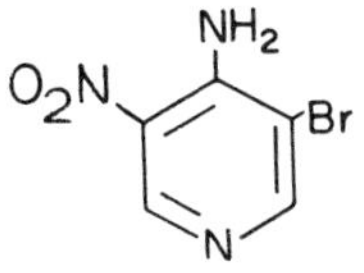   (9.68) 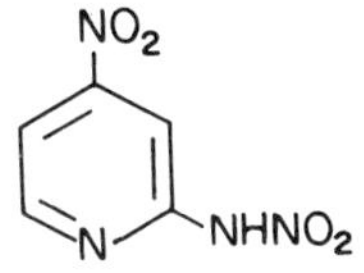   (9.69)

TABLE 9.4

RELATIVE RATES OF REARRANGEMENT
OF x-METHYL-2-NITRAMINOPYRIDINES

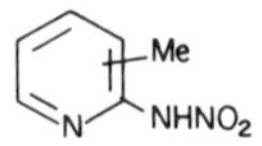

(9.70)

Substituent in 9.70	k_3	k_5
H	1	1
3-Me	—	5.1
4-Me	2.3	0.34
6-Me	10	2.8

Nitration of pyridines via rearrangement of nitramines has been studied. 4-Amino-3-bromopyridine forms, with nitric acid, the nitramine **9.67** in 85% yield; **9.67** rearranges in sulfuric acid to 4-amino-3-bromo-5-nitropyridine (**9.68**) in 97% yield (62RC967). The rearrangement of 2-nitramino-4-nitropyridine (**9.69**) to 2-amino-3,4-dinitropyridine is accompanied by the formation of 4-nitro-2-pyridone (64T81), subsequently shown to arise via nitrosation (79T2895).

A further study of 2-nitraminopyridine and its 3-, 4-, 5-, and 6-methyl-, 5-bromo-, and 5-nitro derivatives showed that in sulfuric acid, free nitronium ions are produced (added mesitylene becomes nitrated). The relative rates of nitration at the 3- and 5-positions of the methyl derivatives are shown in Table 9.4. The results for the 4-methyl compound show that a normal S_EAr mechanism is not involved. 5-Methyl-2-nitraminopyridine seemed to react by a different route since a considerable amount of 5-

TABLE 9.5

REARRANGEMENT OF SUBSTITUTED
2-NITRAMINOPYRIDINES

Substituent	$\log k_2$ (3)	$\log k_2$ (5)
4-OMe	−0.81	−2.19
4-Me	−2.28	−2.10
H	−3.17	−2.22
4-Br	−2.86	−3.04
4-CO$_2$H	−2.98	−3.16
3-D-4-Me		−2.87

TABLE 9.6
Yields (%) of 5-Nitro Isomer from Rearrangement of Nitramines

X	9.71	9.72	9.73
H	90	90	50
Me	60	74	20
OMe	<4	<4	<4
CO_2H	>95	75	>95
Br	40	75	20
Cl	40	78	30

[a] Compounds **9.71–9.73** in 92 wt% H_2SO_4.

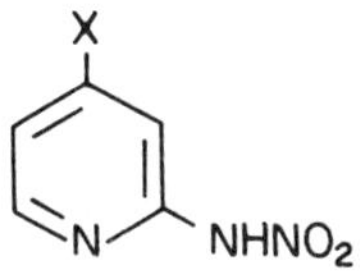

(9.71) 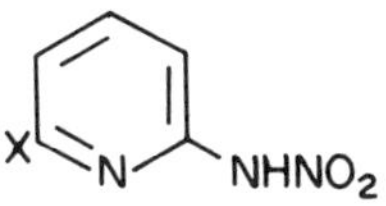 (9.72) 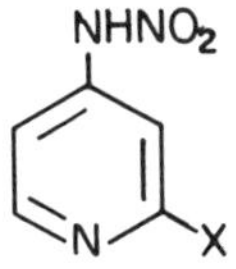 (9.73)

methyl-2-pyridone was formed. Values of $10^2 \, k_2$ at 30°C for different substituents were 3.87 (H), 18.0 (3-Me), 2.0 (4-Me), 13.3 (6-Me), 0.00861 (3-NO_2), and 2.50 (5-Br) (79T2895). Rearrangement in 92 wt% H_2SO_4 and measurement of isotope effects indicated that the rate-determining step occurs prior to formation of the appropriate 3- and 5-nitro complexes.

Rate coefficients were determined for nitration by rearrangement of monoprotonated nitraminopyridines (Table 9.5) (82AJC2035). Rearrangement of 4- and 6-substituted 2-nitraminopyridines (**9.71, 9.72**) at room temperature, and 2-substituted 4-nitraminopyridines (**9.73**) at 75°C gave the yields of 5-nitro derivative shown in Table 9.6 (82AJC2025). For **9.71** and **9.73**, only in the case of X = CO_2H was 5-substitution preferred to 3-substitution, and for **9.72** only in the case X = OMe was 3-substitution preferred to 5-substitution (82AJC2025).

B. Pyridones and Hydroxypyridines

Yields in nitration of alkoxy- and hydroxypyridines have been tabulated (70RCR627). Nitration of 3-hydroxypyridine (**9.74**) gives 74% 2- and only 1% 6-substitution with no 4-substitution detected (68JOC478); this result was confirmed and shown to be qualitatively consistent with charge densities (69BAU1452). If, however, the 2-position is blocked with a methyl or chloro substituent (**9.75**), then both 4- and 6-substitution occur, the former being dominant (68JOC478; 69BAU1452). This suggests that the electrophile coordinates (hydrogen bonds) with the hydroxy group and hence preferentially substitutes ortho to it, a view supported by the decreasing 4- : 6-substitution ratio with increasing temperature, and the complete absence of 4-substitution in the nitration of 3-methoxy-2-methylpyridine (69BAU1452). 2-Substitution of 3-hydroxypyridine could cause the hydrogen of the hydroxy group of lie toward the 4-position, so accounting for the dominance of 4- over 6-nitration in these compounds. Nitration of 2-phenyl-3-hydroxypyridine (**9.76**) occurs primarily at the para position of the phenyl ring (88%) (70BAU1791,2244). Both 3-hydroxy- and 3-methoxypyridine are nitrated as their conjugate acids and at identical rates [70JCS(B)114].

(**9.74**)   (**9.75**, X = Cl, Me)   (**9.76**)

Nitration of 2-pyridone takes place on the free base. Interestingly, the rate of 5-nitration increases more with increasing acidity than does that of 3-nitration: Nitration in sulfuric acid at 80°C gave 100% of 3-substitution, whereas in 20% oleum at 0°C it gave 17% 3- and 83% 5-substitution. The reason for this is not yet clear. The Arrhenius data for the total nitration rate at the 3- and 5-positions are acidity dependent, the activation energies decreasing from ~36.5 kcal mol^{-1} at H_0 −6.5, to 9.7 kcal mol^{-1} at H_0 −10, the log A values decreasing from ~21 to 2.7 sec^{-1} [68JCS(B)1477; 71TL2211; 72J(P2)1953]. Both 3-methyl-2-pyridone and 5-methyl-2-pyridone are nitrated as the free base (the former is the more reactive) as is 1,5-dimethyl-2-pyridone [68JCS(B)1477]. 6-Hydroxy-2-pyridone (**9.77**), its N-methyl- (**9.78**), and its 6-methoxy derivative (**9.79**) all undergo nitration in the 3-position as the free base at or near the encounter rate. Hence the relative rates of ~1 : 0.7 : 0.3 have limited significance [70JCS(B)114]. Nitration of 6-phenyl-2-pyridone has been suggested to take place on the hydroxypyridine tautomer (**9.80**) in order to account for the predominant 3-nitration (through hydrogen bonding to the electrophile), since halogenation goes primarily in the 5-position, but this is most unlikely. As for nitration of 3-hydroxypyridine, the ratio of 3- : 5-substitution decreased with increasing temperature. Acetyl nitrate *at 90°C [N.B. This is explosive!]* gave both 3- and 5-isomers, whereas fuming nitric acid gave the 3,*para*-dinitro product (75CHE1242).

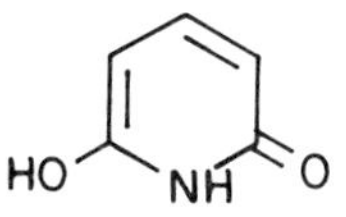

(9.77) HO-NH-O
(9.78) HO-NMe-O
(9.79) MeO-NH-O
(9.80) Ph-N-OH

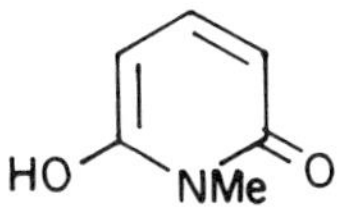

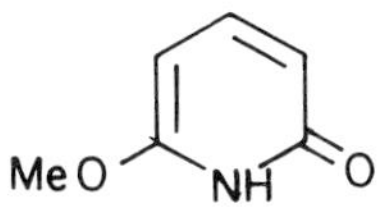

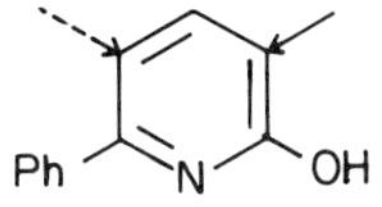

Nitration of 4-pyridone takes place on the free base, except at high acidity where the protonated species is involved [68JCS(B)1477]. 4-Pyridone is less reactive than 2-pyridone toward nitration (Table 9.7), a result which contrasts with hydrogen exchange where the reactivities are identical Table 9.1). As expected, nitration of 3-nitro-4-pyridone also occurs on the free base [68JCS(B)1477]. From this work, partial rate factors of 10^{-10} to 10^{-12} were calculated for the monocations derived from 4-pyridone and 2- and 4-methoxypyridine; deactivation by NH^+ in these systems was estimated as $>10^{13}$ [68JCS(B)1477].

C. Pyridine N-Oxides

Pyridine N-oxide is nitrated in the 4-position as the free base (**9.81**) [64CI(L)1577] and f_p has been estimated as between 2.1×10^{-3} and 4×10^{-6} [66JCS(B)870; 67JCS(B)1213], up to 95% yields of 4-nitropyridine N-oxide have been reported (51RTC581). 4-Nitration on the free base also occurs for the 3,5-dimethyl and 2,6- and 3,5-dichloro derivatives (**9.82–9.84**) [67JCS(B)1213].

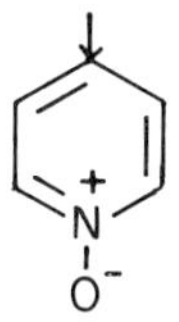

(9.81) 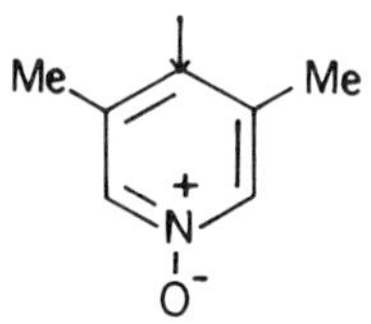 (9.82) 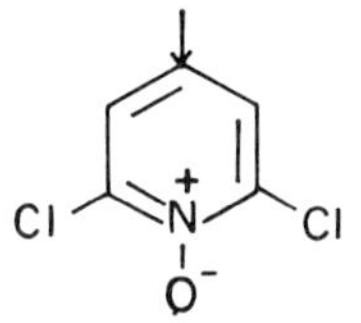 (9.83) 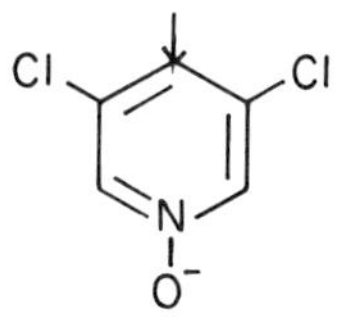 (9.84)

However, when electron-donor groups, located ortho or para to the N-oxide function, sufficiently increase the pK_a value for protonation, nitration can then occur on the conjugate acid at the 3-position (see also section 2.D). This pattern was observed for compounds **9.85–9.87** and for **9.88**, which nitrates as the conjugate acid but in the phenyl ring. Thus, partial rate factors for **9.88** are comparable with those in **9.55–9.57** [67JCS(B)1213]; 68JCS(B)862]. For 3,5-dimethoxypyridine N-oxide, the strong activation at the 4-position, and especially at the 2-position, causes

TABLE 9.7
Standard Log Nitration Rate Coefficients

Substituent	T (°C)	Positions	Species charge	$-H_0$ range	$\dfrac{d[\log k_2 \text{ (stoich)}]}{\delta(-H_0)}$	$-\log k_2^{0''}$
Substituted pyridines						
2-OMe-3-Me	25	5	+	9.1–10.0	−0.58	7.98
2,6-(OMe)$_2$	58.5	3,5	+	5.7–6.8	2.17	4.73
3-OMe	60	2	+	8.1–9.3	−0.41	8.09
3,5-(OMe)$_2$	25	2,6	+	7.8–8.7	2.00	4.75
3,5-(OMe)$_2$-2-NO$_2$	51	6	0	7.3–7.7	1.12	2.26
4-OMe	110	3,5	+	7.4–8.5	−0.34	10.2
3-OH	40	2	+	9.0–10.0	−0.79	7.66
2-NMe$_2$	30	3	+	6.7–8.2	2.01	2.68
						1.77
2-NMe$_2$-3-NO$_2$	30	5	0	7.2–8.2	1.67	−3.71
2-NMe$_2$-5-NO$_2$	30	3	0	6.9–8.0	1.59	−4.86
4-NMe$_2$	50	3,5	+	5.7–7.8	1.49	2.98
4-NMe$_2$-3-NO$_2$	50	5	0	7.2–8.2	1.61	−5.05
2-Ph	25	2',6'	+	7.0–7.6	2.47	5.15
		3',5'				4.45
		4'				4.14
2,6-Cl$_2$	120	3	0			9.01
Substituted pyridones						
2-Pyridone	25, 40	3	0	6.4–8.6	1.04, 0.91	0.37
		5			1.20, 1.08	0.85

3-Me-2-pyridone	31–44	5	0	4.8–8.4	0.85–1.12	−1.70
5-Me-2-pyridone	35	3	0	6.3–7.4	1.03	−1.83
1,5-Me₂-pyridone	29, 40	3	0	5.8–8.1	0.98, 1.21	−2.08
6-OH-2-pyridone	27.5	3	0	5.7–7.0	1.19	−3.43
N-Me-6-OH-2-pyridone	27.5	3	0	5.8–6.9	1.23	−2.79
6-OMe-2-pyridone	27.5	3	0	6.0–7.4	1.29	−2.92
4-Pyridone	86–158	3,5	+	4.5–7.6	1.36–1.53	1.69
	86–110			7.4–9.2	−0.53	9.76
3-NO₂-4-pyridone	157	5	0	5.0–5.3	1.30	6.26

Substituted pyridine N-oxides

2,6-Me₂	80	4	0	5.9–7.6	0.73	2.38
2-OMe	70	5	+			8.68
2,6-(OMe)₂	25	3,5	+	7.5–8.4	2.50	7.48
2,4,6-(OMe)₃	25	3,5	+	7.5–8.3	2.75	6.11
2-Ph	25	2′,6′	+	6.6–7.3	2.41	5.59
		3′,5′				4.17
		4′				4.51

Pyridinium cations

1-Me-2-NHMe	25	3	+	7.2–8.8	2.46	5.95
		5				5.60
1-Me-4-NMe₂	25	3,5	+	6.8–7.4	1.08	2.18

[a]Average values.

substitution at the 2-position on the conjugate acid. Further nitration of 3,5-dimethoxy-2-nitropyridine occurs in the 6-position and probably also involves the conjugate acid [67JCS(B)1213]. In contrast to the nitration of **9.88**, 2,3′-dipyridyl *N*-oxide (**9.89**) undergoes nitration in the N-oxide-containing ring; the 5-nitration observed (75CHE352) suggests that the conjugate acid is involved. By contrast, 2,3′-dipyridyl *N*′-oxide (**9.90**) could not be nitrated at all (75CHE352).

(9.85)

(9.86)

(9.87)

(9.88)

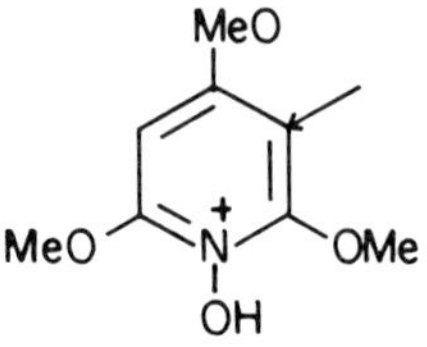

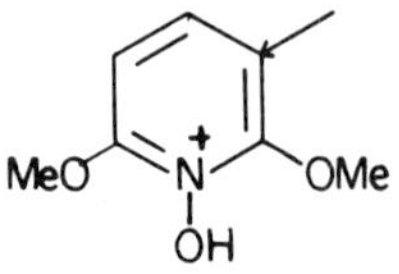

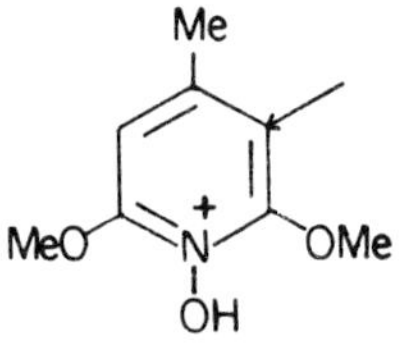

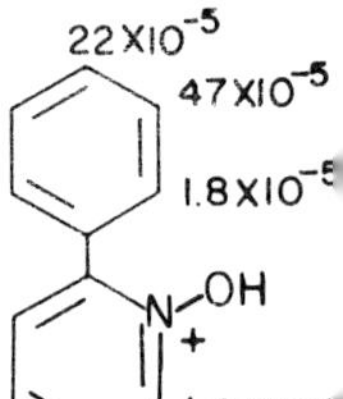

The presence of a strongly electron-supplying dimethylamino group in the 2- or 4-positions (cf. **9.91** and **9.92**) produces 5- and 3-nitration, respectively (accompanied by side reactions) [72JCS(P2)1940], consistent with the involvement, as expected, of the conjugate acids. Attempted nitration of the 3-isomer (**9.93**) gave only side reactions.

Nitration of 2-substituted-3-hydroxypyridine *N*-oxides (**9.94**) gave ~50% yield of 4-nitro derivatives, as also did 2,6-dimethyl-3-hydroxypyridine *N*-oxide (70BAU2440).

(9.89)

(9.90)

(9.91)

(9.92)

(9.93)

(9.94, R = Me, Cl)

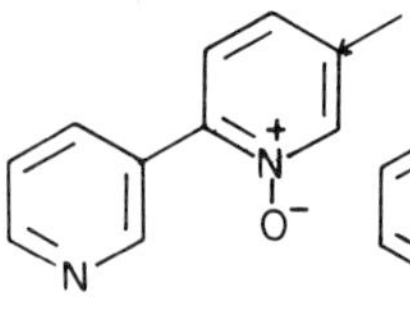

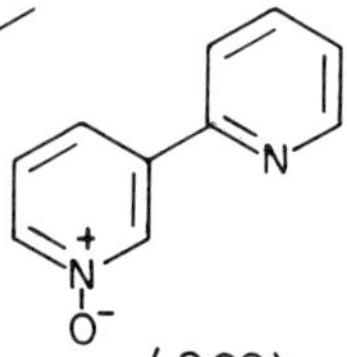

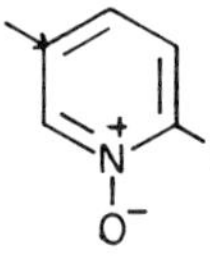

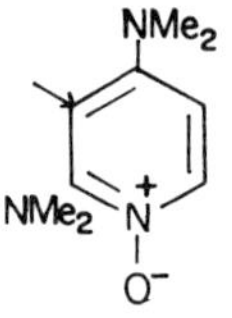

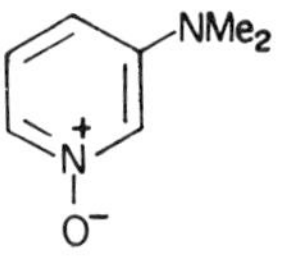

Yields in nitrations of a range of hydroxy- and alkoxypyridine *N*-oxides have been tabulated (70RCR627) as have the activation parameters for various substituted pyridine *N*-oxides [67JCS(B)1213, 67JCS(B)1235].

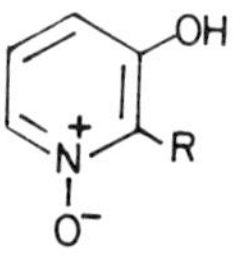

D. 2-Pyrone

The 35% yield of 5-nitro product from nitration of 2-pyrone by nitronium tetrafluoroborate may result from adduct formation, but a normal

TABLE 9.8
CALCULATED FREE-BASE RATE COEFFICIENTS FOR NITRATION OF PYRIDINE N-OXIDES[a]

Substituent	k_2 (M^{-1} sec^{-1})
H	0.2
3,5-Me$_2$	3.0[b]
2,6-Cl$_2$	8.1 × 10^{-5}
3,5-Cl$_2$	3.6 × 10^{-4} [b]
2,6-(OMe)$_2$	8.7
3,5-(OMe)$_2$	1.5 × 10^6
2,6-Me$_2$-4-OMe	6.3 × 10^4
2,4,6-(OMe)$_3$	1.7 × 10^3
3,5-(OMe)$_2$-2-NO$_2$	0.13

[a] At 25°C [67JCS(B)1213].
[b] In the original paper, these figures were transposed because of a typographical error.

mechanism could also be involved (69JHC313). Nitration of 6-phenyl-2-pyrone takes place at the 3-position of the free base, or at higher acidities, at the para position of the phenyl ring of the conjugate acid (66JOU1113).

E. ARSABENZENE

Nitration gives an overall yield of ~8% mononitro derivatives in the proportion 35% 2- and 65% 4-nitroarsabenzene (81JOC881).

F. SUMMARY OF KINETIC DATA

Standard nitration rates, determined by the method outlined in Chapter 3, are shown in Table 9.7 [75JCS(P2)1600, 75JCS(P2)1624].

Rate coefficients have also been calculated for nitration of pyridine N-oxides as the free bases, at 25°C (Table 9.8) [67JCS(B)1213].

5. Halogenation

A. PYRIDINES

Chlorination and bromination of pyridines have been studied under a variety of conditions, particularly in the gas phase. The positional substitution sequence by electrophilic halogenation is 3 > 5 > 4 [67MI1(164)]. A good method for obtaining a high yield of 3-bromopyridine is to brominate in oleum, the latter oxidizing the hydrogen bromide by-product (62RTC864). The reactive substrate under these conditions is the $C_5H_5N^+SO_3^-$ coordination compound. Hydrogen bonding between the oxygen and the 2(6)-hydrogen has been postulated to account for the 2(6)-bromination observed with some derivatives (72BSF2466).

Rates of bromination of several methylpyridines by hypobromous acid in aqueous perchloric acid at 25°C (Table 9.9) have been measured. Rate–acidity profiles and comparison with model compounds showed that the conjugate acids were the reacting species (74JOC3481). From this work the partial rate factor for bromination of the 3-position of the pyridinium ion was estimated as $2-6 \times 10^{-13}$, which gives a σ_3^+ value of ~2.0.

Chlorination of 2-aminopyridine gives 5-chloro and 3,5-dichloro derivatives in equal amounts when the medium is 17 wt% H_2SO_4, but 98 and 6%, respectively, in 72 wt% H_2SO_4. This selectivity was ascribed to differences in rates of chlorination of protonated versus nonprotonated species (76JOC93). For the more basic 2-aminopyridine chlorination presumably occurs on the monocation throughout the acidity region but the less basic 5-chloro derivative chlorinates via the free base at low acidity.

Bromination of 2-aminopyridine, 2-N,N-dimethylaminopyridine, and their 5-substituted derivatives (**9.95**) has been studied in detail [7OJCS(B)117]. All react as the free bases and logarithms of the true second-order rate coefficients (corrected to take account of the equilibrium con-

TABLE 9.9

RELATIVE RATES AND PARTIAL RATE FACTORS FOR BROMINATION OF PYRIDINIUM IONS[a]

Substituents	Relative Rate	f_3	f_5
1-Me-2,6-(OMe)$_2$	7.8×10^{-3}	2.3×10^{-2}	—
1-Me-2-OMe	1.8×10^{-7}	—	—
2,4,6-Me$_3$	3.0×10^{-7}	9.0×10^{-7}	—
1,2,4,6-Me$_4$	2.3×10^{-7}	6.9×10^{-7}	—
2,6-Me$_2$	2.8×10^{-9}	8.4×10^{-9}	—
1,2,6-Me$_3$	2.8×10^{-9}	8.4×10^{-9}	—
2,4-Me$_2$	4.1×10^{-9}	1.5×10^{-8}	9.8×10^{-9}
1,2,4-Me$_3$	3.6×10^{-9}	—	—
4-Me	3.1×10^{-11}	9.3×10^{-11}	—
1,4-Me$_2$	4.6×10^{-11}	1.4×10^{-10}	—
2-Me	3.1×10^{-11}	1.0×10^{-10}	8.6×10^{-11}
1,2-Me$_2$	3.4×10^{-11}	—	—

[a] With HOBr in HClO$_4$ at 25°C.

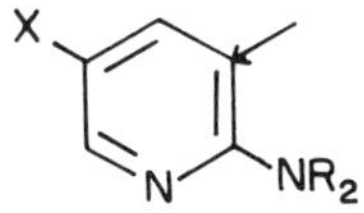

(9.95, X = Me,Cl, Br,NO$_2$; R = H,Me 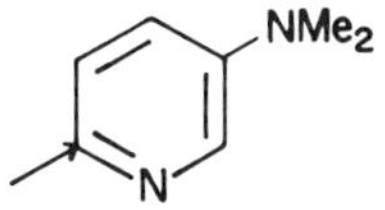)   (9.96 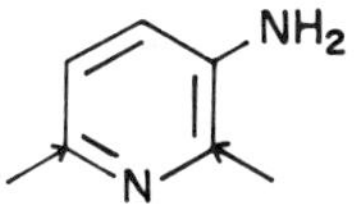)   (9.97)

centration of free base) are given in Table 9.10. From the rates of bromination of the corresponding anilines, values of σ_m^+ for the aza substituent (i.e., the 3-position in pyridine) were deduced as 0.60 (0.56), 0.63 (0.63), and 0.71 (0.58) from the 5-bromo, 5-chloro, and 5-nitro compounds, respectively (values for the dimethylamino compounds in parentheses). Comparison with values obtained in the gas phase show that these value are affected by hydrogen bonding to the pyridine nitrogen.

Bromination of 3-aminopyridine (**9.97**) and 3-N,N-dimethylaminopyridine (**9.96**) with 2,4,4,6-tetrabromocyclohexa-2,5-dienone shows a remarkable difference in orientation. Thus, **9.96** is brominated exclusively at the 6-position, whereas **9.97** gives 2- and 6-monosubstitution together with 2,6-disubstitution. This has been attributed to steric hindrance, which follows from the size of the reagent [73JCS(P1)68]. Both 2-amino- and 2-N,N-dimethylamino-pyridines are brominated solely in the 5-position; in view of the previous results, some 3-derivative might also have been expected from the former.

Bromination of methylpyridines in oleum gives moderate yields of the 3-bromo derivatives **9.98–9.100** (65RTC951). Appreciable (7%) disubstitution occurs with **9.100**.

Bromination of 2,6-dimethylpyridine in oleum at 50°C gave 62% of the 3,5-dibromo derivative, whereas iodination gave 50% of the 3-iodo deriv-

TABLE 9.10

VALUES OF LOG k_0 FOR BROMINATION OF SUBSTITUTED PYRIDINES[a]

Substituent	log k_2^0
2-NH$_2$-5-Cl	4.76
2-NH$_2$-5-Br	4.79
2-NH$_2$-5-NO$_2$	2.53
2-NMe$_2$-5-Cl	6.09
2-NMe$_2$-5-Br	6.31
2-NMe$_2$-5-NO$_2$	4.27

[a] At 25°C.

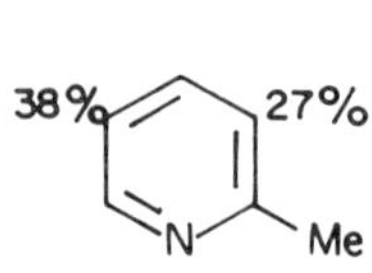

(9.98) — pyridine with 38%, 27%, Me at 2-position
(9.99) — pyridine with 15%, Me, 7%
(9.100) — pyridine with (7%), 45%, Me

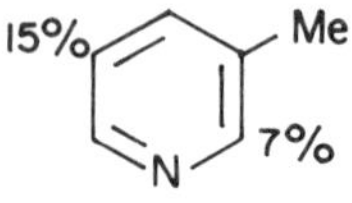

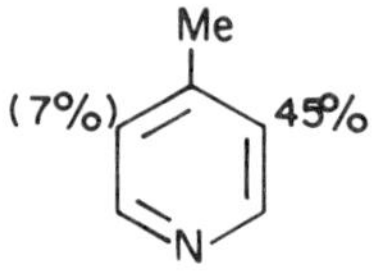

ative. 2,4,6-Trimethylpyridine gave 87% of the 3,5-dibromo derivative, whereas iodination gave only 5.5% of the 3,5-diiodo derivative, the main product (60%) being 3-iodo-2,4,6-trimethylpyridine (70RC779).

B. PYRIDONES AND HYDROXYPYRIDINES

Iodination of 2-benzyl-2-hydroxypyridine (**9.101**) occurs at the 6- and subsequently at the 4-position. Under similar conditions, the corresponding N-oxide gave only the 4,6-diiodo product. This does not necessarily prove that the N-oxide function is more effective in activating the 4-position; steric hindrance to 6-substitution may be involved (cf. 75CHE478). Whereas both aminomethylation and diazonium coupling were easier for the N-oxide than for **9.101**, the 4- : 6-isomer ratio for both reactions was similar for both pyridine and its N-oxide (75CHE478).

Bromination of 2-pyridone in aqueous solution occurs via the tautomer **9.102** at pH < 6 and via the conjugate anion at pH > 6; attack occurs preferentially at the 3-position in 2-pyridone but at the 5-position in the anion. These conclusions, deduced from rate–acidity profiles, were confirmed by using model compounds. Below pH 6, **9.102** and **9.103** are of similar reactivity, whereas **9.104** reacted 2300 times slower. Above pH 6, however, **9.102** becomes much more reactive than **9.103** (82JA4142).

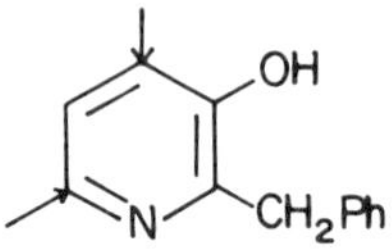

(9.101) 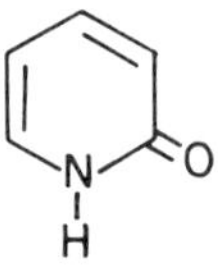 (9.102) 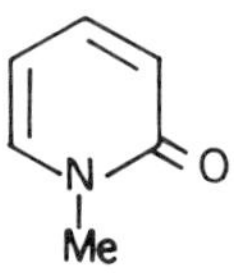 (9.103) 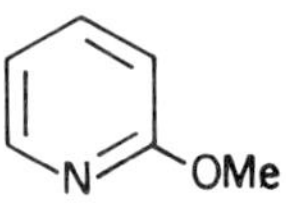 (9.104)

Bromination of the 5-monobromo compounds gave similar results, except that 5-bromo-2-pyridone now becomes much more reactive than its N-methyl analogues above pH 4, which follows from the lower pK_a of the bromo derivative (82JA4142).

4-Pyridone likewise reacts mainly as the neutral pyridone tautomer at pH < 6 and via the conjugate anion at pH > 6, 4-methoxypyridine being unreactive by comparison over the whole pH range. Facile dibromination of 4-pyridone occurs because at most pH values the lower pK_a of the monobromo derivative more than compensates for its intrinsically lower reactivity (83CJC2556).

Bromination of 3-hydroxypyridine takes place readily in aqueous solution at 20°C giving a 50% yield of the 2,4,6-tribromo derivative (50RTC1281); this is indicative of the greater ease of halogenation compared to nitration and sulfonation.

C. PYRIDINE N-OXIDES

Pyridine N-oxide is unreactive toward iron-catalyzed bromination at 110°C (55JA2902), but silver sulfate-catalyzed bromination in sulfuric acid at 200°C gives a 10% yield of 2- and 4-bromination in the ratio 1:2 (61TL32). With bromine in oleum the main product is 3-bromopyridine N-oxide (60%) together with the 2,5-dibromo (~35%), and 2,3- and 3,4-dibromo compounds (~5%) (62T227). Presumably the N-oxide function is here complexed with sulfur trioxide, which causes deactivation and β-orientation. Bromination in acetic anhydride also gives 3-substitution (35%); an addition–elimination mechanism has been proposed (65JPJ62).

Bromination of 3-bromopyridine N-oxide in 90% sulfuric acid gives a low yield of 2,3-, 3,4-, and 2,5-dibromo products; that is, bromination takes place α and γ to the N-oxide function. A suggestion that this arises because the 3-bromine takes over the orientation from the N-oxide (62T227) is probably incorrect. Under such conditions, pyridine N-oxide itself appears to brominate as the free base and bromination as free base is still more probable for the bromo derivative.

This view is supported by the bromination in 90% sulfuric acid of 2-bromopyridine N-oxide, which gives 2,4- and 2,6-dibromopyridine N-oxide in a 4.5:1 ratio. Thus, bromination has taken place α and γ to the N-oxide function, but meta to bromine, so the N-oxide group retains control of the orientation. Moreover, in oleum, bromination of 2-bromopyridine N-oxide yields 2,3- and 2,5-dibromopyridine N-oxides; reaction now occurs on the protonated N-oxide species. Similar bromination of 4-bromopyridine N-oxide gives 3,4-dibromopyridine N-oxide (84%) and 3,4,5-tribromopyridine (10%) (62T227). Use of α-lithiopyridine N-oxide utilizes activation of the 2(6)-positions relative to the 4-position (the C—Li bond being more polar than the C—H bond). Hence, chlorination and bromination give 2,6-dihalogenated products (72JHC1367). 2-Bromo- and 2,6-dibromopyridine *N*-oxide can both be obtained by mono- or bis(chloromercuriation) followed by bromodemercuriation (72JHC1367).

5-Benzyl-3-hydroxypyridine *N*-oxide (**9.105**) is iodinated in the 4- and 6-positions (cf. other reactions described in, Section 6) (74BAU2023), as is 3-hydroxypyridine *N*-oxide (66RC1875).

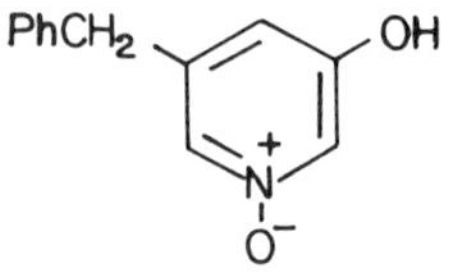

(9.105)

6. Other Reactions

A. METALLATION

Pyridine can be lithiated in the 3-position by *n*-butyllithium in 70% yield (40JA446; 48JA1037), and in similar yields at other positions by lithiodebromination (46JA103; 51JOC1485; 52JA6289; 55RTC1003). Mercuriation of pyridine gives up to 70% of the 3-mercuri derivative (23CB2223; 32JCS1263), and the reaction is facilitated by electron-supplying groups (37USP2085063). Mercuriation of pyridine *N*-oxide goes into the 2- and 3-positions (2:1 ratio) possibly due to coordination of the (very large) electrophile with oxygen (58RTC340); 62RTC124). With mercuric acetate in acetic acid at 100°C, a mixture of 2-, 3-, and 2,6-mercuriated *N*-oxides in the ratio 8:2:1 is obtained (62RTC124), the 3-isomer suggesting some involvement of the protonated oxide.

B. ALKYLATION AND ACYLATION

Neither pyridine nor pyridine *N*-oxide can be alkylated or acylated [but see 67MI1(162) for early reports]. 2-Pyridone and 3-hydroxypyridine can be carboxylated ortho and para to the hydroxy group under basic conditions (24CB1161; 54CA1337i, 54JOC510).

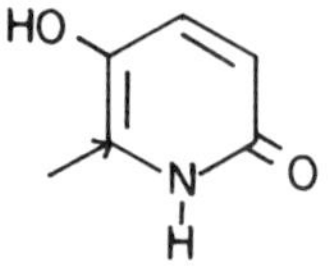

(9.106) 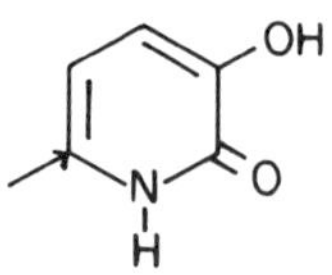   (9.107)

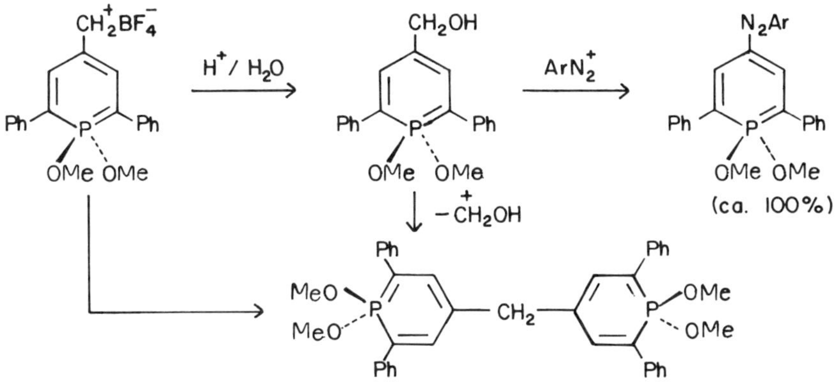

SCHEME 9.1. Electrophilic substitution of λ^5-phosphorins.

Aminomethylation is possible in hydroxypyridine N-oxides; thus, for example, **9.105** is substituted at the 2- and 6-positions (74BAU2023). 3-Hydroxypyridine N-oxide itself is aminomethylated at the 2-position. 2,6-Disubstituted 3-hydroxypyridine N-oxides react at the 4-position (72JOU416). Hydroxy-2-pyridones **9.106** and **9.107** are substituted at the 6-position (71BAU2222).

As in hydrogen exchange, nitration, and desilylation, acetylation of arsabenzene (**9.12**) shows ortho and para orientation, with the 2:3:4-isomer ratio of 40:<1:300 (78TL2537). Steric hindrance from the arsenic d orbitals presumably accounts for the lower reactivity of the 2- compared to the 4-position (cf. hydrogen exchange, Section 2.G). In acetylation, arsabenzene is more reactive than mesitylene, though the reactivity of the latter is reduced through steric hindrance.

1,1-Dimethoxy-2,6-diphenyl-λ^5-phosphorin-4-yl tetrafluoroborate undergoes the alkylation and diazonium coupling reactions shown in Scheme 9.1 in high yield [73AG(E)753].

C. DIAZONIUM COUPLING

Pyridine is not nearly reactive enough to undergo direct azo-coupling, but 3-phenylazopyridine has been prepared via the prior addition of three molecules of sodium sulfite to the pyridine ring (69TL4855).

Azo-coupling of 5-benzyl-3-hydroxypyridine N-oxide (**9.105**) goes only into the 2-position, probably because of steric hindrance to 6-substitution (74BAU2023).

D. SULFONATION AND SULFENYLATION

With oleum in the presence of mercuric sulfate, pyridine gives a 75% yield of the 3-sulfonic acid (42CB1108; 43JA2233; 58RTC963). The methylpyridines are sulfonated at the appropriate β-position, but in lower yields, namely (60%) (2-methylpyridine), (23%) (3-methylpyridine), (40%) (4-methylpyridine) (43JA2233). The lower yields probably arise from oxidative side reactions involving the methyl groups, and this is supported by the failure of 2,4-dimethyl- and 2,6-dimethylpyridine to sulfonate at all. By contrast, 2,6-di-*t*-butylpyridine sulfonates quite readily when reacted under aprotic conditions in liquid SO_2 (37% yield, 4 hr, $-35°C$); this is attributed to steric hindrance to formation of the deactivated pyridine-SO_3 complex (53JA3865; 66JA986). Hydroxy- and aminopyridines undergo sulfonation at the expected positions.

Pyridine N-oxide appears to be rather less reactive than pyridine, because sulfonation is carried out under conditions of high acidity where the protonated species dominates. Under conditions that give 70–75% of pyridine-3-sulfonic acid, pyridine N-oxide gives 40–45% of pyridine-3-sulfonic acid N-oxide, 2.5% of the corresponding 4-isomer, and 45% recovery of starting material (55JA2902; 59RTC586). It is also possible that pyridine N-oxide-SO_3 complexes are formed and rearrangement of such complexes has been proposed as the rate-determining step for sulfonation of both pyridine and its N-oxide (75CHE745).

Sulfenylation of 2-pyridone with sulfur dichloride takes place at the 5-position to give a 13% yield of 2,2′-dihydroxy-5,5-dipyridyl sulfide (**9.108**). Neither the 3-hydroxypyridine nor 4-pyridone reacted (64ACS269).

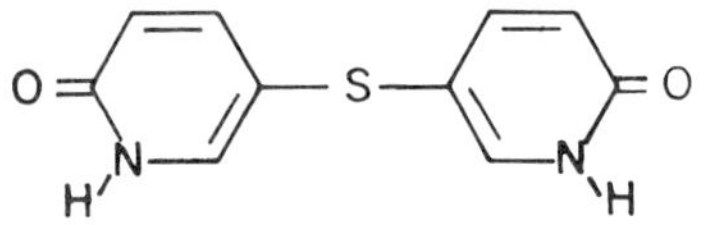

(9.108)

E. DEMETALLATION

Demetallation of 2-substituted pyridines (2-$PyMR_3$) by water or lower alcohols is thought to proceed via the formation of an intermediate anion, the mechanism thus being analogous to the mechanism for base-catalyzed hydrogen exchange. Relative rates for cleavage were Me_3Si (1.0), Me_3Ge (1×10^{-4}), Me_3Sn (22). For silanes, the relative reactivity order was $SiMe_2H > SiMe_3 > SiPh_3 > SiEt_3$. Compounds with methyl substituents at the 3- and 5-positions were less reactive than those with methyl substituents at the 4- and 6-positions [68JCS(B)450, 68JC765, 68JC878,1008, 68JOM(13)113].

7. Side-Chain Reactions

A. Pyrolysis of Esters

a. *Pyridine*

This gas-phase reaction provides the only method currently available for determining σ^+ values for the non-hydrogen-bonded pyridine free base. The method requires neither assumptions, extrapolations, nor approximations; the ρ factor of the reaction is sufficiently small that the reactivities of the pyridine and benzene derivatives can be compared directly. The method was first introduced for determining the electrophilic reactivity of pyridine [62JCS4881; 71JCS(B)2382] using 1-arylethyl acetates (**9.109**). Subsequent determinations used 1-aryl-1-methylethyl acetates. (**9.110**) and 1-arylethyl methyl carbonates (**9.111**) [79JCS(P2)228].

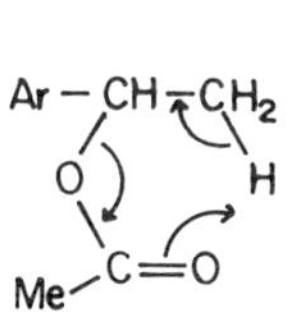

(9.109, Ar = Ph, 2–, 3–, 4–Py) 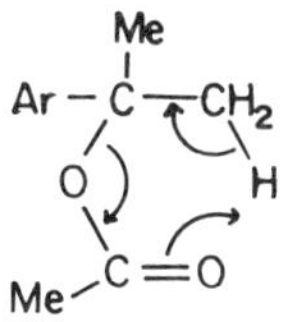   (9.110, Ar = Ph, 2–, 3–, 4–Py) 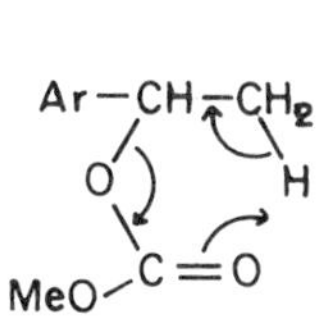   (9.111, Ar = Ph, 2–, 3–, 4–Py)

The stoichiometry is 2.0 for the two former reactions and 3.0 for the latter, in which the initially formed methylcarbonic acid decomposes instantaneously. Of these three reactions the first is the best documented. The reaction data and the σ^+ values obtained are given in Table 9.11.

TABLE 9.11
Electrophilic Reactivity of Pyridine via Pyrolysis of Esters

Esters	T (K)	ρ	σ^+ 2	3	4
1-Arylethyl acetates (**9.109**)	625	−0.63	0.78	0.295	0.865
1-Arylethyl methyl carbonates (**9.111**)	600	−0.71	0.76	0.30	0.85
1-Aryl-1-methylethyl acetates (**9.110**)	550	−0.743	0.89	0.285	0.86

With the exception of the result for the 2-position in the pyrolysis of 1-aryl-1-methylethyl acetates, the other data are in excellent agreement. They show that the electrophilic reactivity of the pyridine positions relative to benzene is benzene > > 3-pyridine > > 2-pyridine > 4-pyridine.

This order is that required by theory, and discussion of the magnitude of the values in relation to those obtained in solution, and theoretical calculations, is deferred to Section 9. The abnormally low reactivity of the 2-position in pyrolysis of the bulkier ester (**9.110**) may reasonably be attributed to a steric effect of some kind. While it is easy to propose various effects relating to the degree of coplanarity of the bulkier carbocationic transition state and the aromatic ring, it is not certain that decreased coplanarity with an electron-deficient ring would in fact result in reduced reactivity. Alternative effects are (1) attraction of the side-chain methyl hydrogens to the nitrogen atom, favoring coplanarity; (2) steric interaction between the side-chain methyl group and the nitrogen lone pair, inhibiting coplanarity; and (3) electrostatic repulsion between the nitrogen lone pair and the oxygen, reducing coplanarity.

Each of these factors is avoidable in the 1-arylethyl compounds.

b. *Pyridine N-Oxide*

The gas-phase pyrolysis of 1-arylethyl acetates has also been used to determine σ^+ values of 0.81 and 0.016, respectively, for the 3- and 4-positions of pyridine *N*-oxide [75JCS(P2)277]. These show that the 4-position is the most reactive, and little less reactive than that of a position in benzene; further discussion is deferred to Section 9.

For the 2-position, no valid results were obtained, due to an exceptionally fast elimination which yielded 2-acetylpyridine instead of the expected 2-vinylpyridine *N*-oxide; the former was shown not to arise from isomerization of the latter. This elimination is probably accelerated by a neighboring group effect (these have since been identified in other eliminations, 83MI1), which induces an acetyl migration followed by hydride transfer (for example, see Scheme 9.2). An alternate mechanism involves

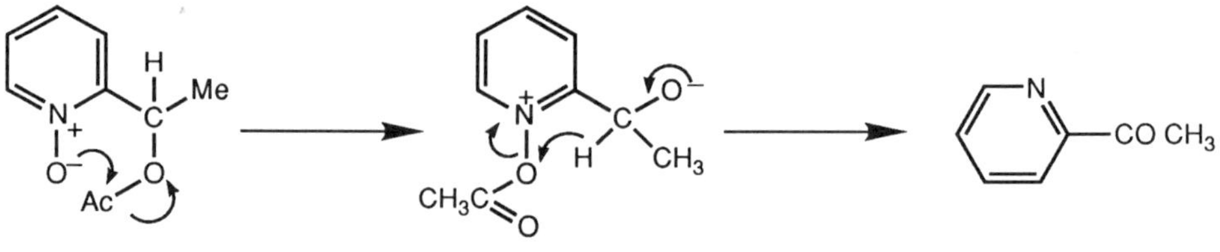

SCHEME 9.2. Possible mechanism for formation of 2-acetylpyridine from 2-(1-acetoxymethyl)pyridine *N*-oxide.

prior oxygen migration to the side chain to give **9.112**, which, being a 1,1-diol derivative, would eliminate very rapidly. Elimination from 2-(1-hydroxyethyl)pyridine N-oxide occurred at almost the same rate as from the ester, possibly by successive migration of a proton and a hydride ion, though the similarities in rates indicates a common rate-determining step for both reactions.

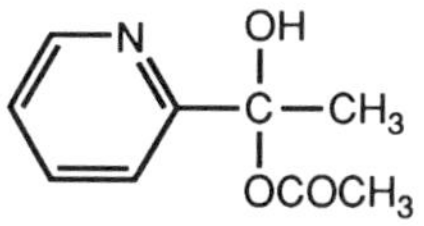

(9.112)

B. SOLVOLYSIS OF 1-ARYL-1-METHYLETHYL CHLORIDES

a. *Pyridine*

The side-chain carbocation method for determining heterocyclic reactivities, first used in the pyrolysis above, has been applied in solution using the solvolysis of 1-aryl-1-methylethyl chlorides (ArCMe$_2$Cl) (73CJC1620, 73JOC2657). Results obtained (Table 9.12) show clear solvent dependence. The values for the 2-position agree well with those determined in the gas phase (Table 9.11), but the values for the other positions are ~0.25 σ units more positive in solution. This has been attributed to hydrogen bonding in the 3- and 4-substituted derivatives; at the 2-position this bonding would be sterically inhibited by the bulky tertiary group, leading to consistency of the gas-phase and solution data [73CJC1620; 74MI1(222)]. Activation parameters indicated that the solvolyses are entropy controlled, so solvation factors are important. Hydrogen bonding

TABLE 9.12

ELECTROPHILIC REACTIVITY OF PYRIDINE VIA SOLVOLYSIS OF 1-ARYL-1-METHYLETHYL CHLORIDES

Solvent	T (°C)	σ^+		
		2	3	4
55% aq. acetone	25.5	0.72	0.47	1.20
45% aq. acetone	40	0.72	0.45	1.14
Methanol	25	0.73	0.57	1.13
Aq. ethanol	75	0.75	0.54	1.16

and differential solvation must be principal causes of the wide variation in σ values for pyridine from various determinations [see e.g., 64AHC(3)209; 66JCS(B)937; 70AJC203; 72AJC431]. Significantly, of all the determinations in solution, those using carbon tetrachloride as solvent (no hydrogen bonding) gave the least positive overall values [i.e., 0.63 (2), 0.33 (3), 0.66 (4)] [66ZN(A)1906]. Values for σ_3 between 0.27 and 0.40 have been obtained in chloroform or solvent-free systems.

b. *Pyridine N-Oxide*

From solvolysis of 1-aryl-1-methylethyl chlorides in 80% aqueous ethanol σ^+ values for the 2- and 4-positions of pyridine *N*-oxide have been determined as 0.68 and 0.45, respectively (73JOC2657). The value for the 4-position is again more positive than that obtained in the gas phase and the discrepancy is about twice that found for pyridine. This is precisely the expectation for the effect of hydrogen bonding, which is much stronger in the case of the *N*-oxide.

8. Transmission of Substituent Effects in Pyridine

In contrast to the heteroatom in five-membered heterocycles (Chapter 6), a pyridine nitrogen atom should be able to transmit electronic effects in a manner similar to a CH group in benzene. Difficulty attends assessment of such transmission abilities by comparison of the reactivities of pyridines and benzenes because the Hammett ρ factor obtained will be a measure of both the transmission ability (64JCS627) and the transition-state charge. Because pyridine is less reactive, the transition state for its reactions should be nearer to the Wheland intermediate than the corresponding transition states for reactions of the corresponding benzene

TABLE 9.13

ρ FACTORS FOR HYDROGEN EXCHANGE IN SULFURIC ACID

Compounds	ρ
Dimethylanilines	−3.2
Anilines	−3.3
Phenols	−4.9
Benzenes	−7.5
Pyridinium ions	−12.9

compounds, according to the principles given by Norman and Taylor [65MI1(298)]. Early work showed that conjugative effects in pyridine appeared to be greater as a result of attachment to a more electron-deficient system (59JA1935). A more detailed analysis of substituent effects for hydrogen exchange in sulfuric acid (Table 9.13; the pyridine data are derived from these in Table 9.1) showed clearly that the ρ factor parallels the reactivity of the parent aromatic as required by the reactivity–selectivity effect, and thus a very large value applies to the unreactive pyridinium ions [74JCS(P2)1294].

These results do not, as noted above, give any direct indication of the difference in transmission ability of =N versus =CH—. For this, one must compare the effects of substituents in **9.113** with **9.114**, though even here the results are blurred because of the dual transmission pathway. It is difficult to make the comparison using a conventional electrophilic substitution (except by prelabeling, e.g., as in detritiation) because substitution at the desired point is likely to be a minor component and difficult to measure accurately. The side-chain carbocation method has therefore been used.

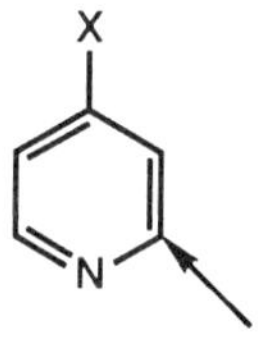

(9.113) 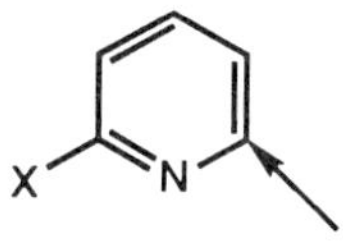   (9.114)

An extensive set of data has been obtained by Noyce and Virgilio (73JOC2660) for solvolysis of 1-aryl-1-methylethyl chlorides (**9.115–9.119**) in 80% aqueous ethanol at 70°C, for which the ρ factor for benzenoid compounds is -4.0. The log k/k_0 values are given in Table 9.14, along with δ log k/k_0 values, the differences between the observed values and those calculated from the standard σ^+ values and the ρ factor of -4.0. Thus a positive value means that the compound is more reactive

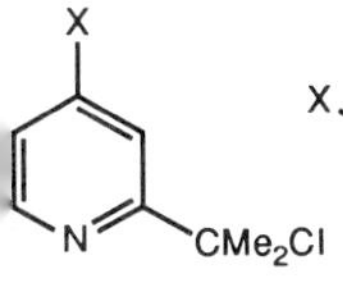

(9.115) 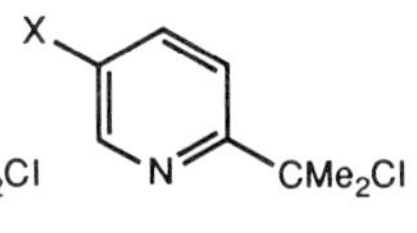   (9.116) 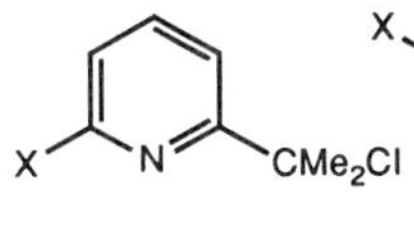   (9.117) 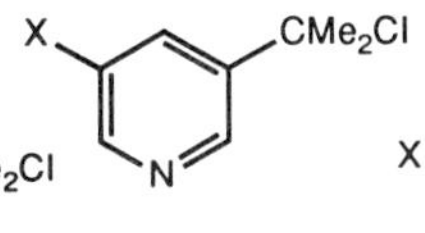   (9.118) 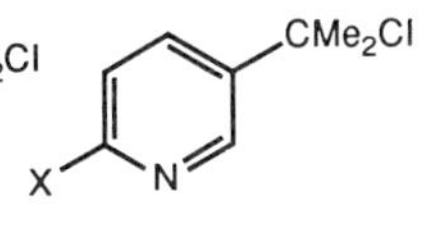   (9.119)

TABLE 9.14
Kinetic Data for Solvolysis of Substituted 1-(x-Pyridyl)-1-methylethyl Chlorides in 80% Ethanol

Compounds	Substituents	log k/k_o	δ log k/k_o
2-Pyridyl	4-Me	0.22	−0.04
(9.115, 9.116, 9.117)	5-Me	1.15	−0.09
	6-Me	0.62	0.36
	4-Cl	−1.45	0.14
	5-Cl	−0.13	0.32
	6-Cl	−0.85	0.74
	5-OMe	3.67 ± 0.3	0.5
	6-OMe	1.003	1.19[a]
	6-OEt	1.01	1.20
	6-Ph	0.23	0.67
	4-Cl-5-OMe	1.81	0.313
3-Pyridyl	5-Me	0.22	−0.04
(9.118, 9.119)	6-Me	1.29	0.05
	5-Br	−1.22	0.40

[a] Sign incorrect in original paper.

than expected, and vice versa. These interesting results were not analyzed in the original paper. The complexity of the observed effects includes the following factors.

(1) Methyl groups in positions 4 and 5 in either series all activate less than they should.

(2) Chloro groups in positions 4 and 5 in the 2-pyridyl compounds and 5-bromo in the 3-pyridyl compound all deactivate less than they should.

(3) Methoxy groups in positions 5 and 6 activate *much* more than they should.

(4) Substituent effects are not additive.

(5) Compounds **9.117** are more reactive than the corresponding compounds **9.115**, and *much* more reactive than predicted.

The last result strongly suggests that hydrogen bonding must be a major factor involved. Thus the presence of the bulky substituents at the 2- and 6-positions will inhibit hydrogen bonding so that the 6-substituted compounds (**9.117**) will be less hindered, and therefore more reactive, than the 4-substituted isomers (**9.115**). Compounds **9.117** will also be exceptionally reactive relative to the unsubstituted compound, for which hydrogen bonding will be greater. To assess this factor, rates of gas-phase pyrolysis of 1-(2-pyridyl)ethyl acetates (**9.120**) have been measured (Table 9.15) [79JCS(P2)624]; the δ log k/k_0 values are derived the same way

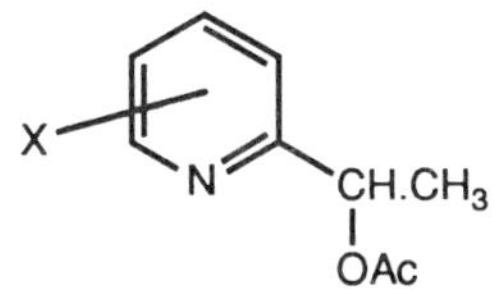

(9.120, X = 4–, 5–, and 6– Substituents)

as those in Table 9.16 and for qualitative comparison with the latter they must be multiplied by 6.6 (the ratio of the respective ρ factors).

Here the 4-Me substituent behaves normally, but as in the solvolysis, 5-Me is less activating while 5-Cl is less deactivating than predicted by comparison with their effects in benzene. Clearly a factor operates which is solvent independent, and it has been suggested that *the inductive effect is less readily transmitted in pyridine than in benzene* [79JCS(P2)624]. This would also account for the fact that all of the alkoxy substituents in both sets of work are more activating than predicted. The discrepancy in reactivity between the 4- and 6-methyl-substituted compounds is much greater in the solvolysis (0.40) than in the pyrolysis (0.013), the latter value becoming 0.086 after correction for ρ factor difference. This strongly supports the view that the abnormally high reactivity of all of the 6-substituted compounds in solvolysis is due to inhibition of hydrogen bonding. Nevertheless, there appears to be an additional factor at work, since even in the gas phase both the Me and OEt substituents activate more from the 6- than from the 4-position. Since the discrepancy is greater for OEt, which has the larger $+M$ effect, it may be that nitrogen transmits conjugative effects more effectively than carbon. Alternatively, electron-donor mesomeric effects may be more drawn into play in the electron-deficient nitrogen-containing system. Firmer conclusions, however, can come only from additional studies of substituent effects in the gas phase.

TABLE 9.15
KINETIC DATA FOR PYROLYSIS OF 1-(2-PYRIDYL)ETHYL ACETATES (**9.120**)

X	H	4-Me	5-Me	6-Me	5-Cl	4-OEt	6-OEt
log k/k_o at 650 K	0	0.061	0.137	0.074	−0.033	0.027	0.047
log k/k_o	—	0	−0.038	0.013	0.032	0.030[a]	0.50[a]

[a] Estimated, since *m*-OEt has not been measured for 1-arylethyl acetates.

9. Comparison of Theoretical Calculations of the Reactivity of Pyridine and Pyridine N-oxide with Observed Data

A. PYRIDINE FREE BASE

The σ^+ values for the pyridine free base determined in the gas phase are as shown in **9.121** [62JCS4881; 71JCS(B)2382]. These show that all positions are deactivated, the reactivity order being benzene >> 3-Py >> 2-Py >> 4-Py, and that the 3-position is about as deactivated as a meta-position in chlorobenzene whereas the 2- and 4-positions are about as deactivated as the ortho- and para-positions in nitrobenzene (62JCS4881).

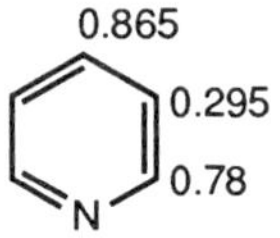

$$\begin{array}{c} 0.865 \\ 0.295 \\ 0.78 \end{array}$$

(9.121)

Various theoretical calculation of pyridine reactivity have been carried out, usually involving the Hückel or Extended Hückel method (59JA1935; 62JCS4881; 65CCC355; 67T2513; 68CCC3138). However, their validity depends critically on the parameters used for the coulomb and resonance integrals, and use of auxilliary inductive parameters for the 2- and 6-carbon adjacent to the electronegative nitrogen. For the coulomb integral α_N ($= \alpha_C + h\beta$) it is generally agreed that h should be 0.5 (51JCP1323; 56JCS272), while the resonance integral β_{CN} should be the same as β_{CC} (52QR63; 56JCS272). Analysis of nucleophilic and free-radical substitution reactions of pyridine, and substitution in quinoline and isoquinoline

TABLE 9.16

π DENSITIES AND LOCALIZATION ENERGIES FOR ELECTROPHILIC SUBSTITUTION OF PYRIDINE[a]

	π Density	$L_r^+/-\beta$
2-Position	0.957	2.686
3-Position	0.982	2.596
4-Position	0.951	2.698

[a] (62J4881).

[59JCS3451; 60T(8)23], indicated that the coulomb integral for the 2(6)-carbons $\alpha_{C'}$ ($= \alpha_C + h'\beta$) requires a value of 0.085 for h'. Experimental σ^+ values are plotted against the π densities calculated with these parameters (Table 9.16) in Fig. 9.1. The quality of the correlation is quite remarkable, and indeed is unparalleled for any other electrophilic substitution of any aromatic; all other sets of parameters give poorer correlations.

Because electron-supplying conjugative effects are not important in pyridine, the molecule is one for which a correlation of reactivities with both π densities and localization energies could be expected. Localization energies were calculated with the requirement that the auxilliary inductive effect is only partially operative at localized centers (42JA900), and using a value of δ of 0.43 ($\alpha_{C'} = \alpha_C + \delta h\beta$) indicated by reactions of quinoline and isoquinoline [59JCS3451; 60T(8)23]. The correlation with σ^+ values (Fig. 9.2) is again excellent.

B. Hydrogen-Bonded Pyridine

From the solvolysis of 1-aryl-1-methylethyl chlorides, average σ^+ values for the hydrogen-bonded molecule are 0.73 (2), 0.51 (3), and 1.16 (4) and these values should apply reasonably well in protic solvents. Any

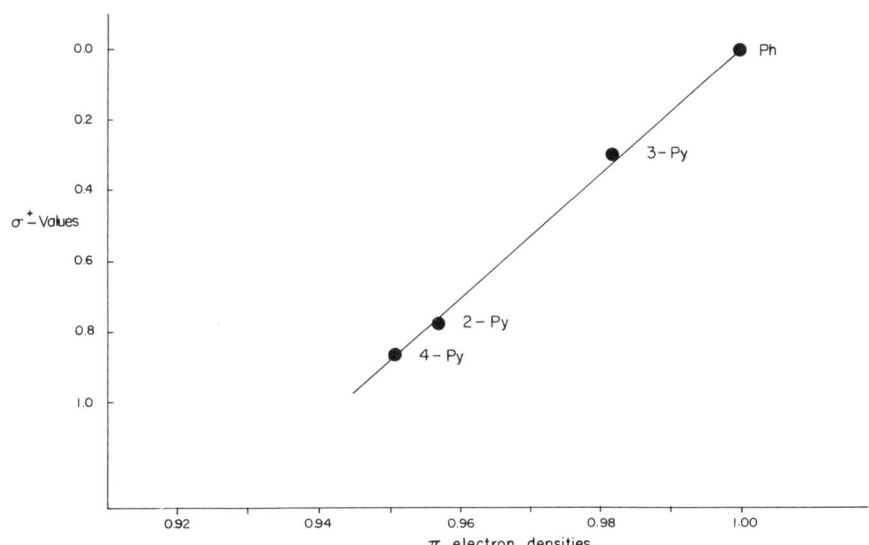

FIG. 9.1. Plot of experimental σ^+ values versus π densities for pyridine.

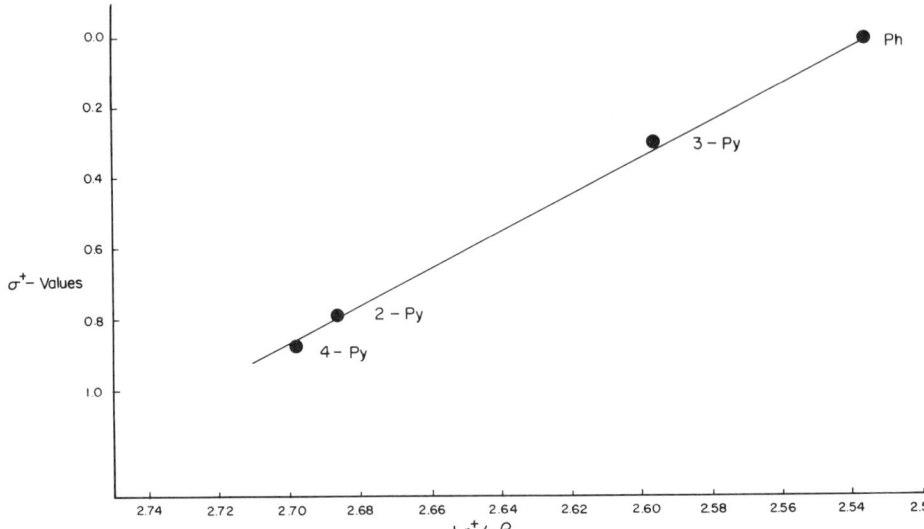

FIG. 9.2. Plot of σ^+ values versus localization energies for electrophilic substitution in pyridine.

attempts to correlate them with theoretically calculated parameters is at present impractical because of the differential positional and substrate hydrogen bonding.

C. Pyridinium Cation

Localization energies have been calculated for the pyridinium ion as follows: -2.85β (2-position), -2.62 (3-position), -2.84 (4-position) (59JA1935). These show the expected greater deactivation relative to pyridine (cf. Table 9.16), coupled with 3-orientation. However, the values will be rather inaccurate because no auxiliary inductive parameter (most important here) was used, and also the calculations do not take account of hydrogen bonding. A value of σ_m^+ for NH^+ has been determined from hydrogen exchange of substituted phenols and the pyridine analogues as 1.85 [68JCS(B)866]; the value is less positive than the σ_m values of 2.02 and 2.1 [64AHC(3)209; 72JCS(P2)671]. The differences reflect solvation effects, or, more probably, the higher reactivity of the hydroxy compounds with consequent attenuated deactivation by nitrogen.

D. PYRIDINE N-OXIDE FREE BASE

The σ^+ values for the 3- and 4-positions of pyridine N-oxide free base determined in the gas phase are shown in **9.122** [75JCS(P2)277]. These show that the 4-position is much more reactive than the 3-position and is almost as reactive as a position in benzene. For this molecule, $+M$ electron release will be substantial, so quantitative correlations with both π densities *and* localization energies will not be possible. The small ρ factor for pyrolysis means that the transition state is nearer to the ground state and so a correlation with π densities should apply. This is indeed the case, the values (Table 9.17) [59JA1935; 75JCS(P2)277] predicting the positional order 2 > benzene ≥ 4 > 3; CNDO/2 calculations also produce the same positional order (benzene excluded) (71TL387). Because of the polarizability of pyridine N-oxide, localization energies predict that, in reactions of high demand for resonance stabilization of the transition states, both the 2- and 4- positions should be activated.

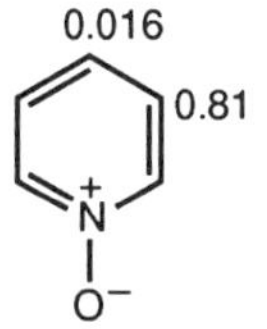

(9.122)

E. HYDROGEN-BONDED PYRIDINE N-OXIDES

Hydrogen bonding substantially reduces the reactivity of pyridine N-oxide so that σ^+ values determined for the 2- and 4-positions from solvol-

TABLE 9.17

π DENSITIES AND LOCALIZATION ENERGIES FOR ELECTROPHILIC SUBSTITUTION OF PYRIDINE N-OXIDE[a]

	π Density	$L_r^+/-\beta$
2-Position	1.011	2.34
3-Position	0.987	2.58
4-Position	0.999	2.42
Benzene	1.000	2.54

[a] [59JA1935; 75JCS(P2)277].

ysis of 1-aryl-1-methylethyl chlorides (73JOC2657) are 0.68 and 0.45, respectively. These indicate substantial deactivation. Comparison of the reactivity of the 4-position in this reaction with that in pyrolysis of the 1-arylethyl acetates indicates that hydrogen bonding reduces the former reactivity by 0.44 σ units. This is not unreasonable since, for example, in trifluoroacetic acid the reactivity of the para-position of anisole is reduced by 0.2 σ units as a result of hydrogen bonding (73MI1).

Because hydrogen bonding here is at oxygen rather than nitrogen, the reactivities of the 2- and 4-positions should be more adversely affected than that of the 3-position. This may partially account for the value of σ_m^+, determined from hydrogen exchange of hydroxy compounds, being 0.8 which is the same as in the gas phase. However, this value was determined from comparison of much more reactive compounds, which has probably attenuated it somewhat (see also Section 3), and indeed good evidence supports this view [68JCS(B)866].

F. Protonated Pyridine N-Oxides

The value of σ_m^+ for N^+OH has been determined as above by comparison of the reactivities of substituted pyridines and benzenes, giving 2.1 (phenoxides) and 2.3 (phenols) [68JCS(P2)866] and these are similar to values of σ_m (2.25, 2.3) [64AHC(3)209; 72JCS(P2)671]. Localization-energy calculations [59JA1935] gave values of L_r^+ as -2.70β (2), -2.58β (3), and -2.67β (4), which predict that protonation should lower the reactivity only of the 2- and 4-positions (cf. Table 9.17). However it appears from comparison of the reactivity of the 3-position that this is also substantially lowered.

G. Pyrylium and Thiopyrylium Ions

The values of σ_m^+, again determined from phenols and phenoxides, are 3.0 to >3.7 (O^+) and 3.2 to >3.7 (S^+) [68JCS(B)866]. Positively charged O^+ and S^+ are by far the most deactivating substituents known.

H. Methyl-Substituted 2-Pyridones

π-Density calculations for a variety of mono- and dimethyl-substituted 2-pyridones indicate the positional reactivity order 3 > 5 > 4 > 6, regardless of the location of the substituents (68CCC394).

I. COMPARISON OF STANDARD DATA FOR NITRATION AND HYDROGEN EXCHANGE

The large amount of rate data for hydrogen exchange and nitration of substituted pyridines under the defined standard conditions has been compared [78JCS(P2)613]. No simple correlation is found (Fig. 9.3) (78JCS(P2)613). Polarizability, differential hydrogen bonding and solvation effects can, as described in the foregoing discussion, affect different reactions in considerably different ways. It follows that there is no *unique* set of reactivity parameters applicable to all reactions, and the concept of an absolute "reactivity scale" toward electrophilic substitution is unattainable. The reactivity produced by a substituent depends on the electrophile involved, and upon the conditions.

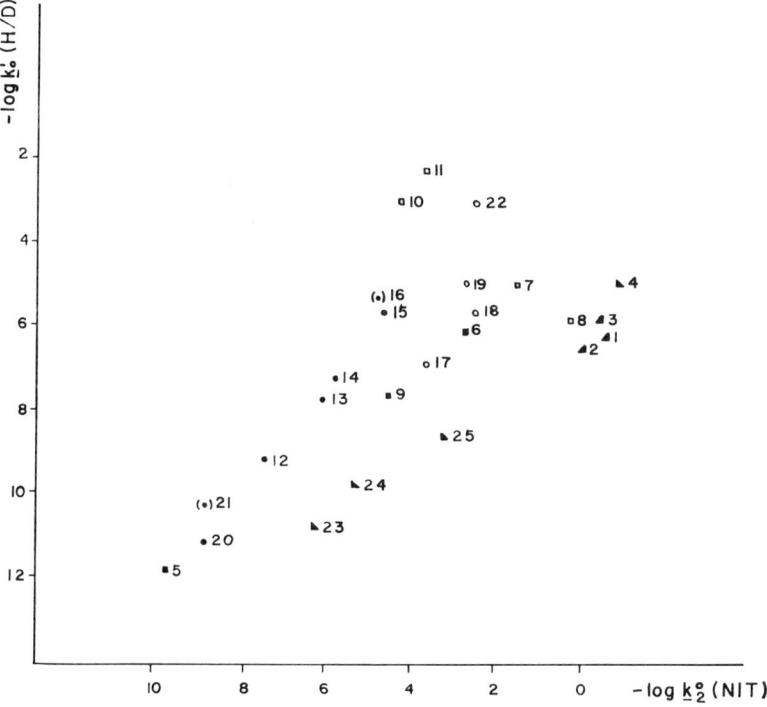

FIG. 9.3. Plot of rate data for hydrogen exchange versus rate data for nitration of substituted pyridines.

CHAPTER 10

Six-Membered Rings: Electrophilic Substitution in the Azines

BY M. ROSS GRIMMETT*

1. Reactivity of the Monocyclic Azines

This chapter is concerned with the electrophilic substitution of pyridazine, pyrimidine, pyrazine, and their derivatives. Aspects of this topic have been reviewed previously [72AHC(14)99; 74AHC(16)1] and the general chemistry of the monocyclic azines has been surveyed in *Comprehensive Organic Chemistry*, Vol. 4 (1979) and *Comprehensive Heterocyclic Chemistry*, Vol. 3 (1984).

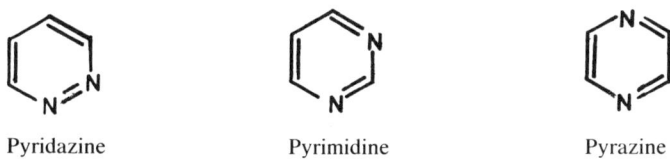

Pyridazine Pyrimidine Pyrazine

The fact that none of these reports has emphasized the physical aspects of electrophilic substitution in the series reflects the paucity of quantitative studies, and the low reactivity of these compounds in the presence of electrophiles. Few kinetic studies have been reported and the regiochemical effects of substituents have seldom been quoted in quantitative form. The present chapter brings together those quantitative results that are available, and collates data on substituent effects. One worthwhile field of study would appear to be the application to the azines of Taylor's method involving thermolysis of esters [75JCS(P2)277, 75JCS(P2)1783].

The resistance of pyridine toward electrophilic substitution is well known, and this reluctance to react is enhanced in the protonated species. Incorporation of extra annular nitrogen atoms further decreases this reactivity with electrophiles; each aza substituent has an electron-withdrawing effect similar to a nitro group. One estimate of the σ_m^+ value for an aza substituent ($=$N—) is 0.65, and 1.95 for the protonated species ($=$N$^+$H—) [68JCS(B)1484] (see also Chapter 9, Sections 9.A–9.C). Val-

*Department of Chemistry, University of Otago, Dunedin, New Zealand

ues have been calculated for σ. For example, Perrin (65JCS5590) determined the Hammett substituent constants [—N=, σ_o (0.56), σ_m (0.60), σ_p (0.83); —N$^+$H=, σ_o (3.21), σ_m (2.18), σ_p (2.42); =N$^+$(O)$^-$, σ_m (1.48), σ_p (1.35)] by comparison of the dissociation constants of pyridinecarboxylic acids and benzoic acids. Although there are variations in the values of the constants [e.g., 0.73 for σ_m (=N—)] [66JCS(B)937, 66QR75; 68JCS(B)864], they provide a useful measure of the comparative effects of the annular nitrogens. Additivity of the values allows one to calculate the influence of two or more such substituents on the ring. Thus, using Perrin's figures, the total σ constant for the =N—N$^+$H= group of pyridazinium at C-5 is computed to be 3.01, a combination of σ_m for =N$^+$H— (2.18) and σ_p for =N— (0.83).

Exocyclic substituents will also modify reactivity through their steric and electronic effects. Substitution will occur readily only in those azines that possess one or more activating groups, and the ease of attack will decrease as the number of annular nitrogens increases (although π-density calculations suggest that there is little difference in π deficiency between the 2-positions of pyrazine and pyridine) [72AHC(14)99]. As a general rule, those azines with adjacent nitrogens are less prone to electrophilic attack at nitrogen than when the aza functions are remote from each other, while azine N-oxides seem to be only marginally more reactive toward electrophilic substitution than the nonoxygenated analogues.

2. Acid-Catalyzed Hydrogen Exchange

In view of the known resistance of pyridine and its N-oxide and methiodide to acid-catalyzed tritiation and deuteration (64CPB1384; 67MI1), any analogous exchange in the azines can only be slow. Acid-catalyzed hydrogen exchange has been studied by Katritzky and co-workers as a prospective means of comparing the reactivity of a wide range of heteroaromatics in electrophilic substitutions [73JCS(P2)1065, 73JCS(P2)1077; 74JCS(P2)1294] (See Chapter 2). All hydrogen exchange data for such compounds can be summarized by the equation

$$\log f = \rho \sum \sigma^+ - \rho' \sum \pi\sigma^+ + 1.6$$

where f is the partial rate factor defined under standard conditions (H_0 = 0; T = 100°C) [74JCS(P2)1294]. The $\sum \sigma^+$ term sums the substituent constants and $\sum \pi\sigma^+$ allows for interactions between substituents. This empirical equation, with values of ρ and ρ' of 6.3 and 2.1, respectively, gives the best fit for the points and has considerable predictive power, though its theoretical basis is doubtful. Unfortunately, complicated extrapolations are frequently needed to bring temperature and acidity to the standard conditions. Standardized rate coefficients calculated for acid-catalyzed hydrogen exchange under these conditions obey the Hammett equation with a ρ value of -7.5 [73JCS(P2)1077].

A. PYRIDAZINES

An attempt to study the kinetics of acid-catalyzed deuterium exchange for a number of substituted pyridazines was partially frustrated by the instability of some of the azines in concentrated acid. Although 3,6-dihydroxypyridazine and its 4-chloro and 4-methyl analogues exchanged to the extent of about 50%, the 3,6-dimethoxy, 3-amino-6-methoxy, 3-amino-6-chloro, and 4-methoxy compounds were either hydrolyzed or decomposed by the acid. Only 4-aminopyridazine and pyridazin-4(1H)-ones gave satisfactory kinetic results [68JCS(B)873]. It had been reported earlier that acid-catalyzed exchange would not occur with pyridazin-3(2H)-one or its 1-oxide even at 100°C in 98% sulfuric acid (67CPB1411).

Katritzky and Pojarlieff [68JCS(B)873] demonstrated that 4-aminopyridazine exchanges as the cation at C-5 (at 186°C; $H_0 = 0.8$–3.3), the exchange competing with hydrolysis to pyridazin-4(1H)-one. In both of the pyridazinones exchange occurs at C-5 in the free base, while at lower acidity all positions can exchange. At pD 2.1, H-3 of pyridazin-4(1H)-one exchanges most rapidly ($k = 3.91 \times 10^{-5}$ sec^{-1}), and H-6 ($k = 1.47 \times 10^{-5}$ sec^{-1}) exchanges slightly faster than H-5 ($k = 9.98 \times 10^{-6}$ sec^{-1}). The surprising occurrence of acid-catalyzed exchange at C-5 in pyridazin-3(2H)-one may be due to reaction via a covalently hydrated species [68JCS(B)873], as with pyrimidin-2(1H)-one *(vide infra)*.

2-H-Cyclopenta[d]pyridazine undergoes exchange at the 5- and 7-positions (69JA924). Protonation at these sites rather than at nitrogen is aided by the benzenoid nature of the transition state (Scheme 10.1) (cf. indolizine, Chapter 8, Section 2.B).

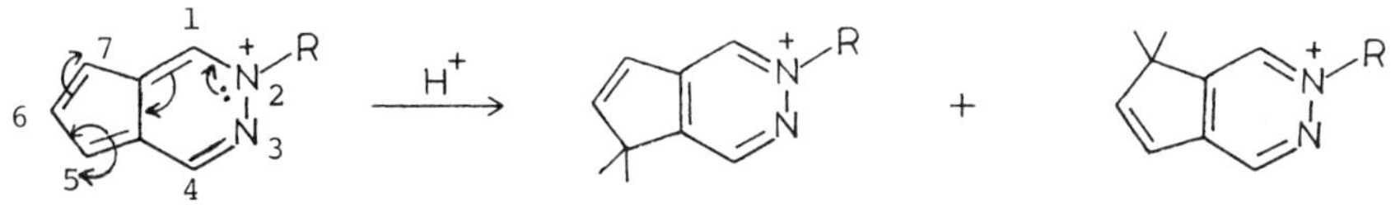

SCHEME 10.1

B. Pyrimidines

The free base forms of 2-amino- and 6-amino-2,4-dimethylpyrimidines can undergo hydrogen–deuterium exchange at the 5-position in dilute acid (pD > 3.7). The latter compound can also exchange as the conjugate acid, but both are hydrolyzed in concentrated acid. Activating substituents are necessary for reaction at any measurable rate. Second-order rate coefficients determined at 107°C were $1.43 \times 10^{-2} \, M^{-1} \, \text{sec}^{-1}$ for 2-aminopyrimidine, $9.78 \, M^{-1} \, \text{sec}^{-1}$ for 6-amino-2,4-dimethylpyrimidine, and $1.65 \times 10^{-4} \, M^{-1} \, \text{sec}^{-1}$ for 6-methyl-2,4-diaminopyrimidine [68JCS(B)1484]. However, in the case of pyrimidin-2(1H)-one, the rate profile suggested reaction of the free base in the pD range 0.27–5.00, although the value of the second-order rate coefficient was 10^4 times higher than that for 2-pyridone [67JCS(B)1226]. This implies that the two reactions go by different mechanisms. An indication that the addition of water is involved in the exchange of pyrimidin-2(1H)-one came from the relatively steep slope of the rate profile in the H_0 region. Thus it was inferred that the covalent hydrate **10.1** (Scheme 10.2), formed in small equilibrium quantity in a fast reversible reaction, was undergoing slow deuterium exchange [68JCS(B)1484]. Support for this mechanism came from a study of the 1,2-dihydro-1,3-dimethyl-2-oxopyrimidinium ion (**10.2**), in which the methyl groups were expected to retard the rate of formation of any cova-

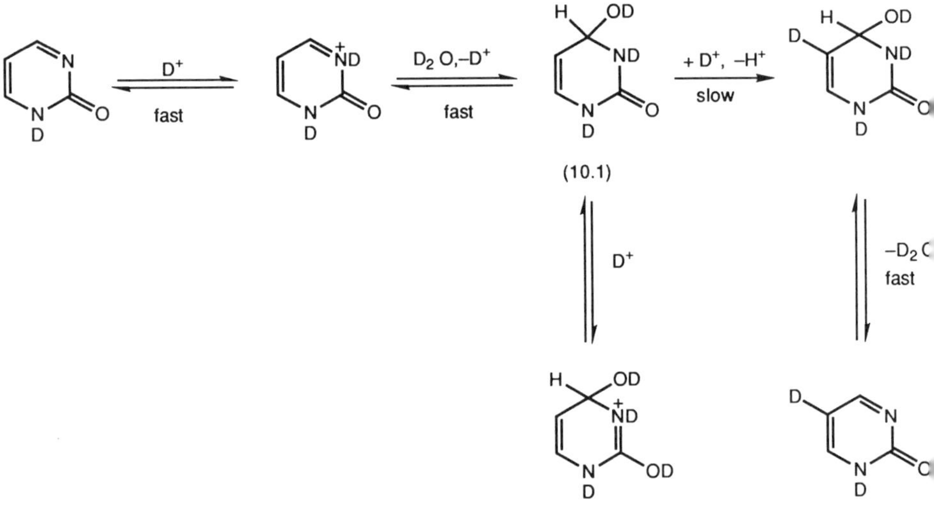

SCHEME 10.2

lent hydrate. The bimolecular rate coefficient (80 M^{-1} sec^{-1}) for the pseudo-base of **10.2** is somewhat greater than that (50 M^{-1} sec^{-1}) for **10.1**. Additional methyl groups, as in 1-methyl- and 4,6-dimethylpyrimidin-2(1H)-one, have two partially cancelling effects in that they decrease the amount of covalent hydrate in the equilibrium mixture, but at the same time they increase its reactivity.

Exchange at C-2 in 1-methylpyrimidin-4(1H)-one is believed to involve ylide formation at that position (71T953).

Although attempts to exchange deuterium for ring hydrogens in the 1,4-diethylpyrazinium dication led only to decomposition, exchange does occur in the corresponding pyrimidinium species (**10.3**), for which the half-lives of exchange at C-2 and C-5 at 78.2°C in CF_3CO_2D were 12 and 29 hr, respectively. Neither of the equivalent H-4 and H-6 atoms of **10.3**, nor any of the ring hydrogens of 1-ethylpyrimidinium tetrafluoroborate, was subject to measurable exchange under similar conditions. These observations prompted Curphey (65JA2063) to conclude that two distinct mechanisms must be operative, an addition–elimination process for C-5 deuteration, and exchange via an ylide at C-2.

(10.2)   (10.3)   (10.4)

With more than one electron-releasing substituent present, electrophilic subtitution becomes much easier. Thus, uracils are deuteriated at C-5 when boiled with 6 M DCl (70JHC903), and 1,3,6-trimethyluracil readily undergoes acid-catalyzed H–D exchange at C-5 in D_2O. Although this process may involve direct electrophilic substitution, covalent hydration across the 5,6-bond followed by elimination cannot be ruled out (66CJC335). Indeed, uracil itself, existing largely as structure **10.4** in aqueous media [66JCS(B)565], is subject to slow deuterium exchange when dissolved in D_2O–DMSO; the rate is increased by acid or base (66JOC175). In D_2SO_4, exchange in both uracil and 1,3-dimethyluracil varied little with concentration of the catalyzing acid, showing reactions to occur on the free bases. Log exchange rate under standard conditions were determined as -1.23 and -0.37, respectively [78JCS(P2)613].

Tritium is lost from the 5-position of tritiated pyrimidine nucleosides. For example, at the 5-position of uridine, tritium has a half-life of 12 hr in 0.5 M H_2SO_4 at 100 °C, compared to 2.4 hr for uracil (64MI1; 67B3576).

Initial attack in all these species is considered to be at C-6 by a nucleophile such as water, (i.e., covalent hydration in implicated).

In pyrimidine *N*-oxides, deuterium exchange is facilitated more than in the nonoxygenated analogues, with the effective rate constant for exchange at C-2 being an order of magnitude higher in the oxide (83CHE1003). Exchange in 5-hydroxypyrimidine *N*-oxide (**10.5**, R = H) and the 4,6-dimethyl derivatives (**10.5**, R = Me) has been examined at pD 0.5–10.0 at 160°C. The anionic and neutral forms are the most reactive and the reaction shows a conversion from the normal S_E2 mechanism at high acidity to the base-catalyzed S_E1 mechanism (see below) at pD > 6. The greater electron supply in the dimethyl compound permitted exchange to occur partially on the protonated species (84BAU2469).

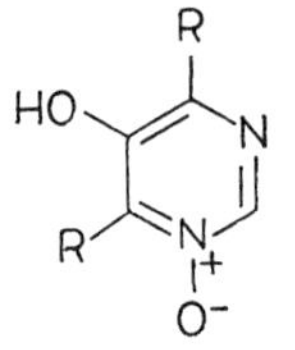

R = H, Me

(10.5)

C. Pyrazines

Pyrazine is partially deuteriated in D_2SO_4 at 230°C (63SA1473).

D. Triazines

Acid-catalyzed deuterium exchange at C-3 in 1,2,4-triazine occurs at 100°C in neutral D_2O and in 0.02 *M* DCl with the same rate coefficient ($k_{obs.}$ = 14 × 10^{-5} sec^{-1}). At higher acidity (0.2 *M* DCl) the exchange rate rapidly falls to zero (72JHC995; 73T2495). Clearly, substituents at C-5 are involved in the reactions since a 5-methyl group decreases the rate 20-fold ($k_{obs.}$ = 6.9 × 10^{-6} sec^{-1}); a 5-phenyl group prevents reaction, probably for steric reasons, although the insolubility of the phenyltriazine in D_2O must have been a factor. That a 6-methyl group has little effect can be deduced from the rates for 6-methyl-1,2,4-triazine (13 × 10^{-5} sec^{-1}) and the 5,6-dimethyl derivative (9.0 × 10^{-6} sec^{-1}). Thus, a covalently hydrated species [not protonated as in the pyrimidin-4(1*H*)-ones (71T953)] must be involved since exchange will not take place under more acidic conditions. Only those 1,2,4-triazines that can be covalently hydrated are able to exhibit H-3–D-3 exchange in neutral D_2O (73T2495).

While it is clear that 1,3,5-triazine can be covalently hydrated, no evidence of deuterium exchange has been presented (75JOU2691), and this follows because each position is ortho and para to nitrogen, which produces strong $(-M, -I)$ electron withdrawal at these sites.

3. Base-Catalyzed Hydrogen Exchange

The base-catalyzed exchange mechanism involves proton abstraction by base to give a carbanion (hydrogen or deuterium bonded to the solvent), which removes deuterium from the solvent (Chapter 2, Section 2.A). Positions adjacent to the annular nitrogen atoms are abnormally unreactive, and this is believed to be due either to decreased s character of the CH bond next to the sp^2-nitrogen atom, or to electrostatic repulsive interaction between the electron pairs on nitrogen and the developing carbanion causing reduced acidity of the C-3—H or C-6—H bonds (69JA5501) (see also Chapter 9, Section 3). Results to extended Hückel molecular orbital calculations suggest that the second factor may be more important (69JA2590).

All of the diazines are more reactive than pyridine (Table 10.1; cf. Tables 9.2 and 9.3), with the annular nitrogens activating the ring in the order para $\simeq$ meta >> ortho. Log partial rate factors are ortho, 1.31; meta, 2.43; and para, 2.46. These were calculated assuming additive effects of nitrogen on exchange rates. Zoltewicz (69JA5501) found no correlation between the rates of H–D exchange for all positions in the azines and ^{13}C- or ^{1}H-NMR chemical shifts, but Adam et al. claimed good correlations of total $(\sigma + \pi)$-electron densities with these measurements for a variety of heterocycles, including the azines (67T2513).

A. Pyridazines

Base-catalyzed exchange in pyridazine, reacting in 1% $NaOD–D_2O$ (64CPB1384) or in NaOMe–MeOD (68MI1), occurs more readily at C-4 and C-5 than at C-3 and C-6. This follows from a combination of acceleration of the reaction by electron withdrawal, and unfavorable juxtaposition of lone pairs for 3- and 6-substitution. In the corresponding N-oxide the order is 6 >> 5 > 3 $\simeq$ 4, with exchange at C-3 being slightly faster than at C-4 (64CPB1384). The order arises from the $-I$ effect of the positive pole, base-catalyzed exchange being particularly susceptible to inductive effects. These orders parallel the reactivities of methyl hydrogen atoms in methylpyridazines and their N-oxides (67CPB2000). Attack at the acti-

TABLE 10.1
RATE DATA FOR BASE-CATALYZED HYDROGEN–DEUTERIUM EXCHANGE

Compound	Positions	MeOD–MeONa at 164.6°C[a] k_2 ($M^{-1}sec^{-1}$)	$k_{rel.}$
Pyridazine	3,6	1.4×10^{-3}	438
	4,5	2.1×10^{-3}	656
Pyrimidine	2	1.2×10^{-4}	38
	4,6	3.9×10^{-4}	122
	5	5.6×10^{-3}	1750
Pyrazine	2,3,5,6	3.1×10^{-4}	97
Pyridine	2,6	3.2×10^{-6}	1
	3,5	3.0×10^{-5}	9.4
	4	3.8×10^{-5}	11.9

Compound	Positions	MeOD–MeOK (0.6 M)[b] $-\log k$ (140°C)	E (kcal mol^{-1})	log A (sec
Pyridazine	3,6	4.1		
	4,5	2.9		
Pyrimidine	2	5.1		
	4,6	4.5		
	5	3.5–4.0		
Pyrazine	2,3,5,6	4.7–5.0	21.9	6.6
3,6-(MeO)$_2$-pyridazine	4,5	2.6	14.8	4.5
2,4-(MeO)$_2$-pyrimidine	5	3.9	24.0	8.8
2,4,6-(MeO)$_3$-pyrimidine	5	3.1	30.2	12.9

		MeOD–MeOK (0.1 M)[b]		
Pyrimidine N-oxide	2	2.4	20.5	11.7
	4	7.4		
	6	3.8	21.6	10.7
Pyrazine N-oxide	2,6	3.2		
	3,5	7.4	23.0	9.5

[a] Data from (69JA5501).
[b] Data from (74CHE1387).

vated C-6 of pyridazine N-oxide is 100 times faster than in pyridine N-oxide [in which the sequence of base-catalyzed hydrogen exchange is 2 >> 3 > 4 (69JA5501)], and occurs even at room temperature. If the positively charged N-oxide nitrogen and the other ring nitrogen act independently, then one would expect exchange to occur most easily at C-6. That H-5 exchanges much faster than H-3 even though C-3 may be more electron deficient is probably again a function of the proximity of the latter position to the pyridine-type nitrogen. The exchange results for the N-oxide contrast with its NMR spectrum, which shows that H-4 is more shielded than H-5 in neutral organic solvents (63YZ523), and also with the order of nucleophilic reactivity (5 > 3 > 6 > 4) (67CPB2000), though here conjugative effects are more important.

When Kawazoe et al. (67CPB2000) studied the base-catalyzed H–D exchange at the exocyclic methyls of methylpyridazines and their N-oxides, they also found (1) some ring deuteriation occurring (e.g., some exchange at C-5) in 3-methylpyridazine; (2) in 3- and 4-methylpyridazine N-oxides exchange occurs at C-6, and to a lesser extent at C-5; (3) some exchange at C-6 in 5-methylpyridazine N-oxide; and (4) partial deuteration at C-5 in 6-methylpyridazine N-oxide. Thus, the methyl groups have little effect on the previously observed regiochemistry of substitution.

Exchange in the more activated 3-hydroxypyridazine [which exists mainly as the 3(2H)-one] occurs initially at C-3, and then to some extent at C-5. In the corresponding N-oxide the order is C-6 > C-4. These last results suggest some activation of the 4- and 6-positions by a "phenolic" hydroxyl group (67CPB1411).

B. Pyrimidines

Comparative results for some of the diazines discussed (*vide supra*) have also been obtained by workers in the Soviet Union (67MI2; 68MI1), who have reported results relating to pyridine, quinoline, pyrimidine, pyrazine, and some N-oxides. The relative rates are comparable, although the rate coefficients they have obtained are smaller than those of Zoltewicz by factors of up to 5. Thus, for pyrimidine, relative rates have been quoted as $k_2 : k_4 : k_5 = 1.0 : 3.2 : 48.0$ (69JA5501) and $k_{2,4,6} : k_5 = 1 : 15$ (67MI2). The relatively high reactivity of the 5-position presumably arises from unfavorable juxtaposition of lone pairs for 2-, 4-, and 6-exchange. Carbon-5 is the only position not subject to the adjacent lone-pair effect. In pyrimidine N-oxide [and pyrazine N-oxide (70JOC3467)] the position α to the oxide function exchanges most readily, and at least 100 times as rapidly as in pyridine N-oxide (72JOC4188). Results for some 5-substituted pyrimidine oxides are listed in Table 10.2. In pyrimidine N-oxide C-6 is 26 times as reactive as C-2 (adjacent lone-pair effect), and both positions are less reactive than C-2 of pyrazine N-oxide (*vide infra*), confirming that the nonoxygenated nitrogen atom is much more effective in facilitating hydrogen exchange when it is meta to the exchanging position than when it is para, due to the adjacent lone pair effect. A linear relationship exists between log k_{H-6} and σ_I for the 5-substituted pyrimidine N-oxides, implying that the inductive effect of a substituent ortho to the exchanging proton is the controlling factor in the process (see Chapter 2, Section 2.A). No such correlation holds for H-2 exchange rates. In

TABLE 10.2
SECOND-ORDER RATE COEFFICIENTS FOR BASE-CATALYZED DEUTERIUM–HYDROGEN EXCHANGE IN 5-SUBSTITUTED PYRIMIDINE N-OXIDES[a]

5-Substituent	k_{H-6} (M^{-1} min^{-1})	k_{H-2} (M^{-1} min^{-1})
H	0.047	0.0018
Me	0.098	0.0017
Br	540	[b]
MeO	4.7	0.0031
Me$_2$N	0.17	[b]

[a] (72JOC4188).
[b] No exchange was observed within 14 days in 0.1 M NaOD.

strongly basic media (0.3 M NaOD), covalent hydration competes with the exchange process (72JOC4188). Other comparative results quoted for pyrimidine N-oxide exchange are 2 > 6 > 4 > 5, and appear to be somewhat contradictory in that C-2 is here more reactive than C-6 (68MI1).

Exchange is more facile in some pyrimidine derivatives. In 1-methylpyrimidin-4(1H)-one (**10.6**), exchange at C-2 proceeded at a rate independent of the concentration of base (66JHC440; 69JOC589). A mechanism with this characteristic would be rate-limiting attack by OD$^-$ on a cationic species (**10.7**) to form an ylide intermediate (Scheme 10.3). The rate equation, obtained by applying the steady-state approximation to the concentration of (**10.7**), is

$$dP/dT = k_1 k_2 [D_2O][\mathbf{10.6}]/(k_{-1} + k_2)$$

and if $k_{-1} \gg k_2$,

$$dP/dT = k_2[D_2O][\mathbf{10.6}]/k_{-1} = k_{D_2O} k_2[\mathbf{10.6}]/K_a$$

An approximate value for k_2 [$(9 \pm 4) \times 10^6$ M^{-1} sec^{-1} at 70°C] was obtained by means of the model compound **10.8**. Using this rate coeffi-

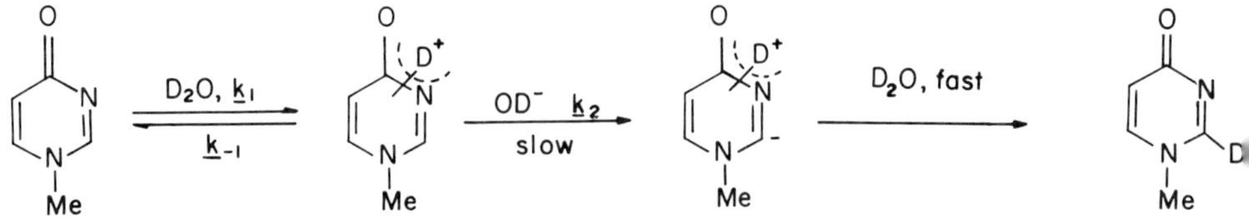

SCHEME 10.3

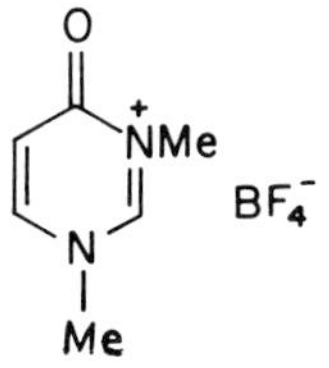

(10.8)

cient in the above equation gave a value of $(4.6 \pm 0.2) \times 10^{-5}$ sec^{-1}, which is satisfactorily close to the observed first-order rate coefficient $[(9.0 \pm 0.5) \times 10^{-5}$ sec$^{-1}]$ for the reaction of **10.6**. In contrast to the quaternary salt (**10.8**), 4-methoxy-1-methylpyrimidinium would not react under the same conditions [66JHC440; 71T953; 74AHC(16)1].

The change to this mechanism for exchange at C-2 in **10.6** from the direct proton-removal mechanism which takes place in 1-methyl-4-pyridone can be accounted for in terms of the relative acidity of the free base (e.g., **10.6**) and the corresponding onium salt (e.g., **10.7**) in each case. In a series of 1-alkyl-4-pyrimidinones the exchange rates for different alkyl substituents decreased in the order 1-methyl > 1-benzyl > 4-pyrimidinone (66JHC440). Ylide intermediates may also be involved in the biosynthesis of uridylic acid (71T953).

In uridine and cytidine (Scheme 10.4), deuterium exchange occurs at C-5 after initial attack at C-6 by a nucleophile (D_2O or OD^-). When that nucleophile is able to abstract the C-6 proton to form a delocalized anion,

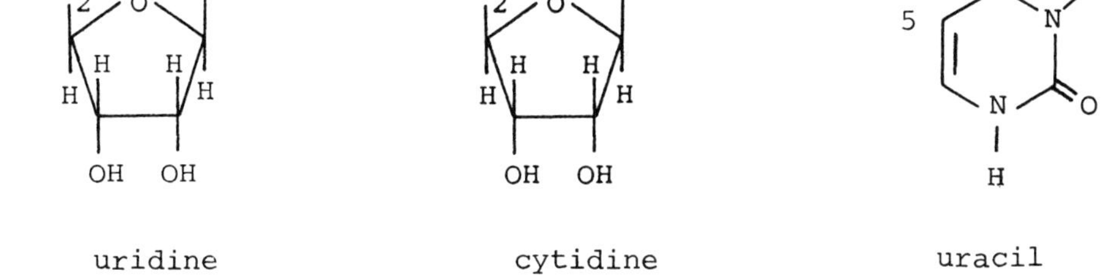

uridine   cytidine   uracil

SCHEME 10.4

some exchange at that position can also be observed (70CCC1991, 70CCC2003; 73JA1628). Thus, treatment of uridine with a threefold excess of NaOD in D_2O at 95°C for 2.5 hr resulted in 4% C-6 deuteriation, 24% C-5 deuteriation, and 20% decomposition (70CCC2003). The use of deuteriated base in DMSO-d_6 increased the extent of exchange at C-6 and decreased the proportion of hydrolysis (73JA1628).

Base-catalyzed protiodedeuteriation of 5-deuterio-1,3-dimethyluracil (Scheme 10.4) probably involves the addition of water across the 5,6-bond, initiated by hydroxyl attack at C-6. While a similar mechanism may apply with 5-deuterio-1-methyluracil, direct exchange with the anion is also a possibility; the anion is almost certainly implicated in the case of the 5-deuterio-3-methyl isomer (70JHC903).

C. Pyrazines

Pyrazine exchanges more readily than pyridine (Table 10.1) (68MI1; 69JA5501) with the exchange in 0.6 M MeOK–MeOD being first order in methoxide (68MI1); the exchange of an α-hydrogen is about 100 times faster in pyrazine than in pyridine.

A similar difference in reactivity is evident in the N-oxides: pyrazine N-oxide is even more reactive than pyrimidine or pyridine N-oxide. Substituents in the 3-position appear to act mainly through their inductive effects (Table 10.3). Log k_2 varies linearly with pK_a, and the carbanion mechanism probably operates (there is, however, doubt about the mode of substitution at C-6 in pyrazine N-oxides) (70JOC3467). All four hydrogens of pyrazine N,N'-dioxide are exchanged in MeOD at 65°C (68MI2).

TABLE 10.3

SECOND-ORDER RATE COEFFICIENTS FOR BASE-CATALYZED DEUTERIUM–HYDROGEN EXCHANGE IN 3-SUBSTITUTED PYRAZINE N-OXIDES[a]

3-Substituent	k_{H-6} (M^{-1} min^{-1})	k_{H-2} (M^{-1} min^{-1})
H	0.16	0.16
CN	–[b]	280
Cl	0.99	75
OMe	0.021	4.3
NMe$_2$	4.5 × 10^{-4}	0.033

[a](70JOC3467).
[b]Hydrolysis occurs before H-6 exchange.

TABLE 10.4
SECOND-ORDER RATE COEFFICIENTS FOR BASE-CATALYZED
DEUTERIUM–HYDROGEN EXCHANGE AT C-3 IN 5-SUBSTITUTED
1,2,4-TRIAZINES[a]

5-Substituent	$10^3 k_2$	Reagents
H	1.00	NaOD–D_2O (0.016–0.157 M)
Me	0.20	NaOD–D_2O (0.016–0.157 M)
H	0.02	NaOD–D_2O–DMSO (0.16–0.67 M)
Ph	0.30	NaOD–D_2O–DMSO (0.16–0.67 M)

[a] (73T2495).

D. 1,2,4-TRIAZINES

In NaOD–D_2O (or with added DMSO) 1,2,4-triazine and its 5- or 6-methyl and 5-phenyl derivatives exchange H-3 for deuterium with an opposite reactivity order to that observed under acid catalysis (*vide supra*) (73T2495). The observation that the 5-phenyl derivative reacts 15 times faster than the unsubstituted parent (Table 10.4; comparison was made in D_2O–DMSO because of solubility problems) clearly indicates that the carbanion mechanism operates in basic media rather than a process involving covalent hydration.

4. Nitration

Attempts to nitrate the parent diazines and triazines have been uniformly fruitless, since not only are the bases inactive, but their cations are even less reactive (50JCS3236; 51JCS2323; 57CB1837). Aryl-substituted azines nitrate preferentially in the aryl ring, with the protonated azine acting mainly as a meta director [47CRV279; 51JCS2323; 65BCJ777; 71JCS(C)1945; 73MI2; 75MI1; 78JCR(S)133; 79H745; 81JCR(S)104; 82CJC2668].

The nature of the nitrating agent can affect overall product distribution. With mixed acids 4-phenylpyrimidine gives a 2 : 3 mixture of *o*- and *m*-nitrophenyl products; nitric acid in trifluoroacetic anhydride gives an ortho : meta : para-product ratio of 45 : 29 : 26 ; nitronium acetate specifically attacks the pyrimidine ring giving 2,4-diacetoxy-1,3,5-trinitro-6-phenyl-1,2,3,4-tetrahydropyrimidine (67CJC1431). Although there is no convincing explanation for these results, addition-elimination may be involved with trifluoroacetic anhydride, with N-nitration and perhaps addition to the free base implicated in acetic anhydride. When the aryl substituent attached to the azine is thienyl or furyl, nitration can occur in both α- and β-positions of the five-membered ring, with β-attack predominating [78JCR(S)133; 81JCR(S)104; 82CJC2668].

As noted in Chapter 9 attempts to apply the general equation devised to relate partial rate factors in hydrogen-exchange reactions to correlations of nitration rates in five- and six-membered heterocycles (including azines) have proved unsuccessful [74JCS(P2)1294; 75JCS(P2)1600, 75JCS(P2)1624, 75TL1395].

A. Pyridazines

When electron-releasing substituents are present the chances of annular nitration are greatly improved, although apparent amino group activation can be a result of nitramine rearrangement. There are a number of instances in which such nitramine intermediates have either been identified or suspected (50JCS3236; 60CPB999, 60JGU1531; 65JHC67). The formation of a mixture of 6-nitro, 4-nitro, and 4,6-dinitro products on nitration of 3-methoxy-5-methylpyridazine has been reported. The second nitro group is probably introduced by reaction of the mononitro free base. Even the presence of two chloro groups in 4,5-dichloro-2-methylpyridazin-3(2H)-one does not prevent nitration at C-6 (84MI1).

Rather more information is available for nitration of pyridazine N-oxides. As in other series the product orientation depends on reagent. Thus, while mixed acids convert pyridazine N-oxide (62CPB643), its 3-methyl (63YZ934) and 3,6-dimethyl (61CPB149; 62YZ253) derivatives into 4-nitro products (22, 27, and 54% yields, respectively), acetyl nitrate transforms the same compounds, respectively, into a mixture of 3- (33%) and 5- (<1%) nitropyridazine N-oxides (63CPB342), 3-methyl-5-nitropyridazine N-oxide (12%) (64CPB228), and 3,6-dimethyl-5-nitropyridazine N-oxide (25%) (64CPB228; 75CPB923). With methoxypyridazine N-oxides, too, there are marked differences. In nitric–sulfuric acid mixtures, the 3-methoxy compound reacted at C-4 and C-6 (60CPB550, 60YZ712), whereas 3,6-dimethoxypyridazine N-oxide gave the 4-nitro product (84%) (55YZ966). In contrast, acyl nitrates induced 5-nitration (11%) in the 3-methoxy oxide; the dimethoxy oxide resisted nitration (64CPB228). The different products may be a consequence of protonated and free base forms of the pyridazines reacting in the different media (*vide infra*), but initial N-nitration cannot be ruled out with acyl nitrates. Nitration in the 5-position can also occur with 4-hydroxypyridazine N-oxides on treatment with mixed acids (75CPB1879).

Most of the comparative work relates to the use of mixed nitric–sulfuric acids. In this medium, nitration takes place mainly at the 4- and 6-positions. Whereas 3-, 5-, and 6-methyl groups attack at C-4 [yields of 27% (63CPB726, 63YZ934), 18% (63CPB35), and 87% (63CPB29), respectively], a 4-methyl group inhibits the overall rate of nitration by blocking the most reactive 4-position (63CPB35). The added electron release provided in 3,4-dimethylpyridazine N-oxide allows minor (9%) 6-nitration (63CPB726, 63YZ934). If pyridazine N-oxides follow the nitration behavior of pyridine N-oxides (which operate through the free base), the preference for 4-nitration might not be unexpected. Nitration at C-6 only seems to occur to any extent when the 4-position is blocked and when added activation is provided by a methoxy substituent. Thus, 3-methoxypyridazine N-oxide heated to 50°C with mixed acids for 1 hr gave a mixture of 4-nitro (34%) and 4,6-dinitro (6%) products. Another worker (60YZ712) reported isolation of a small amount (5%) of the 6-mononitro derivative under similar conditions. The dinitro compound was also formed by nitration of 3-methoxy-4-nitropyridazine N-oxide (60CPB550; 62YZ1005).

Alkoxy and hydroxy groups at C-3 and C-6 usually promote 4-nitration under mild conditions (63CPB29, 63CPB726) even in the presence of chloro groups (62CPB934, 62YZ253; 63CPB29; 73JHC551). Nitration in the 5-position occurs with 4-hydroxypyridazine N-oxide (75CPB1879). It is difficult to arrive at any conclusive assessment of the effectiveness of substituents from the qualitative results available. Indeed, the variable yields reported in the literature may be a consequence of the use of severe conditions.

B. PYRIMIDINES

As with the 1,2-diazines, electron-releasing groups facilitate nitration. Two such activating groups at C-2 and at C-4 or C-6 usually guarantee successful nitration at C-5, and there are examples where only one such substituent is necessary [e.g., pyrimidin-2(1H)-one] [63M11; 69JHC593; 71JCS(B)1], but not the corresponding 4-isomer [71JCS(B)1]. Under conditions of both high and low acidity, pyrimidin-2(1H)-one and uracils nitrate as the free base species; the cations are quite unreactive.

Two activating substituents are even better at promoting 5-nitration as in the uracils [52MI3; 55JCS211; 58CPB482; 60LA(633)158; 64JCS4769; 65JMC187; 71JCS(B)1, 71JCS(C)1945; 77JOC3821] and the corresponding O-alkyl derivatives (53JA5758; 61MI1), but hydrolysis is a complicating factor with the latter compounds. Katritzky *et al.* [71JCS(B)1] found that the kinetics of nitration of 2,4-dimethoxypyrimidine could be followed only over the first 20% of reaction in 98% sulfuric acid, and not at all at lower acidity for this reason. Uridine and 2-deoxyuridines usually suffer extensive glycoside ring-rupture during nitration (77JOC3821), while hydrolysis of a halogen substituent may also accompany nitration [71JCS(B)1].

The similarities in rate of nitration for the uracils suggest that they are all nitrated in the dioxo forms [log k_2(fb) at 40–50°C at H_0 −10.26: uracil, 1,3; 1,3-dimethyluracil, 1.7; 6-methyluracil, 1.9]. 2,4-Dimethoxypyrimidine is slower by a factor of ∼10^4. Whereas pyrimidin-2(1H)-one and its 1-methyl derivative also react as the free bases, the rates are much lower than those of the corresponding uracils [log k_2(fb) at 115°C at H_0 −9.27 for pyrimidin-2(1H)-one is 8.1; for the 1-methyl derivative the value is 8.8], and 100 times slower than 3-methyl-2-pyridone. These results reflect the deactivating effect of a second annular nitrogen and the presence of only one oxo function [71JCS(B)1].

4,6-Dihydroxypyrimidines require even less vigorous nitrating conditions than uracils, possibly because one oxygen function must retain its phenolic character and be more electron-releasing than the lactam structure generally preferred by hydroxyl substituents in the 2-, 4-, and 6-positions of pyrimidine (38CB87; 51JCS96; 56JCS2312, 56JOC177; 61JOC4504; 62MI1, 62RTC443; 64AJC1309; 76JHC1141).

Interpretation of nitration results for aminopyrimidines is again complicated by the formation of nitramines, which can rearrange to give products of nuclear nitration. Only nitramines appear to form with monoaminopyrimidines (O1CB1234); both 5-nitro and 5-nitramino products have been obtained from some diaminopyrimidines with mixed acids [O1 CB3362; 5OCI(L)353; 63CB2977]. The nitramines formed at lower temperatures rearrange to the 5-nitro products when heated in concentrated sulfuric acid. A variety of aminohydroxypyrimidines have also been successfully nitrated at C-5 (55MI1; 61JCS1298, 61JOC526; 64JCS3204).

The introduction of a third activating substituent facilitates nitration even more. Temperatures below 50°C with mixed acids are frequently successful with various combinations of alkyl, amino, and hydroxy substituents (O1CB3362; 57JCS2146; 61JOC455; 65JOC3153), and barbituric acids react very readily indeed, even to the extent of *ipso*-substitution at C-5 [O1LA(315)259; O5LA(339)37].

As long as two other activating groups are present, a halogen function seldom prevents nitration, even though it may retard the reaction (63JCS4186; 64JCS460, 64JCS4769; 66CB2997, 66JMC573). Thiol and alkylthio groups are usually too readily oxidized for their general use as activating substituents in pyrimidine nitration (49JCS2490; 78CB1006). Indeed, nitric acid converts 2-ethylthio-4-chloropyrimidine into 2,4-dihydroxy-5-nitropyrimidine, presumably by prior replacement or modification of the original substituents [05JBC435; 51LA(572)217].

Attempts to nitrate pyrimidine N-oxide were unsuccessful (59JCS525), but 2,6-diamino-4-ethylaminopyrimidine N-oxide formed the 5-nitro derivative (79AJC2049).

C. Pyrazines

Pyrazine itself has yet to be nitrated; at least two electron-donating groups are necessary for successful reaction (53JA5517; 56JA4071). An hydroxyl function directs the nitro group into the para position (56JA4071), or ortho if the para position is blocked (53JA5517; 75MI1). 2,5-Dimethylpyrazine N,N'-dioxide was unaffected by treatment at 50°C with sodium nitrate in sulfuric acid (58JOC1603).

D. Triazines

Attempts to nitrate the parent rings have, not surprisingly, been singularly unsuccessful; the strongly acidic reagents may preferentially destroy the ring [63AG(E)309]. The reported nitration of 3-amino-6-methyl-1,2,4-triazine in the 5-position [64CA(60)8031] has been criticised by Neunhoeffer (84MI2) on the grounds that the starting material was the 5-methyl isomer, and nitration must have occurred at C-6. That this reaction occurs at all is remarkable in view of the resistance of the much more highly activated 1,2,4-triazin-3,5($2H,4H$)-dione (83CCC2676).

E. Borazapyridines

Nitration of 2,3-dimethyl-4-ethyl-3,2-borazapyridine (**10.9**) and the 5-ethyl isomer (**10.10**) with N-nitropicolinium tetrafluoroborate goes exclusively meta to the N-Me group in the yields shown (75ACS457). This orientation may be attributed to conjugative electron release from the N-Me group, readily seen from the canonical form **10.11**.

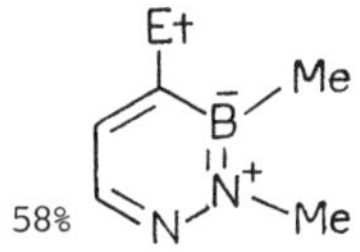

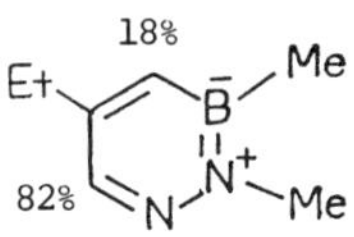

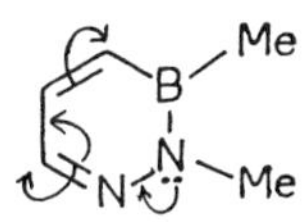

5. Halogenation

This reaction frequently takes place with ease even in the absence of electron-releasing groups. It may involve prior complex formation, and several azine–halogen adducts have been isolated. The situation clearly resembles that observed in the pyridine series, in which the decomposition of similar complexes leads to halogenation at relatively low temperature.

A. Pyridazines

Pyridazine forms complexes readily with bromine and iodine. The latter is stable above 200°C (70CC188), and activating groups permit some annular halogenation. Thus, 3-hydroxypyridazine N-oxide (78CPB3884) and 5-hydroxy-3,6-dimethylpyridazine N-oxide (75CPB923) are brominated at C-4, whereas 4-amino-6-chloro-3-ethoxypyridazine reacts at the 5-position (70CPB1680). 3,6,-Dimethylpyridazine N,N'-dioxide is, however, resistant (73YZ59). 3-Substituted pyridazine N-oxides tend to undergo electrophilic substitution preferentially para to the oxide function, and then next to it (67CPB1411), which accords with the conjugative electron release ($+M$) from the oxygen. When 3-aminopyridazine N-oxide is brominated, the oxygen is apparently protonated, this being followed by concurrent nucleophilic bromination and elimination of water to give 3-amino-6-bromopyridazine (83JOC1064).

Polychlorinated pyridazines may be obtained under vigorous conditions from di- and trichlorinated pyridazines [66CI(L)904; 69USP3466283; 76URP388556].

B. Pyrimidines

Pyrimidine forms 4-bromopyrimidine when the hydrochloride is heated with bromine at 160°C (or at 130°C in nitrobenzene) (73JHC153), the process being preceded by a vigorous reaction at lower temperature (57CB1837; 58AG571). It is likely that N-bromo compounds and perbromides are implicated in these reactions, which occur β to the ring nitrogens, and they are not conventional electrophilic aromatic substitutions.

If alkyl substituents are present, C-5 halogenation can only compete with side-chain halogenation when there are electron-releasing substituents adjacent to one or both nitrogen atoms. Thus, 4-methylpyrimidine is halogenated on methyl in basic media, but the presence of an amino group at C-2 leads to considerable 5-bromination (60CB2405). A methyl group at C-2 appears to be both more activating and more activated than one at the 4-position. 2-Methylpyrimidine gave 5-bromo-2-tribromomethylpyrimidine [54CI(L)1203; 60AC(R)351], and a 4-phenyl group still allowed 5-bromination [73JHC153, 73JHC409; 78CHE1132].

The potentially tautomerizable amino and hydroxy substituents α or γ to a ring nitrogen facilitate nuclear halogenation most markedly, with the highly activated substrates being readily chlorinated and brominated at C-5 even in aqueous solution [46JA453; 60JOC1916; 63JCS1276; 64TL2093; 65RTC1101; 67JCS(C)1922; 71BAU2108; 83JOC1064]. Direct iodination of 2- and 4-amino, but not 2-hydroxypyrimidines, gives >50% yield of the 5-iodo products (84S252), and 4-hydroxypyrimidine [pyrimidin-4(3H)-one] iodinates in alkaline solution (62CCC2550). Fluorine in acetic or anhydrous hydrofluoric acids gives 5-fluorinated derivatives of pyrimidin-2(1H)-ones (77CCC2694).

Groups already present at C-5 can block halogenation [e.g., 4,5-dihydroxy- (64JCS1001) or 5-phenylpyrimidine (77JCS(P1)1862)], but in 5-hydroxypyrimidine the OH group is phenolic enough to activate adjacent carbons. Thus, 5-hydroxy-4-phenylpyrimidine readily forms the 6-iodo derivative (83CHE1008, 83CHE1012).

The mechanism of bromination of pyrimidin-2(1H)-one (**10.12**; R = H) [72CC1032; 74CJC451; 77JCS(P1)1862, 77JOC3670; 78CJC2970; 80JOC2072; 81JOC4172), pyrimidin-4(3H)-one (**10.19**) (79JOC3256; 81JOC4172), and their N-methyl derivatives has been examined in detail. For **10.12** (R = H), the data support a rate-determining attack at high acidity (pH < 2) by Br$_2$ upon a covalent hydrate (**10.14**), which, as an enamine, is rapidly brominated at C-5 to give **10.15**. Compound **10.15** is capable of further hydration to yield **10.16**. Slow, acid-catalyzed conversion of **10.15** or **10.16** into the 5-bromopyrimidinone (**10.17**) follows (Scheme 10.5). The reaction is anything but simple, and in the presence of excess bromine, the dibromo product (**10.18**) is formed (74CJC451; 78CJC2970). N-Methyl groups assist the reaction (74CJC451). At lower acidity (pH > 4) the formation of the pseudo-base appears to be rate limiting. This change occurs because at high acidity the reversal of **10.14** to the cation **10.13** is fast relative to bromine attack; at low acidity the reverse is true. Results obtained between pH 2 and 4 are consistent with a slow changeover in mechanism (80JOC2072). Furthermore, a separate kinetic study of pseudo-base formation and decomposition yielded rate coefficients in good agreement with those derived from the bromination study (78CJC2970; 80JOC2072).

A parallel study of the bromination of pyrimidin-4(3H)-one (**10.19**) and its N-methyl derivatives (79JOC3256; 81JOC4172) again points to an addition–elimination process involving observable cationic intermediates

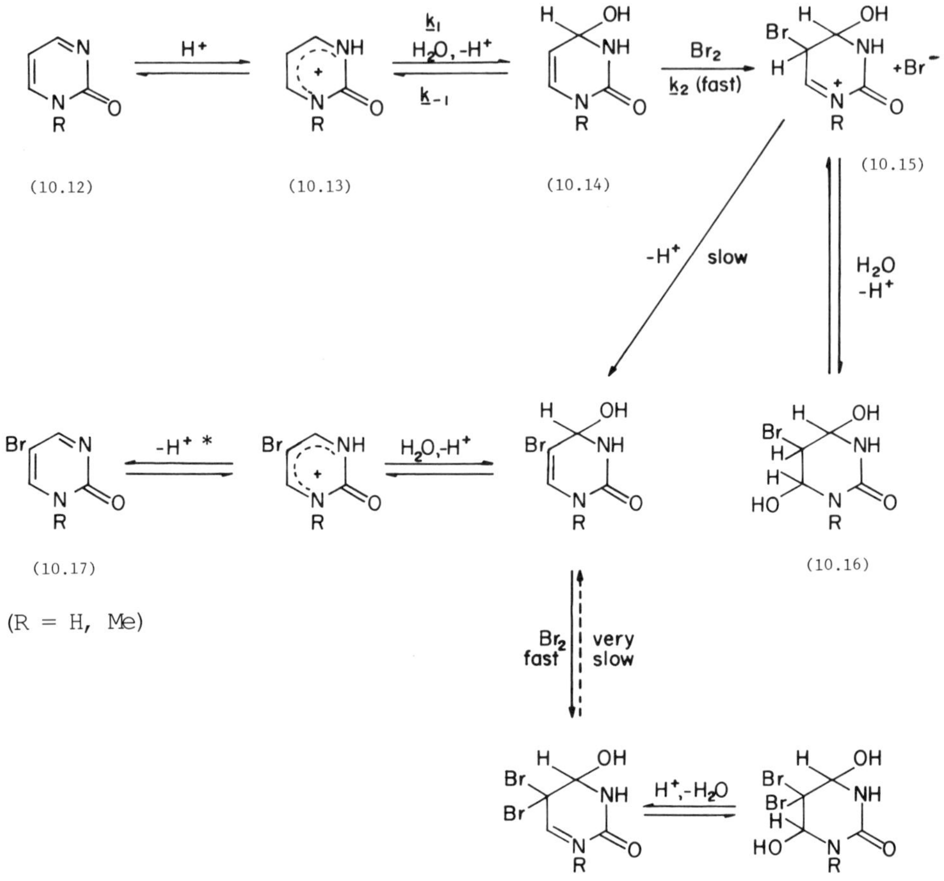

SCHEME 10.5. *This step is irreversible when R = Me.

(Scheme 10.6). The kinetics of dehydration of these showed unusual acidity dependence (for the parent compound $k_{obs.} = 18.7 \times 10^{-3}$ sec^{-1} at $H_0 = 0.1$ and 8.03×10^{-3} sec^{-1} at $H_0 = -0.89$) in that the rates decreased with increasing acidity (slope of log $k_{obs.}$ against H_0 was in the region 0.36–0.43). Usually such dehydrations are acid catalyzed and their rates increase with increasing acidity. Thus, for the pyrimidin-2(1*H*)-ones (74CJC451) and uracils (77JOC3670) the comparable slopes are about -1. Apparently the intermediates in the pyrimidin-4(3*H*)-ones are dehydrated as cations while the uracil and pyrimidin-2(1*H*)-one pseudo-bases dehydrate as neutral molecules (79JOC3256).

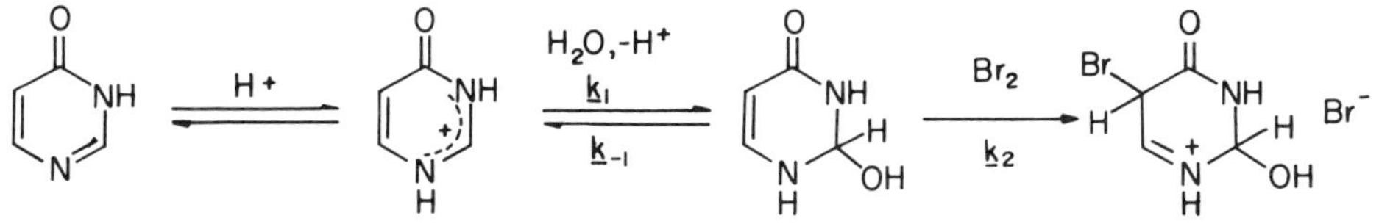

SCHEME 10.6

Although in aqueous solution pyrimidin-4(3*H*)-one (**10.19**) exists only to a very limited extent (~0.0003%) as its covalent hydrate (**10.21**), at pH < 5, attack by bromine is on this very reactive species (**10.21**) ($k_2 \approx 10^9$ M^{-1} sec^{-1}). At pH > 2.5, formation of **10.21** from **10.20** is rate limiting; below pH 2.5, bromine attack on **10.21** is rate limiting.

The results obtained in these studies demonstrate dramatically the effects of covalent hydrates on reactivity. For direct bromination of **10.19**, a rate coefficient of $k_2 \ll 5$ M^{-1} sec^{-1} is indicated, but by virtue of it reacting as its covalent hydrate, it has an apparent value of 2.9×10^3, an enhancement of >>580 times. The enhancement is even greater (>>10^4) for the 2-isomer (**10.12**, R = H), which exhibits a greater degree of covalent hydration (~0.05%).

When there is more than one activating substituent as in the uracils and cytosines, 5-halogenation becomes more facile [46JA453, 46JA1039; 60JOC1583, 60ZOB899; 63JCS1276; 64CCC121; 67JCS(C)1922; 71BAU2108, 71JA3309; 72JOC329; 74JCS(P1)2095; 79JOC4385; 81RTC267, 81S701; 84S252, 84TL3325], to the extent that a group other than hydrogen (e.g., chloromethyl) may be displaced [25LA(441)192]. The mechanistic complications of two or more tautomerizable substituents follow.

Products of both addition and substitution are formed, particularly in aqueous solution, and 5,5-dihalogenated products are common. Overall, uracil (**10.23**, R^1 = R^2 = H) bromination parallels that of **10.12** (R = H) (74CC535; 79CJC626; 80JOC830; 81RTC267), in that addition products, including covalent hydrates, can form rapidly, followed by slower acid-catalyzed dehydration to 5-bromouracils (Scheme 10.7). Variations in second-order rate coefficients with acidity are apparent from Table 10.5. The rates remain fairly constant at low pH and then rise rapidly; 1,3-dimethyluracil (**10.23**, R^1 = R^2 = Me) rates scarcely vary with pH. These results can be interpreted as demonstrating that substrates with at least one NH group react predominantly as anions (**10.24**) at higher pH values. All substrates follow the sequence **10.23** → **10.25** → **10.26** → **10.27** at high acidity, but those which can form anions (e.g., **10.24**) increasingly follow the alternative path as the pH increases (Scheme 10.7) (80JOC830).

TABLE 10.5
SECOND-ORDER RATE COEFFICIENTS FOR BROMINATION OF URACILS[a]

Uracil		1,3-Dimethyluracil	
$k_2^{obs.}$ (M^{-1} sec^{-1})	pH	$k_2^{obs.}$ (M^{-1} sec^{-1})	pH
64.8	0.11		
86.6	1.26	19.9	1.03
142	2.13	20.1	2.25
1,580	3.05		
2,780	3.30	19.0	4.15
115,000	4.95		

[a](80JOC830).

Derivatives of barbituric acid (**10.28**) are so reactive in the presence of halogens that it is often difficult to obtain the 5-monohalogenated derivatives [21CB1035; 22CB3400; 74JCS(P1)2095; 79JOU357], 5,5-dihalo compounds being obtained. Fluorination of 5-substituted barbiturates with perchloryl fluoride was shown to be first order in each component. Elec-

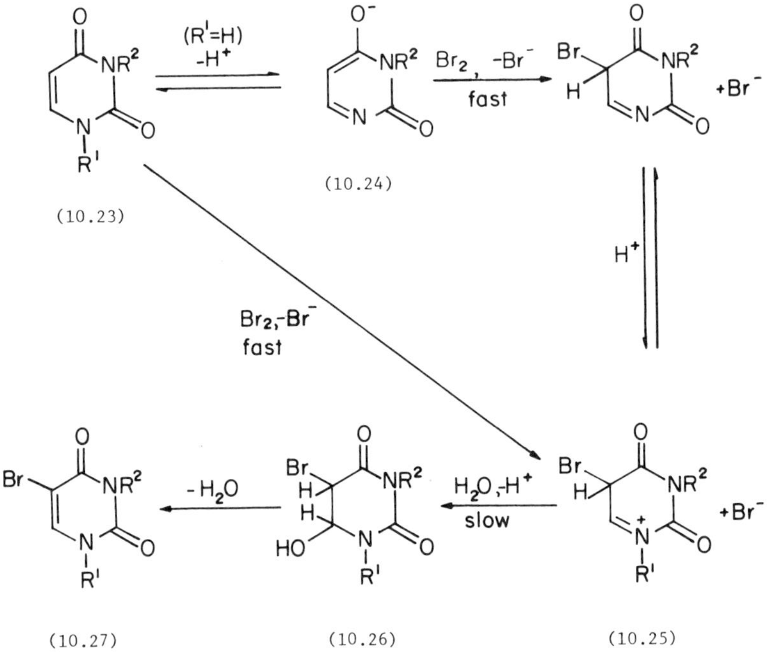

(10.23) (10.24) (10.27) (10.26) (10.25)

SCHEME 10.7

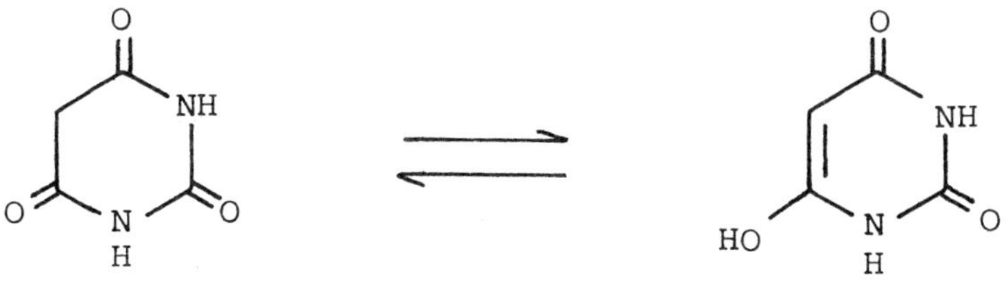

(10.28)

tron-donating substituents increased the rate, and rate coefficients correlated well with σ_p in line with other aromatic halogenations (79JOU357).

Although substituted pyrimidine N-oxides can brominate at C-5 (78CHE1132), the isolated oxide function is seldom sufficiently activating to overcome the inherent unreactivity of pyrimidine (83JOC1064). The failure of 2-, 4-, and 5-methoxypyrimidine N-oxides to undergo bromination (except under forcing conditions) demonstrates the insufficient activation of methoxy and oxide groups. Amino functions at the 2- and 4-positions assist 5-bromination (83JOC1064).

C. PYRAZINES

Those examples of halogenation of pyrazine and its simple alkyl derivatives which are heterolytic rather than homolytic probably involve an addition–elimination process [64JOC415; 66CI(L)1721; 72AHC(14)99], and many of the facile electrophilic substitutions may arise after initial formation of perhalides in solution (61JOC2360). It is possible to achieve selectivity by careful choice of reagents and solvents. With $SOCl_2$–DMF at 20°C, 2-alkylpyrazines are chlorinated exclusively at C-3; with $POCl_3$–PCl_5, the 6-chloro products predominate [72JCS(P1)2004].

As with the other diazines, pyrazines that are activated by strong electron donors substitute readily, the incoming halogen being directed into positions ortho and para to the activating group. Thus 2-aminopyrazine gives a 91% yield of the 3,5-dibromo product (83JOC1064); under milder conditions the major product is that monobrominated at C-5 with traces of the 3-isomer (82JHC673). Similar behavior has been noted with 2-dimethylamino- (83JOC1064), 2-amino-6-phenyl- (80JHC143), 2-amino-5-phenyl- (88% yield of 3-bromo product) (78JHC665), and 2-amino-3-carboxamidopyrazine (96% yield of 5-bromo product) [71JCS(C)3727]. A 2-hydroxy group also assists 3- and 5-bromination (50JA4071; 78JHC665) even if chloro and phenyl groups are also present in the ring. Thus, 6-chloro-2-hydroxy-5-phenylpyrazine gave an 87% yield of the 3-bromo derivative, but 2-hydroxy-3,5-diphenylpyrazine would not react because both the 3- and 5-positions were blocked (56JA4071). Neither 2-methoxypyrazine nor its *N*-oxide can be brominated, but when two such groups act in concert as in 3-methoxypyrazine *N*-oxide or 3-aminopyrazine *N*-oxide, halogenation takes place at C-6. Similar arguments explain the more ready halogenation of 2-aminopyrazine than its *N*-oxide, in which the directing effects of the substituents are in opposition (83JOC1064; 84H1195). In contrast to its effect in the pyridines, an N-oxide function is insufficient by itself to promote annular bromination, even with 2,3-dimethylpyrazine *N,N'*-dioxide (72CHE1153).

D. TRIAZINES

Direct halogenation of 1,2,4-triazine (and the tetrazines) has not been described, while reported chlorinations and brominations of 1,3,5-triazine were probably not electrophilic (54JA632). Even when strongly $+I$ or $+M$ groups are present, triazines frequently resist nuclear halogenation.

3-Amino-6-methyl-1,2,4-triazine can be converted into its 5-bromo derivative [64CA(60)8031], while 3,5-dihydroxy-1,2,4-triazine ("6-azauracil") is attacked at C-6 in a manner reminiscent of the 5-bromination of uracil (*vide supra*) (61JOC1118). Formation of such 6-halogenated products seems to require either a 3-amino or a 5-oxo substituent. Thus, 3-amino-1,2,4-triazine 1- and 2-oxides (but not the 3-chloro, 3-methoxy, or 3-methylthio compounds) are readily brominated and chlorinated at the 6-position (77JOC3498; 78JOC2514), with some concurrent deoxygenation that appears to be a consequence of reaction with HBr (78JOC2514).

In aqueous sulfuric acid 3,5-dihydroxy-1,2,4-triazine is brominated 10^{10} times more slowly than uracil, but with some mechanistic similarities. Provided that there is at least one NH group present, bromine can react with the deprotonated anionic species analogous to **10.24** (76JOC4004).

6. Other Electrophilic Substitutions

A. DIAZO COUPLING

Like pyridine, the azines are not normally attacked by diazonium cations, but since coupling reactions seldom involve extensive protonation of the substrate, the reactions become viable provided that sufficient numbers of activating groups are present, and if reactive diazonium salts are used. Pyridazin-3,4(1*H*,2*H*)-diones are among the few pyridazines that couple; compounds in which an amino group replaces the 3-oxo function fail to react (70JPR591).

In contrast, various combinations of substituents allow pyrimidines to couple at C-5, and sometimes at C-6. A survey (83AJC1659) of the reported structural requirements for successful reaction [44JCS315; 48BSF688; 56CI(L)1453; 64JCS1001; 66JPS568; 73CPB1327, 73JCS-(P1)1089; 77JCS(P1)1985; 79BAU633; 81JHC1639; 82CHE297; 83AJC-1659] concluded that diazo coupling at the 5-position needs two strongly electron-releasing groups at C-2 and C-4 (or C-6), with an amino or hydroxy group at the 4-position mandatory. Even then reaction is not guaranteed. Electrophilic substitution may be accompanied by exocyclic coupling at 2-, 4-, or 6-amino, -thiol, or -methyl groups (80H1753). Hurst (83AJC1659) commented that a report of C-2 coupling in 5-hydroxy-4,6-dimethylpyrimidine (79BAU633) is incorrect; attack must have been at a methyl group, and there are other examples from the earlier literature that require revision. If the 5-position already contains an amino, hydroxy, or acetate group (but not methoxy, methythio, or acetamido), coupling at the 6-position can take place, or at C-2 if C-6 is blocked (81JHC1639; 82CHE297). It has been suggested that these reaction are of the addition–elimination type and not so widely applicable as originally supposed (83AJC1659).

The few examples of diazo coupling reported for pyrazines may not all be annular coupling (28G679; 30G298; 52JCS4870).

B. Nitrosation

Nitrosation, which frequently involves sites comparable in reactivity to those that undergo diazo coupling, is known only in the pyrimidines. Structural requirements are stringent, with three activating substituents at the 2-, 4-, and 6-positions usually necessary [44JCS315; 54YZ674; 58LA(612)158; 60MI1; 64JCS1001; 66LA(691)142; 71TL851; 78H247; 82CPB3392]. Both 4,6-diamino- and 4,6-dihydroxypyrimidines, however, gave the blue 5-nitroso compounds, although the 2,4-isomers were unreactive (44JCS315). Exocyclic nitrosation may occur concurrently or exclusively.

C. Sulfonation

Whereas 2-aminopyrimidine (59JA5166) and some other activated derivatives (54JCS4206, 54JGU2212; 56JA401; 61JOC3863; 63JOC1994) each give the 5-sulfonic acids, 2-aminopyrazine will not react (45JA802).

Nothing is known of the reactivities of the pyridazines, and triazines are cleaved before substitution by the usual sulfonating agents [63AG(E)309].

D. ACYLATION

No kinetic studies of acylation have been made. The minimum requirements for pyrimidines to take part in Reimer–Tiemann reaction at C-5 have been evaluated as one hydroxy and two methyl groups (60JOC1906). Reported failures of the reaction when sufficient activation appeared to be present (57JCS4845) may have been a result of ring cleavage. Pyrimidines with three powerfully activating substituents can also be converted into aldehydes by the Vilsmeier procedure (65M1567; 71CPB215, 71CPB1526, 71JHC445; 73CPB260); hydroxy groups are displaced by chloro after formylation (65M1567). There are also some examples of 5-acylation of 6-aminouracils heated in acetic anhydride [23CB2482; 58LA(612)173; 72JOC578].

E. ALKYLATION

A few examples of Mannich reactions exist in the series. Pyridazine N-oxides, substituted at the 3- or 5-position by hydroxy, are alkylated at either of the vacant ring positions, although activated methyl groups may react preferentially (68CPB939; 73CPB1510; 75CPB923).

Uracils aminoalkylate at C-5 (56YZ230,234; 60JA991), whereas the ortho- and para-directing effect of the phenolic hydroxy group in 5-hydroxypyrimidine and its 4-phenyl derivative assisted formation of the 6-aminomethyl products (82CHE297; 83CHE1008). When the 6-position is blocked, 2-aminoalkylation can occur (79BAU633). Triarylmethyl cations 5-alkylate 2- and 4-hydroxypyrimidines, uracil (**10.4**), and barbituric acid (**10.28**) (especially in the presence of Lewis acids) (74JOC587, 74JOC591). Barbituric acids have been alkylated at C-5 with vinylpyridine (62JOC174) and 6-hydroxy-2,4-diaminopyrimidine has been methylated at the 5-position by MeI/NaOMe (62JCS3172).

F. METALLATION

In pyrimidine, the most likely metallation site for lithiation is C-5 (cf. base-catalyzed hydrogen exchange, Section 2), but even two methoxy groups in the 4- and 6-positions failed to allow lithium–hydrogen exchange at $-40°C$ in this system (65JCS6695). At lower temperature ($< -70°C$), however, 5-pyrimidinyllithium derivatives can be prepared by halogen–lithium exchange (56JA2136; 59T225; 65ACS1741). Gronowitz (65ACS1741) showed that, at very low temperature (-105 to $-110°C$), halogen–lithium exchange predominates with n-butyllithium, but at higher temperature additions occur to the azomethine groups of 5-bromopyrimidine. Above $-30°C$, n-butyllithium adds directly accross the 3,4-bond of the heterocycle. Halogen–metal exchange in pyrimidine decreases in facility in the order C-5 > C-4(6) > C-2 (65ACS1741).

Pyridazine and pyrazine have no ring positions equivalent to C-5 of pyrimidine, but even so there have been a number of metallated derivatives of pyrazines made either by lithium–hydrogen (68JOC1333; 71RTC513; 74JOC3598) or lithium–halogen exchange (65JHC209). Similar pyridazine reactions are believed to be a result of addition reactions (65ACS1652).

CHAPTER 11

Compounds Containing Two or More Six-Membered Rings

1. Introduction

A. SURVEY OF HETEROCYCLES CONSIDERED

A very large number of compounds qualify for inclusion in this chapter; however, quantitative data are available so far for only a few of them. This partly reflects the relative difficulty in carrying out electrophilic substitution on these heterocycles, the rings of which are mainly π deficient. Nevertheless, there appears to be a substantial number of molecules that should be reasonably reactive, but which have not been investigated, and there seems to be scope for considerable further work in this area. We commence this chapter with an overview of the compounds to be considered.

a. *Compounds Containing One Nitrogen Atom*
  i. *Benzo-Annelated Pyridines* (**11.1–11.3**).

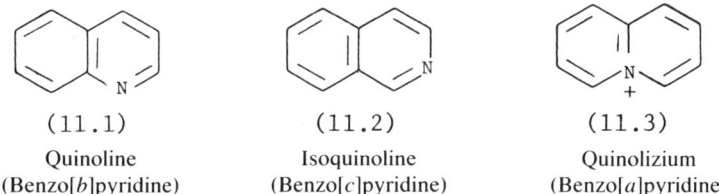

(11.1)
Quinoline
(Benzo[*b*]pyridine)

(11.2)
Isoquinoline
(Benzo[*c*]pyridine)

(11.3)
Quinolizium
(Benzo[*a*]pyridine)

  ii. *Benzo-Annelated Quinolines, Isoquinolines, and Quinoliziums* (**11.4–11.12**).

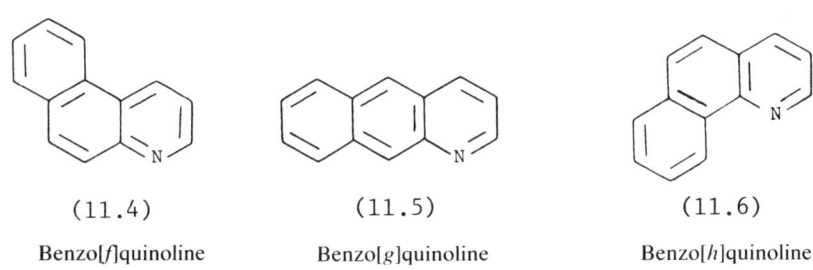

(11.4)
Benzo[*f*]quinoline

(11.5)
Benzo[*g*]quinoline

(11.6)
Benzo[*h*]quinoline

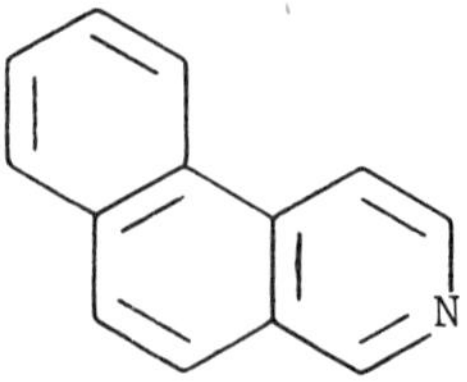

(11.7) Benzo[f]isoquinoline

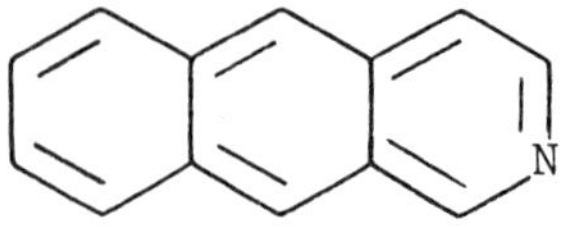

(11.8) Benzo[g]isoquinoline

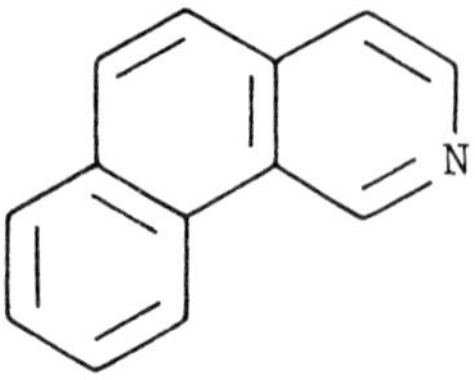

(11.9) Benzo[h]isoquinoline

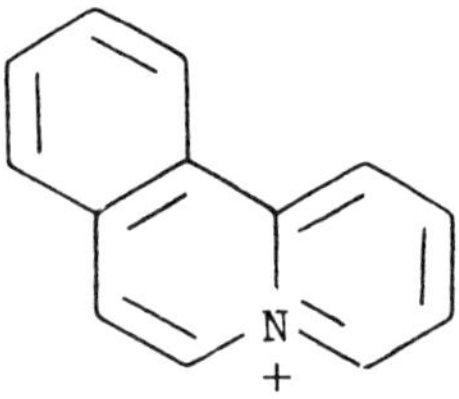

(11.10) Benzo[d]quinolizium

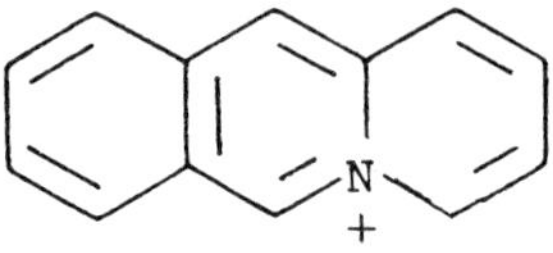

(11.11) Benzo[c]quinolizium (Acridizinium)

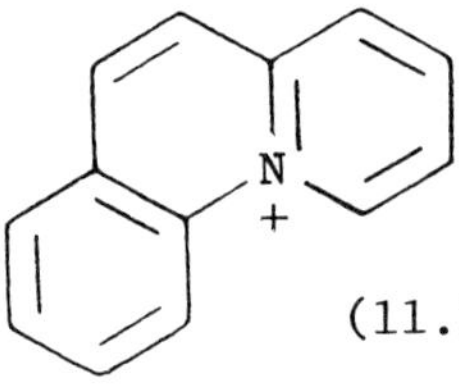

(11.12) Benzo[b]quinolizium

iii. *Dibenzo-Annelated Pyridines* **(11.13, 11.14)**.

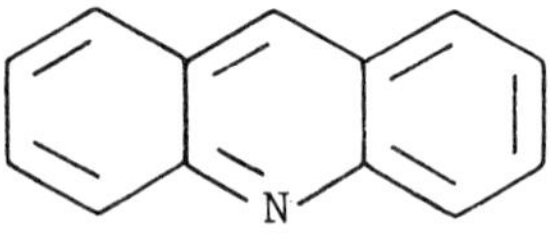

(11.13) Acridine (Dibenzo[b,e]pyridine)

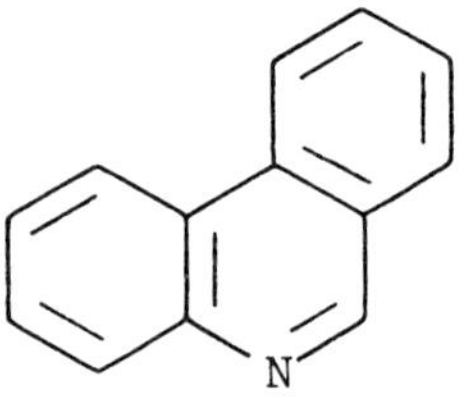

(11.14) Phenanthridine (Dibenzo[b,d]pyridine)

b. *Compounds Containing Two Nitrogen Atoms*

i. *Benzo-Annelated Pyrazines, Pyrimidines, and Pyridazines* **(11.15–11.18)**.

Sec. 1.A] INTRODUCTION 355

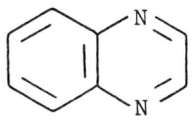 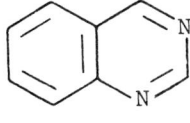 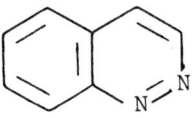 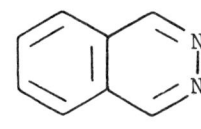

(11.15)   (11.16)   (11.17)   (11.18)

Quinoxaline   Quinazoline   Cinnoline   Phthalazine
(Benzo[b]pyrazine)   (Benzo[d]pyrimidine)   (Benzo[c]pyridazine)   (Benzo[d]pyridazine)

ii. *Dibenzo-Annelated Pyrazine and Pyridazine* **(11.19, 11.20)**.

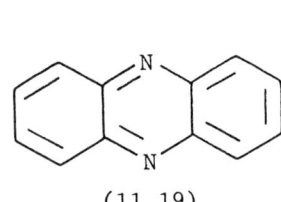 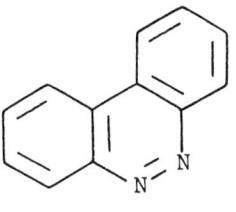

(11.19)   (11.20)

Phenazine   Benzo[c]cinnoline
(Dibenzopyrazine)   (Dibenzopyridazine)

iii. *Pyrido-Annelated Pyridines (Naphthyridines)* **(11.21–11.26)**.

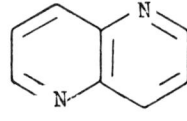 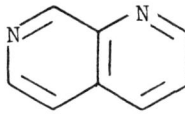 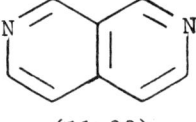

(11.21)   (11.22)   (11.23)

1,5-Naphthyridine   1,7-Naphthyridine   2,7-Naphthyridine
(Pyrido[3,2-b]pyridine)   (Pyrido[3,2-c]pyridine)   (Pyrido[4,3-c]pyridine)

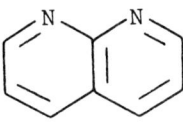

(11.24)   (11.25)   (11.26)

1,8-Naphthyridine   1,6-Naphthyridine   2,6-Naphthyridine
Pyrido[2,3-b]pyridine   (Pyrido[3,2-c]pyridine)   (Pyrido[4,3-c]pyridine)

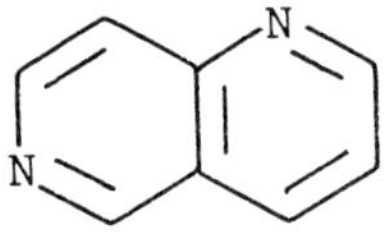

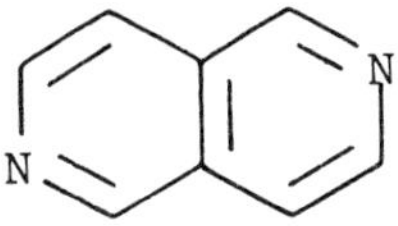

iv. Pyrido-Annelated Quinolines and Isoquinolines (e.g., 11.27–11.30).
There are a large number of compounds in this category, with 10 isomers for **11.27**, six for **11.28**, four for **11.29**, and eight for **11.30**.

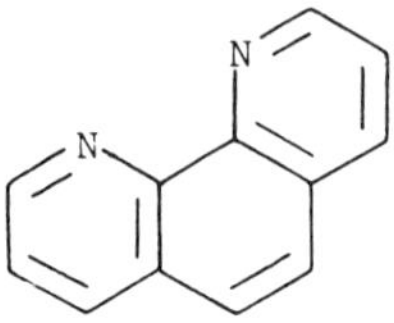

(11.27)
1,10-Phenanthroline
(Pyrido[3,2,-*h*]quinoline)

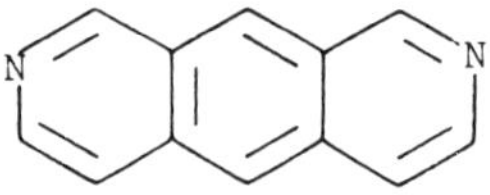

(11.28)
2,7-Anthrazoline
(Pyrido[4,3,-*g*]isoquinoline)

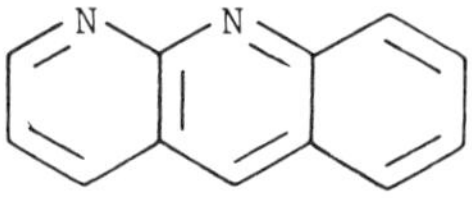

(11.29)
1,9-Anthrazoline
(Pyrido[2,3-*b*]quinoline)

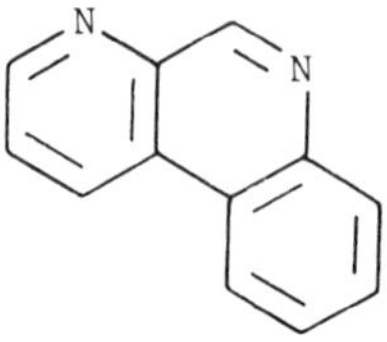

(11.30)
Pyrido[2,3-*c*]quinoline

v. Benzo-Annelated Quinoxalines, Quinazolines, Cinnolines, and Phthalazines (e.g., 11.31–11.34.
There are 10 isomeric compounds of this type, of which four are shown.

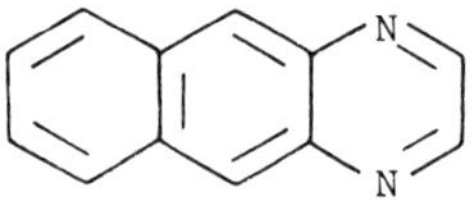

(11.31)
Benzo[*g*]quinoxaline

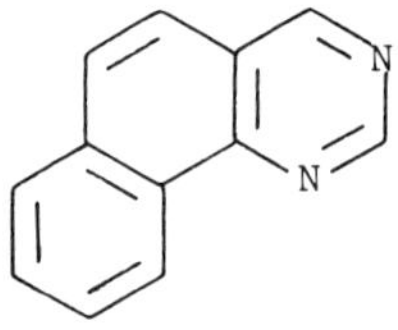

(11.32)
Benzo[*h*]quinazoline

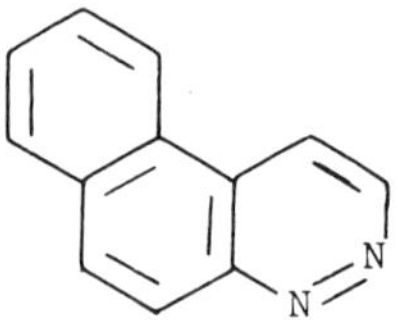

(11.33)
Benzo[*f*]cinnoline

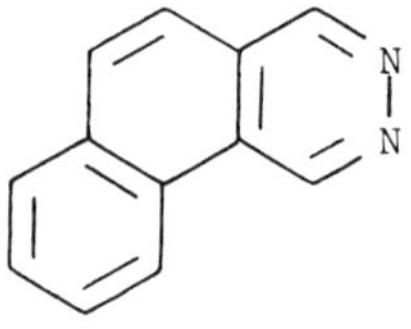

(11.34)
Benzo[*h*]phthalazine

Additionally, derivatives exist with nitrogen at bridgeheads, but they are unreactive with no reports of electrophilic substitution on them.

c. Compounds Containing Three Nitrogen Atoms

i. Dipyrido-Annelated Pyridines (e.g., 11.35).
In addition to 1,9,10-anthyridine, eight other isomers exist in this category.

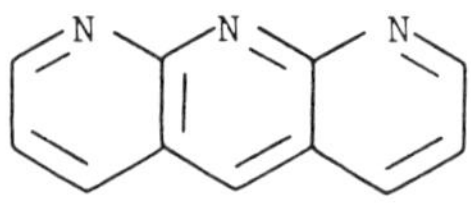

(11.35)
1,9,10-Anthyridine
(Pyrido[2,3-*b*]-1,8-naphthyridine)

ii. *Pyrido-Annelated Pyrazines, Pyrimidines, and Pyridazines (11.36–11.38).* Benzo derivatives of these compounds are possible but reports of them are very rare. Compounds with bridgehead nitrogen are so unreactive as to be of no concern here.

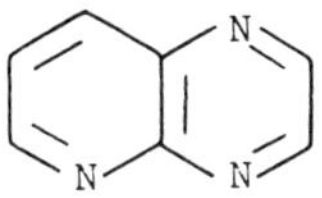

(11.36) Pyrido[2,3-*b*]pyrazine

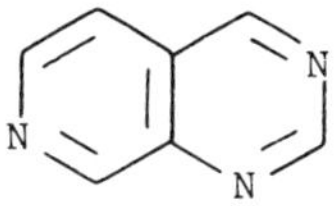

(11.37) Pyrido[3,4-*d*]pyrimidine

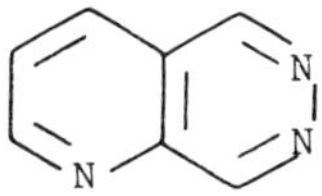

(11.38) Pyrido[2,3-*d*]pyridazine

d. *Compounds with Four or More Nitrogen Atoms*

There are 12 isomers for compounds containing two rings with two nitrogens in each (excluding those compounds with nitrogen bridgeheads). All are very unreactive, and the only one of significant interest is pteridine (**11.39**).

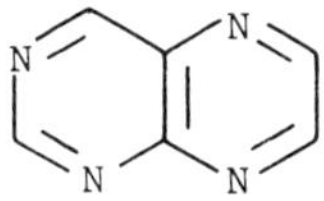

(11.39) Pteridine

e. *Hydroxy Derivatives "Ones" of Compounds* **11.1–11.39** *with Hydroxy Group Conjugated with Nitrogen (e.g., **11.40, 11.41**)*

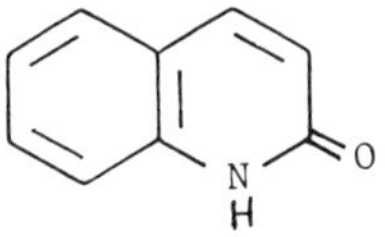

(11.40) 2-Quinolone

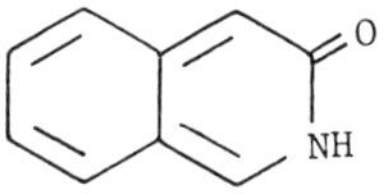

(11.41) 3-Isoquinolone

f. *N-Oxide Derivatives of Compounds* **11.1–11.39** *(e.g., **11.42**)*

More than one oxide function can occur within a given molecule.

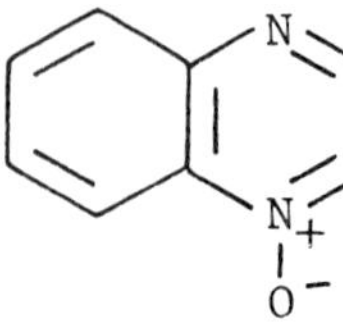

(11.42)
Quinoxaline 1-Oxide

g. *Compounds Containing Other Group VB Elements*

A number of derivatives of compounds in classes discussed in Sections 1.A.a–1.A.d are known, containing arsenic or phosphorus in place of nitrogen, but there are no reports of quantitative electrophilic substitution.

h. *Compounds Containing Boron and Nitrogen (e.g., 11.43)*

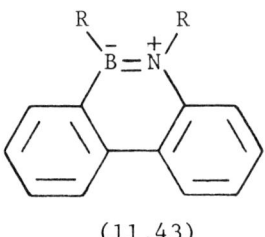

(11.43)
A Borazaphenanthrene

i. *Benzo-Annelated Pyrylium Ions (e.g., 11.44, 11.45)*

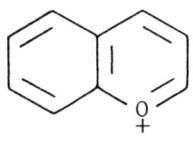 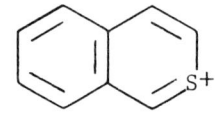

(11.44)　　　　　　(11.45)
1-Benzopyrylium　　2-Benzothiopyrylium

j. *Benzo-Annelated Pyrones (11.46–11.48)*

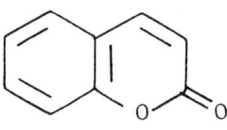 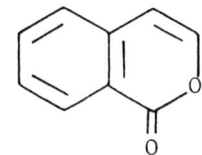 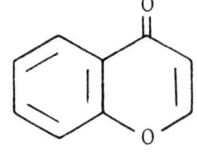

(11.46)　　　(11.47)　　　(11.48)
Coumarin　　Isocoumarin　　Chromone

B. REACTIVITY PATTERNS

In general the reactivity of nitrogen-containing compounds will be governed by the following factors.

(1) Reactivity will decrease as the number of nitrogen atoms increases. Molecules with nitrogens concentrated in one ring will tend to be more reactive than those in which the nitrogens are more evenly distributed. Thus benzo[g]quinoxaline (**11.31**) should be more reactive than 1,9-anthrazoline (**11.29**).

(2) The reactivity will be subject to the same factors that produce the differential positional reactivities in poly(carbo)cyclic aromatics. Thus, α-naphthalene-like positions will be kinetically more reactive than β-naphthalene-like ones though subject in some reactions to greater steric hindrance. Anthracene-like molecules will be more reactive than phenanthrene- or naphthalene-like ones. Thus, benzo[g]quinoline will be more reactive than benzo[h]quinoline or quinoline.

(3) An aza analogue with nitrogen occupying a reactive position in the carbocycle will be less reactive than an aza analogue in which a less reactive site is occupied. Thus, quinoline is less reactive than isoquinoline, and acridine (**11.13**) is expected to be less reactive than benzo[g]quinoline (**11.5**).

(4) Bond-order differences are of crucial importance, the deactivation by nitrogen across a bond of low order being much less than that across a bond of high order. This substantially affects the positional reactivities in molecules with nitrogen occupying β-naphthalene-like positions [e.g., isoquinoline (**11.2**), benzo[f]isoquinoline (**11.7**), 1,7-naphthyridine (**11.22**), and 2,7-anthrazoline (**11.28**)]. The bond-order effect also indirectly accounts for the fact that the position of highest electron density in the neutral isoquinoline species is the 4-position, i.e., *in the ring containing the nitrogen, consequently this is the preferred site for substitution of the free base*. There appears to be considerable misunderstanding of this point in the heterocyclic literature [see, e.g., 81HC(381)32; 85MI1, for examples] and it is dealt with more fully in Section 7. Sites meta to nitrogen in molecules such as **11.7**, **11.22**, and **11.28** should likewise all be relatively reactive.

(5) Deactivating conjugative interactions that are weak because of unfavorable location of the double and single bonds in the transition state become stronger in benzo derivatives because one ring will remain benzenoid in the transition state. Thus, whereas the 5-position is poorly conjugated with the 1-nitrogen in quinoline (substituent interaction factor A_f = 3.6 [81JCS(P2)1153]) leading to substantial 5-substitution, in acridine (**11.13**) the 1-position is much better conjugated with the 10-nitrogen (A_f = 10.9) so that little 1-substitution is observed.

(6) Compounds with bridgehead nitrogen atoms between two six-membered rings are necessarily cationic and are therefore very unreactive. The positions in these molecules that are best conjugated with the positive nitrogen [e.g., the 4- and 8-positions in acridizinium (**11.11**)] will be the most strongly deactivated.

(7) Hydroxy derivatives (which normally exist predominently as carbonyl compounds or ketones) will be much more reactive than the parent compounds.

(8) N-Oxides will be more reactive than their parents. The activating effect of the oxide function should vary according to the extent of bond fixation within a molecule. Thus, in isoquinoline *N*-oxide the reactivity of the 1-position relative to that in isoquinoline will be raised much more than will that of the 3-position.

(9) Compounds will be very much less reactive under conditions in which nitrogen is protonated and the positional order may be altered.

Similar factors account for the relative reactivity of oxygen-containing compounds. Thus 1-benzopyrylium (**11.44**) is very unreactive, but more reactive than pyrylium. The least reactive sites will be the 2- and 4-positions and in the benzenoid ring it will be the 7-position because of the high stability of the *p*-quininoid structure (**11.49**). For 2-benzopyrylium, the least reactive sites will be the 1- and then the 6-position, the latter arising from the high stability of structure **11.50**; **11.50** is more stable than **11.49**, so the 6-position of **11.50** will be less reactive than the 7-position of **11.49**.

(11.49)   (11.50)

Coumarin (**11.46**), iscoumarin (**11.47**), and chromone (**11.48**) should each be more reactive than the corresponding pyrone, and consideration of conjugative effects indicates substitution to be preferred at the 3-position of **11.46** and **11.48**, and at the 4-position of **11.49**. Isocoumarin (**11.47**), in which the oxygen lone pair is less readily delocalized into the benzenoid ring, should be more reactive than the other two isomers. Substituents in these molecules will have very marked directional effects because of significant bond fixation.

2. Acid-Catalyzed Hydrogen Exchange

A. QUINOLINES AND ISOQUINOLINES

Deuteriation of quinoline with sulfuric acid of various strength and at a range of temperatures indicated that the positional reactivity order for quinoline was $8 > 5, 6 > 7 > 3$ [67CPB826; 72BAU2029], and the order $8 > 5 > 6 > 7$ has been confirmed in a kinetic study [71JCS(B)4]. The reactivity order shows that the conjugate acid is involved (since the 5-position is the most reactive in the free base; see Section 7) and this was confirmed for exchange at each of those positions by the slope of the rate–acidity profile. However, at $D_o = 0.5$, both the 2- and 3-positions were *more* reactive than the other positions, which is not in agreement with an A-$S_E 2$ mechanism on either the free base or conjugate acid. The high reactivity of the 2-position was attributed to base-catalyzed exchange of the conjugate acid, exchange next to a positive pole being very rapid by such a mechanism (see 2-exchange in pyridine, Chapter 9, Section 3). This mechanism is however unable to account for the high reactivity of the 3-position under these conditions since it should be substantially less than in the 2-position, which was not the case. It therefore appeared possible that exchange at the 3-position occurs via yet another mechanism, the rate-determining formation of a covalent hydrate by water at higher acidity or by OD$^-$ at lower acidity (Scheme 11.1). The slow first step is understandable because of the loss of the aromaticity which is implied, and is in agreement with the low concentration of OD$^-$ under the conditions used.

Exchange in isoquinoline showed a marked difference in rate–acidity profile between the 1- and 4-positions on the one hand, and the 5- and 8-positions on the other (Fig. 11.1). The 5- and 8-positions clearly exchange as the conjugate acids, but the decrease in rate for the 1-position with increasing acidity, together with it being more reactive than the 5- and 8-positions at low acidity or weak basicity, show that a base-catalyzed mechanism on the conjugate acid applies. For exchange at the 4-position at this acidity a mechanism similar to that in Scheme 11.1 was considered probable [71JCS(B)4]. However, the conclusion is less certain in the isoquinoline case because in the free base the 4-position of isoquinoline is expected to be the most reactive (see Section 7).

Exchange in the hydroxy derivatives of quinoline and isoquinoline has been studied. For exchange at the 3-position of 4-quinolone, the rate is roughly invariant with acidity below $H_0 - 7$ but increases rapidly at higher acidity. This rate–profile slope alteration is due to a change-over from reaction on the free base to reaction on the conjugate acid and relative

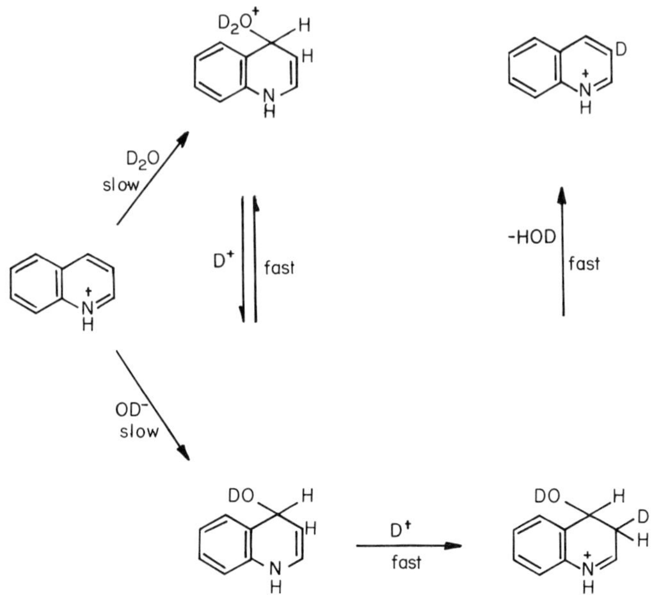

SCHEME 11.1. Mechanism for 3-exchange in quinoline under neutral conditions.

reactivities of these two species was calculated to be $10^{7.3}$. At high acidity, the slope of the rate–acidity profile (0.53) was similar to that of 4-methoxyquinoline, which also reacts as the conjugate acid. The relative rates of exchange at the 3-, 6-, 8-, and 5-positions of the 4-quinolone conjugate acid (i.e., the 4-hydroxyquinolinium cation) were found to be $1 : 10^{-2.5}$:

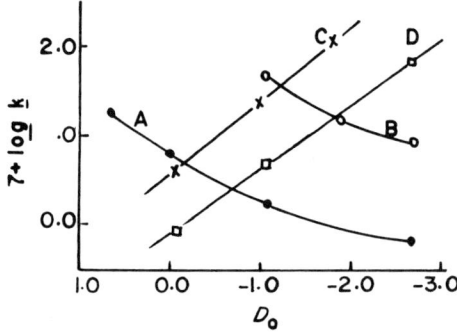

FIG. 11.1. Rate–acidity profile for hydrogen exchange in isoquinoline at the A,1-; B,4-; C,5-; and D,8-positions.

10^{-4} : 10^{-5} [67JCS(B)1226]. The order arises because the 6- and 8-positions are conjugated with the 4-hydroxy group, and the 5- and 8-positions are fairly reactive in the parent quinolinium cation. The 8-position is here less reactive than the 6-position because 4,8(1,5)-conjugation is unfavorable (see Section 7). By contrast, for 2-quinolone, the positional reactivity order (for reaction on the conjugate acid at acidity greater than H_0 -6) is $8 > 6 > 5 > 3$ ($> 4, 7$); the relative reactivity for the first four positions is $1.0 : 0.2 : 0.03 : 0.03$ (68BAP453, 68CPB715). Here conjugation between the 2- and 8-positions is better than between the 4- and 8-positions in 4-quinolone. Conversely, 2,3-conjugation is poor (due to the low 2,3-bond order), hence the 3-position is relatively unreactive. The 5-, 6-, and 8-positions of 2-quinolone were 90–15,000 times more reactive than those in 4-quinolone (**11.51** and **11.52** show the rate coefficients in hr^{-1} at H_0 -9 and 110°C): this is a direct consequence of the better conjugation between the 2-position and the benzenoid ring compared to the 4-position and the benzenoid ring. For example, the substituent interaction factors A_f [81JCS(P2)1153] are 7.3 and 10.9 for **11.53** and **11.54**, respectively.

(11.51)   (11.52)   (11.53)   (11.54)

Comparison of the exchange data for 4-quinolone and 4-pyridone indicated that benzo annelation increases the reactivity of the latter by $\sim 10^2$ for the free base and $>10^5$ for the conjugate acid [67JCS(B)1226]. Again, the factor is greatest for the less reactive system, in keeping with the reactivity–selectivity principle.

3-Hydroxyquinoline underwent exchange at the 5-, 6-, and 8-positions 10-fold faster than quinoline under the same conditions, and exchange at the 7-position became observable. In this work, a 2-methyl substituent was found to increase the reactivity of the 5- and 8-positions 10-fold and of the 6-position by sixfold (72BAU2029). These values are somewhat surprising because the 2,6-conjugative interaction is much the strongest (cf. substituent effects in detritiation of naphthalene [68JCS(B)1112]); perhaps the mechanism is not the same under all conditions.

For exchange in 1-hydroxyisoquinoline with 61 wt% D_2SO_4 at 180°C, the positional reactivity order was $4 > 5 \geq 7 > 8 > 3 > 6$ (68CPB715), which is almost exactly the order for the free base of isoquinoline (Section 7).

However, the conjugate acid is almost certainly the exchanging species in 3-hydroxyisoquinoline, so this agreement is probably fortuitous and may reflect near cancellation of the effects of the OH substituent and N-protonation.

For exchange in 4-hydroxyisoquinoline by 94 wt% D_2SO_4 at 145°C, the positional order was 3 > 5 > 8 > 6; the corresponding rate coeffecents k (sec^{-1}) were 1.3×10^{-2}, 1×10^{-2}, 2×10^{-3}, and 2.1×10^{-4} (72CHE1495). Here the 5- : 8-rate ratio is not significantly different from that evident in Fig. 11.1, which again demonstrates the poor conjugation between the 4- and 8-positions. The observation of 6-exchange in 4-hydroxyisoquinolinium cation suggests that exchange at the 6-position should also be observable with isoquinoline itself; the failure to do so may reflect the complexity of the NMR signals [71JCS(B)4]. The increments in exchange rate produced by a 3-methyl substituent in the 4-hydroxyisoquinolinium cation at the 5-, 8-, and 6-positions were 6.3-, 8-, and 8-fold, respectively (72CHE1495). Again the smaller effect at the 5-position is surprising as it is the only one conjugated with the methyl group.

Rates of exchange over a wide range of acidity have been measured for 6-hydroxyquinoline (**11.55**), 6-methoxyquinoline (**11.56**), and 6-hydroxy-1-methylquinoline (**11.57**) [71JCS(B)11]. In the alkaline region, 6-hydroxyquinoline gave a rate versus acidity slope of 1.3, and the same slope value was found for 2-naphthol, indicating that both undergo exchange by reaction of the anion with D_3O^+. Over the weakly acidic range (pD 4–2) all three compounds give rate profiles with slopes near zero, indicating that each exchanges as a neutral species reacting with D_3O^+. However, **11.56** was much less reactive than either **11.55** or **11.57** showing that it is the zwitterionic form of 6-hydroxyquinolinium that undergoes reaction. In the strongly acidic region (H_0 −1 to −4), all three compounds have similar exchange rates, and rate versus acidity slopes of 0.7, showing that each exchanges as the cation.

(11.55)   (11.56)   (11.57)

Rate profiles have been obtained for exchange in 6- and 7-aminoquinolines at the 5- and 8-positions, respectively [71JCS(B)11]. These rate profiles are fairly similar and show the expected changes in slope corresponding to the first and second protonations at the appropriate pK values.

B. Quinoline and Hydrogen Isoquinoline N-Oxides

The positional order for exchanging quinoline (Q) N-oxide is 8 > 5,6 > 7 > 3 [67CPB826; 70CPB203; 71JCS(B)4], the same as in quinoline. For isoquinoline (iQ) N-oxide, the order 5 > 8 has been obtained. Both N-oxides react (at the positions mentioned) as their conjugate acids [71JCS(B)4]. The relative reactivities of the 5- and 8-positions in the N-oxides are 8-iQ > 8-Q > 5-Q = 6-Q > 5-iQ.

Comparison between the reactivity of the N-oxides and the parent compounds (Table 11.1) shows that formation of the N-oxides increases the reactivity by ~10-fold, except that at the 8-position of quinoline, where the reactivity remains the same. The difference between the result for the 8-position and the others probably again reflects the poor 1,5-conjugative interaction. The relatively small incremental activation for the other positions reflects the generally poor transmission of conjugative effects between one ring and the other in naphthalene-like molecules [68JCS-(B)1112].

3-Methylisoquinoline N-oxide showed exchange at the 8,5- and then at the 6,7-positions, while the less reactive 6-chloro compound gave only 5,8-exchange (67CPB826).

C. Chromone and Thiachromone

Both these molecules exchange as the free bases at the 3-, 6-, and 8-positions [71JCS(B)11]. These are the sites conjugated with the cyclic oxygen and sulfur atoms, and the 3-sites are much more reactive than the 6- and 8-sites because the benzenoid aromaticity is undisturbed for exchange at the 3-position. 4-Hydroxycoumarin undergoes exchange at the 3-position with D_2O in acetone (68CJC1949). Standard log exchange rates for the above compounds are given in Table 11.1 [73JCS(P2)1065].

3. Base-Catalyzed Hydrogen Exchange

In the rate–acidity profile studies described in Section 2, some compounds were found to undergo exchange on the protonated substrate, via the base-catalyzed mechanism (ylide intermediate). This mechanism has also been found to apply to 2-exchange in quinoline 1-oxide (67CPB826) and for exchange in 1,10-phenanthroline with D_2O at 250°C. The percentage of deuterium incorporation after 24 hr is given in formula **11.58**, which shows the expected pattern of reactivity at positions relative to a protonated nitrogen of $\alpha > \gamma > \beta$.

TABLE 11.1
STANDARDIZED REACTIVITY DATA FOR ACID-CATALYZED HYDROGEN EXCHANGE

Substituent	T (°C)	Position	Species charge	$-H_o$ range	$\dfrac{d[\log k_2(\text{stoich})]}{d(-H_o)}$	$-\log k_2^a$
Quinolines and Isoquinolines						
Quinoline	245	5	+	−0.0–2.3	0.45	11.68
Quinoline	245	6	+	−0.3–2.3	0.52	12.09
Quinoline	245	7	+	1.0–3.3	0.83	13.72
Quinoline	180	8	+	3.4–5.9	0.83	11.86
Quinoline	245	8	+	−0.3–2.3	0.78	11.19
Quinoline	245	3	+	2.3–3.3	0.68	13.06
4-OMe-Q	90	3	+	5.4–9.0	0.72	9.52
6-OH-Q	50	5	+	1.5–3.9	0.81	4.97
6-OH-Q	50	5	0	−(2.7–0.6)	−0.06	−0.60
6-OH-Q	50	5	−	−(12.7–10.8)	1.34	−13.86
6-OMe-Q	50	5	+	1.7–3.9	1.14	6.08
6-OMe-Q	180	5	+	−(0.8–0.3)	0.76	6.79
6-OH-Q	180	5	0	−(4.5–1.4)	0.00	3.50
6-OH-1-Me-Q	50	5	+	1.1–2.5	0.65	4.76
6-OH-1-Me-Q	50	5	0	−9.2–1.1	−0.06	−1.67
6-NH$_2$-Q	35	5	+(min)	−1.7–1.1	−0.13	−0.23
6-NH$_2$-Q	35	5	+(maj)	−(3.2–1.7)	0.53	−0.22
6-NH$_2$-Q	50	5	+	−(4.0–1.9)	0.56	−0.03
6-NH$_2$-Q	50	5	0	−7.0–4.5	−0.33	−1.29

Compound					
7-NH$_2$-Q	35	8	+(min)	0.0–1.2	0.00	−0.41
7-NH$_2$-Q	35	8	+(maj)	−1.2–0.0	0.74	−0.26
7-NH$_2$-Q	50	8	+	−(3.4–1.9)	0.88	−0.59
7-NH$_2$-Q	50	8	0	−(7.5–4.5)	0.00	−3.26
2-Quinolone	110	3	+	4.9–6.6	0.51	8.90
2-Quinolone	110	5	+	4.9–6.6	0.76	10.41
2-Quinolone	110	6	+	4.9–6.6	0.75	9.49
2-Quinolone	110	8	+	4.9–6.6	0.60	7.86
4-Quinolone	90	3	+	5.9–9.0	0.54	7.61
4-Quinolone	90	3	0	−0.2–4.4	−0.13	2.09
4-Quinolone	124	3	0	0.6–4.3	0.00	2.75
Isoquinoline	180	5	+	4.1–5.7	0.93	11.79
Isoquinoline	245	5	+	0.2–1.7	1.00	11.55
Isoquinoline	180	8	+	4.1–6.3	0.90	13.02

Quinoline and Isoquinoline N-Oxides

Compound						
iQ 2-oxide	180	5	+	3.5–5.7	0.64	10.25
iQ 2-oxide	180	8	+	4.7–5.7	0.69	12.32
Q 1-Oxide	180	5,6	+	4.7–5.5	0.58	11.53
Q 1-Oxide	180	8	+	3.5–5.5	0.56	10.27

Chromone and Thiachromone

Compound						
Chromone	180	3	0	1.9–5.4	0.00	6.71
Chromone	180	6	0	1.9–5.4	0.00	7.94
Chromone	180	8	0	2.8–5.4	0.00	8.50
Thiachromone	180	3	0	1.8–5.4	0.00	6.95
Thiachromone	180	6,8	0	1.8–5.4	0.00	8.89

"Average values.

Rate coefficients k (sec^{-1}) for exchange of quinoline and 1,5-naphthyridine as neutral species with NaOEt/EtOD at 191°C are shown in **11.59** and **11.60** (73JA3928). Quinoline is more reactive than pyridine by factors

(11.58) — 99, 60, 74, 5

(11.59) — 94×10^{-5}, 30×10^{-5}, 45×10^{-6}, 45×10^{-6}

(11.60) — 930×10^{-5}, 220×10^{-5}, 260×10^{-6}

of 1.45, 1.15, and 2.29 at the 2-, 3-, and 4-positions, respectively. This modest rate enhancement is consistent with the $-I$ effect of the benzo substituent, and the established fact that the $-I$ effect of benzo annelation is greater at the 1- than at the 2-position. Thus, for example, the 2:1-rate ratio for base-catalyzed exchange in naphthalene is ~2.0 [72MI(267)]. Comparison of **11.59** with **11.60** showed the effect of the second nitrogen to be 9.8 (4-position), 7.0 (3-position), and 5.8 (2-position), consistent also with the inductive effect. Because the 3- and 8-positions are formally meta to nitrogen, the seven-fold difference in reactivity between them was considered to be too small for the reactivity of the 8-position to have been significantly affected by unfavorable interactions between the σ lone pair on the (neutral) nitrogen and the carbanion in the transition state (73JA3928). However the 2-:1-rate ratio for naphthalene should also be taken into account, and it seems probable that there is in fact a small but significant unfavorable interaction between the two lone pairs. Some log rate coefficients for exchange under the conditions stated are gathered in Table 11.2 (74CHE1397). Comparison with Tables 9.2 and 9.3 (74CHE1397) indicate that benzo annelation produces a similar activation at the 4-position of pyridine to that in the reaction with NaOEt given above, but the 2-position is activated about twice as much. Annelation produces a much bigger increase in reactivity in the N-oxides.

Isoquinoline is more reactive at the 1-position than quinoline is at the 2-position by approximately the same factor by which the 1-position of naphthalene is more reactive than the 2-position; this is true also for the N-oxides. Comparison of the rate data for quinoline and quinoxaline (**11.15**), and the data in **11.59**, suggests that replacement of =CH— by =N— accelerates the exchange at the 2- and 3-positions by factors of 135 and 20, respectively. The low reactivity of quinoxaline N-oxide seems anomalous and may be in error.

TABLE 11.2
RATE DATA FOR BASE-CATALYZED HYDROGEN EXCHANGE BY MeOK IN MeOD[a]

Compound	Position	MeOK (M)	T (°C)	−log k
Quinoline	2	0.6	140	6.3
	4	0.6	140	5.8
Isoquinoline	1	0.6	140	6.0
	4	0.6	140	5.2
Quinoxaline	2,3	0.6	140	5.0
Quinoline N-oxide	2	0.1	50	5.5
Isoquinoline N-oxide	1	0.1	50	4.2
Quinoxaline N-oxide	3	0.1	50	6.1
Quinoxaline N,N'-dioxide	2,3	0.1	50	3.5

[a](74CHE1397).

Finally, the log rate of deuteriation at 50°C of the quinazoline derivative **11.61** has been determined as −5.3. For the derivatives with —CH$_2$—CH$_2$— or —CH=CH— *peri* bridges, the corresponding values were −5.4 and −5.2, respectively (77CHE1235).

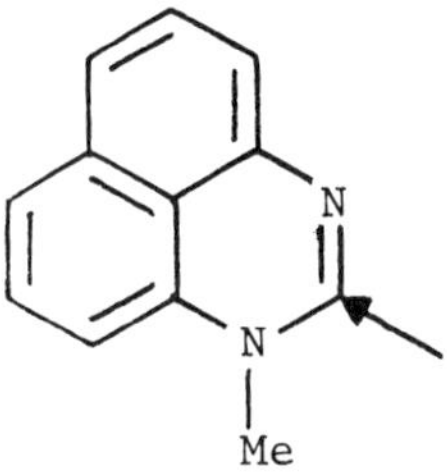

(11.61)

4. Nitration

A. COMPOUNDS CONTAINING ONE NITROGEN ATOM

Nitration of quinoline with nitric acid or metal nitrates in acetic anhydride gives the 3-nitro derivative, together with some of the 6- and 8-isomers (40JA1640; 57JCS2521); the 3-derivative (~30%) is also obtained by reaction with tetranitratotitanium(IV) [74JCS(P1)1751]. These reactions almost certainly involve the free base, for which the expected positional reactivity order is 5 > 8 = 6 > 3; all of these positions are of closely similar reactivity (Section 7). Steric hindrance may reduce the amount of *peri*-(5,8)-substitution. The predominance of 3-substitution may, however, arise under some (but not all) conditions from the formation of a 1,2-dihydroquinoline intermediate (**11.62**) (57JCS2521).

Nitration of quinoline with nitric acid/sulfuric acid gives mainly the 5- and 8-derivatives (40JA1640; 57JCS2521) and studies of rate–acidity profiles show that the quinolinium ion is the reacting species [62CI(L)1057]. This was confirmed by the log A value of 9.0 sec^{-1}, which is typical of reactions between two positively charged species (63JCS4204), and by the parallel rate–acidity profiles for nitration of all the four species, quinoline, isoquinoline, and their N-methyl perchlorates [63CI(L)1283; 64T89]. The overall relative rates for these four species were 1.0 : 13.9 : 0.27 : 15.4 showing the typically higher reactivity of isoquinoline compared to quinoline [see Section 1.B, paragraph (3)], the low reactivity of the 1-methylquinolinium cation here may be a solvation rather than an electronic effect [63CI(L)1283].

The quinolinium ion is at least 10^5 times more reactive in nitration than the pyridinium ion (63JCS4204). The ratio of the reactivity of quinolinium to benzene and to naphthalene has been derived as 1.2×10^{-7} and 1×10^{-10}, respectively [63CI(L)1283; 68ZC201]. A detailed study of the products of nitration of quinoline by nitric acid in 80 wt% sulfuric acid at 25°C yielded the partial rate factors shown in **11.63** [71JCS(B)1254] so that the positional reactivity order for the quinolinium ion is $5 > 8 > 6 > 7 > 3$, which is almost exactly that which applies to the free base, although the relative differences are much greater (Section 7). Likewise, partial rate factors for nitration on the isoquinolinium ion are given in **11.64** (64MI2), and the differences in the i-Q/Q-rate ratios for the 5- and 8-positions (52 and 7, respectively) show the effect of conjugation between nitrogen and the 5-position in quinoline and the 8-position in isoquinoline. Nitration of isoquinoline with nitric acid in acetic anhydride almost certainly involves the free base and consequently gives the 4-derivative in 14% yield (72OPP9); the 4-position is the most reactive site (see Section 7).

(11.62)

(11.63) 1.74 × 10^{-7}, 5.58 × 10^{-9}, 4.57 × 10^{-12}, 4.75 × 10^{-12}, 1.47 × 10^{-7}

(11.64) 90 × 10^{-7}, 10 × 10^{-7}

Nitration of 5-nitroquinoline also goes via the cation, as shown by the rate relative to quinoline of 4.6×10^{-7}, about the same as that (1.8×10^{-7}) which applies between nitrobenzene and benzene [79CI(L)28]. Reaction on the conjugate acid of 5-nitroquinoline would be expected because of the rather poor 1,5-conjugative interaction in benzenoid systems and thus the relatively small effect of the 5-nitro group on the basicity of quinoline; moreover the σ lone-pair density on nitrogen can only be reduced inductively. Nitration of all of the nitroquinolines gave the positional yields shown in Table 11.3 (79CPB2627). The results are as expected based on the electronic effects of the nitro group.

Nitration of the quinoline derivatives 1,2,3,4-tetrahydroacridine (**11.65**) and 2,3-dihydro-1*H*-cyclopenta[b]quinoline (**11.66**) goes into the 5-and 8-positions as it does in 7,8,9,10-tetrahydrophenanthrene (**11.67**) (65-JPJ645).

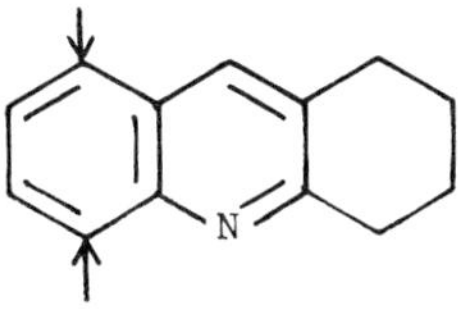

(11.65) 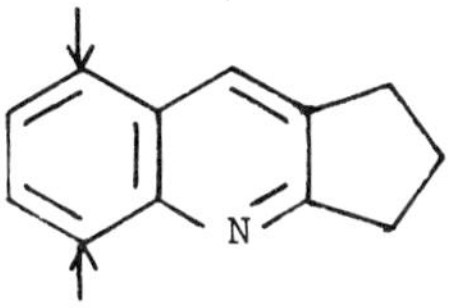   (11.66) 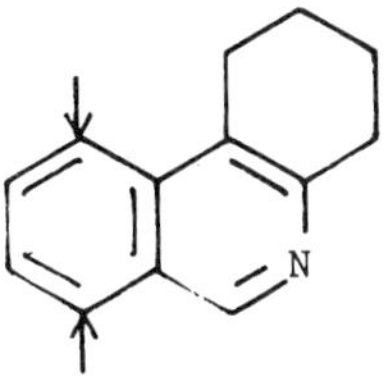   (11.67) 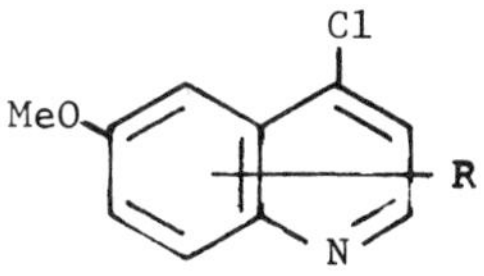   (11.68)

High yields of 5- and 8-nitro derivatives have been reported for nitration of quinoline with strongly activating groups at the 6-position (which direct into the 5-position). Electron-withdrawing groups at the 4-position

TABLE 11.3
PERCENTAGE YIELDS IN NITRATION OF NITROQUINOLINES

Substituent	Yields at position			
	5	6	7	8
2-NO$_2$	40	—	—	33
3-NO$_2$	22	—	—	41
4-NO$_2$	46	—	—	18
5-NO$_2$	—	—	35	—
6-NO$_2$	—	—	—	31
7-NO$_2$	14	10	—	8
8-NO$_2$	—	38	—	—

of quinoline also gives good yields of 5- and 8-nitro derivatives, possibly because reaction takes place on the free base. This seems particularly probable for reaction of 4,7-dichloroquinoline, which gave 91% of the 8-nitro derivative. Percentage yields for nitration of variously substituted 4-chloro-6-methoxyquinolines (**11.68**) for different R groups were H, 83; 2-Me, 58; 7-Cl, 42; 7-F, 63; and 7-CF_3, 40 [69JCS(C)1369].

Nitration of 4-hydroxyquinoline (**11.69**,4-methoxyquinoline (**11.70**), and 1-methyl-4-quinolone (**11.71**) (as the conjugate acids) gave the partial rate factors shown [71JCS(B)1493]. Thus (cf. **11.63**) OH activates the 6- and 8-positions by factors of 3300 and 29.5, respectively; the corresponding values for the OMe substituent are 1600 and 23. This confirms the earlier findings from hydrogen exchange that substituent effects operate more effectively in naphthalenoid systems between the 1,7(2,8)- than between the 1,5-positions [68JCS(B)1112]. The corresponding substituent interaction factors A_f are 3.6 and 7.3 [81JCS(P2)1153].

1.84×10^{-5} OH (**11.69**) 4.34×10^{-6}

8.84×10^{-6} OMe (**11.70**) 3.34×10^{-6}

8.14×10^{-6} OH (**11.71**) 3.85×10^{-7}

It is surprising that no 3-substitution was found in 4-hydroxyquinoline because the high 3,4-bond order would lead one to expect an activation of $\sim 10^5$–10^6 (A_f factor for the 1,2-interaction is 18.2). The failure to detect 8-nitration of **11.71** must reflect considerable steric hindrance from the N-methyl group (cf. hydrogen exchange of chromone, Sect. 2.C). Earlier work has shown 4-quinoline to nitrate in the 3-position with nitric acid (49JCS1367) and in the 6-position with fuming nitric acid (49JCS255). 3-Methyl-4-quinolone, in which steric hindrance to 8-substitution is much less, gave mainly 6- with some 8-substitution (50JCS2092). 6-Methyl- and 6-methoxy-4-quinoline gave the corresponding 5-nitro products in 81 and 78% yields, respectively [69JCS(C)1369]. There is considerable bond fixation in quinolones, which causes strong activation across the 5,6-bond.

Nitration of 3-hydroxyquinoline in nitric acid/sulfuric acid gave the 5- and 7-derivatives in ~ 90 and 10% yields, respectively. The latter is due to the very strong 2,6(3,7)-interaction in naphthalenoid systems (A_f factor 10.9). Nitration of 3-hydroxyquinoline in acetic acid probably involves a neutral species (free base or zwitterion) and gives 70% of the 4-derivative (74CHE699), the 4-position being very strongly activated as a result of the high 3,4-bond order.

Sec. 4.A] NITRATION 373

Nitration of various quinolizium species has been studied. The quinolizium pseudo-base **11.72** nitrated readily but gave complex products (63JCS2203). 1-Hydroxyquinolizium (**11.73**) gave the 2-nitro- and 2,4-dinitro derivatives (64JCS3030); the lack of 4-nitro product demonstrates the field/inductive effect of the pole. Quinoliz-4-one (**11.74**) dinitrates with nitric acid but gives the 1- and 3-nitro derivatives in a ratio of 2 : 5 with cupric nitrate in acetic anhydride (64T1051).

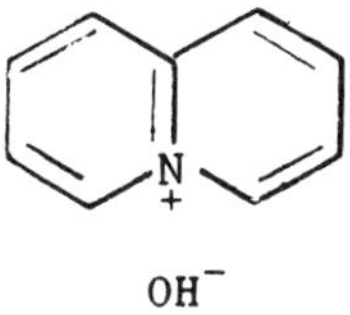

(11.72) 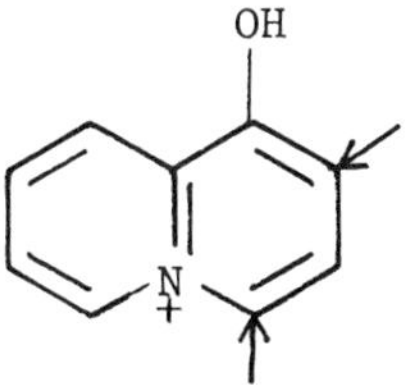   (11.73) 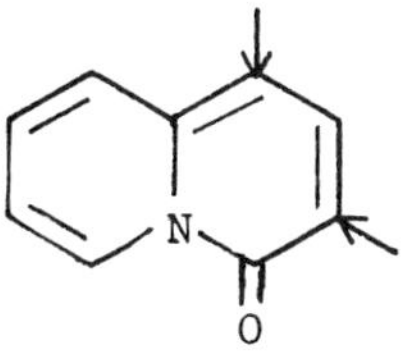   (11.74) 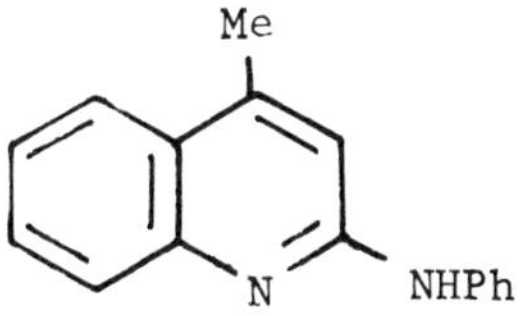   (11.75)

Nitration of 2-anilino-4-methylquinoline (**11.75**) with nitric acid in 50 wt% H_2SO_4 gives equal amounts of the *o*- and *p*-nitrophenyl derivatives, but in 85 wt% acid the *p*-nitrophenyl and 6-*p*-dinitro derivatives are formed (the latter arising from the strong 2,6-conjugative interaction). Acetyl nitrate gave mainly *o*-nitrophenyl and *o*,*p*-dinitrophenyl derivatives (70CPB2094).

Nitration of phenanthridine gives the isomer distribution shown in **11.76** (52JCS2156), which demonstrates two features. First, the reactive sites are those which are not conjugated with nitrogen, as expected. More interesting, however, is the fact that the 3-, 1-, 10-, and 8-positions correspond to the 7-Q, 5-Q, 5-iQ, and 7-iQ positions, respectively, in quinoline and isoquinoline, and give exactly the relative isomer yields for these positions predicted by their σ^+ values (Section 7). Phenanthridin-6-one (**11.77**) gave the expected 2- and 4-nitro derivatives in a ratio of 6 : 1 (70JHC313,597). The 1,10-dicarbonyl compound **11.78** gave the 4-nitro derivative in 96% yield, with dinitration occurring in 45% yield at the positions shown [81CA(95)80688].

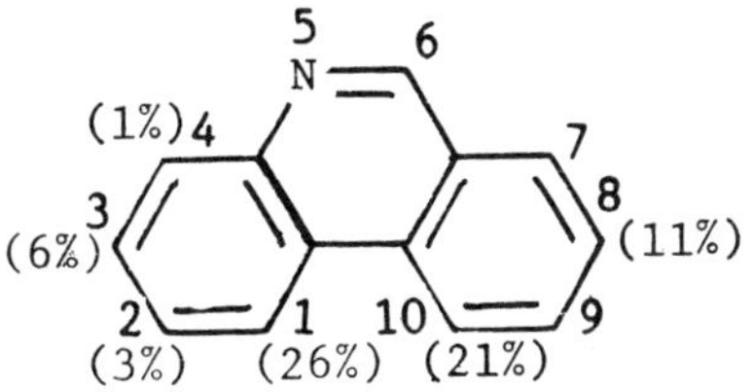

(11.76) 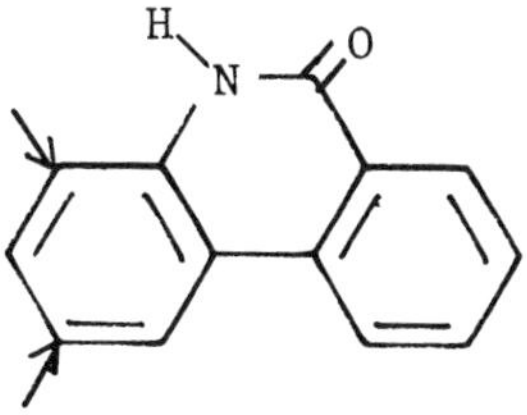   (11.77)

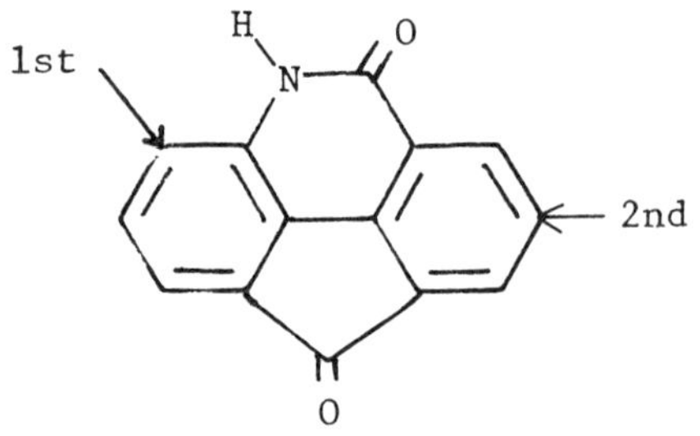

(11.78) 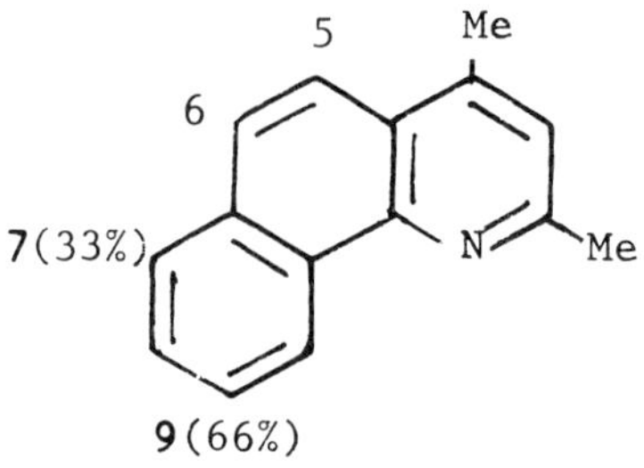   (11.79)

Nitration of 2,4-dimethylbenzo[*h*]quinoline gives the yields at the 7- and 9-positions shown in **11.79** (77MI1), neither position being conjugated with nitrogen. The reason for the higher yield of the 9-isomer is not obvious, especially since in phenanthrene the positional order in nitration (9 > 1 > 3 > 2 > 4) predicts the 7-position of **11.79** to be most reactive.

Nitration with mixed acid of benzo[*g*]isoquinoline and various methyl derivatives gave 5- and 9-substitution, with overall yields as shown in **11.80**. Just as anthracene is very reactive (positional order 9 >> 1 > 2), here too the reactivity was such that, even at 0°C, dinitration could not be prevented (81CHE1217). 9-Substitution is somewhat surprising, since the 9-position is conjugated with nitrogen, but presumably the expected 5,6-disubstitution is prevented by steric hindrance.

R_1	R_2	R_3	Yield
H	H	H	70
H	H	Me	36
H	Me	H	72
H	Me	Me	48
Me	Me	H	5
Me	Me	Mw	56

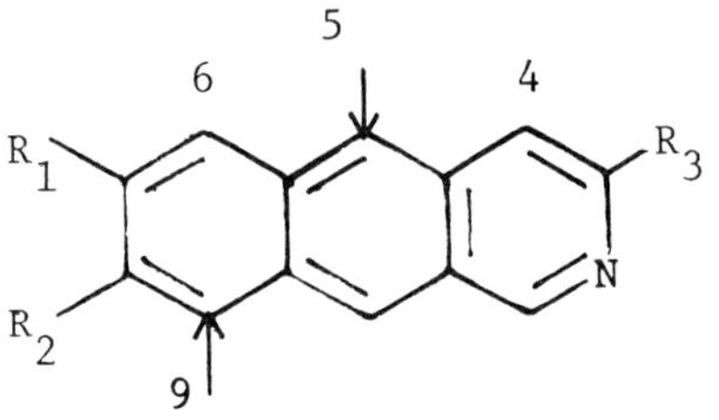

(11.80)

The high reactivity of anthracene-like systems is shown by the 190-fold relative reactivity of acridine to quinoline (57JCS2521); the isomer distribution is given in **11.81** (38CB808); the reason for the relative lack of 1-substitution (cf. 5-substitution in quinoline) is given above [Section 1.B, paragraph (5)]. 9-Aminoacridine also nitrates rapidly at 0°C to give 47% of the 2,4-dinitro derivative (49JCS1008). Likewise, acridizinium (**11.11**) nitrated at the 10-position in 65.5% yield at −5°C, and this position corresponds to the 5-position in isoquinolinium (74JOC1157).

(11.81) structures with positions: 1(5%), 2(13%), 3(1%), 4(25%)

(11.82) 54×10^{-7} and 6×10^{-7} partial rate factors for 1-hydroxyisoquinolinium

(11.83) adduct structure with H, NO$_2$, and OBz

Nitration of quinoline and isoquinoline N-oxides has been studied. The latter compound gave an acidity dependence (in 76–83 wt% sulfuric acid) indicative of reaction on the protonated species, 1-hydroxyisoquinolinium. Partial rate factors (**11.82**) were derived [63CPB1326; 64CI(L)1577; 66JCS(B)870]. Comparison of rate factors for isoquinoline N-oxide with those in **11.64** indicates that the protonated N-oxide is less reactive than the protonated quinoline (i.e., the inductive effect of the N-hydroxyl group outweighs its conjugative effect, which is in accordance with other evidence) [see, e.g., 76AHC(20)1].

2-Methoxyisoquinolinium was nearly four times less reactive than 2-hydroxyisoquinolinium (**11.82**), suggesting that solvation factors are particularly important here. For quinoline N-oxide, the situation is more complicated. With nitric acid/sulfuric acid at 0°C, the 5- and 8-isomers are the main products, with only a trace of the 4-isomer being obtained, whereas at higher temperature (60–100°C), 4-substitution is dominant [50JPJ22; 68JCS(B)316]. Rate–acidity profiles showed that 5- and 8-substitution involved the protonated species and 4-substitution involved the free base. The suggestion [see 68JCS(B)316] that free-base substitution would have the higher activation energy is erroneous. At higher temperature, acidity function changes cause the solution to be less acidic [see, e.g., 69JA6654] and it is this, rather than the temperature per se, which brings about increased reaction via the free base.

With acyl nitrates (e.g., benzoyl nitrate), quinoline N-oxide gives 40% 3-substitution (together with some 6-substitution). A widely quoted [e.g., 82HC(382)447] mechanism for this which is *bimolecular* in nitrating species (59CPB267) has been criticized [75JCS(P2)277]: Acyl nitrates, in benzenoid chemistry, are known to form adducts in which the nitronium ion attaches itself initially to the ring site of highest electron density (75MI3), shown by the free base nitration to be the 4-position in quinoline N-oxide. These adducts then undergo 1,2-migration of the nitro group. For quinoline N-oxide, an adduct such as **11.83** could form (aided by retention of aromaticity), followed by 1,2-nitro group migration and benzoic acid elimination to give the 3-nitro derivative [75JCS(P2)277]. An alternative explanation, favored by one of us, is addition of the benzoyl nitrate as $PhCO_2^-$ and NO_2^+ at the 4-position of the ring and at the N-oxide oxygen, followed by electrophilic nitration at the 3-position and the loss of benzoic acid.

The nitrations of a wide range of substituted quinoline N-oxides under various conditions [82HC(382)447] show the usual pattern of substituent effects superimposed upon the pattern resulting from the use of nitric acid/sulfuric acid at low temperature (5,8-positions), weaker mixtures of these acids at high temperature (4-position), or acyl nitrates (3-position). Nitration of acridine N-oxide by nitric acid/sulfuric acid occurs in the 5-position (60JCS3367).

B. Compounds Containing More Than One Nitrogen Atom

Only ring carbonyl derivatives of naphthyridines have been nitrated. The products **11.84–11.87** were obtained in high yield (75%) from the corresponding precursors (56JCS212; 60JCS1794; 72G253), with substitution occurring at the expected sites.

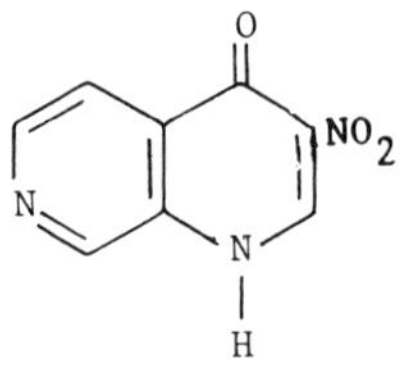

(11.84) 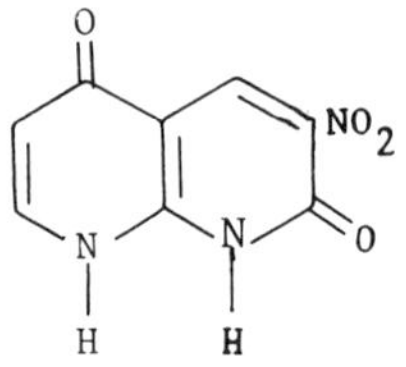   (11.85) 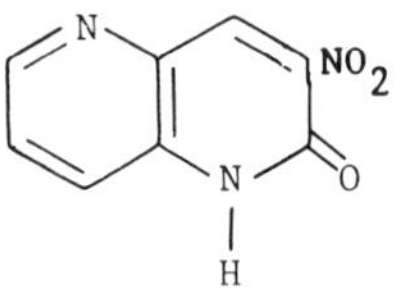   (11.86) 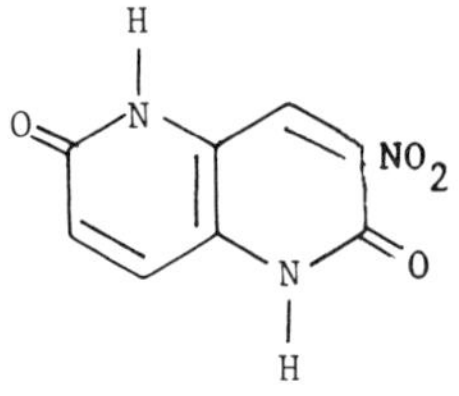   (11.87)

Nitration of phenanthrolines by nitric acid in oleum gives the yields shown in **11.88–11.91** (73RC2255). It is probable that the diprotonated species are involved, except for the 1,10-isomer (**11.90**), in which juxtaposition of positive charges is unfavorable and this may account for the higher yield here. The low yield from the 1,8-isomer (**11.89**) follows, because each of the 5- and 6-positions is conjugated with one of the nitrogens. Likewise, for the 1,7-isomer (**11.88**), the 5-position is conjugatively deactivated by both nitrogens. These conjugative deactivations are greater here than in, for example, quinoline, for reasons given in Section 1.B, paragraph (5). Hence, substitution occurs in the 6-position.

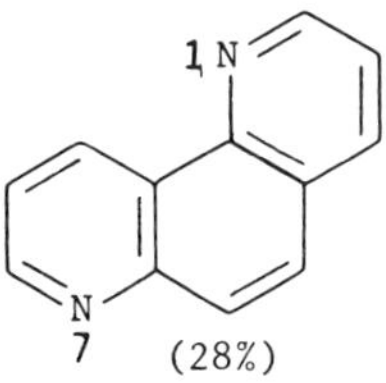

(11.88) 7-N, (28%)
(11.89) 8-N, (6.5%), (6.5%)
(11.90) 1,10-, 3-position (75%)
(11.91) 4,7-, 2-position (17%)

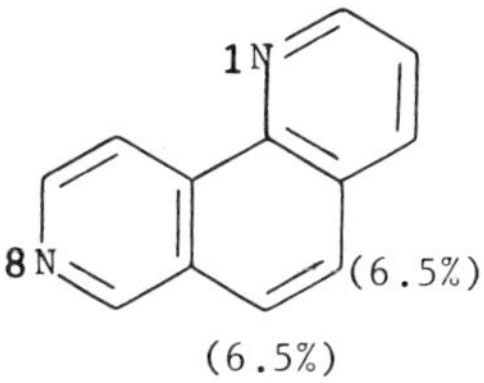

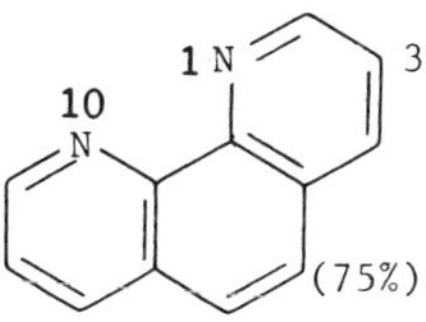

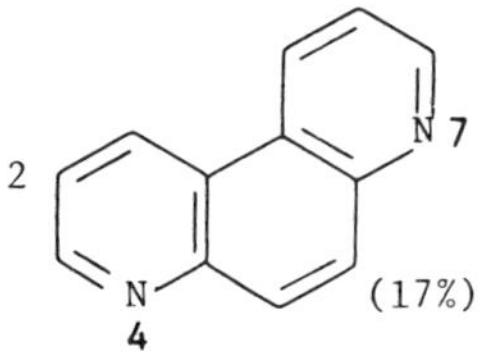

With acetyl nitrate, the 1,10- (**11.90**) and 4,7-phenanthrolines (**11.91**) gave the 3- and 2-nitro products, respectively; this was attributed to an addition–elimination mechanism (73RC2255), though it is possible here that direct nitration occurs on the free base. Nitration of 3,4-dihydro-4-methyl-3-oxo-4,7-phenanthroline (**11.92**) occurs in the 2-position (58-JCS825). This is the site that can conjugate with the nitrogen lone pair and substitution there involves the least interruption of benzenoid character in the transition state.

Quinazoline (**11.16**) nitrates in the 6-position, and because theoretical calculations predict 8- > 6-substitution (the 5- and 7-positions are each conjugatively deactivated by both nitrogens), it has been assumed that reaction must involve the hydrated quinazolinium cation (**11.93**) (47JOC405; 49JCS1367). However, these calculations may not take sufficient account of inductive deactivation of the 8-position; the deficiency of the calculations in this respect will be particularly marked if no auxiliary inductive parameter is used for the bridgehead carbon between the 1- and 8-positions.

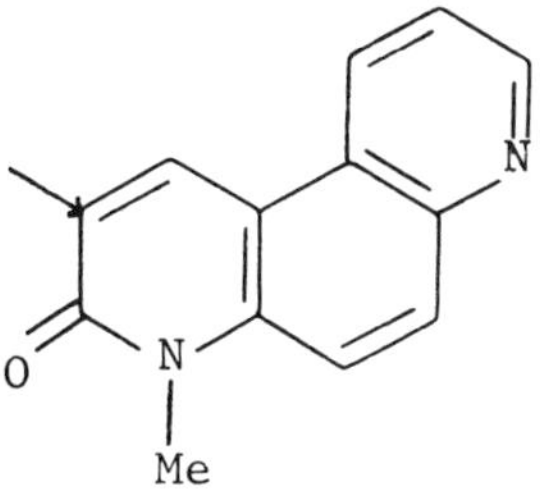

(11.92)

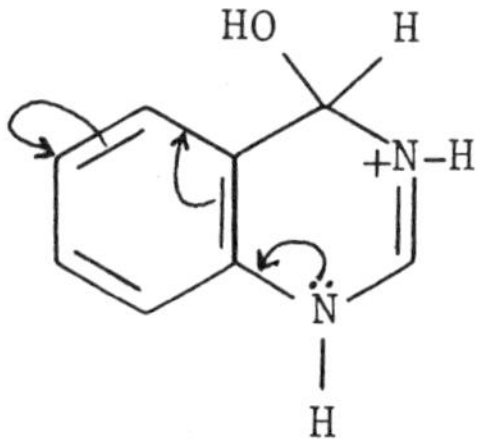

(11.93)

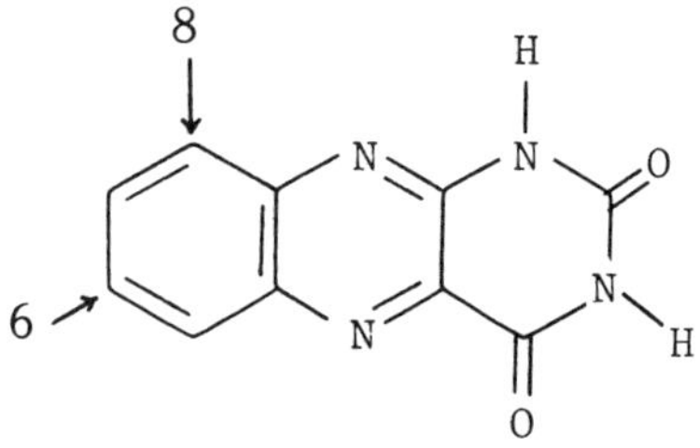

(11.94)

Quinoxaline (**11.15**) is nitrated by nitric acid in oleum to give the 5-nitro (1.5%) and 5,6-dinitro (24%) products (57JCS2518; 59JA6297). Mesomeric donor ($+M$) activating groups in the 6-position direct nitration into the 5-position (58CPB566; 59JA6297). In the 5-position, $+M$ groups direct into the 6- and 8-positions as expected, producing only the dinitro products (55NKZ311). Because the high 5,6-bond order results in the reactivity of the 6-position being raised to that of the 8-position, no mononitration occurs. With fuming nitric acid and sulfuric acid at 20°C, alloxazine (**11.94**) gives the 6- and 8-nitro derivatives in a ratio of 2:3. More severe conditions led to dinitration (69JGU1599). 2,3-Dihydroquinoxaline and its 1,4-dimethyl derivative are nitrated in the 6-position and then dinitrated in high yield in the 6,7-positions (62JCS1170). Quinoxalin-2-one is nitrated at the 7-position in acetic acid, and in the 6-position (58CPB566) in sulfuric acid (57JCS2518), suggesting that the free base and cation, respectively, are involved under these conditions.

Rate–acidity profile studies showed cinnoline (**11.17**) to undergo nitration as the conjugate acid (protonated at N-2). Cinnoline and its 2-methyl cation are nitrated at similar rates, and gave roughly equal amounts of the 5- and 8-nitro isomers [68JCS(B)312]. Comparison with the 2-isoquinolinium cation indicates that unprotonated nitrogen of cinnolinium deactivates by $\sim 2 \times 10^2$-fold, affecting the 5-position more than the 8-position, as expected. Values of $10^4 \, k_2$ at 80°C in 81.2 wt% sulfuric acid were cinnolinium, 3.21; 2-methylcinnolinium, 7.35; quinolinium, 141; 1-methylquinolinium, 45.7; isoquinolinium, 933; and 2-methylisoquinolinium, 1514. N-Methylation of quinoline produces a 3.1-fold rate decrease, whereas N-methylation of isoquinoline produces a 1.7-fold rate increase. It is notable that the effect in cinnoline (2.2-fold increase) is almost exactly the product of these individual effects. For 4-hydroxycinnolinium, the positional partial rate factors are as shown in **11.95** [71JCS(B)1493]; conjugative activation of the 3-, 6-, and 8-positions is evident.

Benzo[*c*]cinnoline (**11.20**) is nitrated at the 1- and 4-positions in a ratio of 4:1 (56 and 15% yields) as predicted by calculations (48MI1; 49JCS971; 54JA5807; 57JCS2521; 62JCS2454, 62JCS4384, 62JCS4860). The rate difference arises because **11.96** is benzenoid and **11.97** is not. The 1,10-, 2,9-, and 3,8-dimethyl derivatives are nitrated at the 4-, 1-, and 4-positions, respectively [53LA(581)117; 62JCS4860], which follows from the expected combination of substituent and bond-order effects. 1-Nitrobenzo[*c*]cinnoline is nitrated at the 10-position, but the 4-nitro isomer will not undergo dinitration (62JCS4384).

(11.95)   (11.96)   (11.97)

TABLE 11.4
VARIATION IN ISOMER DISTRIBUTION WITH ACIDITY IN NITRATION OF CINNOLINE[a]

Wt%H$_2$SO$_4$	Nitro compound (%)			Total yield (%)
	5-	6-	8-	
64.4	8.3	48.1	43.6	70.5
70.05	10.7	45.5	43.9	79.5
81.4	17.4	20.2	62.4	85.9
90.0	22.4	4.7	72.9	79.7

[a] At 80°C.

Nitration of some of the corresponding N-oxides have been studied. With nitric acid/sulfuric acid, cinnoline 1-oxide gave 90% 4-substitution and 2% 5-substitution (63CPB268; 64CPB1090); the 4- and 5-positions are both conjugated with the oxygen. Cinnoline 2-oxide gave by contrast an unusual mixture of 5-, 6-, and 8-nitro derivatives (64CPB1090; 66CPB816); a detailed kinetic study showed that both free base and cation are being nitrated under the same conditions [68JCS(B)316]. The variation in isomer distribution with acidity (Table 11.4) indicated that nitration as the free base gave the 6-isomer whereas nitration as the cation gave the 5- and 8-isomers.

Although nitration of quinoxaline N-oxide fails, that of phenazine N-oxide is very easy, which again demonstrates the high reactivity of anthracene-like systems. At 0°C, nitric acid/sulfuric acid gives the 1- and 3-isomers; under more drastic conditions the 1,7- and 3,7-dinitro isomers are produced (Scheme 11.2) (54CPB283; 58CPB77); these nitrations are believed to involve the cation formed by protonation at the 10-nitrogen [67AG(E)608].

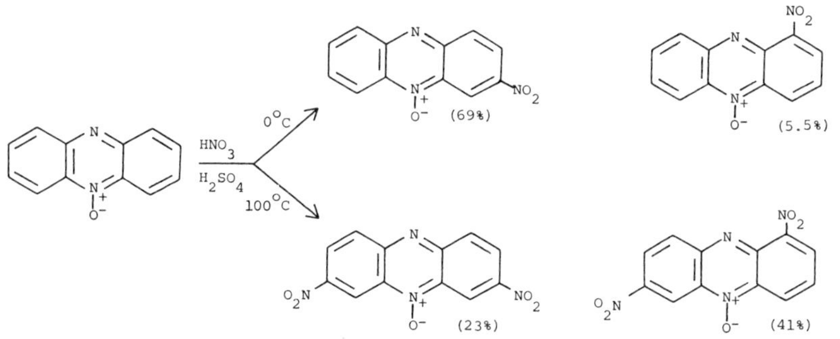

SCHEME 11.2. Isomer distribution in nitration of phenazine N-oxide.

Orientation in the nitration of the 5-oxide of benzo[c]cinnoline (**11.20**) depends upon conditions. With nitric acid alone it gives 69% of the 9-nitro 5-oxide and 13.5% of the 2-nitro 5,6-dioxide (62JCS2454). But with nitric acid/sulfuric acid, the products appear to be the 1- and 7-nitro 5-oxides in 43 and 26% yields, respectively. One group gives the yields as 43 and 26%, respectively (62JCS2454), whereas another finds the latter to be the major product (62JCS4384). 2,9-Dimethylbenzo[c]cinnoline 5-oxide and the 3,8-dimethyl isomer are nitrated at the 1- and 4-positions, respectively, as expected (62JCS4384).

C. XANTHYLIUM SALTS

Nitration of 9-phenylxanthylium perchlorate (**11.98**, X = O) and the sulfur and selenium analogues (**11.98**, X = S, Se) by nitric acid/sulfuric acid has shown a rather unusual isomer distribution (73CPB1272; 74CPB21, 74CPB27). For X = O, mononitration in the phenyl ring was observed, the ratio of meta to para products being 4.5 : 1. However, for X = S or Se, the ratios become only 1.25 and 1.5, respectively. Dinitration of the selenium compound occurred in the presence of excess nitric acid, the second nitro group entering the 4-position. The relative reactivity order for different heteroatoms X was deduced as S > Se > O.

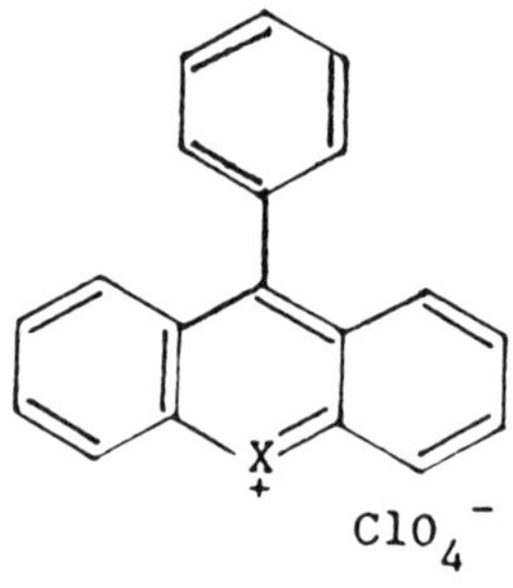

(**11.98**, X = O, S, Se)

D. BORAZA COMPOUNDS

Nitration of 3,4-dimethyl-4,3-borazoisoquinoline (**11.99**) by N-nitropicolinium tetrafluoroborate in acetonitrile goes 82% into the 1-position (75ACS457). This orientation arises because the other main canonical form for **11.99** has a lone pair on nitrogen (N-3) and this is readily delocalized to the 1-position. The corresponding lone pair on nitrogen in the other canonical form of 10-methyl-10,9-borazophenanthrene (**11.100**) is likewise readily delocalized to the 6- and 8-positions, which causes very large partial rate factors of 937,000 and 2,060,000, respectively for its nitration by nitric acid/acetic anhydride (68JA1924).

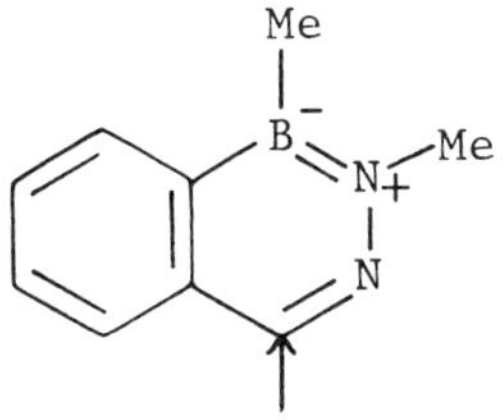

(11.99) 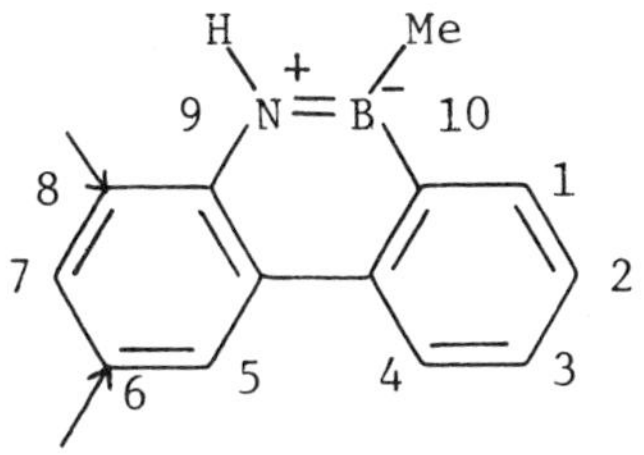   (11.100)

E. Summary of Kinetic Data

Standard log nitration rates, determined by the method given in Chapter 2, are shown in Table 11.5 [75JCS(P2)1600].

TABLE 11.5
Standard Log Nitration Rate Coefficients for Cations

Compound	T (°C)	Positions	$-H_o$ range	$d[\log k_2(\text{stoich})]/d(-H_o)$	$-\log k_2^o$
Quinoline	25	5	7.1–8.7	2.19	6.28
		8			6.36
4-Quinolone	25	6	7.6–8.4	2.45	4.69
		8			5.32
1-Me-4-quinolone	25	3	7.1–8.4	2.28	5.67
		6			4.88
4-OMe-quinoline	25	6	7.7–8.3	2.50	5.10
		8			5.51
Isoquinoline	80	5	4.7–5.4	1.88	5.46
		8			5.46
2-Me-isoquinolinium	80	5	4.3–5.5	1.94	5.23
		8			5.23
2-OH-isoquinolinium	25	5	6.9–8.0	2.19	5.53
		8			6.58
2-OMe-isoquinolinium	25	5	6.9–8.0	2.22	6.08
		8			7.13
Cinnoline	80	5	5.9–6.8	2.17	7.66
		8			7.63
2-Me-cinnolinium	80	5	6.0–6.5	2.26	7.28
		8			7.28
4-Cinnolone	25	6	7.7–8.4	2.07	6.91
		8			7.07
Cinnoline 2-oxide	80	5	5.1–6.6	1.35	7.55
		8			6.98

5. Halogenation

A. COMPOUNDS CONTAINING ONE NITROGEN ATOM

Consideration of the halogenation of quinoline and isoquinoline is complicated by uncertainties relating to the mechanisms of reaction of the free base; this applies particularly to quinoline.

Under conditions in which nitrogen acquires a positive charge, quinoline gives mainly 5- and 8-substitution and isoquinoline gives 5-substitution. For example, in bromination of quinoline in $H_2SO_4/AgSO_4$, the yields were 28% of 5-bromo- and 29% of 8-bromoquinoline, together with 43% of 5,8-dibromoquinoline (60JCS561); under the same conditions isoquinoline gives mainly the 5-bromo derivative [59MI1(199)]. Chlorination and iodination (in $H_2SO_4/AgSO_4$) of quinoline also gives 5- and 8-substitution [63CI(L)1840; 64CI(L)1753]. The quinoline–aluminum chloride complex (e.g., **11.101**) also undergoes bromination to give the 5-bromo derivative in 46% yield, together with the 5,6- (3%) and 5,8-dibromo derivative (8%). The yield of the dibromo compound is markedly increased in the presence of excess bromine (64JOC329). The corresponding isoquinoline–aluminum chloride complex is brominated in the 5-position (76%); reaction with two equivalents of bromine gives 55% of the 5,8-dibromoisoquinoline, and with three equivalents a mixture of 5,7,8- and 5,6-8-tribromoisoquinoline (overall the results indicate that the 6-position is more reactive than the 7-position in quinoline, the reverse being true in isoquinoline; see also Section 7) (64JOC329).

Under neutral conditions, the positional reactivity order for the halogenation of quinoline appears to be $3 > 6 > 8$, whereas isoquinoline gives mainly 4-substitution. For isoquinoline, the fact that reaction occurs on the free base is adequate explanation for the change in orientation, since, contrary to common belief, the 4-position is shown by calculations and gas-phase studies of reactivity to be the most reactive in the neutral isoquinolines (see Section 11.G.); indeed, more reactive than benzene.

For quinoline, reaction on the neutral species would require that 5- and 8-substitution would be sterically hindered, since σ^+ values (Section 7) predict the order $5 > 8 = 6 > 3$. However, it seems improbable that steric hindrance is that severe and, moreover, the order for the unhindered 3- and 6-positions is the wrong way around. An alternative explanation is that bromination occurs on an intermediate addition product. This takes account of the fact that the 3-, 6-, and 8-positions may all conjugate with a lone pair on nitrogen. The order $3 > 6 > 8$ would then precisely follow from such conjugation, since resonance to the 6- and 8-positions requires interruption of the benzenoid conjugation, and resonance is always relayed better to para (8) than to the ortho (6) positions. Support for such a mechanism was provided in N-cyanoquinolinium ion, which undergoes bromination under reaction conditions in which the pseudo-base (**11.102**) is formed; a tetrahydroquinoline intermediate was formed, which underwent elimination in acidic media to give 3-bromoquinoline in 58% yield (62JCS291).

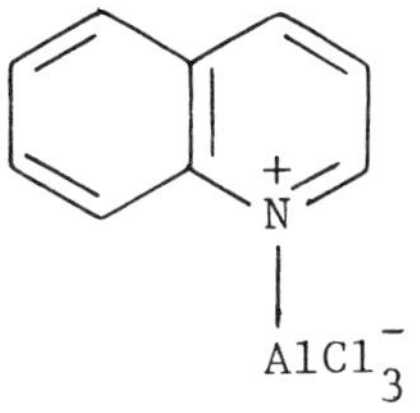

(11.101)

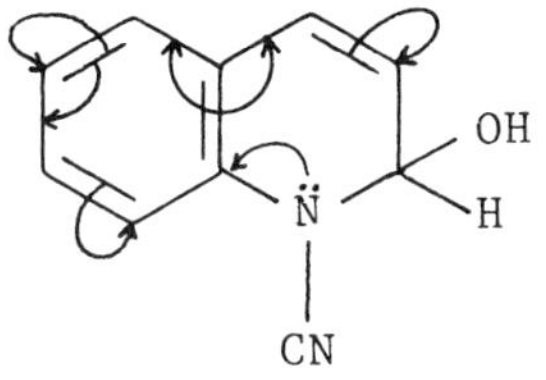

(11.102)

Quinoline and bromine react readily below 100°C to give a $C_9H_7NBr_2$ complex, which in the presence of pyridine as base gives >80% of 3-bromoquinoline together with 2% of 3,6-dibromoquinoline [59CI(L)1449]. Likewise, 6- and 8-bromoquinolines give the 3,6-dibromo derivatives in 62 and 70% yield, respectively (62JOC1318). 6-Nitroquinoline with bromine/pyridine in carbon tetrachloride gives 70% of the 3-bromo derivative [66LA(699)98]. The orientation under these conditions has been explained in terms of attack on quinoline by the quinoline–bromine complex acting as electrophile [59CI(L)1449], but it is difficult to satisfactorily account for the observed orientation in these terms, and in any event the electrophile would more likely be the pyridine–bromine complex.

Bromination with bromine in thionyl chloride or sulfur monochloride and pyridine also gave 3-bromoquinoline (65%) and 4-bromoisoquinoline from the parent heterocycles (60JA4430).

The difficulties in interpreting these results are compounded by the fact that the hydrochlorides with bromine in nitrobenzene give 81% of 3-bromoquinoline and 76% of 4-bromoisoquinoline (73JHC409). The mechanism proposed here involved formation (by a mechanism not entirely clear but presumably involving the equilibrium concentration of free base) of a 1,2-dibromo complex, which then underwent bromination at the 3-position (i.e., a second molecule of the electrophile is needed). Again this mechanism is not necessary to account for 4-bromination of isoquinoline, since the latter would result from direct substitution of the much more reactive free base in equilibrium with the hydrochloride.

Chlorination of quinoline at 160–190°C without solvent gave polysubstitution with proportions of products implying the reactivity order 3, 4 >

6, 8 > 7; substitution probably also occurs at other sites since there were unidentified products (70JOC171). In view of the occurrence of 4-substitution it is not clear if electrophilic or radical substitution is involved.

Chlorination and bromination of 8-methoxyquinoline with N-halosuccinimides occurred, as expected, in the 5-position, but iodination took place in the 7-position (also strongly activated by the methoxy group). 8-Hydroxyquinoline behaved similarly; the preference for 7-iodination may suggest steric hindrance at the *peri*-position. However, under basic conditions, 8-hydroxyquinoline (but not the 8-methoxy analogue) underwent halogenation (as the anion) giving 7-chlorination and 7-bromination, but 5-iodination (72JOC4078). Halogenation of 3-hydroxyquinoline goes 76% into the 4-position, strongly activated because of the high 3,4-bond order; likewise, 4-hydroxyisoquinoline is halogenated in the 3-position (71-BAU395,400).

Quinoline N-oxide with bromine and water gave a small yield (3–4%) of the expected 4-bromo derivative (47JPJ87). However, with bromine and chloroform in acetic anhydride, a 60% yield of the 3,6-dibromo derivative was obtained (61CPB414), a result which rather parallels the nitration by acyl nitrates. It is possible that in bromination, formation of an adduct (corresponding to **11.83**) followed by 1,2-rearrangement is involved, rather than the adducts originally proposed (61CPB414). Bromination of quinoline N-oxide by bromine–thallium triacetate in acetic acid at 50°C gave 4-bromoquinoline N-oxide (60%); the 2-methyl, 3-bromo, and 2-cyano N-oxides gave 63, 95, and 22% yields, respectively, of the corresponding 4-bromo compounds (79H475).

Reports of bromination of phenanthridine (**11.14**) are contradictory, claiming (1) a 40% yield of 2-bromophenanthridine (55JA6379) and (2) a positional reactivity order of 10 > 4 > 2, the isomer yields being 6, 4, and 0.7%, respectively, with ~1% of dibromo products (69AJC1105). These reports both contrast with nitration, in which positional order is 1 > 10 > 8 > 3.

In acetic acid, acridine is brominated to give the 3-bromo and 3,7-dibromo derivatives in ~75% yield (54JCS4142). This finding again contrasts with nitration (**11.81**). Chlorination with $SOCl_2$ goes into the 9-position, presumably because of the lower steric hindrance compared to bromination (83PHA83). Bromination of acridine N-oxide goes, as does nitration, into the 10-position (45%) (60JCS3367).

Quinolizin-4-one is brominated at the 3- and then at the 1-position, thus following the pattern for nitration (cf. **11.74**) (65T945); the 1- and 3-positions are conjugated with the lone pair on nitrogen. 1-Hydroxyquinolozinium (**11.73**) is brominated in the 2-position in 80% yield (63JCS2203), and 3-hydroxyquinolizinium in the 4-position (65JOC526); these orientations arise from strong activation across bonds of high order. Likewise, 2-hydroxyquinolizinium is brominated at the 1-position in 67% yield (64JCS2760). 3-Alkyl-1-aminoquinolizinium salts undergo bromination in the expected 2- and 4-positions [68JCS(C)1088].

B. Compounds Containing More Than One Nitrogen Atom

Bromination of 1,5-naphthyridine in fuming sulfuric acid gives up to 15% of the 3-bromo derivative, depending upon time and temperature, together with rather less of the 3,7-dibromo derivative (63RC1589). Similar results were obtained on bromination of various naphthyridines with bromine in CCl_4 in the presence of pyridine: The results are shown in **11.103–11.106**. The corresponding yields of dibromo products were 10,

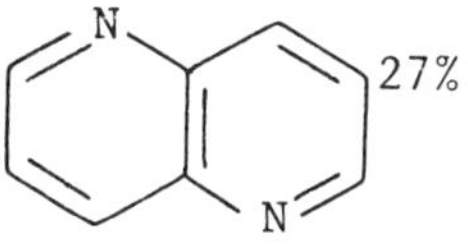

(11.103) 27%   (11.104) 25%, 23%   (11.105) 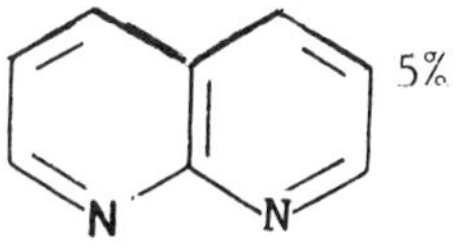   (11.106) 5%

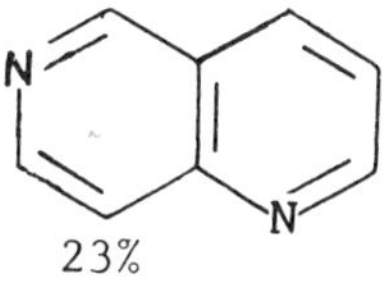

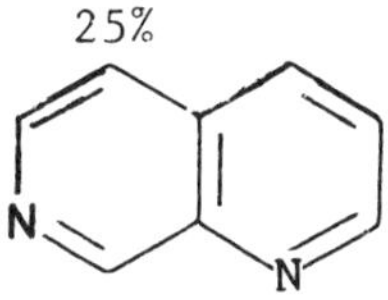

11, 2, and 0.5%. The low yields for 1,8-naphthyridine may stem from its complexing properties (68JOC1384). A modification of this (Eisch) procedure dispenses with pyridine and uses nitrobenzene as a solvent, resulting in higher yields (**11.107** and **11.108**) (76JHC961). The yields are markedly dependent upon whether the hydrobromide or the hydrochloride is used. This was true also for dibromination, which gave yields of 46% (X = Br) or 6.5% (X = Cl) for 1,7-naphthyridine (and 30% for 1,8-naphthyridine). Both 1,5-naphthyridin-2-one and -4-one are brominated at the expected 3-position (56JCS212), as was 1,6-naphthyridin-4-one (65JHC393).

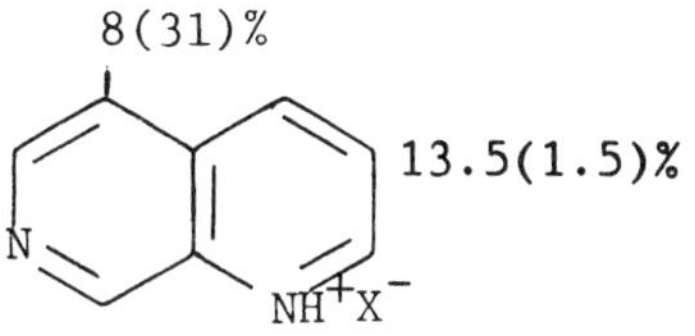

(11.107, X = Br, (Cl)) 8(31)%, 13.5(1.5)%

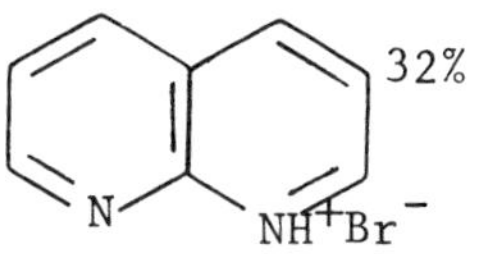

(11.108) 32%

Bromination of 4-(3*H*)-quinazoline (**11.109**, R = H), its 3-methyl derivative (**11.109**, R = Me), and 1,4-dihydro-1,3-dimethyl-4-oxoquinazolinium perchlorate (**11.110**) occurred in each case at the 6-position, no 8-substitution being observed (though the 6-bromo compounds were slowly converted into the 6,8-dibromo derivatives). At pH <2, all three compounds exist predominantly as cations: at pH 0.29 the relative rates for the Me_2, Me, and H derivatives were 4.68 : 1.54 : 1. The mechanisms are thought to involve attack of bromine on covalent hydrates (76JOC838).

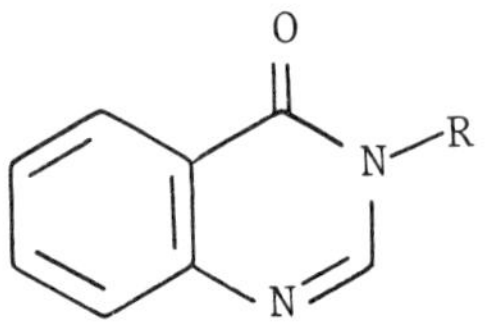

(11.109, R = H, Me) 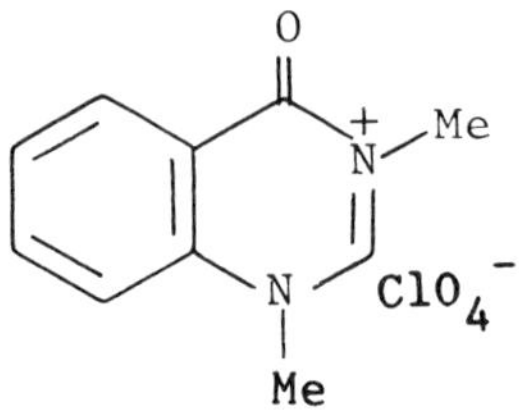   (11.110) 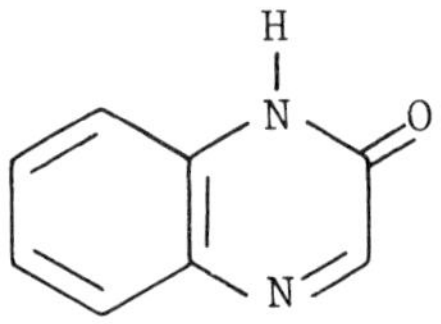   (11.111)

Quinoxaline-2,3-dione and its 1,4-dimethyl analogue are brominated by bromine/sulfuric acid/silver sulfate to give the 6,7-dibromo derivatives, paralleling the result for nitration (62JCS1170). Chlorination of 2-quinoxalinone (**11.111**) in acetic acid goes in high yield (95%) in what was stated to be the 7-position (63MI2). However, this should surely be the 6-position, since (1) the NH group conjugatively activates the 6- and 8-positions; (2) the C=O group conjugatively deactivates the 5- and 7-positions; (3) the =N— atom cannot conjugatively deactivate any of these; this latter factor also accounts for the high yield: (4) the melting point and rate coefficient for methoxydechlorination of the product differed significantly from the values determined for an authentic sample of the 7-chloro compound.

Bromination occurs in the 5- and 6-positions for 1,7- 1,8- 1,10- and 4,7-phenanthrolines (74RC2145). However, with $Br_2/SOCl_2$, 1,10-phenanthroline gave initially the 3- and 5-monobromo derivatives and eventually 3,5,6,8-tetrabromo-1,10-phenanthroline (78JPR172).

Bromination with bromine/sulfuric acid/silver sulfate of benzo[*c*]cinnoline gives the 1- and 4-monobromo derivatives in a ratio of 2.3 : 1 [79JCS(P1)1503], contradicting an earlier report [67JCS(C)1638] in which the 4-isomer was said to dominate. The lower orientation compared with that in nitration has been attributed to steric hindrance to bromination at the 1-position [79AHC(24)151].

Bromination of 1,3,6-triazacycl[3,3,3]azine (**11.112**, R = X = H) occurs at the 4-position and then at the 7- and 9-positions. Calculations predict this (73ACS2421) but the reason can be easily seen by comparison of the transition states for 4- and 7-substitution (**11.113** and **11.114**). Aromaticity is created in both transition states, so reaction should be rapid. However, the triazanaphthalene 10π-ring in **11.113** is more stable than the tetrazanaphthalene 10π-ring in **11.114**. Bromination of the 4-cyano-3-methyl-substituted cyclazine occurs at the 7- and 9-positions (72ACS624), and of 9-ethoxycarbonyl-2-methyl-1,3,4,7-tetrazacycl[3,3,3]azine (**11.115**) at the 6-position (73ACS2421).

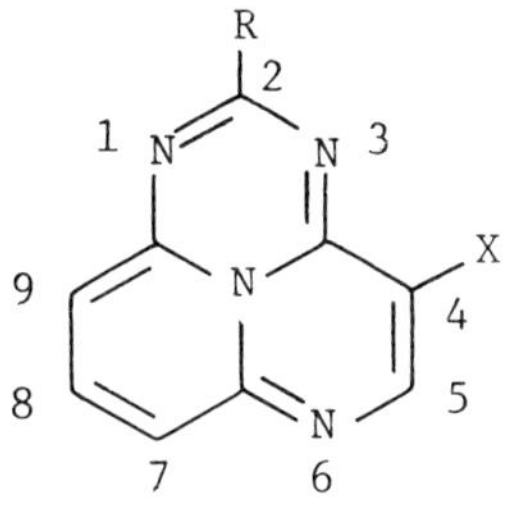

(11.112) 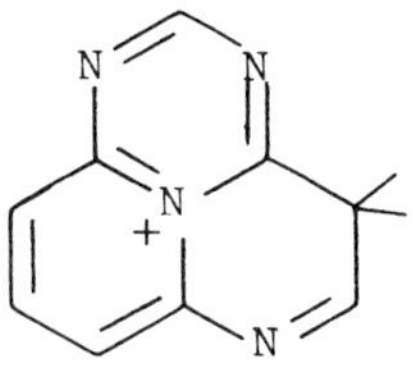   (11.113) 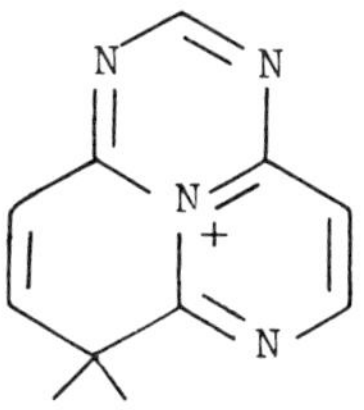   (11.114) 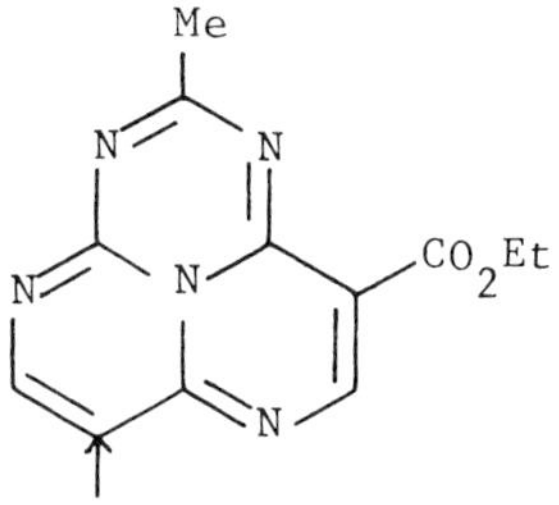   (11.115)

C. Boraza Compounds

Bromination of 4-methyl-4,3-borazaisoquinoline (**11.116**) is now believed (66ACS1448) to occur at the 1-position and not at the 8-position, as previously reported (66JA358). The 1-, 6-, and 8-sites are conjugated with the N-3-nitrogen lone pair, but delocalization into the 6- and 8-positions involves loss of benzenoid structure. The conjugative activation of the 1-position evidently outweighs strong conjugative and inductive deactivation of the 1-position by the 2-nitrogen.

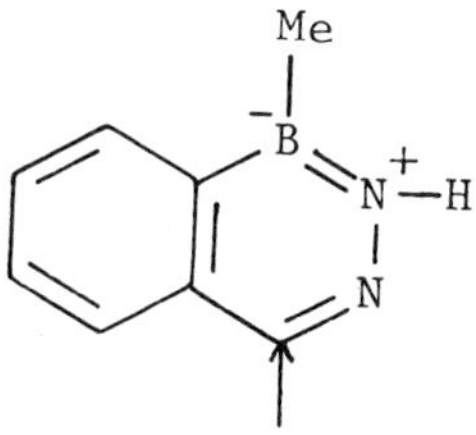

(11.116)

6. Other Electrophilic Substitutions

A. MERCURIATION

Quinoline and isoquinoline are mercuriated in the 3- and 4-positions, respectively (31JPJ542). The latter orientation arises because the 4-position is the most reactive for isoquinoline free base (Section 7). For quinoline, the situation is less clear, but, as in all other reactions of the free base, 3-substitution occurs. While this may involve adduct formation, it is nevertheless surprising that a variety of reagents that might be expected to have different tendencies toward adduct formation all give 3-substitution.

Chloromercuriation of quinoline N-oxide takes place at the 8-position (53YZ823); this also applies to acetoxymercuriation (81% yield); minor amounts of 3-, 5-, 6-, and 7-substitution also occur. Coordination between mercury and oxygen is the most probable cause of 8-substitution (58RTC340; 62RTC124; 69CPB906).

B. SULFONATION

Quinoline is sulfonated by sulfuric acid at the 5-, 6-, 7-, and 8-positions; the cation is obviously involved. The isomer yields are temperature dependent (1888JPR258), as they are for naphthalene. At 100°C, the 8-isomer predominates [1870LA(155)311; 1882CB683], and a 60% yield can be obtained with 20% oleum at 150°C (61USP2950283). The amount of 5-isomer increases with increasing temperature (1882CB1979; 1887CB731) and the yield can be increased by using a mercury catalyst (54USP-2689850).

With 20% oleum at 200°C, 3-hydroxyquinoline and 4-hydroxyisoquinoline undergo sulfonation in the 5- and 8-positions in 61 and 57% yield, respectively (72BAU406). These results contrast with the behavior of these molecules in nitration (Section 4.A), halogenation (Section 5.A), and diazonium coupling (Section 6.C), and demonstrate the usual high steric hindrance to sulfonation. However, under basic conditions, the very powerful activation by O^- is sufficient to produce the normal 4- and 3-substitution, respectively, in yields of 80 and 60% (72BAU404).

Sulfonation of acridizinium ion (**11.11**) occurs at the 10-position in 82% yield (66JOC565). Phenanthridin-6-one sulfonates exclusively at the 2-position at 150°C (57JA5479), steric hindrance presumably preventing 4-substitution.

Quinoxaline-2,3-dione is sulfonated in the 6-position (66BRP1043042), a 6-methyl substituent directs into the 7-position, and a 5-methyl into the 6- and 7-positions (64JAP26975). 1,10-Phenanthroline is sulfonated mainly at the 5- and then at the 3-position (61AC867). Phenazine (**11.19**) sulfonates at the 2-position under severe conditions (50G651). Coumarin (**11.46**) sulfonates in the 3- and 6-positions, and if these are blocked in derivatives, sulfonation takes place at the 8-position (23CB480; 28JIC433; 57JIC35,45, 57JOC884); 2-methylchromones sulfonate in the 6- and 8-positions (56JOC1104).

C. Miscellaneous Electrophilic Substitutions

Quinoline is hydroxylated by electrophilic hydroxyl at the 3-position in 6% yield [54JBC(208)741]. Methylation by methanol/alumina at 450°C gives 41% of 3-methylquinoline (61BRP845562); once again under neutral conditions 3-substitution dominates. Moreover it is hard to believe that either of these results is due to adduct formation.

Diazonium coupling of 3-hydroxyquinoline and 4-hydroxyisoquinoline occur at the expected 4- and 3-positions, respectively (72BAU452). Formylation of 6-methoxy-3-methylbenzo[d,e]cinnolines (**11.117**) takes place at the 7- and 9-positions (3.3 and 33%, respectively) for R = H. However, for R = Me, no 9-formylation occurs, the yield of 7-formylation is increased to 30%, and 5.5% of N-formylation occurs (81JOU2183). The difference between these results appears to reflect a steric effect.

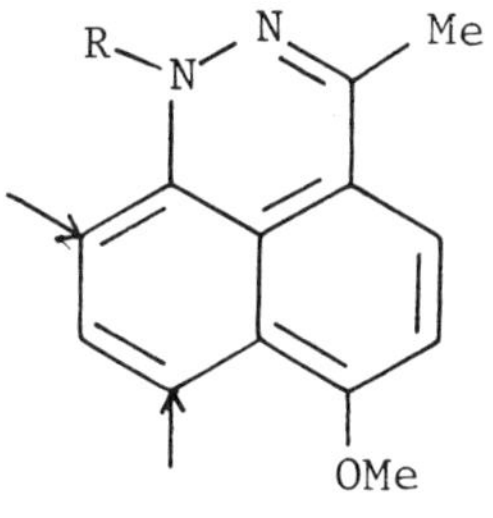

(11.117)

7. Side-Chain Reactions: Pyrolysis of 1-Arylethyl Acetates

The pyrolysis of 1-arylethyl acetates has led to the quantitative determination of the electrophilic reactivity of all positions of the quinoline and isoquinoline free bases [71JCS(B)2382; 75JCS(P2)1783)]. The results have demonstrated a number of major points concerning the transmission of electronic effects in naphthalene-like systems; a summary of the conclusions is given here. The results are shown in terms of σ^+ values in Scheme 11.3, along with the corresponding values for pyridine and naphthalene obtained under the same conditions; no extrapolations or assumptions are used or necessary in this work. The results show the following points.

(1) The positional reactivity order in quinoline is $5 > 6 = 8 > 3 > 7 \gg 2 > 4$.

(2) The positional reactivity order in isoquinoline is $4 > 5 = 7 > 8 > 6 \gg 3 > 1$.

(3) The 5-position in quinoline and the 4-position in isoquinoline are both *more* reactive than a position in benzene. All other positions are less reactive than a position in benzene.

(4) All positions are more reactive than the corresponding position in pyridine.

(5) All positions are less reactive than the corresponding position in naphthalene.

(6) The least reactive sites in each molecule are, as expected, those conjugated with nitrogen. The sites conjugated with nitrogen in the benzenoid ring of isoquinoline are significantly less reactive than the sites conjugated with nitrogen in the benzenoid ring of quinoline. This arises because the substituent interaction factors A_f [81JCS(P2)1153] are larger in the former case (i.e., the values for **11.118–11.121** are 10.9, 7.3, 7.3, and 3.6, respectively). This result is exactly paralleled by the effects of the chloro substituent in protiodetritiation of naphthalene [68JCS(B)1112] (Scheme 11.4). Here the $+M$ effect of the halogen is better relayed from the 2-position than from the 1-position. Thus, overall deactivation is *less* in the former case, and there is almost an exact parallel between the differential effects between the 5- and 7-position on the one hand, and the 6- and 8-positions on the other, in the two systems. The relative effects of the methyl substituent in detritiation of naphthalene also parallel closely the effects of the nitrogens in quinoline and isoquinoline and clearly the same factors govern the transmission of the electronic effects [75JCS(P2)1783].

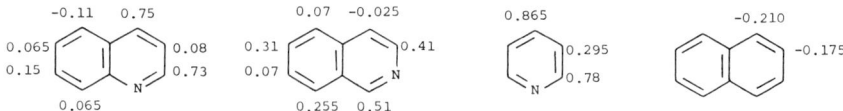

SCHEME 11.3. Values of σ^+ determined from pyrolysis of 1-arylethyl acetates at 625 K.

(11.118)   (11.119)   (11.120)   (11.121)

(7) In the pyridinoid ring of quinoline the conjugated positions are less reactive than predicted on the basis of additivity of the effects of the =N— substituent in pyridine, superimposed upon the reactivity of naphthalene. Additivity predicts that the σ^+ values should be 0.645 (4-position) and 0.605 (2-position). Thus nitrogen deactivates more in quinoline than it does in pyridine, which follows because conjugation between the 1,4- or 1,2-position in quinoline does not involve such a large loss of aromaticity as is the case in pyridine. Again the results for the chloro substituent in hydrogen exchange show an exact parallel. The $+M$ effect should operate better in naphthalene than in benzene and thus the observed values of f_o^{Cl} and f_p^{Cl} for benzene are 0.035 and 0.161 (61JCS2388), smaller than the corresponding values for 1-chloronaphthalene shown in Scheme 11.4.

Comparison of the reactivity of the 1- and 3-positions in isoquinoline with those of the corresponding position in naphthalene shows that nitrogen deactivates much less strongly across the 2,3-bond than across the 1,2-bond ($\sigma_{1\text{-iQ}}^+ - \sigma_{1\text{-nap}}^+ = 0.72$; $\sigma_{3\text{-iQ}}^+ - \sigma_{2\text{-nap}}^+ = 0.585$). This is because the order of the 2,3-bond is much lower than that of the 1,2-bond, so conjugative effects in particular are much less effectively relayed across the 2,3-bond. This again is exactly parallel to the effect of the chloro substituent in hydrogen exchange (protiodetritiation) of naphthalene (Scheme 11.4), in which the poorer transmission of the $+M$ effect of chlorine across the 2,3-bond compared to the 1,2-bond causes a 4-fold greater deactivation across the former. Likewise, in the hydrogen exchange of naphthalene, methyl at position 2 activates the 1- and 3-positions 300 and

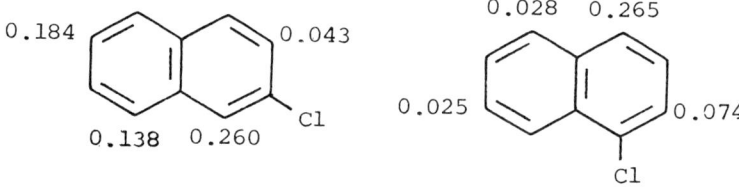

SCHEME 11.4. Deactivating effects (rates relative to the corresponding position in naphthalene) of chloro substituents in acid-catalyzed hydrogen exchange.

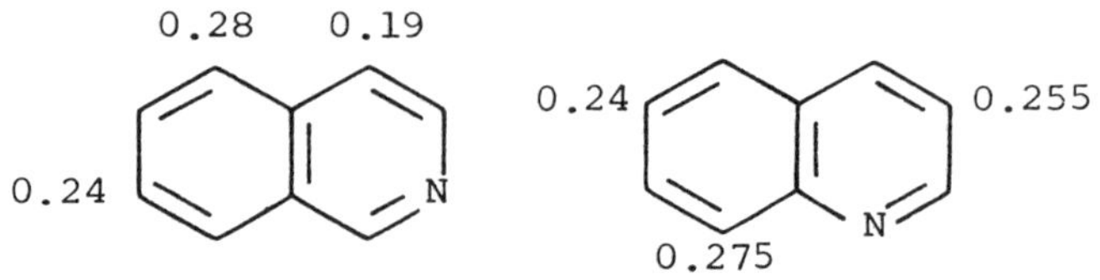

SCHEME 11.5. Net deactivation at the nonconjugated positions of quinoline and isoquinoline relative to naphthalene, in terms of σ^+ values.

3.6 times, respectively, the corresponding values for the effect of methoxy being 2.2×10^4 and 35.9 [75JCS(P2)1783].

(8) Positions not conjugated with nitrogen have (with the exception of the 4-position in isoquinoline) an almost identical reactivity in each molecule ($\sigma^+ \cong 0.07$). However, to analyze the effect of nitrogen upon each position more precisely, the reactivity relative to the corresponding position in naphthalene needs to be considered, and the net deactivations are shown in Scheme 11.5. Again there is a remarkable parallel with the effects of the chloro substituent at nonconjugated positions in hydrogen exhange of naphthalene (Scheme 11.6). Positions that are most deactivated by nitrogen, and show the greatest (secondary) relay of the $-M$ effect, are the least deactivated by the chloro substituent, because the relay of the $+M$ effect of chlorine is correspondingly greatest. The parallel is exact except that the reactivity of the 3- and 6-positions in quinoline (or the corresponding positions in 1-chloronaphthalene) should be marginally reversed.

The important conclusion is that the hydrogen exchange results confirm that there is very poor secondary relay of conjugative effects between the 4-position in isoquinoline and the nitrogen at position 2. As a result, the 4-position is the most reactive site in isoquinoline. Thus, contrary to expectations, *the most reactive site is in the ring containing the nitrogen.* This has long been predicted by calculations (Section 8), but in general

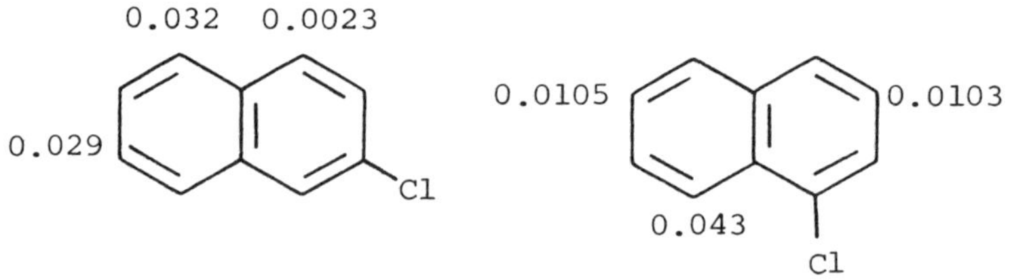

SCHEME 11.6. Deactivating effects (rates relative to the corresponding position in naphthalene) of chloro substituents in acid-catalyzed hydrogen exchange.

workers have been convinced that this could not possibly be true, and alternative and complex mechanisms have been sought to explain the *generally observed* 4-substitution in isoquinoline free base. The reason for the high reactivity of the 4-position has been attributed to the high energy of structures **11.122** and (especially) **11.123**. Since it is difficult to place a positive charge at either of the positions indicated, the deactivation of the 4-position by secondary relay will be correspondingly difficult [75JCS(P2)1783].

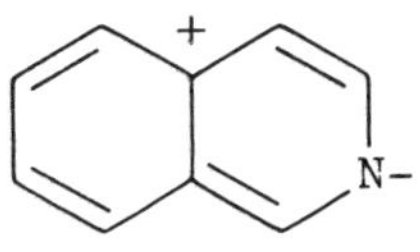

(11.122) 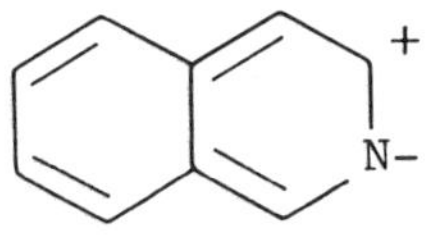   (11.123)

(9) The values of $\Sigma \sigma^+$ for quinoline and isoquinoline are 1.73 and 1.60, respectively. Overall, therefore, isoquinoline should be more reactive than quinoline and this follows from nitrogen occupying the more reactive α-naphthalene-like position in quinoline. Regarding the individual rings, the benzenoid ring is the most reactive in quinoline, and the pyridinoid ring is the most reactive in isoquinoline. Thus, the difference in reactivity of the benzenoid and pyridinoid rings is greater in quinoline than in isoquinoline. This explains, for example, why the benzenoid ring always opens in oxidation of quinoline, whereas both rings may be opened in isoquinoline; the ease of oxidation depends upon electron availability within a given ring.

(10) The σ^+ values show that the position in quinoline should be between ~10 and 2×10^6 times less reactive than naphthalene toward nitration. The observed value is ~10^{10}, confirming that, under these conditions, nitration takes place on the conjugate acid and not on the free base [71JCS(B)2382]. Likewise, the partial rate factors for nitration of the 5- and 8-positions in isoquinoline require σ^+ values of 0.77 and 0.92, confirming that reaction takes place on the conjugate acid [75JCS(P2)1783].

8. Theoretical Calculations of Reactivity

A. Summary of General Methods

A very large number of papers describe theoretical calculations of the electrophilic reactivities of nitrogen-containing heterocycles. Most have used the Hückel method to calculate π densities and, in some cases, localization energies. A few have used more sophisticated methods [e.g., CNDO/2 (74MI2), SCF (68BSB181, 68BSB191), and MINDO/2 [73JCS(P2)179]]. In some cases, frontier electron densities have been used as indexes of reactivity (e.g., 54BCJ423). None of these methods appears to have produced results significantly more meaningful than the Hückel method.

The main problem is that no method is yet able to take the structure of the transition state (and its solvation) into account. This is particularly important for polycyclic molecules, since the extent of conjugative interaction will vary quite markedly with the nature of the reaction. π Densities give the reactivity order appropriate to a reaction, with a transition state close to the ground state, whereas localization energies give the order appropriate for a transition state near to the Wheland intermediate. For a given heterocycle the positional orders produced by these two methods may differ considerably. An additional problem is that only for quinoline and isoquinoline have the positional orders been quantitatively and completely determined. Therefore, the theoretical predictions are considered (below) in some detail only for these molecules. For all other molecules, Table 11.6 gives the references for various calculations, and the results here can be summarized generally by the fact that positions conjugated with nitrogen are found, not surprisingly perhaps, to have a lower π density than those which are not [indeed it has been shown that π densities are a direct measure of the extent of conjugative interactions [75JCS(P2)1783]].

The precise values of the π densities will vary according to the values chosen for the coulomb and resonance integrals; however, there seems to be general agreement that $\alpha_N = \alpha_C + 0.5\beta$, and that $\beta_{CN} = \beta_{CC}$ (i.e., $h = 0.5$, $k = 1.0$). For more meaningful results an auxiliary inductive parameter (h') for carbon adjacent to nitrogen is necessary.

Some π densities for a representative set of azaphenanthrenes and which demonstrate some of the general points are shown in Scheme 11.7; these were calculated with an extended set of auxiliary inductive parameters of 0.17, 0.055, and 0.002 for carbons α, β, and γ to nitrogen (62T507). The following main features are evident.

(1) With but two exceptions (see 4, below) positions conjugated with nitrogen (indicated in Scheme 11.7) have lower charge densities than adjacent positions.

(2) The differences between the densities at the 1- and 3-positions in 2-azaphenanthrene and between the densities of the 2- and 4-positions in 3-azaphenanthrene show the effect of bond fixation (the lowest energy canonicals are shown in Scheme 11.7). The greater effectiveness of the $-M$ effect across the higher order bonds results in a lower π density.

(3) π Densities in the rings most remote from the nitrogen are relatively

TABLE 11.6
MOLECULES FOR WHICH MOLECULAR ORBITAL CALCULATIONS HAVE BEEN MADE[a]

Heterocycle	References
Quinoline	52JCP1554; 53JCS2406; 54BCJ423; 57JCS2521; 62T507; 65CCC355; 68BSB181; 71JCS(B)2382; 73JCS(P2)179; 74CHE953; 74MI2; 75JCS(P2)1783
Isoquinoline	52JCP1554; 53JCS2406; 57JCS2521; 59JCS3451; 62T507; 65CCC355; 68BSB181; 74CHE953; 75JCS(P2)1783
Acridine	47TFS87; 50CR(231)1146; 52JCP1554; 53JCS2406; 57JCS2521; 62T507; 63CCC2089; 65CCC355; 68BSB181; 74CHE953, 74MI2
Benzo[f]quinoline	57JCS2521; 62T507; 65CCC355; 68BSB181; 74CHE953
Benzo[g]quinoline	50CR(231)1146; 53JCS2406; 62T507; 65CCC355; 74CHE953
Benzo[h]quinoline	62T507; 65CCC355; 68BSB181; 74CHE953
Benzo[f]isoquinoline	62T507; 65CCC355
Benzo[g]isoquinoline	50CR(231)1146; 62T507; 65CCC355
Benzo[h]isoquinoline	62T507; 65CCC355
Benz[a]acridine	65CCC355
Benz[b]acridine	65CCC355
Benz[c]acridine	65CCC355
Dibenz[b,h]acridine	65CCC355
1- and 7-Azafluoranthene	62T507
Naphtho[2,3-f]quinoline	65CCC355
Naphtho[2,3-g]quinoline	65CCC355
Naphtho[2,3-g]isoquinoline	65CCC355
Phenanthridine	49JCS971; 52JCP1554; 57JCS2521; 62T507; 65CCC335; 68BSB181; 74CHE953
Naphthyridines	49JCS971; 68BS191; 68JOC1384
Quinazoline	49JCS971; 57JCS2521; 65JCP2658; 68BSB191
Quinoxaline	49JCS971; 57JCS2521; 68BSB191
Cinnoline	49JCS971; 57JCS2521
Phenazine	49JCS971; 52JCP1554; 57JCS2521; 68BSB191
Benzo[c]cinnoline	49JCS971; 68BSB191
Phenanthrolines	49JCS971; 57JCS2521; 62T507; 65CCC355; 69T583; 74CHE953

[a] In addition, the papers by Longuet–Higgins and Coulson give π densities for every position in every diaza derivative of anthracene and phenanthrene (47TFS87; 49JCS971).

little affected by it. It matters little if nitrogen is in the 1- or 3-position on the one hand, or the 2- or 4-position on the other.

(4) The highest densities are at the 9- or 10-positions (which correspond to the most reactive positions in phenanthrene), except for 9-azaphenanthrene, in which deactivation by the adjacent nitrogen is clearly

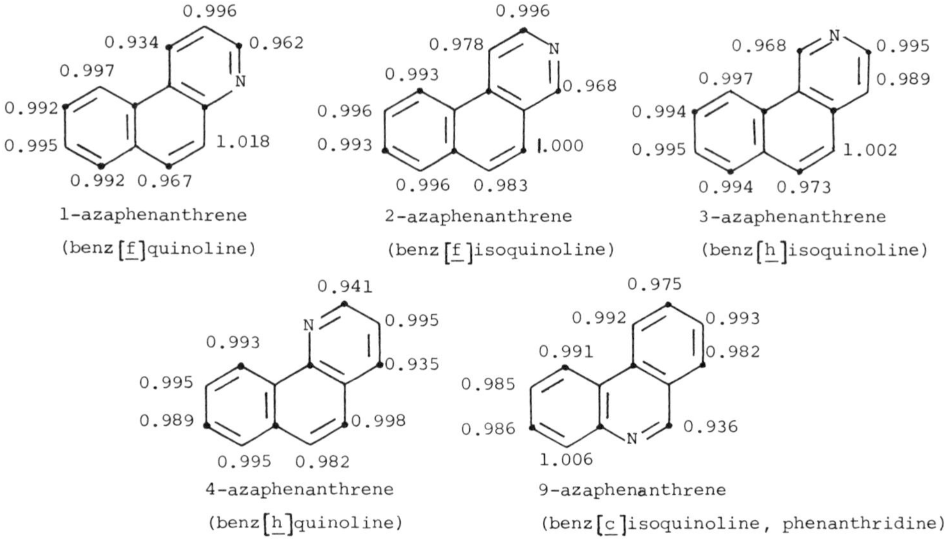

SCHEME 11.7. π Densities for azaphenanthrenes (62T507).

greatest for any position (after allowing for the intrinsically higher reactivity of the 9,10-phenanthrene positions). This exception follows because the 9,10-bond has the highest order in phenanthrene-like molecules.

(5) In 2- and 4-azaphenanthrene, the 10-position (conjugated with nitrogen) actually has a *higher* density than the adjacent nonconjugated 9-position. This is due to the very poor conjugation between the 2- or 4- and 10-positions, and this is the same factor which causes the 5-position in quinoline and the 8-position in isoquinoline to be relatively reactive despite being conjugated with nitrogen. It also accounts for the π density at the 5-position (conjugated) of 9-azaphenanthrene being *higher* than that at the 6-position (nonconjugated).

B. QUINOLINE AND ISOQUINOLINE

π Densities and localization energies have been calculated for both quinoline and isoquinoline [71JCS(B)2382; 75JCS(P2)1783]. The results are shown in Scheme 11.8.

Although the parameters used were very successful in correlating the reactivity of the pyridine free base (Chapter 9), the localization energies are unsatisfactory and grossly overestimate the reactivities of α-naphthalene-like positions. For example, even the 4-position of quinoline is pre-

π-densities

Localization energies

SCHEME 11.8. π Densities and localization energies (-β) for quinoline and isoquinoline.

dicted to be activating. Indeed, most positions are predicted to be equal to or more reactive than benzene. However, these predictions refer to high ρ factor reactions of the free bases, for which no data have yet been obtained. It could well be that for such reactions the high demand for resonance stabilization of the transition state would result in considerably enhanced reactivity at some positions.

π Densities are much better parameters for the data presently available. For quinoline they predict the order 8 > (benzene) > 6 > 3 > 5 > 7 > 4 > 2, whereas the observed order (pyrolysis of 1-arylethyl acetates, Section 7) is 5 > (benzene) > 8 = 6 > 3 > 7 > 2 > 4. The only significant difference is that the 5-position is predicted to be too deactivated, the calculations evidently overestimating the stability of the structure **11.121**. This is of particularly high energy because practically every bond needs to be adversely lengthened or shortened relative to the ground state. The above π-density order is not significantly affected by changing the value for the auxiliary inductive parameters [71JCS(B)2382].

For isoquinoline, π densities predict the order benzene > 5 > 7 > 8 > 6 > 3 > 4 > 1, whereas the observed order (pyrolysis of 1-arylelthyl acetates, Section 7) is 4 > benzene > 5 = 7 > 8 > 6 > 3 > 1.

However, Brown and Harcourt observed that for isoquinolines the positional reactivity order is markedly affected by the value of the auxiliary inductive parameter and if a smaller value for h' of 0.018 is used then the predicted order becomes [60T(8)23] 4 > 5 > benzene > 7 > 8 > 6 > 3 > 1, which is effectively that observed. Notable is the fact that the activation of the 4-position is correctly predicted. The reason for the need for the smaller parameter was not clear originally [75JCS(P2)1783], but almost certainly arises because of the low 2,3-bond order. The greater C-3—N-2 distance will make the effective electronegativity of C-3 less than would be the case in pyridine. It follows from this conclusion that one probably needs a correspondingly greater auxiliary inductive parameter for C-1, but no calculations with independently varying parameters have yet been carried out.

Calculations of π densities, total electron densities, localization energies, or delocalization energies for pyridine and quinoline by the MINDO/2 method give results that are either no better or worse than those calculated by the simple Hückel method [73JCS(P2)179]. It is generally found that for aromatic reactivity the Hückel method outperforms other methods often regarded as superior. *Ab initio* calculations may in due course produce better results, but at present are far too expensive for molecules of this size.

CHAPTER 12

Thiaazepines

An intriguing new class of heteroaromatic compounds containing nitrogen and sulfur in seven-membered rings has been described by Morris and Rees (85CC396, 85CC398; 86CSR1, 86PAC197). The class comprises 1,3,5,2,4-trithiadiazepine (**12.1**), benzo-1,3,5,2,4-trithiadiazepine (**12.2**), and 1,3,5,2,4,6-trithiatriazepine (**12.3**); some related compounds have also been made but these do not contain C—H bonds and are therefore not considered here.

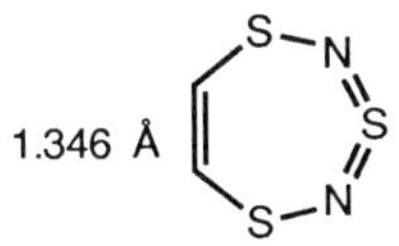

(**12.1**) 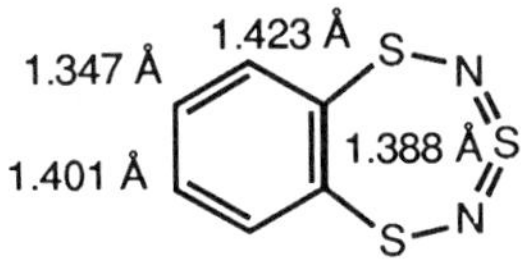   (**12.2**) 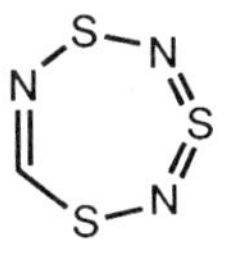   (**12.3**)

Trithiadiazepine (**12.1**) is colorless, planar, and symmetrical, and has bond lengths intermediate between double and single as expected for a 10π aromatic system; the C—C bond length is 1.346 Å. The benzo derivative (**12.2**) is also planar and symmetrical, but is bright yellow, and there is bond alternation in the benzene ring (see **12.2**) akin to that found in naphthalene. Crystal discontinuities have prevented the accurate measurement of the structural details of **12.3**, which is also colorless, but it is presumed to be planar.

Each of **12.1**, **12.2**, and **12.3** undergoes electrophilic substitution, and for **12.1** and its bromo and nitro derivatives, and for **12.3**, the reactivities have been determined quantitatively via acid-catalyzed hydrogen exchange (detritiation) by A. Laws and R. Taylor [89JCS(P2)1911]. The rate–acidity profile for **12.1** in trifluoroacetic acid/acetic acid media shows hydrogen bonding between substrate and solvent. This is similar to all other sulfur-containing heteroaromatics (see Chapter 8, Sections 4.A, 5.A, and 6.A), for which the extent of bonding is, in general, proportional to the number of sulfur atoms present. For the trithiadiazepine (**12.1**), the extent of this bonding is slightly greater than for thienothiophenes, but less than for dithienothiophenes or anisole, and thus also roughly proportional to the number of sulfur atoms present. However, this must be fortuitous here, since hydrogen bonding to nitrogen will be more important.

The partial rate factors corrected for hydrogen bonding, and the derived σ^+ values, are given in Scheme 12.1. These values show trithiadiazepine to be slightly more reactive than the β-position of thiophene, for which the corrected σ^+ value is −0.56 (cf. −0.913 for the α-position). Conjugation between a given carbon and the α- and β-sulfur atoms involves structures **12.4** and **12.5**, respectively, formally analagous to the corresponding structures **12.6** and **12.7** for substitution at the α- and β-positions of thiophene. However, greater bond reorganization from the ground state is required to produce **12.4** than is the case for **12.6**, and, moreover, the sulfur atoms in **12.4** and **12.5** should be less electron releasing than in thiophene due to the adjacent nitrogen atoms. The observed reactivity of less than the average of the positional reactivities in thiophene is therefore quantitatively reasonable. The reactivity greater than benzene is also consistent with *ab initio* calculations of the net atomic charges at C6,7 of −0.189 (85CC398) (cf. −0.063 for benzene) (both at the STO-3G level).

(12.4)   (12.5)   (12.6)   (12.7)

Bromotrithiadiazepine is slightly less susceptible to hydrogen bonding than the parent molecule (**12.1**), as expected, and the partial rate factor

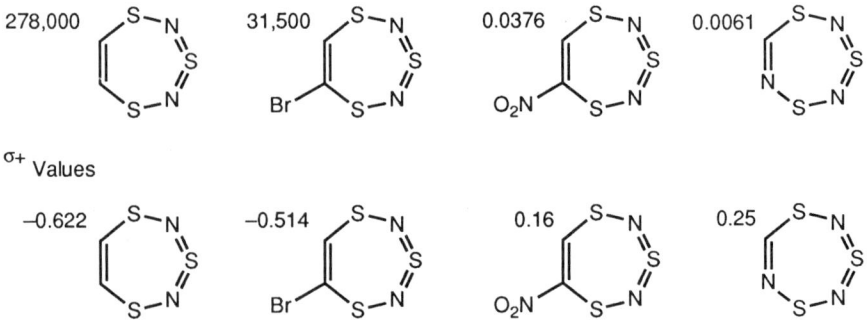

SCHEME 12.1. Partial rate factors for protiodetritiation.

for detritiation (Scheme 12.1) shows it to be ninefold less reactive than **12.1**. This difference compares with a 37-fold deactivation by an ortho-bromine atom in benzene (61JCS2388) and a 6.1-fold deactivation by ortho bromine in naphthalene (in 2-bromo-1-T-naphthalene) [68JCS(B)1112]. The latter differences are due to differences in bond order, the $+M$ effect of bromine (which counteracts the $-I$ effect) being much more effective across a bond of higher order. The effect of bromine in trithiadiazepine is consistent with this explanation, since the C—C bond length (1.35 Å) is close to that of the 1,2-bond in naphthalene (1.36 Å) (cf. 1.40 Å in benzene).

Exchange in nitrotrithiadiazepine could be achieved only in trifluoroacetic acid containing 5% trifluoromethanesulfonic acid (TMSA) and was accompanied by some decomposition, so the partial rate factor and σ^+ value (Scheme 12.1) are *maximum* values. The quantitative deactivation by an ortho-nitro group in benzene in hydrogen exchange is not known, and values from other reactions may be unreliable due to conformation dependence. The value of σ^+ for 2-NO_2 has been determined in the gas phase as 0.84 [71JCS(B)622], which predicts a deactivation of 2.2×10^7. The observed deactivation in trithiadiazepine ($>7.4 \times 10^7$) agrees remarkably well; greater deactivation could be expected because of the higher C—C bond order.

Exchange in trithiatriazepine (**12.3**) could also be achieved only in 5% TMSA in TFA and was likewise accompanied by decomposition. The derived rate parameters (Scheme 12.1) are again *maximum* values. The ortho-aza "substituent" is slightly more deactivating than the ortho nitro, and closely similar effects were predicted earlier from data on pyrolysis of 1-arylethyl acetates [62JCS4881; 71JCS(B)622].

Electrophilic substitution in **12.2** indicates that the 7-position (corresponding to the β-position of naphthalene) is the more reactive, and this follows since this position is para conjugated with sulfur (**12.8**), whereas the 6-position is ortho conjugated (**12.9**). Yields are lower than in substitution of **12.1**, but this does not necessarily indicate that **12.2** is less reactive, since substitution is accompanied by ring-contraction reactions. No quantitative data are yet available.

(**12.8**)   (**12.9**)

REFERENCES

1870LA(155)311	N. Lubavin, *Justus Liebigs Ann. Chem.* **155**, 311 (1870). 4-5B; 11-6B
1882CB683	K. Bedall and O. Fischer, *Chem. Ber.* **15**, 683 (1882). 4-5B; 11-6B
1882CB1979	O. Fischer, *Chem. Ber.* **15**, 1979 (1882). 4-5B; 11-6B
1884CB1558	V. Meyer and H. Kreis, *Chem. Ber* **17**, 1558 (1884). 4–6; 6-4A
1886CB636	A. Biedermann, *Chem. Ber.* **19**, 636 (1886). 6-7A
1886CB644	E. A. von Schweinitz, *Chem. Ber.* **19**, 644 (1886). 6-4A
1886CB660	E. Schleicher, *Chem. Ber.* **19**, 660 (1886). 4-5B; 6-8A
1886CB2623	K. Krekeler, *Chem. Ber.* **19**, 2623 (1886). 4-5B; 6-8A
1887CB731	O. Fischer, *Chem. Ber.* **20**, 731 (1887). 4-5B; 11-6B
1888JPR258	A. Claus, *J. Prakt. Chem.* **37**, 258 (1888). 4-5B; 11-6B
1889G128	L. Balbiano, *Gazz. Chim. Ital.* **19**, 128 (1889). 4-3Eb; 7-6
1892LA(267)172	L. Volhard, *Justus Liebigs Ann. Chem.* **267**, 172 (1892). 6-8A
1894LA(279)217	L. Knorr, *Justus Liebigs Ann. Chem.* **279**, 217 (1894). 4-5B; 7-7
1896CB2560	F. Mehlert, *Chem. Ber.* **29**, 2560 (1896). 4-5B; 6-8A
1900LA(312)237	R. Stoermer, *Justus Liebigs Ann. Chem.* **312**, 237 (1900). 8-2Ab(iii)
01CB1234	S. Gabriel and J. Colman, *Chem. Ber.* **34**, 1234, (1901). 10-4B
01CB3362	S. Gabriel and J. Colman, *Chem. Ber.* **34**, 3362 (1901). 10-4B
01LA(315)259	C. Vogel, *Justus Liebigs Ann. Chem.* **315**, 259 (1901). 10-4B
02CB1633	R. Stoermer, *Chem. Ber.* **35**, 1633 (1902). 8-2Ab(iii)
05JBC435	H. L. Wheeler and L. D. Bristol, *J. Biol. Chem.* **33**, 435 (1905). 10-4B
05LA(339)37	R. Bartling, *Justus Liebigs Ann. Chem.* **339**, 37 (1905). 10-4B
12CB428	F. Friedl, *Chem. Ber.* **45**, 428 (1912). 9-4A
12CB1596	W. J. Hale, *Chem. Ber.* **45**, 1596 (1912). 8-2Ab(iii)
14LA(403)50	W. Steinkopf and M. Bauermeister, *Justus Liebigs Ann. Chem.* **403**, 50 (1914). 6-8A
19LA(426)61	W. Steinkopf and H. Otto, *Justus Liebigs Ann. Chem.* **426**, 61 (1919). 4–6; 6-4A
21CB1035	H. Biltz and H. Wittek, *Chem. Ber.* **54**, 1035 (1921). 10-5B
22CB3400	W. Bock, *Chem. Ber.* **55**, 3400 (1922). 10-5B
22JCS947	I. E. Balaban and F. L. Pyman, *J. Chem. Soc.*, 947 (1922). 7-5A
22LA(430)79	W. Steinkopf, H. Augestad-Jensen, and H. Donat, *Justus Liebigs Ann. Chem.* **430**, 79 (1922). 4–6; 6-4A
23CB480	M. Krüger, *Chem. Ber.* **56**, 480 (1923). 4-5B; 11-6B
23CB2223	G. Sachs and R. Eberhartinger, *Chem. Ber.* **56**, 2223 (1923). 4-2C; 9-6A
23CB2482	H. Biltz and T. Kohler, *Chem. Ber.* **56**, 2482 (1923). 4-3Eb; 10-6D

23CB2498	W. Borsche and B. Schacke, *Chem. Ber.* **56**, 2498 (1923). 8-3A*b*(ii)
24CB555	H. Lindemann, *Chem. Ber.* **57**, 555 (1924). 8-3A*b*(ii)
24CB1161	A. E. Tschitschibabin and A. W. Kirssanow, *Chem. Ber.* **57**, 1161 (1924). 9-6B
24JCS919	R. Forsyth, J. A. Moore, and F. L. Pyman, *J. Chem. Soc.*, 919 (1924). 4-5B; 7-7
24LA(437)14	W. Steinkopf and W. Ohse, *Justus Liebigs Ann. Chem.* **437**, 14 (1924). 4-5B; 6-8A
25LA(441)192	K. Schmedes, *Justus Liebigs Ann. Chem.* **441**, 192 (1925). 10-5B
26CB611	K. von Auwers and H. Mauss, *Chem. Ber.* **59**, 611 (1926). 4-3E*b*; 7-6
26LA(449)63	H. Lindemann and H. Thiele, *Justus Liebigs. Ann. Chem.* **449**, 63 (1926). 8-C*b*(iii)
27JCS2711	G. R. Barnes and F. L. Pyman, *J. Chem. Soc.*, 2711 (1927). 4-5B; 7-7
28G679	C. Gastaldi and E. Princivalle, *Gazz. Chim. Ital.* **58**, 679 (1928). 10-6A
28JIC433	R. N. Sen and D. Chakravarti, *J. Indian Chem. Soc.* **5**, 433 (1928). 4-5B; 11-6B
28JPR(118)33	H. Pauly and E. Arauner, *J. Prakt. Chem.* **118**, 33 (1928). 4–6; 7-5A
29CB226	O. Schmidt-Dumont, *Chem. Ber.* **62**, 226 (1929). 6-7D
30G298	E. Princivalle, *Gazz. Chim. Ital.* **60**, 298 (1930). 10-6A
30JCS2267	N. M. Cullinane, *J. Chem. Soc.*, 2267 (1930). 8-3A*b*(ii)
30RTC552	H. J. den Hertog and J. Overhoff, *Recl. Trav. Chim. Pays-Bas* **49**, 552 (1930). 9-4A
31JCS3283	G. T. Morgan and J. G. Mitchell, *J. Chem. Soc.*, 3283 (1931). 8-3A*b*(ii)
31JPJ542	T. Ukai, *J. Pharm. Soc. Jpn.* **51**, 542 (1931). 4-2C; 11-6A
32JCS1263	N. P. McCleland and R. H. Wilson, *J. Chem. Soc.*, 1263 (1932). 4-2C; 9-6A
32LA(495)166	W. Steinkopf and P. Leonhardt, *Justus Liebigs Ann. Chem.* **495**, 166 (1932). 6-8A
32RTC1054	H. Gilman and E. B. Towne, *Recl. Trav. Chim. Pays-Bas* **51**, 1054 (1932). 4-1D; 6-8A
32RTC1134	I. J. Rinkes, *Recl. Trav. Chim. Pays-Bas* **51**, 1134 (1932). 6-8A
33LA(501)174	W. Steinkopf and T. Höpner, *Justus Liebigs Ann. Chem.* **501**, 174 (1933). 4-5B; 6-8A
34CR(198)2260	C. Courtot, *C. R. Hebd. Seances Acad. Sci.* **198**, 2260 (1934). 4-5B; 8-3A*b*(vi)
34LA(512)136	W. Steinkopf, H. Jacob, and H. Penz, *Justus Liebigs Ann. Chem.* **512**, 136 (1934). 4-5C; 6-8A
34N(L)347	C. K. Ingold, C. G. Raisin, and C. L. Wilson, *Nature (London)* **134**, 347 (1934). 2-1A
34RTC77	J. P. Wibaut and H. E. Jansen, *Recl. Trav. Chim. Pays-Bas* **53**, 77 (1934). 4–6; 7-5A
35JA27	H. S. Harned and W. J. Hamer, *J. Am. Chem. Soc.* **57**, 27 (1935). 3-4C

REFERENCES 405

35JA1763	V. S. Babasinian, *J. Am. Chem. Soc.* **57**, 1763 (1935). 6-3A
35JCS741	S. G. P. Plant, K. M. Rogers, and S. B. C. Williams, *J. Chem. Soc.*, 741 (1935). 4-3E*b;* 8-3A*b*(iv)
36JCS915,1637	C. K. Ingold, C. G. Raisin, C. L. Wilson, *J. Chem. Soc.*, 915, 1637 (1936). 2-1A
36JCS1435	N. M. Cullinane, C. G. Davis, and G. I. Davies, *J. Chem. Soc.*, 1435 (1936). 8-3A*b*(ii)
36JOC146	H. Gilman and R. H. Kirby, *J. Org. Chem.* **1**, 146 (1936). 4-2C; 8-3A*b*(vi)
37JA933	D. E. Worrall, *J. Am. Chem. Soc.* **59**, 933 (1937). 7-4B*a*
37LA(527)237	W. Steinkopf, H. F. Schmitt, and H. Fiedler, *Justus Liebigs Ann. Chem.* **527**, 237 (1937). 6-4A
37USP2085063	C. N. Andersen, U. S. Pat. 2,085,063 (1937). 4-2C; 9-6A
38BCJ95,307,681	M. Koizumi and T. Titani, *Bull. Chem. Soc. Jpn.* **13**, 95, 307, 681 (1938). 8-2A*b*(i)
38BCJ643	M. Koizumi, Y. Komaki, and T. Titani, *Bull. Chem. Soc. Jpn.* **13**, 643 (1938). 2-1A; 8-2A*b*(i)
38CB87	W. Huber and H. A. Holscher, *Chem. Ber.* **71**, 87 (1938). 10-4B
38CB808	K. Lehmstedt, *Chem. Ber.* **71**, 808 (1938). 11-4A
38JA1198	D. E. Worrall, *J. Am. Chem. Soc.* **60**, 1198 (1938). 7-4B*a*
38JA2628	A. Burger, W. B. Wartman, and R. E. Lutz, *J. Am. Chem. Soc.* **60**, 2628 (1938). 4-3E*b;* 8-3A*b*(iv)
38JCS28	A. P. Best and C. L. Wilson, *J. Chem. Soc.*, 28 (1938). 2-1A
38JOC120	H. Gilman, A. L. Jacoby, and H. A. Pacevitz, *J. Org. Chem.* **3**, 120 (1938). 4-2A; 8-3A*b*(vi)
39BCJ353	M. Koizumi, *Bull. Chem. Soc. Jpn.* **14**, 353 (1939). 2-1A.
39BCJ453	M. Koizumi, *Bull. Chem. Soc. Jpn.* **14**, 453 (1939). 8-2A*b*(i)
39CB1470	E. Ochiai and H. Nagasawa, *Chem. Ber.* **72**, 1470 (1939). 7-5A
39JA104	D. E. Worrall and E. Lavin, *J. Am. Chem. Soc.* **61**, 104 (1939). 7-4B*a*
39JA951	H. Gilman, L. C. Cheney, and H. B. Willis, *J. Am. Chem. Soc.* **61**, 951 (1939). 4-2A; 8-3A*b*(vi)
39JA2370	S. Shankman and A. R. Gordon, *J. Am. Chem. Soc.* **61**, 2370 (1939). 3-4C
39JPJ462	E. Ochiai, T. Kakuda, I. Nakayama, and G. Masuda, *J. Pharm. Soc. Jpn.* **59**, 462 (1939). 7-4A*b*
40G1,11	A. Quilico and R. Justani, *Gazz. Chim. Ital.* **70**, 1, 11 (1940). 4-5B; 7-7
40JA446	H. Gilman and S. M. Spatz, *J. Am. Chem. Soc.* **62**, 446 (1940). 4-2A; 9-6A
40JA1640	L. F. Fieser and E. B. Hershberg, *J. Am. Chem. Soc.* **62**, 1640 (1940). 11-4A
40JA2606	H. Gilman, M. W. Van Ess, H. B. Willis, and C. G. Stuckwisch, *J. Am. Chem. Soc.* **62**, 2606 (1940). 4-2A; 8-3A*b*(vi)
40JPJ433	H. Nagazawa, *J. Pharm. Soc. Jpn.* **60**, 433 (1940). 7-4A*b*
41G327	A. Quilico and C. Musante, *Gazz. Chim. Ital.* **71**, 327 (1941). 7-4B*a*
41JA358	W. G. Brown and N. J. Letang, *J. Am. Chem. Soc.* **63**, 358 (1941). 8-3A*b*(i)

41JA879	J. V. Scudi and R. P. Buhs, *J. Am. Chem. Soc.* **63**, 879 (1941). 4-5B; 8-C*b*(vi)
41JA2479	H. Gilman, F. W. Moore, and O. Baine, *J. Am. Chem. Soc.* **63**, 2479 (1941). 4-2A; 8-3A*b*(vi)
41RTC650	I. J. Rinkes, *Recl. Trav. Chim. Pays-Bas* **60**, 650 (1941). 6-3B
42CB1108	E. F. Moller and L. Birkofer, *Chem. Ber.* **75**, 1108 (1942). 4-5B; 9-6D
42G537	C. Musante, *Gazz. Chim. Ital.* **72**, 537 (1942). 7-4B*a*
42JA477	F. F. Blicke and J. H. Burckhalter, *J. Am. Chem. Soc.* **64**, 477 (1942). 4-3B; 6-6
42JA900	G. W. Wheland, *J. Am. Chem. Soc.* **64**, 900 (1942). 9-9A
42JA2444	J. J. Rosenbaum and W. E. Cass, *J. Am. Chem Soc.* **64**, 2444 (1942). 7-4A*a*
42JPJ105	E. Ochiai and Y. Kashida, *J. Pharm. Soc. Jpn.* **62**, 105 (1942). 7-4A*b*
43CB419	O. Dann, *Chem. Ber.* **76**, 419 (1943). 6-3A
43JA1729	H. Gilman and C. G. Stuckwisch, *J. Am. Chem. Soc.* **65**, 1729 (1943). 4-2A; 8-3A*b*(vi)
43JA2233	S. M. McElvain and M. A. Goese, *J. Am. Chem. Soc.* **65**, 2233 (1943). 4-5B; 9-6D
43OSC357	W. Minnis, *Org. Synth., Collect. Vol.* **2**, 357 (1943). 4–6; 6-4A.
44JCS315	B. Lythgoe, A. R. Todd, and A. Topham, *J. Chem. Soc.*, 315 (1944). 4-4A; 10-6A,B
44LA(556)1	N. P. Buu-Hoi, *Justus Liebigs Ann. Chem.* **556**, 1 (1944). 4–6; 6-4A
45JA802	J. Weijlard, M. Tishler, and A. E. Erickson, *J. Am. Chem. Soc.* **67**, 802 (1945). 10-6C
45JA877	H. Gilman and C. G. Stuckwisch, *J. Am. Chem. Soc.* **67**, 877 (1945). 4-2A; 8-3A*b*(vi)
45JA2092	R. Mozingo, S. A. Harris, D. E. Wolf, C. E. Hoffhine, Jr., N. R. Easton, and K. Folkers, *J. Am Chem. Soc.* **67**, 2092 (1945). 6-4A
45PIA(A)343	K. Ganapathi and A. Venkataraman, *Proc.—Indian Acad. Sci., Sect. A* **22A**, 343 (1945). 7-4A*b*
46G131	C. Musante, *Gazz. Chim. Ital.* **76**, 131 (1946). 7-4B*a*
46JA103	H. Gilman and D. S. Melstrom, *J. Am. Chem. Soc.* **68**, 103 (1946). 4-7A; 9-6A
46JA453	J. P. English, J. H. Clark, J. W. Clapp, D. Seeger, and R. H. Ebel, *J. Am. Chem. Soc.* **68**, 453 (1946). 4-6; 105B
46JA1039	J. P. English, J. H. Clark, R. G. Shepherd, H. W. Marson, K. Krapcho, and R. O. Roblin, *J. Am. Chem. Soc.* **68**, 1039 (1946). 10-5B
46JA1871	F. H. Westheimer and M. S. Kharasch, *J. Am. Chem. Soc.* **68**, 1871 (1946). 3-4C
47CA754b	M. Colonna, *Chem. Abstr.* **41**, 754b (1947). 8-C*b*(iii)
47CRV279	I. J. Krems and P. E. Spoerri, *Chem. Rev.* **40**, 279 (1947). 10-4
47HCA2110	B. Prijs, J. Ostertag, and H. Erlenmeyer, *Helv. Chim. Acta* **30**, 2110 (1947). 7-4A*b*

47JA1173	H. M. Priestley and C. D. Hurd, *J. Am. Chem. Soc.* **69**, 1173 (1947). 6-8A
47JA1549	R. C. Clapp, J. H. Clark, J. R. Vaughan, J. P. English, and G. W. Anderson, *J. Am. Chem. Soc.* **69**, 1549 (1947). 4-3B; 6-6
47JA1920	C. R. Neumoyer and E. D. Armstutz, *J. Am. Chem. Soc.* **69**, 1920 (1947). 4–6; 8-3A*b*(iii)
47JA3093	H. D. Hartough and A. I. Kosak, *J. Am. Chem. Soc.* **69**, 3093 (1947). 4-3E*b*; 6-7A
47JCS431	E. Pedley, *J. Chem. Soc.*, 431 (1947). 7-5A
47JCS1631	W. J. Gaudion, W. H. Hook, and S. G. P. Plant, *J. Chem. Soc.*, 1631 (1947). 4-3E*b*; 8-2A*b*(v)
47JOC405	R. C. Elderfield, T. A. Williamson, W. J. Gensler, and C. B. Kremer, *J. Org. Chem.* **12**, 405 (1947). 11-4B
47JPJ87	E. Ochiai and T. Okamato, *J. Pharm. Soc. Jpn.* **67**, 87 (1947). 4–6; 11-5A
47TFS87	C. A. Coulson and H. C. Longuet-Higgins, *Trans. Faraday Soc.* **43**, 87 (1947). 8-B*a*, Table 17; Table 11-6
48BSF688	M. Polonovskii and M. Pesson, *Bull. Soc. Chim. Fr.*, 688 (1948). 10-6A
48JA1037	A. Murray, W. W. Foreman, and W. Langham, *J. Am. Chem. Soc.* **70**, 1037 (1948). 4-2A; 9-6A
48JCS1006	H. H. Hodgson and D. P. Dodgson, *J. Chem. Soc.*, 1006 (1948). 8-C*b*(iii)
48JOC635	W. J. King and F. F. Nord, *J. Org. Chem.* **13**, 635 (1948). 6-7A
48MI1	A. Pullman, *Rev. Sci.* **86**, 219 (1948). 11-4B
48MI2	L. A. Kazitsyna, *Vestn. Mosk. Univ.* **2**, 109 (1947) [*CA* **42**, 3751 (1948)]. 4-5B; 6-8A
48RTC45	G. M. van der Want, *Recl. Trav. Chim. Pays-Bas* **67**, 45 (1948) 8-C*b*(iii)
48RTC309	N. P. Buu-Hoi and N. Hoan, *Recl. Trav. Chim. Pays-Bas* **67**, 309 (1948). 6-7A
48YZ195,197	J. Haginiwa, *Yakugaku Zasshi* **68**, 195, 197 (1948). 4-5B; 7-4A*c*,7
49CA6205f	V. F. Borodkin and T. V. Malkova, *Chem. Abstr.* **43**, 6205f (1949). 4-5B; 8-3A*b*(vi)
49JA1593	R. Wendland, C. H. Smith, and R. Muraca, *J. Am. Chem. Soc.* **71**, 1593 (1949). 4-5B; 8-3A*b*(vi)
49JA2066	H. Gilman and F. J. Marshall, *J. Am. Chem. Soc.* **71**, 2066 (1949). 6-8A
49JCS255	A. Adams and D. H. Hey, *J. Chem. Soc.*, 255 (1949). 11-4A
49JCS971	H. C. Longuet-Higgens and C. A. Coulson, *J. Chem. Soc.*, 971 (1949). 11-4B, Table 6
49JCS1008	A. Hampton and D. Magrath, *J. Chem. Soc.*, 1008 (1949). 11-4A
49JCS1367	K. Schofield and T. Swain, *J. Chem. Soc.*, 1367 (1949). 11-4A,B
49JCS2490	K. J. M. Andrews, N. Anand, A. R. Todd, and A. Topham, *J. Chem. Soc.*, 2490 (1949). 10-4B

49JGU763	A. P. Terentev, S. K. Golnbeva, and L. V. Tsymbal, *J. Gen. Chem. USSR (Engl. Transl.)* **19**, 763 (1949). 8-2A*b*(vi)
49JOC405,638	W. J. King and F. F. Nord, *J. Org. Chem.* **14**, 405,638 (1949). 6-7A
49OS(29)31	K. B. Wiberg and H. F. McShane, *Org. Synth.* **29**, 31 (1949). 6-6
49ZOB531	A. P. Terentev and L. A. Kazitsyna, *Zh. Obshch. Khim.* **19**, 531 (1949). 4-5B; 6-8C.
49ZOB538,1365,2118	A. P. Terentev and L. A. Yanovskaya, *Zh. Obshch. Khim.*, **19**, 538, 1365, 2118 (1949). 4-5B; 6-8D.
50BP625173	K. L. Kreuz and R. T. Sanderson, Br. Pat. 625,173 (1950). 6-5
50BSF466,1278	P. Petitcolas and R. Sureau, *Bull. Soc. Chim. Fr.*, 466, 1278 (1950). 4-5B; 8-2C*b*(vi)
50CI(L)353	D. J. Brown, *Chem. Ind. (London)* **69**, 353 (1950). 10-4B
50CR(231)1146	I. Samuel, *C. R. Hebd. Seances Acad. Sci.* **231**, 1146 (1950). Table 11-6
50G651	S. Maffei, *Gazz. Chim. Ital.* **80**, 651 (1950). 4-5B; 11-6B
50HCA306	H. von Babo and B. Prijs, *Helv. Chim. Acta* **33**, 306 (1950). 7-4A*b*
50JCS1515	A. T. James and E. E. Turner, *J. Chem. Soc.*, 1515 (1950). 8-4B
50JCS2092	A. Adams and D. H. Hey, *J. Chem. Soc.*, 2092 (1950). 11-4A
50JCS3236	S. Dixon and L. F. Wiggins, *J. Chem. Soc.*, 3236 (1950). 10-4, 4A
50JPJ22	E. Ochiai and T. Okamoto, *J. Pharm. Soc. Jpn.* **70**, 22 (1950). 11-4A
50MI1	V. F. Borodkin, *J. Appl. Chem. USSR (Engl. Transl.)* **23**, 803 (1950). 4-5B; 8-3A*b*(vi)
50RTC1281	H. J. den Hertog, F. R. Schepman, J. de Bruyn, and G. Thysse, *Recl. Trav. Chim. Pays-Bas* **69**, 1281 (1950). 4–6; 9-5B
50USP2492644	G. C. Johnson, U.S. Pat. 2,492,644 (1950). 6-4A
50USP2500732	R. K. Abbott, U.S. Pat 2,500,732 (1950). 4-3A; 8-3A*b*(iv)
50USP2500734	R. K. Abbott, U.S. Pat. 2,500,734 (1950). 8-3A*b*(iv)
50USP2527680	L. P. Kyrides and D. G. Sheets, U.S. Pat 2,527,680 (1950). 6-6
51G613	C. Runti, *Gazz. Chim. Ital.* **81**, 613 (1951). 4-3B; 8-2A*b*(v)
51JA2614	A. H. Schlesinger and D. T. Mowry, *J. Am. Chem. Soc.* **73**, 2614 (1951). 8-2A*b*(iv)
51JA5628	F. W. Cagle and H. Eyring, *J. Am. Chem. Soc.* **73**, 5628 (1951). 2-1G*c*(iii)
51JCP1323	P.-O. Löwdin, *J. Chem. Phys.* **19**, 1323 (1951). 9-9A
51JCS96	W. R. Boon, W. G. M. Jones, and G. R. Ramage, *J. Chem. Soc.*, 96 (1951). 10-4B
51JCS2323	B. Lythgoe and L. S. Rayner, *J. Chem. Soc.*, 2323 (1951). 10-4
51JGU1415	A. P. Terentev and L. A. Yanovskaya, *J. Gen. Chem. USSR (Engl. Transl.)* **21**, 1415 (1951). 4-5B; 8-2A*b*(vi)
51JOC1485	H. Gilman and S. M. Spatz, *J. Org. Chem.* **16**, 1485 (1951). 4-7A; 9-6A

REFERENCES 409

51LA(572)217	F. G. Fischer and J. Roch, *Justus Liebigs Ann. Chem.* **572**, 217 (1951). 10-4B
51MI1	O. O. Oraz and J. F. Salellas, *An. Asoc. Quim. Argent.* **39**, 175 (1951). 4–6; 8-3A*b*(iii)
51MI2	M. E. Fondovila, O. O. Orazi, and J. F. Salellas, *An. Asoc. Quim. Argent.* **39**, 184 (1951). 4–6; 6-4A
51MI3	S. Yamashiro, *J. Chem. Soc. Jpn.* **54**, 295 (1951). 8-3A*b*(vi)
51RTC581	H. J. den Hertog and W. P. Combe, *Recl. Trav. Chim. Pays-Bas* **70**, 581 (1951). 9-4C
51USP2571742	R. F. McCleary and J. A. Patterson, U.S. Pat., 2,571,742 (1951). 8-2A*b*(iv)
51ZOB281	A. P. Terentev and L. A. Yanovskaya, *Zh. Obshch. Khim.* **21**, 281 (1951). 4-5B; 6-8D
51ZOB1524	A. P. Terentev and G. M. Kadatskii, *Zh. Obshch. Khim.* **21**, 1524 (1951). 4-5B; 6-8A
52BSF713	P. Cagniant and P. Cagniant, *Bull. Soc. Chim. Fr.*, 713 (1952). 6-4A
52JA766,2285	R. Gaertner, *J. Am. Chem. Soc.* **74**, 766, 2285 (1952). 4-3E*b*; 8-2Ab(v)
52JA2185	R. Gaertner, *J. Am. Chem. Soc.* **74**, 2185 (1952). 4-3E*b*; 8-2A*b*(iv)
52JA2965	C. D. Hurd and K. L. Kreuz, *J. Am. Chem. Soc.* **74**, 2965 (1952). 6-3A
52JA4950	R. Gaertner, *J. Am. Chem. Soc.* **74**, 4950 (1952). 8-2A*b*(iv)
52JA6289	A. Murray and W. H. Langham, *J. Am. Chem. Soc.* **74**, 6289 (1952). 4-7A; 9-6A
52JCP1554	H. H. Jaffé, *J. Chem. Phys.* **20**, 1554 (1952). Table 11-6
52JCS1528,4158	E. A. Braude and J. S. Fawcett, *J. Chem. Soc.*, 1528,4158 (1952). 6-9A, 6-9C
52JCS2156	A. G. Caldwell and L. P. Walls, *J. Chem. Soc.*, 2156 (1952). 11-4A
52JCS2172,2184	V. Gold and F. L. Tye, *J. Chem. Soc.*, 2172, 2184 (1952). 2-1A
52JCS4870	J. J. Gallagher, G. T. Newbold, W. Sharp, and F. S. Spring, *J. Chem. Soc.*, 4870 (1952). 10-6A
52JGU189	A. P. Terentev and G. M. Kadatskii, *J. Gen. Chem. USSR (Engl. Transl.)* **22**, 189 (1952). 4-5B; 6-8A
52JGU1069	L. S. Efros, *J. Gen. Chem. USSR (Engl. Transl.)* **22**, 1069 (1952). 4–6; 8-C*b*(iii), 4B
52JPJ1263	Y. Mizuno, K. Adachi, and K. Nakamura, *J. Pharm. Soc. Jpn.* **72**, 1263 (1952). 8-C*b*(iii)
52MI1	K. Ganapathi and K. D. Kulkarni, *Curr. Sci.* **21**, 314 (1952). 7-4A*b*
52MI2	M. E. Fondovila, O. O. Orazi, and J. F. Salellas, *An. Asoc. Quim. Argent.* **39**, 61 (1952). 4–6; 6-4A
52MI3	D. J. Brown, *J. Appl. Chem.* **2**, 239 (1952). 10-4B
52QR63	R. D. Brown, *Q. Rev., Chem. Soc.* **6**, 63 (1952). 9-9A
53CA10519c	L. A. Kazitsyna, *Chem. Abstr.* **47**, 10519c (1953). 4-5B; 8-2A*b*(v),(vi)
53GEP869490	O. Trosken, Ger. Pat. 869,490 (1953). 4-5B,C; 7-7

53JA3278	D. A. Shirley, M. J. Danzig, and F. C. Canter, *J. Am. Chem. Soc.* **75**, 3278 (1953). 8-2A*b*(iv)
53JA3517	C. D. Hurd and H. J. Anderson, *J. Am. Chem. Soc.* **75**, 3517 (1953). 6-3A, 4A; 10-4C
53JA3697	J. Sicé, *J. Am. Chem. Soc.* **75**, 3697 (1953). 4-2A; 6-3A, 8A
53JA3865	H. C. Brown and B. Kanner, *J. Am. Chem. Soc.* **75**, 3865 (1953). 4-5B; 9-6D
53JA5517	G. Karmas and P. E. Spoerri, *J. Am. Chem. Soc.* **75**, 5517 (1953). 10-4C
53JA5758	J. L. Rabinowitz and S. Gurin, *J. Am. Chem. Soc.* **75**, 5758 (1953). 10-4B
53JCS2406	D. A. Brown and M. J. S. Dewar, *J. Chem. Soc.*, 2406 (1953). Table 11-6
53JGU263	A. P. Terentev and G. M. Kadatskii, *J. Gen. Chem. USSR (Engl. Transl.)* **23**, 263 (1953). 4-5B; 6-8A
53JOC1492	E. Sawicki, *J. Org. Chem.* **18**, 1492 (1953). 8-3A*b*(ii)
53JPJ701	M. Ohta, R. Hagiwara, and Y. Mizushima, *J. Pharm. Soc. Jpn.* **73**, 701 (1953). 7-4C*b*
53LA(581)117	W. Theilacker and F. Baxmann, *Justus Liebigs Ann. Chem.* **581**, 117 (1953). 11-4B
53MI1	L. A. Kazitsyna, *Ushenyl Zap. Mosk. im. H. V. Lomonosova*, No. 131, 5 (1950) [*CA* **47**, 10518c (1953)]. 4-5B; 6-8A
53PI(A)758	K. Ganapathi and K. D. Kulkarni, *Proc.—Indian Acad. Sci., Sect. A* **37A**, 758 (1953). 7-4A*b*
53USP2652405	R. E. Conary and R. F. McCleary, U.S. Pat. 2,652,405 (1953). 8-2A*b*(v)
53YZ823	T. Ukai, Y. Yamamoto, and S. Hirano, *Yakugaku Zasshi* **73**, 823 (1953). 4-2C; 11-6A
53YZ1023	I. Hirao, *Yakugaku Zasshi* **73**, 1023 (1953). 6-4A
53ZOB951	L. S. Efros, *Zh. Obshch. Khim.* **23**, 951 (1953). 8-C*b*(iii)
53ZOB1552	L. S. Efros and R. N. Levit, *Zh. Obshch. Khim.* **23**, 1552 (1953). 8-C*b*(iii)
54AK361	S. Gronowitz, *Ark. Kemi* **7**, 361 (1954). 4-2A; 6-8A
54BCJ423	K. Fukui, T. Yonezawa, and C. Nagata, *Bull. Chem. Soc. Jpn.* **27**, 423 (1954). 11-8A, Table 6
54CA1337i	H. Bojarska-Dahlig and T. Urbanski, *Chem. Abstr.* **48**, 1337i (1954). 9-6B
54CB127	H. Plieninger, *Chem. Ber.* **87**, 127 (1954). 8-2A*b*(v)
54CB1184	H. Scheibler, E. Keintzel, and K. Falk, *Chem. Ber.* **87**, 1184 (1954). 4-5B; 6-8A
54CI(L)1203	A. Holland and R. Slack, *Chem. Ind. (London)*, 1203 (1954). 10-5B
54CPB283	H. Otomasu, *Chem. Pharm. Bull.* **2**, 283 (1954). 11-4B
54JA632	C. Grundmann and A. Kreutzberger, *J. Am. Chem. Soc.* **76**, 632 (1954). 10-5D
54JA2447	E. Campaigne and P. A. Monroe, *J. Am. Chem. Soc.* **76**, 2447 (1954). 6-4A
54JA5775	W. J. Burlant and E. S. Gould, *J. Am. Chem. Soc.* **76**, 5775 (1954). 4-2A; 8-3A*b*(vi)
54JA5807	W. T. Smith and P. R. Ruby, *J. Am. Chem. Soc.* **76**, 5807 (1954). 11-4B

54JA6407	R. G. Johnson, H. B. Willis, and H. Gilman, *J. Am. Chem. Soc.* **76**, 6407 (1954). 4-2A; 8-3A*b*(iv)
54JBC(208)741	B. B. Brodie, J. Axelrod, P. A. Shore, and S. Udenfriend, *J. Biol. Chem.* **208**, 741 (1954). 4-5A; 11-6C
54JCS237	J. Cymerman-Craig and J. W. Loder, *J. Chem. Soc.*, 237 (1954). 6-7A
54JCS2293	I. L. Finar and K. E. Godfrey, *J. Chem. Soc.*, 2293 (1954). 7-8A
54JCS4142	R. M. Acheson, T. G. Hoult, and K. A. Barnard, *J. Chem. Soc.*, 4142 (1954). 4–6; 11-5A
54JCS4206	G. R. Barker, N. G. Luthy, and M. M. Dhar, *J. Chem. Soc.*, 4206 (1954). 4-5B; 10-6C; Table 11-6
54JGU1251	A. P. Terentyev, L. I. Belenky, and L. A. Yanovskaya, *J. Gen. Chem. USSR (Engl. Transl.)* **24**, 1251 (1954). 4–6; 6-4A
54JGU2212	N. V. Khromov-Borisov and R. S. Karlinskaya, *J. Gen. Chem. USSR (Engl. Transl.)* **24**, 2212 (1954). 4-5B; 10-6C
54JOC70	J. Sicé, *J. Org. Chem.* **19**, 70 (1954). 4-2A; 6-8A
54JOC510	O. Baine and G. F. Adamson *J. Org. Chem.* **19**, 510 (1954). 9-6B
54JOC894	J. F. Scully and E. V. Brown, *J. Org. Chem* **19**, 894 (1954). 4-5B; 6-8C
54RTC325	H. C. Beyerman, P. H. Berben, and J. S. Bontekoe, *Recl. Trav. Chim. Pays-Bas* **73**, 325 (1954). 7-5A
54USP2689850	N. Grier, U. S. Pat 2,689,850 (1954). 4-5B; 11-6B
54YZ674	T. Ukai, T. Yamamoto, and S. Kanemoto, *Yakugaku Zasshi* **74**, 674 (1954). 4-4A; 10-6B
54ZOB488	L. S. Efros, L. N. Kononkova, and Y. Eded, *Zh. Obshch. Khim.* **24**, 488 (1954). 4–6; 8-3B
54ZOB1265	A. P. Terentev, L. I. Belenkii, and L. A. Yanovskaya, *Zh. Obshch. Khim.* **24**, 1265 (1954). 8.2A*b*(iv)
55AJC100	R. D. Brown, *Aust. J. Chem.* **8**, 100 (1955). 7-5B, 10
55AK87	S. Gronowitz, *Ark. Kemi* **8**, 87 (1955). 6-4B
55AK343	S. Gronowitz and K. Halvarson, *Ark. Kemi* **8**, 343 (1955). 4-2A; 6-8A
55CA296e	E. Gustak, *Chem. Abstr.* **49**, 296e (1955). 7-4A*a*
55CA8257c	Y. Mizuno and K. Adachi, *Chem. Abstr.* **50**, 8257c (1955). 8-C*b*(iii)
55GEP926249	O. Trösken, Ger. Pat. 926,249 (1955). 4-5B; 7-7
55JA1061	G. E. Wiseman and E. S. Gould, *J. Am. Chem. Soc.* **77**, 1061 (1955). 8-3A*b*(ii)
55JA2902	H. S. Mosher and F. J. Welch, *J. Am. Chem. Soc.* **77**, 2902 (1955). 4-5B; 9-5C,6D
55JA3044	N. C. Deno, J. J. Jaruzelski, and A. Schriesheim, *J. Am. Chem. Soc.* **77**, 3044 (1955). 2-1A
55JA3410	W. E. Truce and F. J. Lotspeich, *J. Am. Chem. Soc.* **77**, 3410 (1955). 4-5C; 6-8A
55JA4066	E. C. Spaeth and C. B. Germain, *J. Am. Chem. Soc.* **77**, 4066 (1955). 4-3E*b*; 6-7A
55JA5939	F. G. Bordwell and H. Stange, *J. Am. Chem. Soc.* **77**, 5939 (1955). 8-2A*b*(iii)

55JA6379	H. Gilman and J. Eisch, *J. Am. Chem. Soc.* **77**, 6379 (1955). 11-5A
55JCS21	N. P. Buu-Hoi and N. D. Xuong, *J. Chem. Soc.*, 21 (1955). 4-3E*b*; 6-7A
55JCS211	S. F. Mason, D. J. Brown, and E. Hoerger, *J. Chem. Soc.*, 211 (1955). 10-4B
55JCS1205	I. L. Finar, *J. Chem. Soc.*, 1205 (1955). 7-8A
55JCS1565	W. Davies, F. C. James, S. Middleton, and Q. N. Porter, *J. Chem. Soc.*, 1565 (1955). 8-2A*b*(iv)
55JCS3275	K. H. Saunders, *J. Chem. Soc.*, 3275 (1955).
55JCS3609,3619,3622	V. Gold and D. P. N. Satchell, *J. Chem. Soc.*, 3609, 3619, 3622 (1955). 2-1A
55JOC657	K. Oita, R. G. Johnson, and H. Gilman, *J. Org. Chem.* **20**, 657 (1955). 4–6; 8-3A*b*(iii)
55LA(593)179	R. Hüttel, H. Wagner, and P. Jochum, *Justus Liebigs Ann. Chem.* **593**, 179 (1955). 4-6; 7-5B
55LA(593)200	R. Hüttel, O. Schäfer, and P. Jochum, *Justus Liebigs Ann. Chem.* **593**, 200 (1955). 7-5B
55MI1	D. J. Brown, *J. Appl. Chem.* **5**, 358 (1955). 10-4B
55NKZ311	J. Adachi, *Nippon Kagaku Zasshi* **76**, 311 (1955). 11-4B
55RTC1003	J. P. Wibaut and L. G. Heeringan, *Recl. Trav. Chim. Pays-Bas* **74**, 1003 (1955). 4-7A; 9-6A
55YZ966	T. Itai and H. Igeta, *Yakugaku Zasshi* **75**, 966 (1955). 10-4A
56ACS879	L. Melander and S. Olsson, *Acta Chem. Scand.* **10**, 879 (1956). 2-1A
56BAU627	Yu. L. Goldfarb, L. V. Antik, and P. A. Konstantinov, *Bull. Acad. Sci. USSR, Div. Chem. Sci. (Engl. Transl.)*, 627 (1956). 4-5B; 6-8A
56CI(L)1453	J. F. W. McOmie and J. H. Chesterfield, *Chem. Ind. (London)*, 1453 (1956). 10-6A
56G797	D. D. M. Casoni, A. Mangini, and R. Passerini, *Gazz. Chim. Ital.* **86**, 797 (1956). 7-4B*c*
56JA401	R. R. Herr, T. Enkoji, and T. J. Bardos, *J. Am. Chem. Soc.* **78**, 401 (1956). 4-5B; 10-6C
56JA578	N. C. Deno and R. Stein. *J. Am. Chem. Soc.* **78**, 578 (1956). 3-4C
56JA2136	B. W. Langley, *J. Am. Chem. Soc.* **78**, 2136 (1956). 4-7A; 10-6F
56JA4071	G. Karmas and P. E. Spoerri, *J. Am. Chem. Soc.* **78**, 4071 (1956). 10-4C,5C
56JCS212	E. P. Hart, *J. Chem. Soc.*, 212 (1956). 11-4B, 5B
56JCS272	R. D. Brown, *J. Chem. Soc.*, 272 (1956). 9-9A
56JCS2312	D. J. Brown, *J. Chem. Soc.*, 2312 (1956). 10-4B
56JCS2743	V. Gold and D. P. N. Satchell, *J. Chem. Soc.*, 2743 (1956). 2-1A
56JCS3911	D. P. N. Satchell, *J. Chem. Soc.*, 3911 (1956). 2-1A
56JCS4114	J. Cymerman-Craig, G. N. Vaughan, and W. K. Warburton, *J. Chem. Soc.*, 4114 (1956). 6-3A
56JCS4257	G. A. Olah, S. J. Kuhn, and A. Mlinko, *J. Chem. Soc.*, 4257 (1956). 6-3A
56JCS4858	C. Eaborn, *J. Chem. Soc.*, 4858 (1956). 4-1C; 6-8A

REFERENCES 413

56JOC177	J. W. Daly and B. E. Christensen, *J. Org. Chem.* **21**, 177 (1956). 10-4B
56JOC584	B. B. Corson, H. E. Tiefenthal, G. R. Atwood, W. J. Heintzelman, and W. J. Reilly, *J. Org. Chem.* **21**, 584 (1956). 4-3A; 8-2A*b*(v)
56JOC1104	D. V. Joshi, J. R. Merchant, and R. C. Shah, *J. Org. Chem.* **21**, 1104 (1956). 4-5B; 11-6B
56LA(598)186	R. Hüttel, O. Schäfer, and G. Welzel, *Justus Liebigs Ann. Chem.* **598**, 186 (1956). 7-5B
56YZ230,234	H. Hirano and H. Yonemoto, *Yakugaku Zasshi* **76**, 230, 234 (1956). 10-6E
56ZOB557	A. P. Terentev, A. N. Kost, and V. A. Smit, *Zh. Obshch. Khim.* **26**, 557 (1956). 4-3A, D; 8-2A*b*(v)
57CB1837	H. Bredereck, R. Gompper, and H. Seiz, *Chem. Ber.* **90**, 1837 (1957). 10-4, 5B
57CJC21	H. J. Anderson, *Can. J. Chem.* **35**, 21 (1957). 6-7D
57JA2188	L. L. Bennett and H. T. Baker, *J. Am. Chem. Soc.* **79**, 2188 (1957). 4-5B; 7-7
57JA3800	W. J. Raich and C. S. Hamilton, *J. Am. Chem. Soc.* **79**, 3800 (1957). 6-3A
57JA5479	H. Gilman and J. J. Eisch, *J. Am. Chem. Soc.* **79**, 5479 (1957). 4-5B; 11-6B
57JCS345	M. J. S. Dewar and D. S. Urch, *J. Chem. Soc.*, 345 (1951). 8.3A*b*(ii)
57JCS2146	W. R. Boon, *J. Chem. Soc.*, 2146 (1957). 10-4B
57JCS2398	J. H. Binks and J. H. Ridd, *J. Chem. Soc.*, 2398 (1957). 4-4B; 8-2A*b*(vi)
57JCS2518	M. J. S. Dewar and P. M. Maitlis, *J. Chem. Soc.*, 2518 (1957). 11-4B, 11.H.1.
57JCS2521	M. J. S. Dewar and P. M. Maitlis, *J. Chem. Soc.*, 2521 (1957). 11-4A,B, Table 6
57JCS3024	I. L. Finar and R. J. Hurlock, *J. Chem. Soc.*, 3024 (1957). 7-4B*c*
57JCS3314	I. L. Finar and G. H. Lord, *J. Chem. Soc.*, 3314 (1957). 4-3E*a*; 7-6
57JCS4845	R. Hull, *J. Chem. Soc.*, 4845 (1957). 10-6D
57JIC35,45	J. R. Merchant and R. C. Shah, *J. Indian Chem. Soc.* **34**, 35, 45 (1957). 4-5B; 11-6B
57JOC507	E. Sawicki and A. Carr, *J. Org. Chem.* **22**, 507 (1957). 8-C*b*(iii)
57JOC884	J. R. Merchant and R. C. Shah, *J. Org. Chem.* **22**, 884 (1957). 4-5B; 11-6B
57MI1	R. Motoyama, E. Imoto, and J. Ogawa, *J. Chem. Soc. Jpn.*, **78**, 962 (1957). 6-8A
57ZOB2599	V. G. Pesin and A. M. Khaletskii, *Zh. Obshch. Khim.* **27**, 2599 (1957). 8-C*b*(iii)
57ZOB3210	N. K. Kochetkov, E. D. Khomutova, M. Ya. Karpeysky, and R. M. Khomotov, *Zh. Obshch. Khim.* **27**, 3210 (1957). 4-3B; 7-6
58AC(R)783	D. D. M. Casoni, *Ann. Chim.* (*Rome*) **48**, 783 (1958). 7-4B*c*
58AG571	H. Bredereck, R. Gompper, and H. Herlinger, *Angew. Chem.* **70**, 571 (1958). 10-5B

414 REFERENCES

58AK269,295	S. Gronowitz, *Ark. Kemi* **13**, 269,295 (1958). 4-2A; 6-8A
58CA14317c	K. Kikuchi, *Chem. Abstr.* **52**, 14317c (1958). Table 8-17
58CPB77	H. Otomasu, *Chem. Pharm. Bull.* **6**, 77 (1958). 11-4B
58CPB482	S. Senda, A. Suzui, M. Honda, and H. Fujimura, *Chem. Pharm. Bull.* **6**, 482 (1958). 10-4B
58CPB566	H. Otomasu and S. Nakajima, *Chem. Pharm. Bull.* **6**, 566 (1958). 11-4B
58DOK(123)295	E. M. Panov and K. A. Kocheskov, *Dokl. Akad. Nauk SSSR* **123**, 295 (1958). 4-2D; 6-8A
58JA4979	H. C. Brown and Y. Okamoto, *J. Am. Chem. Soc.* **80**, 4979 (1958). 5, 5-Ba; 7-4Ab
58JA5329	M. A. Paul, *J. Am. Chem. Soc.* **80**, 5329 (1958). 3-2C
58JA6271	P. W. Alley and D. A. Shirley, *J. Am. Chem. Soc.* **80**, 6271 (1958). 4-2A; 7-8B
58JCS825	W. O. Sykes, *J. Chem. Soc.*, 825 (1958). 11-4B
58JCS3079	M. J. S. Dewar and D. S. Urch, *J. Chem. Soc.*, 3079 (1958). 8-3Ab(ii)
58JGU1288	V. N. Ivanova, *J. Gen. Chem. USSR (Engl. Transl.)* **28**, 1288 (1958). 6-3A
58JGU2214	L. S. Efros and A. V. Eltsov, *J. Gen. Chem. USSR (Engl. Transl.)* **28**, 2214 (1958). 8-4B
58JOC(22)610	E. Sawicki and A. Carr, *J. Org. Chem.* **22**, 610 (1958). 8-Cb(iii)
58JOC(23)1603	C. F. Koelsch and W. H. Gumprecht, *J. Org. Chem.* **23**, 1603 (1958). 10-4C
58LA(612)158	W. Pfleiderer and K. H. Schundehutte, *Justus Liebigs Ann. Chem.* **612**, 158 (1958). 4-4A; 10-6B
58LA(612)173	W. Pfleiderer and G. Strauss, *Justus Liebigs Ann. Chem.* **612**, 173 (1958). 4-3Eb; 10-6D
58MP247	C. MacLean, J. H. van der Waals, and E. L. Mackor, *Mol. Phys.* **1**, 247 (1958). 2-1A
58MI1	A. Grimison and J. H. Ridd, *Proc. Chem. Soc.*, 256 (1958). 7-5A
58RTC340	M. van Ammers and H. J. den Hertog, *Recl. Trav. Chim. Pays-Bas* **77**, 340 (1958). 4-2C; 9-6A; 11-6A
58RTC963	H. J. den Hertog, H. C. van der Plas, and D. J. Buurman, *Recl. Trav. Chim. Pays-Bas* **77**, 963 (1958). 4-5B; 9-6D
58ZOB62	L. S. Efros and A. V. Eltsov, *Zh. Obshch. Khim.* **28**, 62 (1958). 8-Cb(iii)
58ZOB359	N. K. Kochetkov and E. E. Khomutova, *Zh. Obshch. Khim.* **28**, 359 (1958). 7-4Ba, 8A
58ZOB2376	N. K. Kochetkov, E. D. Khomutova, and M. V. Bazilevsky, *Zh. Obshch. Khim.* **28**, 2376 (1958). 4-3B; 7-6
59AJC152	R. D. Brown and B. A. W. Coller, *Aust. J. Chem.* **12**, 152 (1959). Table 8-1
59AK507	L. Melander and P. C. Myhre, *Ark. Kemi* **13**, 507 (1959). 2-1A
59BAU1258	E. N. Safonova, V. M. Belikov, and S. S. Novikov, *Bull. Acad. Sci. USSR, Div. Chem. Sci. (Engl. Transl.)*, 1258 (1959). 6-3B
59BAU1925	Ya. L. Goldfarb, G. I. Gorushkina, and B. F. Federov, *Bull. Acad. Sci. USSR, Div. Chem. Sci. (Engl. Transl.)*, 1925 (1959). 4-2A; 6-8A

59BCJ971	Y. Yukawa and Y. Tsuno, *Bull. Chem. Soc. Jpn.* **32**, 971 (1959). 5; 8-2A*c*(ii)
59CB1944	R. Gompper and O. Christmann, *Chem. Ber.* **92**, 1944 (1959). 7-4A*a*
59CI(L)1449	J. J. Eisch, *Chem. Ind. (London)*, 1449 (1959). 11-5A
59CPB267	E. Ochiai and C. Kaneko, *Chem. Pharm. Bull.* **7**, 267 (1959). 11-4A
59G1539	D. D. M. Casoni, *Gazz. Chim. Ital.* **89**, 1539 (1959). 7-4B*c*
59JA1935	R. A. Barnes, *J. Am. Chem. Soc.* **81**, 1935 (1959). 9-8, 9A,C,D,F, Table 17
59JA5166	W. T. Caldwell and G. E. Jaffe, *J. Am. Chem. Soc.* **81**, 5166 (1959). 4-5B; 10-6C
59JA5509	A. J. Kresge and Y. Chiang, *J. Am. Chem. Soc.* **81**, 5509 (1959). 2-1A
59JA5790	N. C. Deno, P. T. Groves, and G. Saines, *J. Am. Chem. Soc.* **81**, 5790 (1959). 2-1A
59JA6297	F. H. Case and J. A. Brennan, *J. Am. Chem. Soc.* **81**, 6297 (1959). 11-4B
59JCS525	R. R. Hunt, J. F. W. McOmie, and E. R. Sayer, *J. Chem. Soc.*, 525 (1959). 10-4B
59JCS1819	I. L. Finar and G. H. Lord, *J. Chem. Soc.*, 1819 (1959). 4-3E*a*; 7-6
59JCS2299	F. B. Deans and C. Eaborn, *J. Chem. Soc.*, 2299 (1959). 4-1C; 6-8A
59JCS3061	A. Adams and R. Slack, *J. Chem. Soc.*, 3061 (1959). 7-4B*b*
59JCS3451	R. D. Brown and R. D. Harcourt, *J. Chem. Soc.*, 3451 (1959). 9-9A; Table 11-6
59JGU2003	Ya. L. Goldfarb, M. A. Kalik, and M. L. Kirmalova, *J. Gen. Chem. USSR (Engl. Transl.)* **29**, 2003 (1959). 4-2A; 6-8A
59JGU3592	Ya. L. Goldfarb, M. A. Kalik, and M. L. Kirmalova, *J. Gen. Chem. USSR (Engl. Transl.)* **29**, 3592 (1959). 6-8A
59JIC434	P. N. Bhargava and M. Nagabhushanam, *J. Indian Chem. Soc.* **36**, 434 (1959). 7-8A
59JOC187	J. B. Dickey, E. B. Towne, M. S. Bloom, W. H. Moore, H. M. Hill, H. Heynemann, D. G. Hedberg, D. C. Sievers, and M. V. Otis., *J. Org. Chem.* **24**, 187 (1959). 7-4A*b*
59LA(626)83,92	R. Gompper and H. Rühle, *Justus Liebigs Ann. Chem.* **626**, 83, 92 (1959). 4–6; 7-5A
59MI1	P. B. D. de la Mare and J. H. Ridd, "Aromatic Substitution, Nitration and Halogenation," pp. 159, 199. Butterworth, London, 1959. 3; 4–6; 11-5A
59RTC586	M. van Ammers and H. J. den Hertog, *Recl. Trav. Chim. Pays-Bas.* **78**, 586 (1959). 4-5B; 9-6D
59T225	T. L. V. Ulbricht, *Tetrahedron* **6**, 225 (1959). 4-7A; 10-6F
59ZOB535	N. K. Kochetkov and E. D. Khomutova, *Zh. Obshch. Khim.* **29**, 535 (1959). 4-5B; 7-4B*a*, 7
60AC(R)351	M. Robba, *Ann. Chim. (Rome)* **5**, 351 (1950). 10-5B
60ACS219	P. C. Myhre, *Acta Chem. Scand.* **14**, 219 (1960). 3-2A
60AK309	S. Gronowitz and R. Håkansson, *Ark. Kemi* **16**, 309 (1960). 4-2A; 6-8A
60AK363	S. Gronowitz, *Ark. Kemi* **16**, 363 (1960). 4-2A; 6-8A

60AK563	R. A. Hoffman and S. Gronowitz, *Ark. Kemi* **16**, 563 (1960). 6-3A
60BAU1700,1705	B. P. Fedorov and F. M. Stoyanovich, *Bull. Acad. Sci. USSR (Engl. Transl.)* 1700, 1705 (1960). 4-2A; 6-7A, 8A
60BSF793	A. Buzas and J. Teste, *Bull. Soc. Chim. Fr.*, 793 (1960). 4-5B,C; 6-8A
60CA17368i	G. Guardiano, A. Ricca, and A. Quilico, *Chem. Abstr.* **54**, 17368i (1960). 7-4B*a*
60CA24661c	G. Travagli and G. Mazzoli, *Chem. Abstr.* **54**, 24661c (1960). 7-8A
60CB2405	H. Bredereck, W. Jentzsch, and G. Morlock, *Chem. Ber.* **93**, 2405 (1960). 10-5B
60CCC1058	R. Lukes, M. Janda, and K. Kefurt, *Collect. Czech. Chem. Commun.* **25**, 1058 (1960). 4-3B; 6-6
60CPB550	H. Igeta, *Chem. Pharm. Bull.* **8**, 550 (1960). 10-4A
60CPB999	T. Itai and S. Suzuki, *Chem. Pharm. Bull.* **8**, 999 (1960). 10-4A
60G356	P. Pino, F. Piacenti, and G. Fatti, *Gazz. Chim. Ital.* **90**, 356 (1960). 4–6; 7-5B
60HC(14,1–4)	E. Klingsberg *et al.*, *Chem. Heterocycl. Compd.* **14**, Parts I-IV (1960). 9-1
60JA15	E. Hogfeldt and J. Bigeleisen, *J. Am. Chem. Soc.* **82**, 15 (1960). 2-1Gc(iv)
60JA991	J. H. Burckhalter, R. J. Seiwald, and H. C. Scarborough, *J. Am. Chem. Soc.* **82**, 991 (1960). 4-3B,C; 10-6E
60JA1447	H. Wynberg and A. Bantjes, *J. Am. Chem. Soc.* **82**, 1447 (1960). 6-8A
60JA2742	M. S. Grant and H. R. Snyder, *J. Am. Chem. Soc.* **82**, 2742 (1960). 4-5E; 8-2A*b*(vi)
60JA2965	R. W. Taft, *J. Chem. Soc.* **82**, 2965 (1960). 2-1A
60JA4430	E. E. Garcia, C. V. Greco, and I. M. Hunsberger, *J. Am. Chem. Soc.* **82**, 4430 (1960). 4–6; 11-5A
60JA4729	R. H. Boyd, R. W. Taft, A. P. Wolf, and D. R. Christman, *J. Am. Chem. Soc.* **82**, 4729 (1960). 2-1A
60JCS561	P. B. D. de la Mare, M. Kiamud-din, and J. H. Ridd, *J. Chem. Soc.*, 561 (1960). 4–6; 11-5A
60JCS938	M. Martin-Smith and S. T. Reid, *J. Chem. Soc.*, 938 (1960). 8-2A*b*(iii)
60JCS1480	C. Eaborn and R. Taylor, *J. Chem. Soc.*, 1480 (1960). 2-1B*a*
60JCS1794	T. K. Adler and A. Albert, *J. Chem. Soc.*, 1794 (1960). 11-4B
60JCS2461	V. Gold, R. W. Lambert, and D. P. N. Satchell, *J. Chem. Soc.*, 2461 (1960). 2-1A
60JCS3301	C. Eaborn and R. Taylor, *J. Chem. Soc.*, 3301 (1960). 2-1A, B*a*, E*b*, Fe(i); 6-1A; 9-2A
60JCS3367	R. M. Acheson, B. Adcock, G. M. Glover, and L. E. Sutton, *J. Chem. Soc.*, 3367 (1960). 11-4A, 5A
60JGU1531	T. V. Gortinskaya and M. N. Shchukina, *J. Gen. Chem. USSR (Engl. Transl.)* **30**, 1531 (1960). 10-4A
60JOC1583	R. C. Elderfield and R. N. Prasad, *J. Org. Chem.* **25**, 1583 (1960). 10-5B

60JOC1906	R. H. Wiley and Y. Yamamoto, *J. Org. Chem.* **25**, 1906 (1960). 10-6D
60JOC1916	D. G. Crosby and R. V. Berthold, *J. Org. Chem.* **25**, 1916 (1960). 4–6; 10-5B
60LA(633)158	F. G. Fischer, W. P. Neumann, and J. Roch, *Justus Liebigs Ann. Chem.* **633**, 158 (1960). 10-4B
60LA(634)84	H. O. Wirth, O. Königsten, and W. Kern, *Justus Liebigs Ann. Chem.* **634**, 84 (1960). 4–6; 6-4A
60MI1	N. S. Drozdov and V. P. Krylov, *Kinet. Katal., Akad. Nauk SSSR, Sb. Statei*, 248 (1960) [*CA* **58** 422 (1963)]. 4-4A; 10-6B
60MI2	L. Pauling, "The Nature of the Chemical Bond." Cornell Univ. Press, New York, 1960. 6-10A
60RTC737	G. Dallinga and G. Ter Maten, *Recl. Trav. Chim. Pays-Bas* **79**, 737 (1960). 2-1A
60T(8)23	R. D. Brown and R. D. Harcourt, *Tetrahedron* **8**, 23 (1960). 9-9A; 11-8B
60T(10)81	G. Del Re, *Tetrahedron* **10**, 81 (1960). 8-C*a*
60T(10)215	M. K. Bhattacharjee, R. B. Mitra, B. D. Tilak, and M. R. Venkiteswaren, *Tetrahedron* **10**, 215 (1960). 4-3E*b*; 8-2A*b*(v)
60TL(21)12	B. L. Hinman and J. Lang, *Tetrahedron Lett.* **21**, 12 (1960). 2-1A
60YZ712	T. Nakagome, *Yakugaku Zasshi* **80**, 712 (1960). 10-4A
60ZOB899	R. S. Karlinskaya and N. V. Khromov-Borisov, *Zh. Obshch. Khim.* **30**, 899 (1960). 10-5B
60ZOB203	I. I. Grandberg and A. N. Kost, *Zh. Obshch. Khim.* **30**, 203 (1960). 4-3E*b*; 7-6
60ZOB1269	N. K. Kochetkov and E. D. Khomutova, *Zh. Obshch. Khim.* **30**, 1269 (1960). 7-8A
60ZOB2931	I. I. Grandbert, A. N. Kost, and N. N. Zheltikova, *Zh. Obshch. Khim.* **30**, 2931 (1960). 7-8A
60ZOB2942	I. I. Grandberg and A. N. Kost, *Zh. Obshch. Khim.* **30**, 2942 (1960). 7-6
61AC867	D. E. Blair and H. Diehl, *Anal. Chem.* **33**, 867 (1961). 4-5B; 11-6B
61BRP845562	G. A. Collins, Br. Pat. 845,562 (1961). 4-3A; 11-6C
61BSF1534	R. Royer, P. Demerseman, and A. Cheutin, *Bull. Soc. Chim. Fr.*, 1534 (1961). 4-3E*b*; 8-2A*b*(v)
61CPB149	T. Itai and S. Sako, *Chem. Pharm. Bull.* **9**, 149 (1961). 10-4A
61CPB414	M. Hamana and M. Yamazaki, *Chem. Pharm. Bull.* **9**, 414 (1961). 4–6; 11-5A
61CR(252)2419	M. Bercol-Vatteroni, R. C. Moreau, and P. Reynaud, *C. R. Hebd. Seances Acad. Sci.* **252**, 2419 (1961). 6-3A
61JA2154	C. G. Swain and A. S. Rosenberg, *J. Am. Chem. Soc.* **83**, 2154 (1961). 8-2A*b*(i)
61JA2877	A. J. Kresge and Y. Chiang, *J. Am. Chem. Soc.* **83**, 2877 (1961). 2-1A
61JA2934	E. Klingsberg, *J. Am. Chem. Soc.* **83**, 2934 (1961). 7-4B*d*
61JA4956	J. F. Bunnett, *J. Am. Chem. Soc.* **83**, 4956 (1961). 3-4C

61JCS1298	D. J. Brown and J. S. Harper, *J. Chem. Soc.*, 1298 (1961). 10-4B
61JCS247	C. Eaborn and R. Taylor, *J. Chem. Soc.*, 247 (1961). 2-1E*a,b*
61JCS2388	C. Eaborn and R. Taylor, *J. Chem. Soc.*, 2388 (1961). 2-1B*b*; 8-2A*b*(i); 11-7; 12
61JCS2733	I. L. Finar and M. Manning, *J. Chem. Soc.*, 2733 (1961). 4-3E*a*; 7-6
61JCS2769	I. L. Finar and D. B. Miller, *J. Chem. Soc.*, 2769 (1961). 7-5B
61JCS2825	E. R. Ward and W. H. Poesche, *J. Chem. Soc.*, 2825 (1961). 8-C*b*(iii)
61JCS3851	W. J. Barry, *J. Chem. Soc.*, 3851 (1961). 7-7
61JCS4921	C. Eaborn and J. A. Sperry, *J. Chem. Soc.*, 4921 (1961). 4-1C; 6-8A,C; 8-2A*b*(vi), 3A*b*(v)
61JCS4927	R. Baker, C. Eaborn, and R. Taylor, *J. Chem. Soc.*, 4927 (1961). 2-1E*b*
61JCS5077	R. Baker and C. Eaborn, *J. Chem. Soc.*, 5077 (1961). 8-3A*b*(i), Scheme 17
61JOC359,363	E. Campaigne and W. E. Kreighbaum, *J. Org. Chem.* **26**, 359, 363 (1961). 4-3E*b*; 8-2A*b*(v)
61JOC455	J. A. Carbon, *J. Org. Chem.* **26**, 455 (1961). 10-4B
61JOC526	J. J. Fox and D. Van Praag, *J. Org. Chem.* **26**, 526 (1961). 10-4B
61JOC1118	P. K. Chang, *J. Org. Chem.* **26**, 1118 (1961). 10-5D
61JOC2360	H. Gainer, M. Kokorudz, and W. K. Langdon, *J. Org. Chem.* **26**, 2360 (1961). 10-5C
61JOC3863	R. C. Elderfield and R. N. Prasad, *J. Org. Chem.* **26**, 3863 (1961). 4-5B; 10-6C
61JOC4504	S. Inoue, A. J. Saggiomo, and E. A. Nodiff, *J. Org. Chem.* **216**, 4504 (1961). 10-4B
61JSP58	B. Bak, D. Christensen, L. Hansen-Nygaard, and J. Restrup-Andersen, *J. Mol. Spectrosc.* **7**, 58 (1961). 6-10A
61M677	M. Pailer and E. Romberger, *Monatsch. Chem.* **92**, 677 (1961). 8-2A*b*(vi)
61MI1	S. Senda, H. Izumi, M. Kano, and J. Tsubota, *Gifu Yakka Daigaku Kiyo* **11**, 62 (1961) [CA 57, 11194 (1961)]. 10-4B
61M2	A. A. Frost and R. G. Pearson, "Kinetics and Mechanism," p. 145. Wiley, New York, 1961. 3-4H
61M13	V. Gold, *Proc. Chem. Soc.*, 453 (1961). 2-1A
61TL32	H. C. van der Plas, H. J. den Hertog, M. van Ammers, and M. B. Haase, *Tetrahedron Lett.*, 32 (1961). 9-5C
61USP2950283	L. L. Zemphier, *U. S. Pat.* 2,950,283 (1961). 4-5B; 11-6B
61ZOB1919	A. P. Grekov and R. S. Azen, *Zh. Obshch. Khim.* **31**, 1919 (1961). 7-4C*a*
62AK(18)513	S. Gronowitz and B. Gestblom, *Ark. Kemi* **18**, 513 (1962). 4-4A; 6-8A
62AK(19)499	B. Östman, *Ark. Kemi* **19**, 499 (1962). 6-3A
62AK(19)527	B. Östman, *Ark. Kemi* **19**, 527 (1962). 6-3A
62BCJ1420	T. Sone, K. Takahashi, and Y. Matsuki, *Bull. Chem. Soc. Jpn.* **35**, 1420 (1962). 6-3A
62CCC372	M. Janda and F. Dvorak, *Collect. Czech. Chem. Commun.* **27**, 372 (1962). 4-3B; 6-6

62CCC2550	Z. Budesinsky, V. Jelinsky, and J. Prikryl, *Collect. Czech. Chem. Commun.* **27**, 2550 (1962). 4–6; 10-5B
62CI(L)1057	M. W. Austin, M. Brickman, and J. H. Ridd, *Chem. Ind. (London)*, 1057 (1962). 9-4A; 11-4A
62CPB643	T. Itai and S. Natsume, *Chem. Pharm. Bull.* **10**, 643 (1962). 10-4A
62CPB933	T. Itai and S. Sako, *Chem. Pharm. Bull.* **10**, 933 (1962). 10-4A
62JA1658	H. C. Brown and G. Marino, *J. Am. Chem. Soc.* **84**, 1658 (1962). 6-7A
62JA2534	R. L. Hinman and E. B. Whipple, *J. Am. Chem. Soc.* **84**, 2534 (1962). 8-2Ab(i)
62JA3687	G. A. Olah, S. J. Kuhn, S. H. Flood, and J. C. Evans, *J. Am. Chem. Soc.* **84**, 3687 (1962). 3-2C
62JA3778	W. M. Schubert and R. H. Quacchia, *J. Chem. Soc.* **84**, 3778 (1962). 2-1A
62JA3976	A. J. Kresge and Y. Chiang, *J. Am. Chem. Soc.* **84**, 3976 (1962). 2-1Gc(v)
62JA4343	A. J. Kresge, G. W. Barry, K. R. Charles, and Y. Chiang, *J. Am. Chem. Soc.* **84**, 4343 (1962). 2-1A
62JCS291	M. D. Johnson and J. H. Ridd, *J. Chem. Soc.*, 291 (1962). 11-4B, 5A
62JCS1170	G. W. H. Cheeseman, *J. Chem. Soc.*, 1170 (1962). 4–6; 11-4B,5B
62JCS2382	R. Baker, C. Eaborn, and J. A. Sperry, *J. Chem. Soc.*, 2382 (1962). 8-2Ab(i)
62JCS2454	J. W. Barton and M. A. Cockett, *J. Chem. Soc.*, 2454 (1962). 11-4B
62JCS3172	D. J. Brown and N. W. Jacobsen, *J. Chem. Soc.*, 3172 (1962). 4-3A; 10-6E
62JCS4384	J. F. Corbett, P. F. Holt, and M. L. Vickery, *J. Chem. Soc.*, 4384, (1962). 11-4B
62JCS4860	J. F. Corbett, P. F. Holt, and M. L. Vickery, *J. Chem. Soc.*, 4860 (1962). 11-4B
62JCS4881	R. Taylor, *J. Chem. Soc.*, 4881 (1962). 5; 9-7Aa, 9A; 12
62JGU1515	F. N. Stepanov and N. I. Grineva, *J. Gen. Chem. USSR (Engl. Transl.)* **32**, 1515 (1962). 8-Ba
62JOC174	S. L. Shapiro, V. Bandurco, and L. Freedman, *J. Org. Chem.* **27**, 174 (1962). 4-3A; 10-6E
62JOC1318	J. J. Eisch, *J. Org. Chem.* **27**, 1318 (1962). 11-5A
62JOC2026	S. F. Bedell, E. C. Spaeth, and J. M. Bobbitt, *J. Org. Chem.* **27**, 2026 (1962). 4-3A; 8-2Ab(v)
62JOC2282	R. A. Parent, *J. Org. Chem.* **27**, 2282 (1962). 7-4Ab
62JSP124	B. Bak, D. Christensen, W. B. Dixon, L. Hansen-Nygaard, R. Anderson, and M. Schottlander, *J. Mol. Spectrosc.* **9**, 124 (1962). 6-10A
62MI1	L. R. Davidenkov, *Med. Prom-st. SSSR* **16**, 25 (1962)[*CA* **58**, 4565 (1963)]. 10-4B
62MI2	F. Dvorak, *Tech. Publ., Stredisko Tech. Inf. Potravin. Prum.* **161**, 56 (1962) [*CA* **60**, 543a (1964)]. 4-3B; 6-6
62RC967	L. S. Wieczorek and T. Talik, *Rocz. Chem.* **36**, 967 (1962). 9-4A

62RTC124	M. van Ammers and H. J. den Hertog, *Recl. Trav. Chim. Pays-Bas* **81**, 124 (1962). 4-2C; 9-6A; 11-6A
62RTC443	A. R. Katritzky, R. G. Shepherd, and A. J. Waring, *Recl. Trav. Chim. Pays-Bas* **81**, 443 (1962). 10-4B
62RTC864	H. J. den Hertog, L. Van der Does, and C. A. Landheer, *Recl. Trav. Chim. Pays-Bas* **81**, 864 (1962). 4–6; 9-5A
62T227	M. van Ammers, H. J. den Hertog, and B. Haase, *Tetrahedron* **18**, 227 (1962). 4–6; 9-5C
62T507	G. Coppens and J. Nasielski, *Tetrahedron* **18**, 507 (1962). 11-8A, Table 6, Scheme 7
62YZ253,1005	T. Nakagome, *Yakugaku Zasshi* **82**, 253,1005 (1962). 10-4A
63AC(R)1860	A. Ricci, A. Martani, O. Graziani, and M. Letizia, *Ann. Chim. (Rome)* **53**, 1860 (1963). 8-Cb(iii)
63ACS268	L. -B. Agenä, *Acta Chem. Scand.* **17**, 268 (1963). 4-5E; 8-2Ab(vi)
63AG(E)309	C. Grundmann, *Angew. Chem., Int. Ed. Engl.* **2**, 309 (1963). 10-4D, 6C
63AG(E)714	F. Hübenett, F. H. Flock, W. Hansel, H. Heinze, and H. Hofman, *Angew. Chem., Int. Ed. Engl.* **2**, 714 (1963). 4-5B, 6; 7-4Bb, 5B, 7-7
63AHC(1)1	S. Gronowitz, *Adv. Heterocycl. Chem.* **1**, 1 (1963). 6.
63AK191	S. Gronowitz and V. Vilks, *Ark. Kemi* **21**, 191 (1963). 6-4A
63AP035	L. M. Stock and H. C. Brown, *Adv. Phys. Org. Chem.* **1**, 35 (1963). 5.
63B1298	W. Haffert, R. Lundin, and L. L. Ingraham, *Biochemistry* **2**, 1298 (1963). 7-3C
63BSF1651	J. Tirouflet and P. Fournari, *Bull. Soc. Chim. Fr.*, 1651 (1963). 6-3A
63CA(58)2422c,d	V. P. Lopatinskii, E. E. Sirotkina, and M. M. Anosova, *Chem. Abstr.* **58**, 2422c,d (1963). 4-3Eb; 8-3Ab(iv)
63CA(58)5606h	V. P. Lopatinskii, E. E. Sirotkina, and M. M. Anosova, *Chem. Abstr.* **58**, 5606h (1963). 8-2Ab(v)
63CB2977	D. Soll and W. Pfleiderer, *Chem. Ber.* **96**, 2977 (1963). 10-4B
63CCC2089	J. Koutecky and R. Zahradnik, *Collect. Czech. Chem. Commun.* **28**, 2089 (1963) Table 11-6
63CI(L)1283	R. B. Moodie, K. Schofield, and M. J. Williamson, *Chem. Ind. (London)*, 1283 (1963). 11-4A
63CI(L)1840	M. Kiamuddin and A. K. Choudhury, *Chem. Ind. (London)*, 1840 (1963). 11-5A
63CJC274	B. M. Lynch and T. L. Chan, *Can. J. Chem.* **41**, 274 (1963). 7-4Cc
63CJC1540	M. A. Khan, B. M. Lynch, and Y. Y. Hung, *Can. J. Chem.* **41**, 1540 (1963). 7-4Bc
63CJC2380	B. M. Lynch, *Can. J. Chem.* **41**, 2380 (1963). 4–6; 7-5C
63CPB29	M. Ogata and H. Kano, *Chem. Pharm. Bull.* **11**, 29 (1963). 10-4A
63CPB35	M. Ogata and H. Kano, *Chem. Pharm. Bull.* **11**, 35 (1963). 10-4A
63CPB268	I. Suzuki, T. Nakashima, and T. Itai, *Chem. Pharm. Bull.* **11**, 268 (1963). 11-4B

63CPB342	T. Itai and S. Natsume, *Chem. Pharm. Bull.* **11**, 342 (1963). 10-4A
63CPB726	T. Nakagome, *Chem. Pharm. Bull.* **11**, 726 (1963). 10-4A
63CPB1326	I. Suzuki, T. Nakashima, and N. Nagasawa, *Chem. Pharm. Bull.* **11**, 1326 (1963). 11-4A
63GI196	G. Palazzo and G. Corsi, *Gazz. Chim. Ital.* **93**, 1196 (1963). 7-4C*a*
63JA329	B. B. P. Tice, I. Lee, and F. H. Kendall, *J. Am. Chem. Soc.* **85**, 329 (1963). 2-1F*d*
63JA2524	B. C. Challis and F. A. Long, *J. Am. Chem. Soc.* **85**, 2524 (1963). 2-1A; 8-2A*b*(i)
63JA4044	P. Haake and W. B. Miller, *J. Am. Chem. Soc.* **85**, 4044 (1963). 7-3B
63JCS736	G. E. Ficken and D. J. Fry, *J. Chem. Soc.*, 736 (1963). 8-C*b*(iii)
63JCS1276	D. J. Brown and J. S. Harper, *J. Chem. Soc.*, 1276 (1963). 4–6; 10-5B
63JCS1363	T. van Es and O. G. Backeberg, *J. Chem. Soc.*, 1363 (1963). 4-5B, C; 7-4A*a*, 5A, 7
63JCS2032	D. Buttimore, D. H. Jones, R. Slack, and K. R. H. Woolridge, *J. Chem. Soc.*, 2032 (1963). 7-5B
63JCS2203	A. Fozard and G. Jones, *J. Chem. Soc.*, 2203 (1963). 4—6; 11-4A; 11-5A
63JCS3753	A. R. Katritzky and B. J. Ridgewell, *J. Chem. Soc.*, 3753 (1963). 9-2A
63JCS4186	A. Stuart and H. C. S. Wood, *J. Chem. Soc.*, 4186 (1963). 10-4B
63JCS4204	M. W. Austin and J. H. Ridd, *J. Chem. Soc.*, 4204 (1963). 3-4A, G; 11-4A
63JGU223	V. G. Pesin, A. M. Khaletskii, and V. A. Sergeev, *J. Gen. Chem. USSR (Engl. Transl.)* **33**, 223 (1963). 4–6; 8-C*b*(iv)
63JGU1946	F. S. Babichev and V. K. Kibirev, *J. Gen. Chem. USSR (Engl. Transl.)* **33**, 1946 (1963). 8-4B
63JOC1420	R. D. Schuetz, D. D. Taft, J. P. O'Brien, J. L. Shea, and H. M. Mork, *J. Org. Chem.* **28**, 1420 (1963). 8-2A*b*(iv)
63JOC1994	H. M. Gilow and J. Jacobus, *J. Org. Chem.* **28**, 1994 (1963). 4-5B;10-6C
63JOC2262	W. E. Noland, L. R. Smith, and D. C. Johnson, *J. Org. Chem.* **28**, 2262 (1963). 4-3E*b*; 8-2A*b*(iii), (v)
63MI1	L. M. Stempel, G. B. Brown, and J. J. Fox, *Abstr. 145th Meet., Am. Chem. Soc., New York*, 14-0 (1963). 10-4B
63MI2	P. Linda and G. Marino, *Ric. Sci.* **3**, 225 (1963). 4–6; 11-5B
63RC1589	W. Czuba, *Rocz. Chem.* **37**, 1589 (1963). 4–6; 11-5B
63SA1473	H. H. Perkampus and E. Baumgarten, *Spectrochim. Acta* **19**, 1473 (1963). 10-2C
63T465	A. R. Katritzky, A. J. Waring, and K. Yates, *Tetrahedron* **19**, 465 (1963). 2-1A
63T937	R. Taylor and G. G. Smith, *Tetrahedron* **19**, 937 (1963). 5; 6-10B
63YZ523	Y. Kawazoe and S. Natsume, *Yakugaku Zasshi* **83**, 523 (1963). 10-3A

REFERENCES

63YZ934	T. Nakagome, *Yakugaku Zasshi* **83**, 934 (1963). 10-4A
64ACS269	A. Senning, *Acta Chem. Scand.* **18**, 269 (1964). 4-5D; 9-6D
64AHC(3)209	H. H. Jaffe and H. Lloyd Jones, *Adv. Heterocycl. Chem.* **3**, 209 (1964). 9-7B*a*, 9C, F
64AJC1309	C. J. Moye, *Aust. J. Chem.* **17**, 1309 (1964). 10-4B
64BSF1525	P. Cagniant, P. Faller, and D. Cagniant, *Bull. Soc. Chim. Fr.*, 1525 (1964). 4-3E*b*; 8-2A*b*(v)
64CA(60)493d	B. Arventiev and H. Offenberg, *Chem. Abstr.* **60**, 493d (1964). 8-2A*b*(iii)
64CA(60)6721d	I. F. Tupitsyn and N. K. Semenova, *Chem. Abstr.* **60**, 6721d (1964). 9-3
64CA(60)8031d	J. Hadacek and E. Kisa, *Chem. Abstr.* **60**, 8031d (1964). 10-4D, 5D
64CA(60)10670g	R. S. Muravnik, *Chem. Abstr.* **60**, 10670g (1964). 4-5B; 8-C*b*(vi)
64CA(60)15718d	P. G. Lykos, *Chem. Abstr.* **60**, 15718d (1964). Table 8-1
64CA(60)15808e	A. S. Angeloni and M. Tramontini, *Chem. Abstr.* **60**, 15808e (1964). 8-2A*b*(iv)
64CCC121	M. Prystas and F. Šorm, *Collect. Czech. Chem. Commun.* **29**, 121 (1964). 10-5B
64CI(L)1576	A. R. Katritzky, B. J. Ridgewell, and A. M. White, *Chem. Ind. (London)*, 1576 (1964). 9-2D
64CI(L)1577	R. B. Moodie, K. Schofield, and M. J. Williamson, *Chem. Ind. (London)*, 1577 (1964) .9-4C; 11-4A
64CI(L)1753	M. Kiamuddin and M. E. Haque, *Chem. Ind. (London)*, 1753 (1964). 11-5A
64CI(M)207	F. Piacenti, P. Bucci, and P. Pino, *Chim. Ind. (Milan)* **46**, 207 (1964). 4–6; 7-5B
64CPB228	T. Itai and S. Natsume, *Chem. Pharm. Bull.* **12**, 228 (1964). 10-4A
64CPB1090	I. Suzuki, T. Nakashima, N. Nagasawa, and T. Itai, *Chem. Pharm. Bull.* **12**, 1090 (1964). 11-4B
64CPB1384	Y. Kawazoe, M. Ohnishi, and Y. Yoshioka, *Chem. Pharm. Bull.* **12**, 1384 (1964). 9-3; 10-2, 3A
64GER1175683	K. Menzl, Ger. Pat. 1,175,683 (1964). 7-5C
64HCA838	C. Moussebois and F. Eloy, *Helv. Chim. Acta* **47**, 838 (1964). 7-8A
64JA1865	R. A. Olofson, W. R. Thompson, and J. S. Michelman, *J. Am. Chem. Soc.* **86**, 1865 (1964). 7-3B, Scheme 7-4.
64JA2857	J. D. Vaughan, D. G. Lambert, and V. L. Vaughan, *J. Am. Chem. Soc.* **86**, 2857 (1964). 7-5B
64JA4770	R. J. Thomas and F. A. Long, *J. Am. Chem. Soc.* **86**, 4770 (1964). 2-1A, B
64JAP26975	H. Sugiyama, T. Ikeda, and S. Koike, Jpn. Pat. 29675 (1964). 4-5B; 11-6B
64JCS173	N. P. Buu-Hoi, N. D. Xuong, and N. V. Bac, *J. Chem. Soc.*, 173 (1964). 4-3E*b*; 8-2A*b*(v)
64JCS446	M. P. L. Caton, D. H. Jones, R. Slack, and K. R. H. Woolridge, *J. Chem. Soc.*, 446 (1964). 4-2A; 7-8B
64JCS460	R. Lohrmann and H. S. Forrest, *J. Chem. Soc.*, 460 (1964). 10-4B

64JCS627	R. Baker, R. W. Bott, C. Eaborn, and P. M. Greasley, *J. Chem. Soc.*, 627 (1964). 9-8
64JCS1001	J. H. Chesterfield, D. T. Hurst, J. F. W. McOmie, and M. S. Tute, *J. Chem. Soc.*, 1001 (1964). 4-4A; 10-5B, 6A, B
64JCS2760	A. Fozard and G. Jones, *J. Chem. Soc.*, 2760 (1964). 4–6; 11-5A
64JCS3030	A. Fozard and G. Jones, *J. Chem. Soc.*, 3030 (1964). 11-4A
64JCS3114	D. H. Jones, R. Slack, and K. R. H. Woolridge, *J. Chem. Soc.*, 3114 (1964). 7-8B
64JCS3204	D. J. Brown and T. Teitei, *J. Chem. Soc.*, 3204 (1964). 10-4B
64JCS3691	A. Fischer, A. J. Read, and J. Vaughan, *J. Chem. Soc.*, 3691 (1964). 3-2C
64JCS4284	B. D. Batts and V. Gold, *J. Chem. Soc.*, 4284 (1964). 2-1A
64JCS4769	A. Stuart, D. W. West, and H. C. S. Wood, *J. Chem. Soc.*, 4769 (1964). 10-4B
64JGU1265	V. G. Pesin, I. A. Belenkaya-Lotsmanenko, and A. M. Khaletiskii, *J. Gen. Chem. USSR (Engl. Transl.)* **34**, 1265 (1964). 4-5B; 8-Cb(vi)
64JGU1814,2201	Y. K. Yurev and N. K. Sadovaya, *J. Gen. Chem. USSR (Engl. Transl.)* **34**, 1814, 2201 (1964). 4-5B; 6-8B
64JGU2491	V. G. Pesin and E. K. Dyachenko, *J. Gen. Chem. USSR (Engl. Transl.)* **34**, 2491 (1964). 4-3B; 8-Cb(vi)
64JGU3063	V. G. Pesin, V. A. Sergeev, and A. M. Khaletskii, *J. Gen. Chem. USSR (Engl. Transl.)* **34**, 3063 (1964). 8-Cb(iv)
64JOC329	M. Gordon and D. E. Pearson, *J. Org. Chem.* **29**, 329 (1964). 11-5A
64JOC415	W. B. Lutz, S. Lazarus, S. Klutchko, and R. I. Meltzer, *J. Org. Chem.* **29**, 415 (1964). 10-5C
64JOC2678	P. Beak and G. A. Carls, *J. Org. Chem.* **29**, 2678 (1964). 9-2E
64JOC2682	D. W. Mayo, P. J. Sapienza, R. C. Lord, and W. D. Phillips, *J. Org. Chem.* **29**, 2682 (1964). 9-2E
64MI1	R. M. Fink, *Arch. Biochem. Biophys.* **107**, 493 (1964). 10-2B
64MI2	R. B. Moodie, K. Schofield, and M. J. Williamson, "Nitrocompounds," pp. 89–96. Pergamon, Oxford, 1964. 3-4A, C; 11-4A
64MI3	A. I. Shatenshtein, I. O. Shapiro, Yu. I. Ranneva, and A. G. Kamrad, *Reakts. Sposobn. Org. Soedin.* **1**, 232 (1964)[*CA* **62**, 11648 (1965)]. 6-2
64MI4	A. I. Shatenshtein, I. O. Shapiro, Yu. I. Ranneva, N. N. Magdesieva, and Yu. K. Yurev, *Reakts. Sposobn. Org. Soedin.* **1**, 236 (1964) [*CA* **62**:8960 (1965)]. 6-2
64PAC217	J. Bigeleisen, *Pure Appl. Chem.* **8**, 217 (1964). 8-2Ab(i)
64RC1321	L. Achremowicz, T. Batowski, and Z. Skrowarzewska, *Rocz. Chem.* **38**, 1321 (1964). 9-4A
64T81	T. Talik and Z. Talik, *Tetrahedron* **20**, (Suppl. 1), 81 (1964). 9-4A
64T89	R. B. Moodie, K. Schofield, and M. J. Williamson, *Tetrahedron* **20**, Suppl. 1, 89 (1964). 11-4A
64T1051	B. S. Thyagarajan and P. V. Gopalakrishnan, *Tetrahedron* **20**, 1051 (1964). 11-4A

64TL845	H. A. Staab, M.-T. Wu, A. Mannschreck, and G. Schwalbach, *Tetrahedron Lett.*, 845 (1964). 7-3B,C; 8-Cb(ii)
64TL2093	H. C. van der Plas and G. Geurtsen, *Tetrahedron Lett.*, 2093 (1964). 4–6; 10-5B
64TL2117	J. Joan and J. Jones, *Tetrahedron Lett.*, 2117 (1964). 9-4A
65AC(R)1028	A. S. Angiloni and M. Tramontini, *Ann. Chim. (Rome)* **55**, 1028 (1965). 8-2A b(iii)
65ACS1652	I. Crossland and L. K. Rasmussen, *Acta Chem. Scand.* **19**, 1652 (1965). 10-6F
65ACS1741	S. Gronowitz and J. Roe, *Acta Chem. Scand.* **19**, 1741 (1965). 4–6, 7A; 10-6F
65ACS2434	B. Bak, C. H. Christensen, T. S. Hansen, E. J. Pedersen, and J. T. Nielsen, *Acta Chem. Scand.* **19**, 2434 (1965). 7-5C
65AG(E)434	C. F. Kroger and H. Frank, *Angew. Chem., Int. Ed. Engl.* **4**, 434 (1965). 7-5C
65AG(E)435	H. Prinzbach, H. Berger, and A. Luttringhaus, *Angew. Chem., Int. Ed. Engl.* **4**, 435 (1965). 7-3B
65AHC(4)1	A. Albert and W. L. F. Armarego, *Adv. Heterocycl. Chem.* **4**, 1 (1965). 2-1Fe(vii).
65AHC(4)43	D. D. Perrin, *Adv. Heterocycl. Chem.* **4**, 43 (1965). 2-1Fe(vii).
65AJC1377	P. G. E. Alcorn and P. R. Wells, *Aust. J. Chem.* **18**, 1377 (1965). 3-2C
65AJC1513	R. D. Brown, A. S. Buchanan, and A. A. Humffray, *Aust. J. Chem.* **18**, 1513, (1965). 4-1A,B; 6-8A,B,C
65AJC1521	R. D. Brown, A. S. Buchanan, and A. A. Humffray, *Aust. J. Chem.* **18**, 1521 (1965). 4-1B; 6-8A
65AJC1527	R. D. Brown, A. S. Buchanan, and A. A. Humffray, *Aust. J. Chem.* **18**, 1527 (1965). 4-7B; 6-8A
65BAU1391	S. N. Godovikova and Ya. L. Goldfarb, *Bull. Acad. Sci. USSR, Div. Chem. Sci. (Engl. Transl.)*, 1391 (1965). 8-Db(iv)
65BCJ777	Y. Ogata, Y. Izawa, and Y. Kawashima, *Bull. Chem. Soc. Jpn.* **38**, 777 (1965). 10-4
65BSF1466	E. Bisagni, J. P. Marquet, A. Cheutin, and R. Royer, *Bull. Soc. Chim. Fr.*, 1466 (1965). 8-2Ab(iii),(v)
65BSF1473	P. Demerseman, J. P. Lechartier, C. Pene, A. Cheutin, and R. Royer, *Bull. Soc. Chim. Fr.*, 1473 (1965). 4-3Eb; 8-2Ab(v), 4B
65CC46	A. J. Kresge, R. A. More O'Ferrall, L. E. Hakka, and V. P. Vitullo, *J.C.S. Chem. Commun.*, 46 (1965). 2-1A
65CC160	J. A. Elvidge, *J.C.S. Chem Commun.*, 160 (1965). 6-10A
65CC408	K.-H. Wüsch, H. Linke, A. J. Boulton, and A. Rahman, *J.C.S. Chem. Commun.*, 408 (1965). 8-Cb(iii)
65CCC355	R. Zahradnik and C. Parkanyi, *Collect. Czech. Chem. Commun.* **30**, 355 (1965). 9-9A; Table 11-6
65CI(L)1384	P. Bellingham, C. D. Johnson, and A. R. Katritzky, *Chem. Ind. (London)*, 1384 (1965). 9-2C
65CI(L)1728	T. M. Harris and J. C. Randall, *Chem. Ind. (London)*, 1728 (1965). 7-3B
65CJC2117	B. M. Lynch and L. Shiu, *Can. J. Chem.* **43**, 2117 (1965). 7-4Ba
65JA2063	T. J. Curphey, *J. Am. Chem. Soc.* **87**, 2063 (1965). 10-2B

65JCP2658	A. H. Gawer and B. P. Dailey, *J. Chem. Phys.* **42**, 2658 (1965). Table 11-6
65JCS459	P. Hodge and R. W. Rickards, *J. Chem. Soc.*, 459 (1965). 4–6; 6-4B
65JCS1051	M. W. Austin, J. R. Blackborow, J. H. Ridd, and B. V. Smith, *J. Chem. Soc.*, 1051 (1965). 7-4Ad, 4Bc
65JCS5590	D. D. Perrin, *J. Chem. Soc.*, 5590 (1965). 10-1
65JCS6695	M. D. Mehta, D. Miller, and E. F. Mooney, *J. Chem. Soc.*, 6695 (1965). 10-6F
65JCS7283	D. L. Pain and E. W. Parnell, *J. Chem. Soc.*, 7283 (1965). 4-5B; 7-7
65JHC67	W. D. Guither, D. G. Clark, and R. N. Castle, *J. Heterocycl. Chem.* **2**, 67 (1965). 10-4A
65JHC209	A. Hirschberg, A. Peterkofsky, and P. E. Spoerri, *J. Heterocycl. Chem.* **2**, 209 (1965). 4-7A; 10-6F
65JHC393	W. W. Paudler and T. J. Kress, *J. Heterocycl. Chem.* **2**, 393 (1965). 11-5B
65JMC187	A. P. Martinez, W. W. Lee, and L. Goodman, *J. Med. Chem.* **8**, 187 (1965). 10-4B
65JOC526	P. A. Duke, A. Fozard, and G. Jones, *J. Org. Chem.* **30**, 526 (1965) 4–6; 11-5A
65JOC3153	E. C. Taylor and A. McKillop, *J. Org. Chem.* **30**, 3153 (1965). 10-4B
65JOC3373	G. A. Olah, J. A. Olah, and N. A. Overchuk, *J. Org. Chem.* **30**, 3373 (1965). 9-4A
65JOC3457	W. E. Noland, L. R. Smith, and K. R. Rush, *J. Org. Chem.* **30**, 3457 (1965). 8-2Ab(iii), Scheme 4
65JOC4085	J. P. Paolini and R. K. Robins, *J. Org. Chem.* **30**, 4085 (1965). 8-Db(iv)
65JPJ62	M. Yamazaki, Y. Chono, K. Noda, and M. Hamana, *J. Pharm. Soc. Jpn.* **85**, 62 (1965). 9-5C
65JPJ645	E. Hayashi and R. Goto, *J. Pharm. Soc. Jpn* **85**, 645 (1965). 11-4A
65M1567	W. Klotzer and M. Herberz, *Monatsh. Chem.* **96**, 1567 (1965). 4-3Ea; 10-6D
65MI1	R. O. C. Norman and R. Taylor, "Electrophilic Substitution in Benzenoid Compounds," pp. 1, 61, 143, 296, 298. Elsevier, London. 1965. 2-1A; 3;6-1D,10C; 8-Cb(iv); 9-8
65MI2	A. J. Kresge, *Discuss. Faraday Soc.* **39**, 49 (1965). 2-1A
65NKZ99,637,643,647	Y. Matsuki and T. Kanda, *Nippon Kagaku Zasshi* **86**, 99, 637, 643, 647 (1965). 4-3Eb; 8-2Ab(v)
65NKZ853	Y. Matsuki and T. Kanda, *Nippon Kagaku Zasshi* **86**, 853 (1965) [*CA* **65**, 674 (1966)]. 8-2Ab(iii),(iv)
65NKZ1067	Y. Matsuki and F. Shoji, *Nippon Kagaku Zasshi* **86**, 1067 (1965). 4-3Eb; 8-2Ab(iv),(v)
65RTC951	L. van der Daes and H. J. den Hertog, *Recl. Trav. Chim. Pays-Bas* **84**, 951 (1965). 4–6; 9-5A
65RTC1101	H. C. van der Plas, *Recl. Trav. Chim. Pays-Bas* **84**, 1101 (1965). 4–6; 10-5B
65T823	A. Da Settimo and M. F. Saettone, *Tetrahedron* **21**, 823 (1965). 8-2Ab(iii)

65T843	G. Marino, *Tetrahedron* **21**, 843 (1965). 4–6; 6-4A
65T945	B. S. Thyagarajan and G. V. Gopalakrishnan, *Tetrahedron* **21**, 945 (1965). 4–6; 11-5A
65T1665	J. Vaughan, G. J. Welch, and G. J. Wright, *Tetrahedron* **21**, 1665 (1965). 8-3A*c*(ii)
65T1923	A. Da Settimo and M. F. Saettone, *Tetrahedron* **21**, 1923 (1965). 8-2A*b*(iii)
66ACS1448	J. Namtvedt and S. Gronowitz, *Acta Chem. Scand.* **20**, 1448 (1966). 4–6; 11-5C
66AG(E)513	H. Prinzbach, E. Futterer, and A. Lüttringhaus, *Angew. Chem., Int. Ed. Engl.* **5**, 513 (1966). 7-3C
66AHC(6)391	A. N. Kost and I. I. Grandberg, *Adv. Heterocycl. Chem.* **6**, 391 (1966). 4–6; 7-5B
66AHC(7)377	P. Bosshard and C. H. Eugster, *Adv. Heterocycl. Chem.* **7**, 377 (1966). 6
66AJC1909	D. G. Hawthorne and Q. N. Porter, *Aust. J. Chem.* **19**, 1909 (1966). 8-2A*b*(iv)
66BCJ2274	Y. Yukawa, Y. Tsuno, and M. Sawada, *Bull. Chem. Soc. Jpn.* **39**, 2274 (1966). 5.
66BRP1043042	J. Hattori, S. Koike, T. Ozaki, K. Yoshioka, and H. Sugiyama, Br. Pat. 1,043,042 (1966). 4-5B; 11-6B
66BSF3055	P. Cagniant, P. Faller, and D. Cagniant, *Bull. Soc. Chim. Fr.*, 3055 (1966). 8-2A*b*(iv)
66BSF3618	P. Faller, *Bull. Soc. Chim. Fr.*, 3618 (1966). 8-2A*b*(iv)
66CA(65)5429c	H. Offenberg and B. Arventiev, *Chem. Abstr.* **65**, 5429c (1966). 8-2A*b*(iv)
66CB2997	E. Buhler and W. Pfleiderer, *Chem. Ber.* **99**, 2997 (1965). 10-4B
66CCC113	V. Sterba, J. Arient, and F. Navratil, *Collect. Czech. Chem. Commun.* **31**, 113 (1966). 8-C*b*(iii)
66CCC1093	V. Sterba, J. Arient, and J. Slosar, *Collect. Czech. Chem. Commun.* **31**, 1093 (1966). 8-C*b*(iii)
66CHE14	O. P. Shvaika and G. P. Khimisha, *Chem. Heterocycl. Compd. (Engl. Transl.)* **2**, 14 (1966). 7-8A
66CHE210	V. V. Avidon and M. N. Shchukina, *Chem. Heterocycl. Compd. (Engl. Transl.)* **2**, 210 (1966). 8-4B
66CHE413	V. I. Minkin, S. F. Pozharskii, and Yu, A. Ostroumov, *Chem. Heterocycl. Compd. (Engl. Transl.)* **2**, 413 (1966). 7–10
66CHE643	A. I. Shatenshtein, A. H. Kamrads, I. O. Shapiro, Yu. I. Ranneva, and S. A. Hiller, *Chem. Heterocycl. Compd. (Engl. Transl.)* **2**, 643 (1966). 6-2
66CHE686	Yu. K. Yurev., M. A. Galbershtam, and A. I. Kandrov, *Chem. Heterocycl. Compd. (Engl. Transl.)* **2**, 686 (1966). Table 6.14
66CI(L)904	R. D. Chambers, J. A. H. MacBride, and W. K. R. Musgrave, *Chem. Ind. (London)*, 904 (1966). 10-5A
66CI(L)1721	R. D. Chambers, J. A. H. MacBride, and W. K. R. Musgrave, *Chem. Ind. (London)*, 1721 (1966). 10-5C
66CJC335	D. Cook, *Can. J. Chem.* **44**, 335 (1966). 10-2B
66CJC2283	G. Van Zyl, C. J. Bredeweg, R. H. Rynbrandt, and D. C. Neckers, *Can. J. Chem.* **44**, 2283 (1966). 8-2A*b*(iv)

66CPB816	I. Suzuki, T. Nakashima, and N. Nagasawa, *Chem. Pharm. Bull.* **14**, 816 (1966). 11-4B
66JA358	M. J. S. Dewar and J. L. von Rosenberg, *J. Am Chem. Soc.* **88**, 358 (1966). 11-5C
66JA986	H. C. Brown and B. Kanner, *J. Am. Chem. Soc.* **88**, 986 (1966). 4-5B; 9-6D
66JA1569	P. C. Myhre and M. Beug, *J. Am. Chem. Soc.* **88**, 1569 (1966). 3-2A
66JA4263	R. A. Olofson and J. M. Landesberg, *J. Am. Chem. Soc.* **88**, 4263 (1966). 7-3B,C, Scheme 5
66JA4265	R. A. Olofson, J. M. Landesberg, K. N. Houk, and J. S. Michelman, *J. Am. Chem. Soc.* **88**, 4265 (1966). Scheme 7-3(a)
66JA5537	D. G. Lambert and M. M. Jones, *J. Am. Chem. Soc.* **88**, 5537 (1966). 4–6; 7-5A
66JCS(B)521	P. B. D de la Mare, O. D. H. el Dusouqui, and E. A. Johnson, *J. Chem. Soc. B*, 521 (1966). 4–6; 8-3A(iii)
66JCS(B)565	A. R. Katritzky, F. D. Popp, and A. J. Waring, *J. Chem. Soc. B*, 565 (1966). 10-2B
66JCS(B)613	C. Eaborn, P. M. Jackson, and R. Taylor, *J. Chem. Soc. B*, 613 (1966). 2-1A
66JCS(B)727	R. Taylor, *J. Chem. Soc. B*, 727 (1966). 3-2C; 6-3A; 7-4B*c*
66JCS(B)870	J. Gleghorn, R. B. Moodie, K. Schofield, and M. J. Williamson, *J. Chem. Soc. B*, 870 (1966). 9-4C; 11-4A
66JCS(B)937	J. H. Blanck, *J. Chem. Soc. B*, 937 (1966). 9-7B*a*; 10-1
66JCS(C)1908	S. McKenzie, B. B. Molloy, and D. H. Reid, *J. Chem. Soc. C*, 1908 (1966). 4-3E*b*, 4A; 8-5C
66JCP389	J. A. Pople and G. A. Segal, *J. Chem. Phys.* **44**, 389 (1966). 8-5A
66JHC440	G. E. Wright, L. Bauer, and C. L. Bell, *J. Heterocycl. Chem.* **3**, 440 (1966). 10-3B
66JMC551	A. W. Chow, N. M. Hall, J. R. E. Hoover, M. M. Dolan, and R. J. Ferlanto, *J. Med. Chem.* **9**, 551 (1966). 8-2A*b*(iv)
66JMC573	D. E. O'Brien, C. C. Cheng, and W. Pfleiderer, *J. Med. Chem.* **9**, 573 (1966). 10-4B
66JOC70	W. E. Noland and K. R. Rush, *J. Org. Chem.* **31**, 70 (1966). 8-2A*b*(iii)
66JOC175	Y. Inoue, N. Furutachi, and K. Nakanishi, *J. Org. Chem.* **31**, 175 (1966). 10-2B
66JOC565	C. K. Bradsher and J. D. Turner, *J. Org. Chem.* **31**, 565 (1966). 4-5B; 11-6B
66JOC3093	A. C. Cope and W. D. Burrows, *J. Org. Chem.* **31**, 3093 (1966). 4-3A; 8-2A*b*(v)
66JOU1113	N. P. Pusherina, N. D. Dmitrieva, N. N. Malysheva, and R. A. Levina, *J. Org. Chem. USSR (Engl. Transl.)* **2**, 1113 (1966). 9-4D
66JPS568	M. Israel, H. Schlein, C. Maddock, S. Farber, and E. Modest, *J. Pharm. Sci.* **55**, 568 (1966). 10-6A
66LA(691)142	H. Goldner, G. Dietz, and E. Carstens, *Justus Liebigs Ann. Chem.* **691**, 142 (1966). 4-4A; 10-6B
66LA(695)55	H. A. Staab, H. Irngartinger, A. Mannschreck, and M.-T. Wu, *Justus Liebigs Ann. Chem.* **695**, 55 (1966). 7-3B

66LA(699)98	F. Zymalkowski and P. Tinapp, *Justus Liebigs Ann. Chem.* **699**, 98 (1966). 11-5A
66MI1	P. B. Venuto, L. A. Hamilton, P. S. Landis, and J. J. Wise, *J. Catal.* **5**, 81 (1966). 6-5
66MI2	A. I. Shatenshtein, A. G. Kamrad, I. O. Shapiro, Yu. I. Ranneva, and E. N. Zvyagintseva, *Dokl. Chem. (Engl. Transl.)* **168**, 502 (1966). 6-2
66MI3	J. Devanneaux and J. F. Labarre, *J. Chim. Phys. Phys.-Chim. Biol.* **66**, 1780 (1966). 6-10A
66PMH142	J. H. Ridd, *Phys. Methods Heterocycl. Chem.* **1**, 142 (1963). Table 8-1
66QR75	G. B. Barlin and D. D. Perrin, *Q. Rev., Chem. Soc.* **20**, 75 (1966). 10-1
66RC1875	K. Lewicka and E. Plazek, *Rocz. Chem.* **40**, 1875 (1966). 4-6; 9-5C
66T(Suppl 7)49	A. Ur-Rahman and A. J. Boulton, *Tetrahedron, Suppl.* **7**, 49 (1966). 8-C*b*(iii)
66T57	K. J. Morgan and D. P. Morrey, *Tetrahedron* **22**, 57 (1966). 6-3B
66T835	W. Adam and A. Grimison, *Tetrahedron* **22**, 835 (1966). 7-10
66TCA(5)401	N. K. Ray and P. T. Narasimhan, *Theor. Chim. Acta* **5**, 401 (1966). 8-C*a*, 3B
66TL2967	S. Gronowitz and J. Namtvedt, *Tetrahedron Lett.*, 2967 (1966). 8-D*b*(iv)
66ZN(A)1906	R. Joeckle, E. D. Schmid, and R. Mecke, *Z. Naturforsch., A* **21**, 1906 (1906). 9-7B*a*
67ACS1674	B. Bak, C. Dambmann, and F. Nicolaisen, *Acta Chem. Scand.* **21**, 1674 (1967). 8-2A*b*(i)
67ACS2823	S. Gronowitz and N. Gjøs, *Acta Chem. Scand.* **21**, 2823 (1967). 6-3A
67AG(E)178	E. Pfeil and U. Harder, *Angew. Chem., Int. Ed. Engl.* **6**, 178 (1967). 4-3A; 8-2A*b*(v)
67AG(E)608	A. R. Katritzky, C. D. Johnson et al., *Angew. Chem., Int. Ed. Engl.* **6**, 608 (1967). 2-1F*d*; 9-4A; 11-4B
67AJC313	W. H. Cherry, W. Davies, B. C. Ennis, and Q. N. Porter, *Aust. J. Chem.* **20**, 313 (1967). 8-2A*b*(iv)
67B3576	R. Shapiro and R. S. Klein, *Biochemistry* **6**, 3576 (1967). 10-2B
67BAU1120	N. I. Shuikin and B. L. Lebedev, *Bull. Acad. Sci. USSR, Div. Chem. Sci. (Engl. Transl.)*, 1120 (1967) 6-5
67BAU1561	N. I. Shuikin, B. L. Lebedev, V. G. Nikolskii, O. A. Korytina, A. V. Kessenikh, and E. P. Prokofev, *Bull. Acad. Sci. USSR, Div. Chem. Sci. (Engl. Transl.)*, 1561 (1967). 6-5
67BCJ130	Y. Aito, T. Matsuo, and C. Aso, *Bull. Chem. Soc. Jpn.* **40**, 130 (1967). 4–6; 6-4B
67BRP1058468	Smith, Kline, and French Lab., Br. Pat. 1058468 (1967) [*CA* **66**, 115592 (1967)]. 4-3E*b*; 8-2A*b*(v)
67CC352	R. G. Coombes, R. B. Moodie, and K. Schofield, *J.C.S. Chem. Commun.*, 352 (1967). 3-2B
67CC377	W. W. Paudler and L. S. Helmick, *J.C.S. Chem. Commun.*, 377 (1967). 8-D*b*(i)

67CC966	R. B. Temple, C. Fay, and J. Williamson, *J.C.S. Chem. Commun.*, 966 (1967). 9-4A
67CHE723	F. S. Babichev and A. F. Babicheva, *Chem. Heterocycl. Compd. (Engl. Transl.)* **3**, 723 (1967). 4-3E*b*; 8-4B
67CJC897	H. J. Anderson and C. W. Huang, *Can. J. Chem.* **45**, 897 (1967). 4-3E*b*; 6-4B, 5,7D
67CJC1431	B. M. Lynch and L. Poon, *Can. J. Chem.* **45**, 1431 (1967). 10-4
67CJC2227	H. J. Anderson and S. J. Griffiths, *Can. J. Chem.* **45**, 2227 (1967). 6-3B, 4B
67CPB826	Y. Kawazoe and M. Ohnishi, *Chem. Pharm. Bull.* **15**, 826 (1967). 11-2A,B,3
67CPB1411	H. Igeta, M. Yamada, Y. Yoshioka, and Y. Kawazoe, *Chem. Pharm. Bull.* **15**, 1411 (1967). 4–6; 10-2A,3A,5A
67CPB2000	H. Igeta, M. Yamada, Y. Yoshioka, and Y. Kawazoe, *Chem. Pharm. Bull.* **15**, 2000 (1967). 10-3A
67CR(C)(264)1652	R. Vivaldi, H. J. M. Dou, and J. Metzger, *C. R. Hebd. Seances Acad. Sci., Ser. C* **264**, 1652 (1967). 7-4A*b*
67G1286	L. Pentimalli and A. M. Guerra, *Gazz. Chim. Ital.* **97**, 1286 (1967). 8-4B
67G1604	V. Berlini, A. DeMunno, V. Dell'Amico, and P. Pino, *Gazz. Chim. Ital.* **97**, 1604 (1967). 7-4B*a*, 5B
67JA1292	J. L. Longridge and F. A. Long, *J. Am. Chem. Soc.* **89**, 1292 (1967). 2-1A
67JA3358	J. A. Zoltewicz and C. L. Smith, *J. Am. Chem. Soc.* **89**, 3358 (1967). 2-2D; 9-3
67JA4411	A. J. Kresge and Y. Chiang, *J. Am. Chem. Soc.* **89**, 4411 (1967). 2-1G*c*(v)
67JA6218	J. D. Vaughan, G. L. Jewett, and V. L. Vaughan, *J. Am. Chem. Soc.* **89**, 6218 (1967). 4–6; 7-5B
67JCP158	D. P. Santry and G. A. Segal, *J. Chem. Phys.* **47**, 158 (1967). 8-5A
67JCS(B)445	A. C. Ling and F. H. Kendall, *J. Chem. Soc. B*, 445 (1967). 2-1A
67JCS(B)780	R. Taylor, G. J. Wright, and A. J. Homes, *J. Chem. Soc. B*, 780 (1967). 8-3A*c*(ii)
67JCS(B)1204	C. D. Johnson, A. R. Katritzky, B. J. Ridgewell, and M. Viney, *J. Chem. Soc. B*, 1204 (1967). 3-3A, 4A; 9-4A
67JCS(B)1211	C. D. Johnson, A. R. Katritzky, and M. Viney, *J. Chem. Soc. B*, 1211 (1967). 3-3C; 9-4A
67JCS(B)1213	C. D. Johnson, A. R. Katritzky, N. Shakir, and M. Viney, *J. Chem. Soc. B*, 1213 (1967). 3-3C; 9-4C,F, Tables 7, 8
67JCS(B)1219	G. P. Bean, C. D. Johnson, A. R. Katritzky, B. J. Ridgewell, and A. M. White, *J. Chem. Soc. B*, 1219 (1967). 2-1E*a*, F*e*(iii); 9-2B
67JCS(B)1222	G. P. Bean, P. J. Brignell, C. D. Johnson, A. R. Katritzky, B. J. Ridgewell, H. O. Tarhan, and A. M. White, *J. Chem. Soc. B*, 1222 (1976). 2-1F*e*(vi); 9-2D
67JCS(B)1226	P. Bellingham, C. D. Johnson, and A. R. Katritzky, *J. Chem. Soc. B*, 1226 (1967). 2-1F*e*(iv), G*b;* 9-2C; 10-2B; 11-2A

67JCS(B)1235	C. D. Johnson, A. R. Katritzky, and N. Shakir, *J. Chem. Soc. B*, 1235 (1967). 9-4C
67JCS(C)1164	G. W. H. Cheeseman and B. Tuck, *J. Chem. Soc. C*, 1164 (1967). 4–6; 8-3B
67JCS(C)1638	J. F. Corbett, *J. Chem. Soc. C*, 1638 (1967). 11-5B
67JCS(C)1922	D. J. Brown and B. T. England, *J. Chem. Soc. C*, 1922 (1967). 4–6; 10-5B
67JCS(C)2084	M. S. El Shanta and R. M. Scrowston, *J. Chem. Soc. C*, 2084 (1967). 4-3Eb; 8-2Ab(v)
67JOC463	S. Gronowitz, N. Gjøs, R. M. Kellogg, and H. Wynberg, *J. Org. Chem.* **32**, 463 (1967). 4–6; 6-4A
67KKZ63	M. Kuroki, *Kogyu Kagaku Zasshi* **70**, 63 (1967) [*CA* **68**, 12802 (1968)]. 4–6; 8-3Ab(iii)
67M254	M. Scholz and D. Heidrich, *Monatsh. Chem.* **98**, 254 (1967). 6-10A
67M2039	F. Sauter and L. Golser, *Monatsh. Chem.* **98**, 2039 (1967). 4-3Eb; 8-2Ab(v)
67MI1	K. Schofield, "Heteroaromatic Nitro Compounds," pp. 1, 162, 164. Butterworth, London, 1967. 9-1, 5A, 6B; 10-2
67MI2	I. F. Tupitsyn, N. N. Zatsepina, and A. V. Kirova, *Isotopenpraxis* **3**, 136 (1967) [*CA* **71**, 21351 (1969)]. 10-3B
67MI3	S. Clementi, F. Genel, and G. Marino, *Ric. Sci.* **37**, 418 (1967). 6–7, 7C,D
67MI4	P. Linda and G. Marino, *Ric. Sci.* **37**, 424 (1967). 4-3Eb; 6-7D
67MI5	N. I. Shuikin, B. L. Lebedev, and V. G. Nikolskii, *Dokl. Chem. (Engl. Transl.)* **174**, 499 (1967). 6-5
67MI6	E. M. Arnett, "Physico-Chemical Processes in Mixed Aqueous Solvents," p. 106. Heinemann, London, 1967. 3-4H
67NKZ751	Y. Matsuki and I. Ito, *Nippon Kagaku Zasshi* **88**, 751 (1967). 4-3Eb; 8-2Ab(v)
67NKZ755	Y. Matsuki and S. Fusaji, *Nippon Kagaku Zasshi* **88**, 755 (1967) [*CA* **69**, 59020 (1968)]. 8-2Ab(iv)
67T1739	P. Linda and G. Marino, *Tetrahedron* **23**, 1739 (1967). 4-3Eb; 6-7A,C, Table 6.6
67T2513	W. Adam, A. Grimison, and G. Rodriguez, *Tetrahedron* **23**, 2513 (1967). 9-9A; 10-3
67TCA(9)181	M. Gelus, P.-M. Vay, and G. Berthier, *Theor. Chim. Acta* **9**, 181 (1967). 8-Ca
67ZC58	K. Schwetlik, K. Unverferth, and R. Mayer, *Z. Chem.* **7**, 58 (1967). 6-1A
68AC(R)1435	E. C. R. DeFabrizio, *Ann. Chim. (Rome)* **58**, 1435 (1968). 4–6; 8-2Ab(iv)
68ACS63	A. Bugge, *Acta Chem. Scand.* **22**, 63 (1968). 8-5B
68ACS2754	B. Östman, *Acta Chem. Scand.* **22**, 2754 (1968). 6-3A
68AJC939	P. D. Bolton and F. M. Hall, *Aust. J. Chem.* **21**, 939 (1968). 2-1Gc(iv)
68BAP453	G. P. Bean, A. R. Katritzky, and A. Marzec, *Bull. Acad. Pol. Sci., Ser. Sci. Chim.* **16**, 453 (1968). 11-2A
68BSB181,191	G. Leroy, C. Aussems, and F. Remoortere, *Bull. Soc. Chim. Belg.* **77**, 181,191 (1968). 11-8A, Table 6
68C1	R. Taylor, *Chimia* **22**, 1 (1968). 8-3Ac(ii)

68CA(69)85848	N. N. Zatsepina, Yu. L. Kaminskii, and I. F. Tupitsyn, *Chem. Abstr.* **69**, 85848 (1968). 8-C*b*(ii)
68CCC394	M. Holik, V. Skala, and J. Kuthan, *Collect. Czech. Chem. Commun.* **33**, 394 (1968). 9-9H
68CCC3138	J. Kuthan, J. Palecek, J. Prochazkova, and V. Skala, *Collect. Czech. Chem. Commun* **33**, 3138 (1968). 9-9A
68CHE314	L. F. Avramenko, T. A. Zacharova, V. Ya. Pochinok, and Yo. S. Rozum, *Chem. Heterocycl. Compd. (Engl. Transl.)* **4**, 314 (1968). 8-5C
68CJC1949	A. R. Knight and J. S. McIntyre, *Can. J. Chem.* **46**, 1949 (1968). 11-2C
68CPB160	T. Naito, S. Nakagawa, and T. Takahashi, *Chem. Pharm. Bull.* **16**, 160 (1968). 7-4B*b*
68CPB715	Y. Kawazoe and Y. Yoshioka, *Chem. Pharm. Bull.* **16**, 715 (1968). 9-2C; 11-2A
68CPB939	S. Kamiya, G. Okusa, M. Osuda, M. Kumagai, A. Nakamura, and K. Koshinuma, *Chem. Pharm. Bull.* **16**, 939 (1968). 10-6E
68CR(C)(266)714	H. J. M. Dou, G. Vernin, and J. Metzger, *C. R. Hebd. Seances Acad. Sci., Ser C* **266**, 714 (1968). 7-4A*b*
68CR(C)(267)697	M. Robba, J. M. Leconte, and M. Cugnon de Sevricourt, *C. R. Hebd. Seances Acad. Sci., Ser. C* **267**, 697 (1968). 8-D*b*(iv)
68IJQ165	R. B. Hermann, *Int. J. Quantum Chem.* **2**, 165 (1968). 8-2A*a*
68JA418	V. J. Shiner, W. E. Buddenbaum, B. L. Murr, and G. Lamaty, *J. Am. Chem. Soc.* **90**, 418 (1968). 6-9A
68JA1924	M. J. S. Dewar and R. H. Logan, *J. Am. Chem. Soc.* **90**, 1924 (1968). 11-4D
68JA2105	P. C. Myhre, M. Beug, and L. L. James, *J. Am. Chem. Soc.* **90**, 2105 (1968). 3-2A
68JA4633	D. S. Noyce, D. R. Hartter, and F. B. Miles, *J. Am. Chem. Soc.* **90**, 4633 (1968). 6-9A
68JA5939	J. A. Zoltewicz, G. M. Kauffman, and C. L. Smith, *J. Am. Chem. Soc.* **90**, 5939 (1968). 9-3
68JCS(B)312	R. B. Moodie, E. A. Quereshi, K. Schofield, and J. T. Gleghorn, *J. Chem. Soc. B*, 312 (1968). 11-4B
68JCS(B)316	R. B. Moodie, E. A. Quereshi, K. Schofield, and J. T. Gleghorn, *J. Chem. Soc. B*, 316 (1968). 11-4A,B
68JCS(B)370	A. R. Butler and C. Eaborn, *J. Chem. Soc. B*, 370 (1968). Table 6.1
68JCS(B)392	P. Linda and G. Marino, *J. Chem. Soc. B*, 392 (1968). 4–6; 6-4B
68JCS(B)397	E. Baciocchi and L. Mandolini, *J. Chem. Soc. B*, 397 (1968). 4–6; 8-2A*b*(iv)
68JCS(B)450	D. G. Anderson, M. A. M. Brodney, and D. E. Webster, *J. Chem. Soc. B*, 450, (1968). 9-6E
68JCS(B)765	D. G. Anderson and D. E. Webster, *J. Chem. Soc. B*, 765 (1968). 9-6E
68JCS(B)800	R. G. Coombes, R. B. Moodie, and K. Schofield, *J. Chem. Soc. B*, 800 (1968). 3-2B, 3A, 4C; 6-3A
68JCS(B)862	A. R. Katritzky and M. Kingsland, *J. Chem. Soc. B*, 862 (1968). 9-4A,C

68JCS(B)864	G. P. Bean and A. R. Katritzky, *J. Chem. Soc. B*, 864 (1968). 9-2D; 10-1
68JCS(B)866	P. Bellingham, C. D. Johnson, and A. R. Katritzky, *J. Chem. Soc. B*, 866 (1968). 2-1Fe(v), Gb; 9-2C,D,E,9C,E,F,G
68JCS(B)873	A. R. Katritzky and I. Pojarlieff, *J. Chem. Soc. B*, 873 (1968). 2-1Fe(vi); 10-2A
68JCS(B)878,1008	D. G. Anderson and D. E. Webster, *J. Chem. Soc. B*, 878, 1008 (1968). 9-6E
68JCS(B)1112	C. Eaborn, P. Golborn, R. E. Spillett, and R. Taylor, *J. Chem. Soc. B*, 1112 (1968). 6-1D; 7-2C; 8-Ba; 11-2A,B,4A,7; 12
68JCS(B)1397	R. Taylor, *J. Chem. Soc. B*, 1397 (1968). 5-Bb; 6–9, 9A,C,10A
68JCS(B)1402	J. Blatchly and R. Taylor, *J. Chem. Soc. B*, 1402 (1968). 8-3Ac(ii)
68JCS(B)1477	P. J. Brignell, A. R. Katritzky, and H. O. Tarhan, *J. Chem. Soc. B*, 1477 (1968). 3-4A, D; 9-4B
68JCS(B)1484	A. R. Katritzky, M. Kingsland, and O. S. Tee, *J. Chem. Soc. B*, 1484 (1968). 2-1Fe(vii); 10-1, 2B
68JCS(B)1559	R. Taylor, *J. Chem. Soc. B*, 1559 (1968). 8-3Ab(i),(ii),c(ii)
68JCS(C)1088	T. L. Hough and G. Jones, *J. Chem. Soc. C*, 1088 (1968). 11-5A
68JCS(C)2145	G. Berti, A. DaSettimo, and E. Nannipieri, *J. Chem. Soc. C*, 2145 (1968). 8-2Ab(iii)
68JCS(C)2848	G. W. H. Cheeseman and P. D. Roy, *J. Chem. Soc. C*, 2848 (1968). 4–6; 8-3B
68JGU1933	E. N. Zvyagintseva, T. A. Yukushina, and A. I. Shatenshtein, *J. Gen. Chem. USSR (Engl. Transl.)* **38**, 1933 (1968). 6-1A,2, Table 6.1
68JGU1938	Yu. I. Shapiro, L. I. Belenkii, I. A. Romanskii, F. M. Stoyanovich, Ya. L. Goldfarb, and A. I. Shatenshtein, *J. Gen. Chem. USSR (Engl. Transl.)* **38**, 1938 (1968). 6-2
68JGU1944	E. N. Zvyagintseva, L. I. Belenkii, T. A. Yakushina, Ya. L. Goldfarb, and A. I. Shatenshtein, *J. Gen. Chem. USSR (Engl. Transl.)* **38**, 1944 (1968). Table 6.1
68JHC69	D. E. Boswell, J. A. Brennan, P. S. Landis, and P. G. Rodewald, *J. Heterocycl. Chem.* **5**, 69 (1968). 8-2Ab(iii)
68JOC478	R. C. de Selms, *J. Org. Chem.* **33**, 478 (1968). 9-4B
68JOC1087	W. W. Paudler and L. S. Helmick, *J. Org. Chem.* **33**, 1087 (1968). 8-Db(ii)
68JOC1333	G. P. Rizzi, *J. Org. Chem.* **33**, 1333 (1968). 4-2A; 10-6F
68JOC1384	W. W. Paudler and T. J. Kress, *J. Org. Chem.* **33**, 1384 (1968). 4–6; 11-5B, Table 6
68JOC2902	R. M. Kellogg, P. A. Schaap, and E. T. Harper, *J. Org. Chem.* **33**, 2902 (1968). 4–6; 6-4A
68JOM(13)113	D. G. Anderson and D. E. Webster, *J. Organomet. Chem.* **13**, 113 (1968). 9-6E
68JOU2057	S. D. Sokolov and I. M. Yudintseva, *J. Org. Chem. USSR (Engl. Transl.)* **4**, 2057 (1968). 7-4Ba
68M11	I. F. Tupitsyn, N. N. Zatsepina, A. V. Kirova, and Yu. M. Kaputsin, *Reakts. Sposobn. Org. Soedin.* **5**, 601, 613 (1968) [*CA* **70**, 76940, 76944 (1969)]. 10-3A,B,C

68MI2	I. F. Tupitsyn, N. N. Zatsepina, and A. V. Kirova, *Reakts. Sposobn. Org. Soedin.* **5**, 626 (1968) [*CA* **70**, 76942 (1969)]. 10-3C
68MI3	I. O. Shapiro, F. S. Yakushin, I. A. Romanskii, and A. I. Shatenshtein, *Kinet. Katal.* **9**, 1011 (1968). 6-2
68MI4	A. I. Shatenshtein, Ya. L. Goldfarb, I. O. Shapiro, E. N. Zvyagintseva, and L. T. Belenkii, *Dokl. Chem. (Engl. Transl.)* **180**, 577 (1968). 6-2
68MI5	K. J. Armstrong and M. Martin-Smith, *Q. Rep. Sulfur Chem.* **3**, 357 (1968). 8-2Ab(iii)
68MI6	N. Trinajstic and A. Hinchcliffe, *Z. Phys. Chem. Wiesbaden* **59**, 271 (1968). Table 8-17
68T239	J. Kobe, B. Stanovnik, and M. Tisler, *Tetrahedron* **24**, 239 (1968). 8-Db(iv)
68TL421	J. A. Zoltewicz and J. D. Meyer, *Tetrahedron Lett.*, 421 (1968). 9-3
68USP3369897	K.-H. Menzel and R. Putter, U.S. Pat. 3,369,897 (1968). 8-4B
68YZ1289	N. Saito, T. Kurihara, K. Yamanaka, S. Tsuruta, and S. Yasuda, *Yakugaku Zasshi* **88**, 1289 (1968). 7-5B
68ZC201	J. H. Ridd, *Z. Chem.*, 201 (1968. 11-4A
69AC(R)787	P. Finocchiaro, *Ann. Chim. (Rome)* **59**, 787 (1969). 4-3Eb; 8-3Ab(iv)
69AC(R)799	P. A. Di San Filippo and E. C. R. DeFabrizio, *Ann. Chim. (Rome)* **59**, 799 (1969). 4–6; 8-2Ab(iv)
69AJC1105	G. S. Chandler, *Aust. J. Chem.* **22**, 1105 (1969). 11-5A
69AJC1963	B. C. Elmes and J. M. Swan, *Aust. J. Chem.* **22**, 1963 (1969). 4-3Ea; 8-3Ab(iv)
69BAU1452	L. D. Smirnov, R. E. Lokhov, V. P. Lezina, B. E. Zaitsev, and K. M. Dyumaev, *Bull. Acad. Sci. USSR, Div. Chem. Sci. (Engl. Transl.)*, 1452 (1969) 9-4B
69BAU2446	B. L. Lebedev, N. A. Karev, O. A. Korytina, and N. I. Shinkin, *Bull. Acad. Sci. USSR, Div. Chem. Sci. (Engl. Transl.)*, 2446 (1969). 4-3A; 8-2Ab(v)
69BSF1149	R. Phan-Tan-Luu, L. Bouscasse, E. Vincent, and J. Metzger, *Bull. Soc. Chim. Fr.*, 1149 (1969). 7-10
69CA(70)68249	F. S. Babichev, G. P. Kutrov, and M. Y. Kornilov, *Chem. Abstr.* **70**, 68249 (1969). 4-3Eb; 8-4B
69CCC3895	I. Dobas, V. Sterba, M. Vecera, *Collect. Czech. Chem. Commun.* **34**, 3895 (1969). 7-7
69CCC3905	I. Dobas, V. Sterba, and M. Vecera, *Collect. Czech. Chem. Commun.* **34**, 3905 (1969). 7-7
69CHE844	V. Grunstein and A. Struzdin, *Chem. Heterocycl. Compd. (Engl. Transl.)* **5**, 844 (1969). 7-5C
69CPB906	N. Ikekawa, O. Hoshino, and Y. Honma, *Chem. Pharm. Bull.* **17**, 906 (1969). 4-2C; 11-6A
69IJQ33	R. A. Sallavanti and D. D. Fitts, *Int. J. Quantum Chem.* **3**, 33 (1969). Table 8-17
69JA924	A. G. Anderson and D. M. Forkey, *J. Am. Chem. Soc.* **91**, 924 (1969). 10-2A
69JA1113	P. Haake, L. P. Bausher, and W. B. Miller, *J. Am. Chem. Soc.* **91**, 1113 (1969). 7-3B, Scheme 7-4.

69JA2590	W. Adam, A. Grimison, and R. Hoffmann, *J. Am. Chem. Soc.* **91**, 2590 (1969). 10-3
69JA5501	J. A. Zoltewicz, G. Grahe, and C. L. Smith, *J. Am. Chem. Soc.* **91**, 5501 (1969). 10-3, 3A,B,C, Table 1
69JA6654	C. D. Johnson, A. R. Katritzky, and S. A. Shapiro, *J. Am. Chem. Soc.* **91**, 6654 (1969). 2-1Gc(i); 3-4A, 5B; 11-4A
69JA7381	E. A. Hill, M. L. Gross, M. Stasiewicz, and M. Manion, *J. Am. Chem. Soc.* **91**, 7381 (1969). 5; 6–9,9A,C.D; 8-2Ab(vii)
69JA7391	E. A. Hill, M. L. Gross, M. Stasiewicz and M. Manion, *J. Am. Chem. Soc.* **91**, 7391 (1969). 5.
69JCS(B)21	C. Eaborn and P. M. Jackson, *J. Chem. Soc. B*, 21 (1969). 6-10A
69JCS(B)270	D. J. Brown and P. B. Ghosh, *J. Chem. Soc. B*, 270 (1969). 7-3B,C, Table 3
69JCS(C)1369	N. D. Heindel, C. J. Ohnmacht, J. Molnar, and P. D. Kennewell *J. Chem. Soc. C*, 1369 (1967). 11-4A
69JCS(C)1766	K. J. Armstrong, M. Martin-Smith, N. M. D. Brown, G. C. Brophy, and S. Sternhall, *J. Chem. Soc. C*, 1766 (1969). 8-2Ab(iii)
69JCS(C)2189	M. Davis and A. W. White, *J. Chem. Soc. C*, 2189 (1969). 8-Cb(iii)
69JCS(C)2755	I. Brown, S. T. Reid, N. M. D. Brown, K. J. Armstrong, M. Martin-Smith, W. E. Sneader, G. C. Brophy, and S. Sternhell, *J. Chem. Soc. C*, 2755 (1969). 8-2Ab(iii)
69JGU1599	Zh. I. Akselrod and V. M. Berezovskii, *J. Gen. Chem. USSR (Engl. Transl.)* **39**, 1599 (1969). 11-4B
69JGU1816	I. N. Somin, *J. Gen. Chem. USSR (Engl. Transl.)* **39**, 1816 (1969). 4-2B; 8-Cb(v)
69JHC199	J. A. White and R. C. Anderson, *J. Heterocycl. Chem.* **6**, 199 (1969). 7-3B
69JHC207	J. F. Gerster, B. C. Hinshaw, R. K. Robins, and L. B. Townsend, *J. Heterocycl. Chem.* **6**, 207 (1969). 8-Db(iv)
69JHC313	W. H. Pirkle and M. Dines, *J. Heterocycl. Chem.* **6**, 313 (1969). 9-4D
69JHC575	H. J. M. Dou, G. Vernin and J. Metzger, *J. Heterocycl. Chem.* **6**, 575 (1969). 7-4Ab
69JHC593	I. Wempen, H. U. Blank, and J. J. Fox, *J. Heterocycl. Chem.* **6**, 593 (1969). 10-4B
69JHC841	J. H. Finley and G. P. Volpp, *J. Heterocycl. Chem.* **6**, 841 (1969). 7-5B, Table 4
69JOC589	P. Beak and E. M. Monroe, *J. Org. Chem.* **34**, 589 (1969). 10-3B
69JOC1008	D. S. Noyce and G. V. Kaiser, *J. Org. Chem.* **34**, 1008 (1969). 6-9A,C
69JOC1405	J. A. Zoltewicz and G. M. Kauffman, *J. Org. Chem.* **34**, 1405 (1969). 9-3
69MI1	C. K. Ingold, "Structure and Mechanism in Organic Chemistry," 2nd ed. Bell, London, 1969. 3
69MI2	N. N. Zatsepina, Yu. L. Kaminskii, I. F. Tupitsyn, *Reakt. Sposobn. Org. Soedin.* **6**, 753 (1969). 8-2Ab(ii)

69T227	K. M. Biswas and A. H. Jackson, *Tetrahedron* **25**, 227 (1969). 8-2A*a,b*(v)
69T583	B. Tinland, *Tetrahedron* **25**, 583 (1969). Table 11-6
69T4599	S. Clementi and G. Marino, *Tetrahedron* **25**, 4599 (1969). 4-3E*b*; 6-7C,D
69T5777	N. Bodor and M. J. S. Dewar, *Tetrahedron* **25**, 5777 (1969). 3-2C
69TH1	E. F. V. Scriven, Ph.D. Thesis, University of East Anglia (1969). 3-4H
69TL579	P. M. Weintraub and R. E. Bamburg, *Tetrahedron Lett.*, 579 (1969). 7-4C*a*
69TL1909	L. N. Yakhontov, V. A. Azimov, and E. I. Lapan, *Tetrahedron Lett.*, 1908 (1969). 8-D*b*(iv)
69TL3377	A. C. Rochat and R. A. Olofson, *Tetrahedron Lett.*, 3377 (1969). 7-3C
69TL4855	Z. J. Allan, J. Podstata, and Z. Vrba, *Tetrahedron Lett.*, 4855 (1969). 9-6C
69USP3466283	U.S. Pat. 3,466,283 (1969) [*CA* **92**, 12756 (1970)]. 10-5A
69ZN(B)12	H. Güsten, L. Klasinc, and O. Volkert, *Z. Naturforsch., B: Anorg. Chem., Org. Chem., Brochem., Biophys., Biol.* **24**, 12 (1969). 8-4A
70ACS23	A. Skancke and P. N. Skancke, *Acta Chem. Scand.* **24**, 23 (1970). 8-5,5A
70ACS99	N. Gjøs and S. Gronowitz, *Acta Chem. Scand.* **24**, 99 (1970). 4-3E*b*; 6-7A
70AHC(11)177	B. Iddon and R. M. Scrowston, *Adv. Heterocycl. Chem.* **11**, 177 (1970). 8-2A*b*(iii),(v)
70AHC(11)327	B. Iddon and R. M. Scrowston, *Adv. Heterocycl. Chem.* **11**, 327 (1970). 4-3E*b*; 8-2A*b*(v)
70AHC(11)383	R. A. Jones, *Adv. Heterocycl. Chem.* **11**, 383 (1970). 6
70AHC(12)1	N. N. Magdesieva *Adv. Heterocycl. Chem.* **12**, 1 (1970). 6,6-10A
70AJC203	A. D. Campbell, S. Y. Chooi, L. W. Deady, and R. A. Shanks, *Aust. J. Chem.* **23**, 203 (1970). 9-7B*a*
70AK89	S. Olsson, *Ark. Kemi* **32**, 89 (1970). 6-1A; 8-2A*b*(i)
70AK249	S. Gronowitz and E. Sandberg, *Ark. Kemi* **32**, 249 (1970). 8-D*b*(iv)
70BAU1791,2244	L. D. Smirnov, V. I. Kuzmin, V. P. Lezina. and K. M. Dyumaev, *Bull. Acad. Sci. USSR, Div. Chem. Sci. (Engl. Transl.)*, 1791, 2244 (1970). 9-4B
70BAU2440	K. M. Dyumaev, L. D. Smirnov, R. E. Lokhov, and B. E. Zaitsev, *Bull. Acad. Sci. USSR, Div. Chem. Sci. (Engl. Transl.)*, 2440 (1970). 9-4C
70BAU2592	Yu. B. Volkenshtein, I. B. Farmanova, and Ya. L. Goldfarb, *Bull. Acad. Sci. USSR, Div. Chem. Sci. (Engl. Transl.)*, 2592 (1970). 4–6; 6-4A
70BCJ3344	M. Kamiya, *Bull. Chem. Soc. Jpn.* **43**, 3344 (1970). 7–10
70BCJ3496	Y. Kawase and S. Hori, *Bull. Chem. Soc. Jpn.* **43**, 3496 (1970). 4-3E*b*; 8-2A*b*(v)
70BSF1029	R. Royer and L. Rene, *Bull. Soc. Chim. Fr.*, 1029 (1970). 8-2A*b*(iii)

70BSF1483	O. Chalvet, R. Royer, and P. Demerseman, *Bull. Soc. Chim. Fr.*, 1483 (1970). 8-2A*a*, Table 1
70BSF3155	A. Friedmann, D. Bouin, and J. Metzger, *Bull. Soc. Chim. Fr.*, 3155 (1970). 7-4A*b*
70BSF3601	R. Royer and R. Rene, *Bull. Soc. Chim. Fr.*, 3601 (1970). 4-3E*b*; 8-2A*b*(v)
70CC188	J. I. Hoppe and B. R. T. Keene, *J. Chem. Soc., Chem. Commun.*, 188 (1970). 10-5A
70CC641	D. J. Blackstock, A. Fischer, K. E. Richards, J. Vaughan, and G. J. Wright, *J.C.S. Chem. Commun.*, 641 (1970). 3-2C
70CC1024	A. A. El-Anani, C. C. Greig, and C. D. Johnson, *J.C.S. Chem. Commun.*, 1024 (1970). 2-1Fe(vii).
70CCC1991	W. J. Wechter and R. C. Kelly, *Collect. Czech. Chem. Commun.* **35**, 1991 (1970). 10-3B
70CCC2003	W. J. Wechter, *Collect, Czech. Chem. Commun.* **35**, 2003 (1970). 10-3B
70CHE254	L. G. Yudin, A. I. Pavlyuchenko, and A. N. Kost, *Chem. Heterocycl. Compd. (Engl. Transl.)* **6**, 254 (1970). 4-2C; 8-2A*b*(vi)
70CHE614	S. S. Novikov, L. I. Khmelnitskii, O. V. Lebedev, V. V. Serastyanova, and L. V. Yepishina, *Chem. Heterocycl. Compd. (Engl. Transl.)* **6**, 614 (1970). 7-4A*d*
70CHE465	S. S. Novikov, L. I. Khmelnitskii, O. V. Lebedev, V. V. Sevastyanova, and L. U. Yepishina, *Chem. Heterocycl. Compd. (Engl. Transl.)* **6**, 465 (1970). 7-4A*d*
70CHE849	T. P. Sicheva, I. D. Kisleva, and M. N. Shchukina, *Chem. Heterocycl. Compd. (Engl. Transl.)* **6**, 849 (1970). 8-4B
70CJC2006	R. G. Micetich, *Can. J. Chem.* **48**, 2006 (1970). 4-2A; 7-8B
70CPB203	N. Uehara and Y. Kawazoe, *Chem. Pharm. Bull.* **18**, 203, 1970). 11-2B
70CPB1680	M. Yanai, T. Kinoshita, S. Takeda, H. Sadaki, and H. Watanabe, *Chem. Pharm. Bull.* **18**, 1680 (1970). 4-6; 10-5A
70CPB2094	Y. Hamada, T. Ito, and M. Hirota, *Chem. Pharm. Bull.* **18**, 2094 (1970). 11-4A
70G556	S. Clementi and G. Marino, *Gazz. Chim. Ital.* **100**, 556 (1970). 6-7D
70JA1567	P. D. Bolton, C. D. Johnson, A. R. Katritzky, and S. A. Shapiro, *J. Am. Chem. Soc.* **92**, 1567 (1970). 2-1Gc(iv)
70JA2546	G. A. Olah, R. H. Schlosberg, D. P. Kelly, and G. D. Mateescu, *J. Am. Chem. Soc.* **92**, 2546 (1970). 2-1A
70JA6309	A. J. Kresge, S. Slae, and D. W. Taylor, *J. Am. Chem. Soc.* **92**, 6309 (1970). 2-1A
70JA7547	J. A. Zoltewicz and L. S. Helmick, *J. Am. Chem. Soc.* **92**, 7547 (1970). 9-3
70JCS(B)43	P. Linda and G. Marino, *J. Chem. Soc. B*, 43 (1970). 4–6; 6-4A
70JCS(B)114	A. R. Katritzky, H. O. Tarhan, and S. Tarhan, *J. Chem. Soc. B*, 114 (1970). 9-4B
70JCS(B)117	P. J. Brignell, P. E. Jones, and A. R. Katritzky, *J. Chem. Soc. B*, 117 (1970). 4–6; 9-5A
70JCS(B)797	P. F. Christy, J. H. Ridd, and N. D. Stears, *J. Chem. Soc. B*, 797 (1970). 3-2C

70JCS(B)848	A. R. Butler and J. B. Hendry, *J. Chem. Soc. B*, 848 (1970). 4–6; Table 6.4
70JCS(B)852	A. R. Butler and J. B. Hendry, *J. Chem. Soc. B*, 852 (1970). 61A
70JCS(B)1153	S. Clementi, P. Linda, and G. Marino, *J. Chem. Soc. B*, 1153 (1970). 4-3E*b*; 6-4A,7A,D,8A
70JCS(B)1364	R. Taylor, *J. Chem. Soc. B*, 1364 (1970). 4-1C; 6-8A,C
70JCS(B)1570	J. Sierra, M. Ojeda, and P. A. H. Wyatt, *J. Chem. Soc. B*, 1570 (1970). 2-1G*c*(i).
70JCS(B)1692	R. E. Burton and I. L. Finar, *J. Chem. Soc. B*, 1692 (1970). 7–10
70JCS(C)933	G. C. Brophy, S. Sternhell, N. M. D. Brown, I. Brown, K. J. Armstrong and M. Martin-Smith, *J. Chem. Soc. C*, 933 (1970). 8-2A*b*(iii)
70JCS(C)1949	J. Cooper, D. F. Ewing, R. M. Scrowston, and R. Westwood, *J. Chem. Soc. C*, 1949 (1970). 8-2A*b*(iii),(iv)
70JCS(C)2334	N. B. Chapman, K. Clarke, and K. S. Sharma, *J. Chem. Soc. C*, 2334 (1970). 4-3E*b*, 6; 8-3B
70JCS(C)2563	C. F. Candy, R. A. Jones, and P. H. Wright, *J. Chem. Soc. C*, 2563 (1970). 4-3E*a*; 6-7D
70JGU1609	T. A. Yakushina, E. N. Zvyagintseva, V. P. Litvinov, S. A. Ozolin, Ya. L. Goldfarb, and A. I. Shatenshtein, *J. Gen. Chem. USSR (Engl. Transl.)* **40**, 1609 (1970). 6-1A; 8-5A
70JHC313,597	H.-L. Pan and T. L. Fletcher, *J. Heterocycl. Chem.* **7**, 313, 597 (1970). 11-4A
70JHC373	L. H. Klemm, R. Zell, I. T. Barnish, R. A. Klemm, C. E. Klopfenstein, and D. R. McCoy, *J. Heterocycl. Chem.* **7**, 373 (1970). 8-D*b*(iv)
70JHC399	P. E. Sonnet, *J. Heterocycl. Chem.* **7**, 399 (1970). 6-3B
70JHC707	M. D. Coburn, *J. Heterocycl. Chem.* **7**, 707 (1970). 7-4B*c*
70JHC903	D. V. Santi, C. F. Brewer, and D. Farber, *J. Heterocycl. Chem.* **7**, 903 (1970). 10-2B,3B
70JOC171	Y. C. Tong, *J. Org. Chem.* **35**, 171 (1970). 11-5A
70JOC1141	J. D. Vaughan, Z. Mughrabi, and E. C. Wu, *J. Org. Chem.* **35**, 1141 (1970). 7-3B
70JOC1146	E. C. Wu and J. D. Vaughan, *J. Org. Chem.* **35**, 1146 (1970). 7-3B
70JOC1662	A. J. Boulton and R. C. Brown, *J. Org. Chem.* **35**, 1662 (1970). 8-C*b*(iii)
70JOC1718	D. S. Noyce, C. A. Lipinski, and G. M. Loudon, *J. Org. Chem.* **35**, 1718 (1970). 6–9, 9A
70JOC3467	W. W. Paudler and S. A. Humphrey, *J. Org. Chem.* **35**, 3467 (1970). 10-3B,C, Table 3
70JOU2531	L. I. Belenkii, A. P. Yakubov, and Ya. L. Goldfarb, *J. Org. Chem. USSR (Engl. Transl.)* **6**, 2531 (1970). 6-7A
70JPR591	H. G. O. Becker, H. Bottcher, G. Fischer, H. Ruckauf, and S. Saphon, *J. Prakt. Chem.* **312**, 591 (1970). 4-4B; 10-6A
70JPR(312)882	K. Schwetlik and K. Unverferth, *J. Prakt. Chem.* **312**, 882 (1970). 6-1C
70MI1	R. A. Olofson, H. Kohn, R. V. Kendall, and W. P. Piekielek, *160th*

438 REFERENCES

	Natl. Meet. Am. Chem. Soc., Abstr. ORGN 76, (1970). 7-3B
70RC779	W. Drzeniek and P. Tomasik, *Rocz. Chem.* **44**, 779 (1970). 4–6; 9-5A
70RCR627	Zh. I. Akselrod and V. M. Berezovskii, *Russ. Chem. Rev. (Engl. Transl.)* **39**, 627 (1970). 9-1, 4B,C
70T5101	A. R. Cooksey, K. J. Morgan, and D. P. Morrey, *Tetrahedron* **26**, 5101 (1970). 6-3B
70TL1389	S. Clementi, P. Linda, and G. Marino, *Tetrahedron Lett.*, 1389 (1970). 4–6; 6-4B,7C
70TL2793	D. J. Blackstock, J. R. Cretney, A. Fischer, M. P. Hartshorn, K. E. Richards, J. Vaughan, and G. J. Wright, *Tetrahedron Lett.*, 2793 (1970). 3-2C
71ACR240	G. A. Olah, *Acc. Chem. Res.* **4**, 240 (1971). 3, 3-2B
71ACR248	J. H. Ridd, *Acc. Chem. Res.* **4**, 248 (1971). 3
71AHC(13)235	G. Marino, Adv. Heterocycl. Chem. **13**, 235 (1971). 4–6; 6, 6-8A, Table 6.4
71AJC1413	B. E. Boulton and B. A. W. Coller, *Aust. J. Chem.* **24**, 1413 (1971). 7-5B
71AJC2679	C. Decoret and B. Tinland, *Aust. J. Chem.* **24**, 2679 (1971). 6-10A
71BAU395,400	N. A. Andronova, L. D. Smirnov, V. P. Lezina, B. E. Zaitsev, and K. M. Dyumaev, *Bull. Acad. Sci. USSR, Div. Chem. Sci. (Engl. Transl.)* 395, 400 (1971). 11-5A
71BAU1142	Ya. L. Goldfarb, E. I. Novikova, and L. I. Belenkii, *Bull. Acad. Sci. USSR, Div. Chem. Sci. (Engl. Transl.)*, 1142 (1971). 6-3A
71BAU1429	M. S. Shvartsberg, L. W. Bizhan, and I. L. Kotlyarevskii, *Bull. Acad. Sci. USSR, Div. Chem. Sci. (Engl. Transl.)*, 1429 (1971). 4-2A; 7-8B
71BAU2108	V. S. Reznik and Yu. S. Shvetsov, *Bull. Acad. Sci. USSR, Div. Chem. Sci. (Engl. Transl.)* **20**, 2108 (1971). 4–6; 10-5B
71BAU2222	L. D. Smirnov, M. R. Abezov, V. P. Lezina, B. E. Zaitsev, and K. M. Dyumaev, *Bull. Acad. Sci. USSR, Div. Chem. Sci. (Engl. Transl.)*, 2222 (1971). 4-3C; 9-6B
71BAU2687	Ya. L. Goldfarb, E. I. Novikova, and L. I. Belenkii, *Bull. Acad. Sci. USSR, Div. Chem. Sci. (Engl. Transl.)*, 2687 (1971). 4–6; 6-4A
71BSF238	B. Roques, M.-C. Zaluski, and M. Dutheil, *Bull Soc. Chim. Fr.*, 238 (1971). 4–6; 6-4B
71BSF4310	M. Baule, R. Vivaldi, J. C. Poite, H. J. M. Dou, G. Vernin, and J. Metzger, *Bull. Soc. Chim. Fr.*, 4310 (1971). 7-4A*b*, Table 4
71CA(74)7637	F. S. Babichev, G. F. Kutrov, and M. Y. Kornilov, *Chem. Abstr.* **74**, 7637 (1971). 4-7C; 8-4B
71CA(75)140864	S. Kano and T. Noguchi, *Chem. Abstr.* **75**, 140864 (1970). 8-5C
71CB3925	H. Reimlinger *et al.*, *Chem. Ber.* **104**, 3925 (1971). 8-3B
71CC394	J. A. Elvidge, J. R. Jones, C. O'Brien, and E. A. Jones, *J.C.S. Chem. Commun.*, 394 (1971). 8-D*b*(ii)

REFERENCES

71CC421	G. P. Bean, *J.C.S. Chem. Commun.*, 421 (1971).	6-1D; 8-2A*b*(i)
71CC1441	F. Fringuelli, G Marino, G. Savelli, and A. Taticchi, *J.C.S. Chem. Commun.*, 1441 (1971). 4-3E*b;* 6-7B	
71CHE144	E. A. Karakhanov, G. V. Drovyannikova, and E. A. Viktorova, *Chem. Heterocycl. Compd. (Engl. Transl.)* 144 (1971).	8-2A*b*(v)
71CHE377	V. V. Melnikov, M. S. Pevzner, V. V. Stolpakova, and L. F. Khorkova, *Chem. Heterocycl. Compd. (Engl. Transl.)* **7**, 377 (1971). 7–10	
71CHE938	A. I. Shatenshtein, E. A. Kovryhuykh, and I. O. Shapiro, *Chem. Heterocycl. Compd. (Engl. Transl.)* **7**, 938 (1971).	4-2A; 6-8B
71CHE953	E. A. Karakhanov, G. V. Drovyannikova, and E. A. Viktorova, *Chem. Heterocycl. Compd. (Engl. Transl.)* **7**, 953 (1971).	4-3A; 8-2A*b*(v)
71CHE956	E. A. Karakhanov, G. V. Drovyannikova, L. A. Kiseleva, and E. A. Viktorova, *Chem. Heterocycl. Compd. (Engl. Transl.)* **7**, 956 (1971). 8-2A*b*(v)	
71CHE1265	L. I. Belenkii, E. I. Novikova, and Ya. L. Goldfarb, *Chem. Heterocycl. Compd. (Engl. Transl.)* **7**, 1265 (1971) 4-3B; 6-6	
71CHE1401	L. G. Yudin, A. I. Pavlyuchenko, V. A. Budylin, and V. I. Minkin, *Chem. Heterocycl. Compd. (Engl. Transl.)* **7**, 1401 (1971). 4-2c; 8-2A*b*(iii),(vi)	
71CHE1406	A. N. Kost, L. G. Yudin, V. A. Budylin, and M. Abchilaev, *Chem. Heterocycl. Compd. (Engl. Transl.)* **7**, 1406 (1971).	8-2A*b*(iv)
71CHE1443	V. I. Minkin, I. I. Zakharov, and L. L. Popova, *Chem. Heterocycl. Compd. (Engl. Transl.)* **7**, 1443 (1971). 8-C*a*	
71CJC2139	R. Raap, *Can. J. Chem.* **49**, 2139 (1971).	4-2A; 7-8B
71CPB215,216	K. Senga, F. Yoneda, and S. Nishigaki, *Chem. Pharm. Bull.* **19**, 215, 216 (1971). 4-3E*a;* 10-6D	
71CPB1526	S. Nishigaki, K. Senga, and F. Yoneda, *Cham. Pharm. Bull.* **19**, 1526 (1971). 10-6D	
71JA3309	D. Lipkin and J. A. Rabi, *J. Am. Chem. Soc.* **93**, 3309 (1971). 10-5B	
71JA6181	A. J. Kresge, S. G. Mylonakis, Y. Sato, and V. P. Vitullo, *J. Am. Chem. Soc.* **93**, 6181 (1971). 2-1A	
71JCS(B)1	C. D. Johnson, A. R. Katritzky, M. Kingsland, and E. F. V. Scriven, *J. Chem. Soc. B*, 1 (1971). 10-4B	
71JCS(B)4	U. Bressel, A. R. Katritzky, and J. R. Lea, *J. Chem. Soc. B* 4 (1971). 2-1F*e*(vi), 2D; 11-2A,B	
71JCS(B)11	U. Bressel, A. R. Katritzky, and J. R. Lea, *J. Chem. Soc., B*, 11 (1971). 11-2A,C	
71JCS(B)79	G. Marino, S. Clementi, and P. Linda, *J. Chem. Soc. B*, 79 (1971). 4-3E*b*, 6; 8-2A*b*(iv),(v)	
71JCS(B)102	A. R. Butler and J. B. Hendry, *J. Chem. Soc. B*, 102 (1971). 6-3A	
71JCS(B)536	R. Taylor, *J. Chem. Soc. B*, 536 (1971). 8-3A*c*(ii)	
71JCS(B)622	R. Taylor, *J. Chem. Soc. B*, 622 (1971). 12	
71JCS(B)712	F. De Sarlo and J. H. Ridd, *J. Chem. Soc. B*, 712 (1971).	9-4A

71JCS(B)1254	D. H. G. Crout, J. R. Penton, and K. Schofield, *J. Chem. Soc. B* 1254 (1971). 11-4A
71JCS(B)1256	S. R. Hartshorn, R. B. Moodie, and K. Schofield, *J. Chem. Soc. B*, 1256 (1971). 3-2C
71JCS(B)1450	R. Taylor, *J. Chem. Soc. B*, 1450 (1971). 6-10A
71JCS(B)1493	R. B. Moodie, J. R. Penton, and K. Schofield, *J. Chem. Soc. B*, 1493 (1971). 11-4A,B
71JCS(B)2363	A. El-Anani, P. E. Jones, and A. R. Katritzky, *J. Chem. Soc. B*, 2363 (1971). 2-1Ea; 9-2B,D
71JCS(B)2365	A. G. Burton, P. P. Forsythe, C. D. Johnson, and A. R. Katritzky, *J. Chem. Soc. B*, 2365 (1971). 7-2A,4Ba, Scheme 9
71JCS(B)2382	R. Taylor, *J. Chem. Soc. B*, 2382 (1971). 9-7Aa,9A; 11-7,8B, Table 6
71JCS(B)2443	R. G. Coombes and L. W. Russell, *J. Chem. Soc. B*, 2443 (1971). 3-2B
71JCS(B)2454	S. R. Hartshorn, R. B. Moodie, and K. Schofield, *J. Chem. Soc. B*, 2454 (1971). 3-4E
71JCS(C)463	N. B. Chapman, C. G. Hughes, and R. M. Scrowston, *J. Chem. Soc. C*, 463 (1971). 4-3Eb; 8-4B
71JCS(C)1308	N. B. Chapman, C. G. Hughes, and R. M. Scrowston, *J. Chem. Soc. C*, 1308 (1971). 4-3Eb, 6; 8-4B
71JCS(C)1945	J. Clark and Z. Munawar, *J. Chem. Soc. C*, 1945 (1971). 10-4,4B
71JCS(C)2018	G. W. H. Cheeseman, M. Rafiq, P. D. Roy, C. J. Turner, and G. V. Boyd, *J. Chem. Soc. C*, 2018 (1971). 8-3B
71JCS(C)3052	J. Cooper and R. M. Scrowston, *J. Chem. Soc. C*, 3052 (1971). 8-2Ab(iii)
71JCS(C)3405	J. Cooper and R. M. Scrowston, *J. Chem. Soc. C*, 3405 (1971). 8-2Ab(iii)
71JCS(C)3727	A. Albert and K. Ohta, *J. Chem. Soc. C*, 3727 (1971). 10-5C
71JGU1945	T. A. Yakushina, I. O. Shapiro, E. N. Zvyagintseva, V. P. Litinov, S. A. Ozolin, Ya. L. Goldfarb, and A. I. Shatenshtein, *J. Gen. Chem. USSR (Engl. Transl.)* **41**, 1945 (1971). 6-2; 8-2Ab(ii), 5B
71JGU2314	E. N. Zvyagintseva, V. E. Udre, M. G. Vorokov, and A. I. Shatenshtein, *J. Gen. Chem. USSR (Engl. Transl.)* **41**, 2314 (1971). Table 6.1
71JHC51	P. N. Neuman, *J. Heterocycl. Chem.* **8**, 51 (1971). 7-4Cc
71JHC293	M. D. Coburn, *J. Heterocycl. Chem.* **8**, 293 (1971). 7-4Bc
71JHC445	A. A. Santilli, D. H. Kim, and S. V. Wanser, *J. Heterocycl. Chem.* **8**, 445 (1971). 4-3Ea; 10-6D
71JHC849	C. D'Erba, G. Garbarino, and G. Guanti, *J. Heterocycl. Chem.* **8**, 849 (1971). 6-3A
71JHC1101	G. G. Smith and J. A. Kirby, *J. Heterocycl. Chem.* **8**, 1101 (1971). 5-'Bb
71JOC1053	D. W. H. MacDowell and A. T. Jeffries, *J. Org. Chem.* **36**, 1053 (1971). 4-2A; 6-8A
71JOC3084	P. Cohen-Fernandes and C. L. Habraken, *J. Org. Chem.* **36**, 3084 (1971). 8-Cb(iii)
71JOC3087	M. Fraser, *J. Org. Chem.* **36**, 3087 (1971). 8-Db(i)
71JOU1232	Yu. G. Erykalov, A. P. Belokurova, I. S. Isaev, A. I. Rezvukhin,

and V. A. Koptyug, *J. Org. Chem. USSR (Engl. Transl.)* **7**, 1232 (1971). 2-1A
71JOU1803 L. I. Belenkii, E. I. Novikova, I. A. Dyachenko, and Ya. L. Goldfarb, *J. Org. Chem. USSR (Engl. Transl.)* **7**, 1803 (1971). 6-3A
71JOU1835 S. D. Sokolov and I. M. Yudintseva, *J. Org. Chem. USSR (Engl. Transl.)* **7**, 1835 (1971). 4–6; 7-5B
71M837 V. Pirc, B. Stanovnik, and M. Tisler, *Monatsh. Chem.* **102**, 837 (1971). 8-Db(i),(ii)
71MI1 J. G. Hoggett, R. B. Moodie, J. R. Penton, and K. Schofield, "Nitration and Aromatic Reactivity." Cambridge Univ. Press, London and New York, 1971. 3,3-4A,5A
71MI2 H. Suzuki and Y. Tamura, *J. Chem. Soc. Jpn.* **2**, 1021 (1971). 4–6; 6-4A
71PMH55 J. H. Ridd, *Phys. Methods Heterocycl. Chem.* **4**, 55 (1971). 9-1
71RTC513 W. Schwaiger and J. P. Ward, *Recl. Trav. Chim. Pays-Bas* **90**, 513 (1971). 4-2A; 10-6F
71T245 K. J. Morgan and D. P. Morrey, *Tetrahedron* **27**, 245 (1971). 6-3B
71T681 D. Farcasiu, A. Vasilescua, and A. T. Balaban, *Tetrahedron* **27**, 681 (1971). 9-2F
71T851 W. Engewald, M. Mühlstadt, and C. Weiss, *Tetrahedron* **27**, 851 (1971). 8-Ba
71T4171 W. Engewald, M. Mühlstadt, and C. Weiss, *Tetrahedron* **27**, 4171 (1971). 8-2Ab(i), Ba, 5A, Table 2
71T953 P. Beak and R. N. Watson, *Tetrahedron* **27**, 953 (1971). 10-2B,D, 3B
71T4045 L. Klasinc and N. Trinajstic, *Tetrahedron* **27**, 4045 (1971). 8-5
71T4667 S. Clementi, P. Linda, and M. Vergoni, *Tetrahedron* **27**, 4667 (1971). 4-3Eb; 6-7A
71TL387 C. Leibovici and J. Streith, *Tetrahedron Lett.*, 387 (1971). 9-9D
71TL851 F. Yoneda, K. Shinomura, and S. Nishigaki, *Tetrahedron Lett.*, 851 (1971). 4-4A; 10-6B
71TL2211 A. G. Burton, P. J. Halls, and A. R. Katritzky, *Tetrahedron Lett.*, 2211 (1971). 9-4B
71TL3833 G. Ciranni and S. Clementi, *Tetrahedron Lett.*, 3833 (1971). 4-3Eb; 6-7C
72ACS624 O. Ceder, J. A. Andersson, and L.-E. Johansson, *Acta Chem. Scand.* **26**, 624 (1972). 4–6; 11-5B
72ACS1851 N. Gjøs and S. Gronowitz, *Acta Chem. Scand.* **26**, 1851 (1972). 4-3Eb, 6; 6-4A, 7A
72ACS2601 A. Helland and P. W. Skancke, *Acta Chem. Scand.* **26**, 2601 (1972). 8-Da
72AHC(14)17 K. R. H. Woolridge, *Adv. Heterocycl. Chem.* **14**, 17 (1972). 4–6; 7-5B
72AHC(14)99 G. W. H. Cheeseman and E. S. G. Werstiuk, *Adv. Heterocycl. Chem.* **14**, 99 (1972). 10-1, 5C
72AJC431 L. W. Deady and R. A. Shanks, *Aust. J. Chem.* **25**, 431 (1972). 9-7Ba

72AP(305)509	I. Simiti and E. Chindris, *Arch. Pharm. (Weinheim, Ger.)* **305**, 509 (1972). 7-4A*a*
72BAU404	L. D. Smirnov, N. A. Andronova, V. P. Lezina, and K. M. Dyumaev, *Bull. Acad. Sci. USSR, Div. Chem. Sci. (Engl. Transl.)*, 404 (1972). 4-5B; 11-6B
72BAU406	L. D. Smirnov, N. A. Andronova, V. P. Lezina, and K. M. Dyumaev, *Bull. Acad. Sci. USSR, Div. Chem. Sci. (Engl. Transl.)*, 406 (1972). 4-5B; 11-6B
72BAU452	L. D. Smirnov, N. A. Andronova, V. P. Lezina, and K. M. Dyumaev, *Bull. Acad. Sci. USSR, Div. Chem. Sci. (Engl. Transl.)*, 452 (1972). 4-4B; 11-6C
72BAU1166	V. P. Lezina, A. U. Stepanyants, L. D. Smirnov, and K. M. Dyumaev, *Bull. Acad. Sci. USSR, Div. Chem. Sci. (Engl. Transl.)*, 1166 (1972). 9-2C
72BAU1169	V. P. Lezina, A. U. Stepanyants, L. D. Smirnov, and K. M. Dyumaev, *Bull. Acad. Sci. USSR, Div. Chem. Sci. (Engl. Transl.)*, 1169 (1972). 9-C
72BAU2029	V. P. Lezina, A. V. Stepanyants, L. D. Smirnov, N. A. Andronova, and K. M. Dyumaev, *Bull. Acad. Sci. USSR, Div. Chem. Sci. (Engl. Transl.)*, 2029 (1972). 11-2A
72BCJ2534	H. Suzuki and K. Nakamura, *Bull. Chem. Soc. Jpn.* **45**, 2534 (1972). 3-2F
72BSF162	J. C. Poite, J. Roggero, H. J. M. Dou, G. Vernin, and J. Metzsger, *Bull. Soc. Chim. Fr.*, 162 (1972). 7-4A*b*, 4B*b*
72BSF2466	J. Abblard, C. Decoret, L. Cronenberger, and H. Pacheco, *Bull. Soc. Chim. Fr.*, 2466 (1972). 9-5A
72BSF2481	P. Guerret, R. Jacquier, and G. Maury, *Bull. Soc. Chim. Fr.*, 2481 (1972). 8-D*b*(iv)
72BSF3955	T. Q. Minh, F. Mantovani, P. Faller, L. Christiaens, and M. Renson, *Bull. Soc. Chim. Fr.*, 3955 (1972). 8-2A*b*(v)
72CA(76)72617	S. L. Gusinskaya, V. Yu Telly, and N. L. Ovchinnikova, *Chem. Abstr.* **76**, 72617 (1972). 7-8A
72CA(77)101477	V. G. Pesin and L. A. Kaukhova, *Chem. Abstr.* **77**, 101477 (1972). 8-3B
72CA(76)126183	I. Simiti, M. Farkas, and I. Schwartz. *Chem. Abstr.* **76**, 126183 (1972). 7-5A
72CA(77)126512	V. G. Pesin and L. A. Kaukhova, *Chem. Abstr.* **77**, 126512 (1972). 8-3B
72CC77	H.-S. Ryang and H. Sakurai, *J.C.S. Chem. Commun.*, 77 (1972). 4-3E*b;* 8-2A*b*(v)
72CC427	S. Clementi, P. Linda, and G. Marino, *J.C.S. Chem. Commun.*, 427 (1972). 4-3E*a;* 6-1D, 7D; 8-2A*b*(v)
72CC641	J. H. Ridd and E. F. V. Scriven, *J.C.S. Chem. Commun.*, 641 (1972). 3-2E
72CC1032	S. Banerjee and O. S. Tee, *J.C.S. Chem. Commun.*, 1032 (1972). 10-5B
72CHE13	N. N. Magdesieva and V. A. Vdovin, *Chem. Heterocycl. Compd. (Engl. Transl.)* **8**, 13 (1972). 4-2A; 8-2A*b*(iv),(v),(vi)
72CHE18	N. N. Magdesieva, V. A. Vdovin, and L. D. Konyashkin, *Chem. Heterocycl. Compd. (Engl. Transl.)* **8**, 18 (1972). 4-2C; 8-2A*b*(vi)

72CHE541	L. I. Belenkii, G. P. Gromova, and Ya. L. Goldfarb, *Chem. Heterocycl. Compd. (Engl. Transl.)* **8**, 541 (1972). 4–6; 6-4B
72CHE627	N. Salbadols and J. Popelis, *Chem. Heterocycl. Compd. (Engl. Transl.)* **8**, 627 (1972). 8-Db(iv)
72CHE1023	L. M. Alekseeva, G. G. Dvoryantseva, I. V. Persianova, Y. N. Sheinker, R. M. Palei, and P. M. Kochergin, *Chem. Heterocycl. Compd. (Engl. Transl.)* **8**, 1023 (1972). 8-4B
72CHE1153	A. S. Elina, I. S. Musatova, and G. P. Syrova, *Chem. Heterocycl. Compd. (Engl. Transl.)* **8**, 1153 (1972). 10-5C
72CHE1223	N. O. Saldabol, L. L. Zeligman, S. A. Giller, Yu. Yu. Popelis, A. E. Abele, and L. N. Alekseeva, *Chem. Heterocycl. Compd. (Engl. Transl.)* **8**, 1223 (1972). 4-4A; 8-5C
72CHE1242	N. P. Shusherina and T. I. Likhomanova, *Chem. Heterocycl. Compd. (Engl. Transl.)* **8**, 1242 (1972). 9-4B
72CHE1495	V. P. Lezina, A. U. Stepanyants, L. D. Smirnov, N. A. Andronova, and K. M. Dyumaev, *Chem. Heterocycl. Compd. (Engl. Transl.)* **8**, 1495 (1972). 11-2A
72CPB2163	S. Naruto and O. Yonemitsu, *Chem. Pharm. Bull.* **20**, 2163 (1972). 8-2Ab(v)
72CPB2678	Y. Ito, Y. Hamada, and M. Hirota, *Chem. Pharm. Bull.* **20**, 2678 (1972). 9-4A
72CPB2686	Y. Hamada, Y. Ito, T. Mizuno and M. Hirota, *Chem. Pharm. Bull.* **20**, 2686 (1972). 9-4A
72CR(C)(275)49	G. Dana, P. Scribe, and J. P. Girault, *C. R. Hebd. Seances Acad. Sci., Ser. C* **275**, 49 (1972). 4-3Eb; 6-7A,C,D
72CS(2)137	A. Bugge, *Chem. Scr.* **2**, 137 (1972). 8-5A
72G253	S. Carboni, A. DaSettimo, D. Bertini, P. L. Ferrarini, O. Livi, C. Mori, and I. Tonetti, *Gazz. Chim. Ital.* **102**, 253 (1972). 11-4B
72G534	F. Fringuelli, G. Marino, and A. Taticchi, *Gazz. Chim. Ital.* **102**, 534 (1972). 6-9B
72JA5759	H. Kohn, S. J. Benlovic and R. A. Olofson, *J. Am. Chem. Soc.* **94**, 5759 (1972). 7-3B
72JA7448	G. A. Olah, S. Kobayashi, and M. Tashiro, *J. Am. Chem. Soc.* **94**, 7448 (1972). 3-2B
72JCS(P1)265	J. Cooper and R. M. Scrowston, *J.C.S. Perkin Trans. 1*, 265 (1972). 8-2Ab(iii)
72JCS(P1)414	J. Cooper and R. M. Scrowston, *J.C.S. Perkin Trans. 1*, 414 (1972). 4-3A; 8-2Ab(iii),(v)
72JCS(P1)1404	N. B. Chapman, K. Clarke, and A. Manolis, *J.C.S. Perkin Trans. 1*, 1404 (1972). 8-2Ab(iv)
72JCS(P1)2004	A. F. Bramwell, I. M. Payne, G. Riezebos, P. Ward, and R. D. Wells, *J.C.S. Perkin Trans. 1*, 2004 (1972). 4–6; 10-5C
72JCS(P1)2567	V. Calo, F. Ciminale, L. Lopez, F. Naso, and P. E. Todesco, *J.C.S. Perkin Trans. 1*, 2567 (1972). 7-5A; 8B1b(iv)
72JCS(P1)2954	J. A. Hickman and D. G. Wibberley, *J.C.S. Perkin Trans. 1*, 2954 (1972). 8-Ba
72JCS(P2)71	G. Marino and S. Clementi, *J.C.S. Perkin Trans. 2*, 71 (1972). 4-3Eb; 6-7C,D, Table 6.6
72JCS(P2)97	R. Baker, C. Eaborn, and R. Taylor, *J.C.S. Perkin Trans. 2*, 97 (1972). 6-1A, Table 6.1; 8-2Ab(i), 3Ab(i), Scheme 3

72JCS(P2)671	P. Forsythe, R. Frampton, C. D. Johnson, and A. R. Katritzky, *J.C.S. Perkin Trans. 2*, 671 (1972). 9-9C,F
72JCS(P2)766	H. V. Ansell, R. B. Clegg, and R. Taylor, *J.C.S. Perkin Trans. 2*, 766 (1972). 2-1C
72JCS(P2)1111,1116	B. C. Challis and E. M. Millar, *J.C.S. Perkin Trans. 2*, 1111,1116 (1972). 8-2A*b*(i)
72JCS(P2)1618	B. C. Challis and E. M. Millar, *J.C.S. Perkin Trans. 2*, 1618 (1972). 2-1A; 8-2A*b*(i)
72JCS(P2)1625	B. C. Challis and E. M. Millar, *J.C.S. Perkin Trans. 2*, 1625 (1972). 8-2A*b*(i)
72JCS(P2)1654	M. R. Grimmett, S. R. Hartshorn, K. Schofield, and J. B. Weston, *J.C.S. Perkin Trans. 2*, 1654 (1972). 3-4E; 7-4A*d*, 4B*c*
72JCS(P2)1940.	A. G. Burton, R. D. Frampton, C. D. Johnson, and A. R. Katritzky, *J.C.S. Perkin Trans. 2*, 1940 (1972). 9-4A,C
72JCS(P2)1950	G. Bianchi, A. G. Burton, C. D. Johnson, and A. R. Katritzky, *J.C.S. Perkin Trans. 2*, 1950 (1972). 3-3C; 9-4A
72JCS(P2)1953	A. G. Burton, P. J. Halls, and A. R. Katritzky, *J.C.S. Perkin Trans. 2*, 1953 (1972). 3-3C; 9-4B
72JCS(P2)2070	S. Alunni, P. Linda, G. Marino, S. Santini, and G. Savelli, *J.C.S. Perkin Trans. 2*, 2070 (1972). 4-3E*a;* 6-7A
72JCS(P2)2567	V. Calo, F. Ciminale, L. Lopez, F. Naso, and P. Todesco, *J.C.S. Perkins Trans. 2*, 2567 (1972) 4-6.
72JHC849	A. Arcoria, E. Maccarone, G. Musumarra, and G. Romano, *J. Heterocycl. Chem.* **9**, 849 (1972). 6-3A
72JHC995	W. W. Paudler and J. Lee, *J. Heterocycl. Chem.* **9**, 995 (1972). 10-2D
72JHC1157	W. W. Paudler and C. I. Patsy, *J. Heterocycl. Chem.* **9**, 1157 (1972). 4-2A; 8-D*b*(iii)
72JHC1367	R. A. Abramovitch, J. Campbell, E. E. Knaus, and A. Silhankova, *J. Heterocycl. Chem.* **9**, 1367 (1972). 9-5C
72JOC329	D. H. R. Barton, R. H. Hesse, H. T. Toh, and M. M. Pechet, *J. Org. Chem.* **37**, 329 (1972). 10-5B
72JOC578	J. L. Shim, R. Niess, and A. D. Broom, *J. Org. Chem.* **37**, 578 (1972). 4-3E*b;* 10-6D
72JOC2615	D. S. Noyce, C. A. Lipinski, and R. W. Nichols, *J. Org. Chem.* **37**, 2615 (1972). 5,5-B*a;* 6-9A, Table 6.11
72JOC2620,623	D. S. Noyce and H. J. Pavez, *J. Org. Chem.* **37**, 2620, 623 (1972). 6-9C
72JOC3355	W. L. Albrecht, D. H. Gustafson, and S. W. Horgan, *J. Org Chem.* **37**, 3355 (1972). 8-3A*b*(iv)
72JOC4078	H. Gershon, M. W. McNeil, and S. G. Schulman, *J. Org. Chem.* **37**, 4078 (1972). 11-5A
72JOC4188	S. A. Krueger and W. W. Paudler, *J. Org. Chem.* **37**, 4188 (1972). 10-3B, Table 2
72JOU416	K. M. Dymumaev and R. E. Lokhov, *J. Org. Chem. USSR (Engl. Transl.)* **8**, 416 (1972). 4-3C; 9-6B
72JOU1685,1808	I. B. Repinskaya, A. I. Rezvukhin, and V. A. Koptyug, *J. Org. Chem. USSR (Engl. Transl.)* **8**, 1685, 1808 (1972). 2-1A
72JPR(314)603	K. Schwetlik and K. Unverferth, *J. Prakt. Chem.* **314**, 603 (1972). 6-1A,B,C,D

72KFZ22	V. M. Aryozina and M. N. Shchukina, *Khim. Farm. Zh.* **6**, 22 (1972). 8-4B
72MI1	A. I. Shatenshtein, A. G. Kamrad, I. O. Shapiro, Yu. I. Ranneva, and E. N. Zvyagintseva, *Khim. Seraorg. Soedin., Soderzh. Neftyakh. Nefteprod.* **9**, 121 (1972). [*CA* **80**, 59256 (1974)]. 6-1C, 2
72MI2	R. Taylor, *Compr. Chem. Kinet.* **13**, 10, 149, 181, 186, 194, 199, 217, 243, 262, 266, 267, 271, 278, 287, 335, 348 (1972). 2-1A, Gc(v),2B,C; 3; 4-1A,B,2C,3Eb; 6-5,7A,8A; 7-2B; 8-2Ab-(i),(ii),(v),(vi),Cb(ii),3Ac(ii); 11-3
72MI3	I. Simiti, E. Chindris, and I. Schwartz, *Rev. Chim. (Bucharest)* **23**, 460 (1972). 7-5A
72NKK387	T. Keumi, N. Takimi, and Y. Oshima, *Nippon Kagaku Kaishi*, 387 (1972) [*CA* **77**, 151771 (1972)]. 4-3Eb; 8-3Ab(iv)
72OPP9	J. W. Bunting and W. G. Meathrel, *Org. Prep. Proced. Int.* **4**, 9 (1972). [*CA* **76**, 153544 (1972)]. 11-4A
72RTC831	J. M. A. Baas and B. M. Wepster, *Recl. Trav. Chim. Pays-Bas* **91** 831 (1972). 3-2C
72RTC1185	P. Cohen-Fernandes and C. L. Habraken, *Recl. Trav. Chim. Pays-Bas* **91**, 1185 (1972). 7-4Bc
72RTC1383	E. F. Godefroi, H. J. J Loozen, and J. T. J. Luderer-Platje, *Recl. Trav. Chim. Pyas-Bas* **91**, 1383 (1972). 7-6
72T3277	G. Dore, M. Bonhomme, and M. Robba, *Tetrahedron* **28**, 3277 (1972). 8-3B
72TL1755	R. Taylor, *Tetrahedron Lett.*, 1755 (1972). 3-2C
72TL2191	B. N. McMaster, M. C. A. Opie, and G. J. Wright, *Tetrahedron Lett.*, 2191 (1971). 2-1Gc(iii)
72TL2771	I. J. Ferguson, M. R. Grimmett, and K. Schofield, *Tetrahedron Lett.*, 2771 (1972). 7-4Bc
72TL3889	D. S. Noyce and R. W. Nichols, *Tetrahedron Lett.*, 3889 (1972). 6-10C
72TL3893	D. A. Forsyth and D. S. Noyce, *Tetrahedron Lett.*, 3893 (1972). 6-9A,C
72TL5277	G. Casnati, A. Dossena, and A. Pochini, *Tetrahedron Lett.*, 5277 (1977). 4-3A; 8-2Ab(v)
72USP3707480	G. L. Dunn and R. E. Hoover, U.S. Pat. 3,707,480 (1972). 6-3A
73ACH107	I. Simiti and M. Farkas, *Acta Chim. Acad. Sci. Hung.* **76**, 107 (1973). 7-5A
73ACS153	P. Salomaa, A. Kankaanpera, E. Nikander, K. Kaipainen, and R. Aaltonen, *Acta Chem. Scand.* **27**, 153 (1973). 6-1C
73ACS2179	K. E. Stensio, K. Wahlberg, and R. Wahren, *Acta Chem. Scand.* **27**, 2179 (1973). 7-5A
73ACS2257	S. Gronowitz, B. Yom-Tov, and U. Michael, *Acta Chem. Scand.* **27**, 2257 (1973). 8-4B
73ACS2421	O. Ceder and K. Rosen, *Acta Chem. Scand.* **27**, 2421 (1973). 4–6; 11-5B
73AG(E)753	W. Schäfer and K. Dimroth, *Angew. Chem., Int. Ed. Engl.* **12**, 753 (1973). 4-3A; 9-6B
73AJC2725	E. L. Samuel, *Aust. J. Chem.* **26**, 2725 (1973). 8-Cb(iii),(iv)

446   REFERENCES

73BAU2233   Ya. L. Goldfarb, F. M. Stoy, and G. B. Chermanova, *Bull. Acad. Sci. USSR, Div. Chem. Sci. (Engl. Transl.)*, 2233 (1973).   6-4A

73BAU2666   L. I. Belenkii, Ya. L. Goldfarb, and G. P. Gramova, *Bull. Acad. Sci. USSR, Div. Chem. Sci. (Engl. Transl.)* 2666 (1973).   4–6; 6-4B

73BSF1760   J. P. Girault, P. Scribe, and G. Dana, *Bull. Soc. Chim. Fr.*, 1760 (1973).   4-3Eb; 6-7C

73CA(78)58306   M. K. A. Khan, A. Mohammady, and F. Y. Ahmed, *Chem. Abstr.* **78**, 58306 (1973).   8-Cb(iv)

73CA(78)84153   M. S. Ogii, V. I. Shcherbachenko, S. N. Petrunyan, and N. K. Moshchinskaya, *Chem. Abstr.* **78**, 84153 (1973).   4-3B; 8-3Ab(iv)

73CA(79)105183   M. Hubert-Habart, C. Pene, G. Bastian, and R. Royer, *Chem. Abstr.* **79**, 105183 (1973).   8-2Ab(iii)

73CA(78)111049   N. I. Baranova and V. I. Shishkina, *Chem. Abstr.* **78**, 111049 (1973).   4-5E; 8-3Ab(vi)

73CA(79)105183   M. Hubert-Habart, C. Pene, G. Bastian, and R. Royer, *Chem. Abstr.* **79**, 105183 (1973).   8-2Ab(iii)

73CC300   A. Fischer and D. R. A. Leonard, *J.C.S. Chem. Commun.*, 300 (1973).   3-2F

73CC540   L. A. P. Kane-Maguire and C. A. Mansfield, *J.C.S. Chem. Commun.*, 540 (1973).   4-3A; 6-5; 8-2Ab(v)

73CC836   R. Taylor and T. J. Tewson, *J.C.S. Chem. Commun.*, 836 (1973).   2-1A, B

73CC936   H. V. Ansell and R. Taylor, *J.C.S. Chem. Commun.*, 936 (1973).   2-1C

73CCC1809   R. Frimm, L. Fisera, and J. Kovac, *Collect. Czech. Chem. Commun.* **38**, 1809 (1973).   6-7A

73CHE95   A. M. Simonov, Yu. V. Koshchienko, and T. G. Belenko, *Chem. Heterocycl. Compd. (Engl. Transl.)* **9**, (1973).   4–6; 8-4B

73CHE366   V. M. Aryozina and M. N. Shchukina, *Chem. Heterocycl. Compd. (Engl. Transl.)* **9**, 366 (1973).   8-4B

73CHE447   N. S. Ksenzhek, L. I. Belenkii, and Ya. L. Goldfarb, *Chem. Heterocycl. Compd. (Engl. Transl.)* **9**, 447 (1973).   4-3Eb; 6-7A

73CHE953   T. V. Shchedrinskaya, V. P. Litvinov, P. A. Konstantinov, Ya. L. Goldfarb, and E. G. Ostapenko, *Chem. Heterocycl. Compd. (Engl. Transl.)* **9**, 953 (1973).   4-2A; 8-2Ab(vi)

73CHE1202   S. D. Sokolov, T. N. Egorova, and N. S. Kuryatov, *Chem. Heterocycl. Compd. (Engl. Transl.)* **9**, 1202 (1973).   7-4Ba

73CHE1331   A. M. Gyalmaliev, I. V. Stankevich, and Z. V. Todres, *Chem. Heterocyl. Compd. (Engl. Transl.)* **9**, 1331 (1973).   4-5B, 6; 7–10; 8-Ca, 3B

73CJC1620   T. J. Broxton, G. L. Butt, L. W. Deady, S. H. Toh, R. D. Topsom, A. Fischer, and N. W. Morgan, *Can. J. Chem.* **51**, 1620 (1973).   9-7Ba

73CPB260   F. Yoneda, K. Senga, and S. Nishigaki, *Chem. Pharm. Bull.* **21**, 260 (1973).   4-3Ea; 10-6D

73CPB1272   M. Hori, T. Kataoka, K. Ohno, and T. Toyoda, *Chem. Pharm. Bull.* **21**, 1272 (1973).   11-4C

73CPB1327	K. Ikeda, T. Sumi, K. Yokoi, and Y. Mizuno, *Chem. Pharm. Bull.* **21**, 1327 (1973). 10-6A
73CPB1510	S. Kamiya and G. Okusa, *Chem. Pharm. Bull.* **21**, 1510 (1973). 10-6E
73GER2243015	D. R. Hoff, P. Kulsa, H. H. Mrozik, and E. F. Rogers, Ger. Pat. 2,243,015 (1973). 7-5A
73IJS233	R. P. Dickinson, B. Iddon, and R. G. Sommerville, *Int. J. Sulfur Chem.* **8**, 233 (1973). 4–6; 8-2Ab(iv)
73JA1628	J. A. Rabi and J. J. Fox, *J. Am. Chem. Soc.* **95**, 1628 (1973). 10-3B
73JA3918	C. G. Stevens and S. J. Strickler, *J. Am. Chem. Soc.* **95**, 3918 (1973). 2-1Fe(i), Gc(iii)
73JA3928	J. A. Zoltewicz and A. A. Sale, *J. Am. Chem. Soc.* **95**, 3928 (1973). 11-3
73JA6139	J. H. Bradbury, B. E. Chapman, and F. A. Pellegrino, *J. Am. Chem. Soc.* **95**, 6139 (1973). 7-3B
73JCS(P1)68	G. J. Fox, J. D. Hepworth, and G. Hallas, *J.C.S. Perkin Trans. 1*, 68 (1973). 4–6; 9-5A
73JCS(P1)623	K. Clarke, R. M. Scrowston, and T. M. Sutton, *J.C.S. Perkin Trans. 1*, 623 (1973). 8-2Ab(iii)
73JCS(P1)789	D. Lichtenberg and F. Bergmann, *J.C.S. Perkin Trans. 1*, 789 (1973). 8-Db(ii)
73JCS(P1)1089	S. Bien, D. Amith, and M. Ber, *J.C.S. Perkin Trans. 1*, 1089 (1973). 10-6A
73JCS(P1)1196	K. Clarke, R. M. Scrowston, and T. M. Sutton, *J.C.S. Perkin Trans. 1*, 1196 (1973). 8-2Ab(iii)
73JCS(P1)1766	D. J. Chadwick, J. Chambers, G. D. Meakins, and R. L. Snowden, *J.C.S. Perkin Trans. 1*, 1766 (1973). 6-4B
73JCS(P1)2327	D. J. Chadwick, J. Chambers, H. E. Hargraves, G. D. Meakins, and R. L. Snowden, *J.C.S. Perkin Trans. 1*, 2327 (1973). 6-7A
73JCS(P2)179	J. N. Murrell, W. Schmidt, and R. Taylor, *J.C.S. Perkin Trans. 2*, 179 (1973). 11-8A,B, Table 6
73JCS(P2)253	R. Taylor, *J.C.S. Perkin Trans. 2*, 253 (1973). 9-4A
73JCS(P2)432	J. A. Elvidge, J. R. Jones, C. O'Brien, E. A. Evans, and J. C. Turner, *J.C.S. Perkin Trans. 2*, 432 (1973). 8-Cb(ii)
73JCS(P2)823	O. Rogne, *J.C.S. Perkin Trans. 2*, 823 (1973). 7–7
73JCS(P2)918	B. C. Challis and A. J. Lawson, *J.C.S. Perkin Trans. 2*, 918 (1973). 4-4A; 8-2Ab(iii)
73JCS(P2)1065	A. El-Anani, J. Banger, G. Bianchi, S. Clementi, C. D. Johnson, and A. R. Katritzky, *J C. S. Perkin Trans. 2*, 1065 (1973). 2-1Fd, Gc, Gc(i),(iii),(iv); 9-2H; 10-2; 11-2C
73JCS(P2)1072	A. El-Anani, S. Clementi, A. R. Katritzky, and L. Yakhontov, *J.C.S. Perkin Trans. 2*, 1072 (1973). 6-1D; 8-Db(i); 9-2B
73JCS(P2)1077	S. Clementi and A. R. Katritzky, *J.C.S. Perkin Trans. 2*, 1077 (1973). 2-1Gd; 10-2
73JCS(P2)1250	S. Clementi, P. Linda, and C. D. Johnson, *J.C.S. Perkin Trans. 2*, 1250 (1973). 4-3Eb; 8-2Ab(iv),(v)
737JCS(P2)1675	S. Clementi, P. P. Forsythe, C. D. Johnson, and A. R. Katritzky, *J.C.S. Perkin Trans. 2*, 1675 (1973). 2-1Gc(ci); 6-1D; 7-2A, Table 1

73JCS(P2)1889	J. A. Elvidge, J. R. Jones, C. O'Brien, E. A. Evans, and H. C. Sheppard, *J.C.S. Perkin Trans. 2*, 1889 (1973). 8-Db(ii)
73JCS(P2)2097	S. Clementi, F. Fringuelli, P. Linda, G. Marino, G. Savelli, and A. Taticchi, *J.C.S. Perkin Trans. 2*, 2097 (1973). 4-3Ea,b; 6-7C, 10A,B
73JGU871	P. A. Konstantinov, N. M. Koloskova, R. I. Shupik, and M. N. Volkov, *J. Gen. Chem. USSR (Engl. Transl.)* **43**, 871 (1973). 4-3Ea; 6-7B
73JHC153	T. J. Kress and L. L. Moore, *J. Heterocycl. Chem.* **10**, 153 (1973). 4–7; 10-5B
73JHC409	T. J. Kress and S. M. Costantino, *J. Heterocycl. Chem.* **10**, 409 (1973). 4–6; 10-5B; 11-5A
73JHC551	P. D. Cook and R. N. Castle, *J. Heterocycl. Chem.* **10**, 551 (1973). 10-4A
73JOC829	J. A. Zoltewicz and V. W. Cantwell, *J. Org. Chem.* **38**, 829 (1973). 9-3
73JOC1955	L. F. Miller and R. E. Banbury, *J. Org. Chem.* **38**, 1955 (1973). 8-5C
73JOC2433,3318,3321	D. S. Noyce and S. A. Fike, *J. Org. Chem.* **38**, 2433, 3318, 3321 (1973). 7-9B
73JOC2657	D. S. Noyce, J. A. Virgilio, and B. Bartman, *J. Org. Chem.* **38**, 2657 (1973). 9-7Ba,b, 9E
73JOC2660	D. S. Noyce and J. A. Virgilio, *J. Org. Chem.* **38**, 2660 (1973). 9-8
73JOC3212	G. A. Olah and Y. K. Mo, *J. Org. Chem.* **38**, 3212 (1973). 2-1A
73JOC3316	D. S. Noyce and S. A. Fike, *J. Org. Chem.* **38**, 3316 (1973). 7-9A, Scheme 12
73JOC3762	D. S. Noyce and G. T. Stowe, *J. Org. Chem.* **38**, 3762 (1973). 4-2A; 7-8B, 9A, Scheme 12
73JOU840	S. P. Maltseva, Z. A. Borodulina, and B. I. Stepanov, *J. Org. Chem. USSR (Engl. Trans.)* **9**, 840 (1973). 4-3Ea; 7-6
73JOU1542	L. I. Belenkii, I. B. Karmanova, and Ya. L. Goldfarb, *J. Org. Chem. USSR* **9**, 1542 (1973). 6–6
73JOU2216	Yu. N. Koshelev, A. V., Reznichenko, L. S. Efros, and I. Ya. Kvitko, *J. Org. Chem. USSR (Engl. Transl.)* **9**, 2216 (1973). 4–6; 8-5C
73JPU1237	G. M. Norikova, V. F. Degtyarev, and V. I. Shishkina, *J. Phys. Chem. USSR (Engl. Transl.)* **47**, 1237 (1973). 8-3Ab(ii)
73M1599	O. Hromatka, D. Binder, and K. Eichinger, *Monatsh. Chem.* **104**, 1599 (1973). 6-4A
73MI1	R. Taylor *Aromat. Heteroaromat. Chem.* **1**, 181 (1973). 9-9E
73MI2	Y. Yuki and S. Mouri, *Nagoya Kogyo Daigaku Gakuho* **25**, 419 (1973) [*CA* **82**, 170841 (1975)]. 10-4
73NKK1505	T. Keumi, Y. Maegawa, T. Takegami, and Y. Oshimia, *Nippon Kagaku Kaishi*, 1505 (1973) [*C.A* **79**, 136927 (1973)[. 4-3Eb; 8-3Ab(iv)
73OS1758	H. Heaney and S. V. Ley, *Org. Synth.* **53**, 1758 (1973). 8-2Ab(v)

73RC2255	J. Mlochowski and Z. Skrowaczewska, *Rocz. Chem.* **47**, 2255 (1973). 11-4B
73T413	J. P. Girault, P. Scribe, and G. Dana, *Tetrahedron* **29**, 413 (1973). 4-3E*b*; 6-7A
73T579	E. B. Pedersen, T. E. Peterson, K. Torssell, and S.-O. Lawesson, *Tetrahedron* **29**, 579 (1973). 3-2D
73T669	G. F. Smith and D. A. Taylor, *Tetrahedron* **29**, 669 (1973). 4-5B; 8-2A*b*(vi)
73T971	J. Bergman, J.-E. Bäckvall, and J.-O. Lindström, *Tetrahedrom* **29**, 971 (1973). 4-3E*b*; 8-2A*b*(v)
73T2495	W. W. Paudler, J. Lee, and T. K. Chen, *Tetrahedron* **29**, 2495 (1973). 10-2D, 3D, Table 4
73T3469	M. A. Schroeder and R. C. Makino, *Tetrahedron* **29**, 3469 (1973). 7-3B
73YZ59	I. Suzuki and S. Sueyoshi, *Yakugaku Zasshi* **93**, 59 (1973). 10-5A
74ACH381	I. Simiti and M. Farkas, *Acta Chim. Acad. Sci. Hung.* **83**, 381 (1974). 4–6; 7-5A
74AHC(16)1	J. A. Elvidge, J. R. Jones, C. O'Brien, E. A. Evans, and H. C. Sheppard, *Adv. Heterocycl. Chem.* **16**, 1 (1974). 9-1; 10-1, 3B
74AHC(16)181	J. Ashby and C. C. Cook, *Adv. Heterocycl. Chem.* **16**, 181 (1974). 8-3A*b*(i)
74AHC(17)255	M. J. Cook, A. R. Katritzky, and P. Linda, *Adv. Heterocycl. Chem.* **17**, 255 (1974). 6-10A
74AJC2331	B. E. Boulton and B. A. W. Coller, *Aust. J. Chem.* **27**, 2331 (1974). 7-5A
74AJC2349	B. E. Boulton and B. A. W. Coller, *Aust. J. Chem.* **27**, 2349. 4–6; 8-D*b*(iv)
74BAU232	L. P. Kamshii and V. A. Koptyug, *Bull. Acad. Sci. USSR, Div. Chem. Sci. (Engl. Transl.)*, 232 (1974). 2-1A
74BAU2023	L. D. Smirnov, V. S. Zhuravlev, V. P. Lezina, and K. M. Dyumaev, *Bull. Acad. Sci. USSR, Div. Chem. Sci. (Engl. Transl.)*, 2023 (1974). 4-3C, 4B, 6; 9-5C, 6B,C
74BCJ1267	T. Okuyama, K. Kunugiza, and T. Fueno, *Bull. Chem. Soc. Jpn.* **47**, 1267 (1974). 4–6; 8-2A*b*(iv)
74BSF183	G. Saint-Buf and B. Lobert, *Bull. Soc. Chim. Fr.*, 183 (1974). 8-3A*b*(ii)
74BSF2099	P. Chauvin, J. Morel, P. Pastour, and J. Martinez, *Bull. Soc. Chim. Fr.*, 2099 (1974). 4-2A; 7-4A*b*, 8B
74CA(80)3443	S. Yurugi, T. Fushimi, and A. Miyake, *Chem. Abstr.* **80**, 3443 (1974). 4–6; 8-3B
74CA(81)105154	I. M. Nasyrov, I. U. Numanov, M. Isabaev, and N. Radzhabov, *Chem. Abstr.* **81**, 105154 (1974). 4-5C; 8-2A*b*(vi)
74CA(81)168829	M. M. Sukhoroslova, V. P. Lopatinskii, and V. V. Bochkarev, *Chem. Abstr.* **81**, 168829 (1974). 4-3E*b*; 8-3A*b*(iv)
74CC333	R. N. McDonald and J. M. Richmond, *J.C.S. Chem. Commun.*, 333 (1974). 4–6, Table 6.4
74CC535	S. Banerjee and O. S. Tee, *J.C.S. Chem. Commun.*, 535 (1974). 10-5B

74CC585	R. F. Cookson and A. C. Richards, *J.C.S. Chem. Commun.*, 585 (1974). 4-3B; 7-6
74CHE136	V. K. Polyakov, Z. P. Zapluivechka, and S. V. Tsukerman, *Chem. Heterocycl. Compd. (Engl. Transl.)* **10**, 136 (1974). 6-7A
74CHE166	G. I. Kagan, V. A. Kosobutskii, V. K. Belyakov, and O. G. Tarakanov, *Chem. Heterocycl. Compd. (Engl. Transl.)* **10**, 166 (1974). 8-Ca
74CHE230	L. I. Savranskii, V. A. Kovtunenko, and F. S. Babichev, *Chem. Heterocycl. (Engl. Transl.)* **10**, 230 (1974). 8-4B,5
74CHE516	S. D. Sokolov, T. N. Egorova, and I. M. Yudintseva, *Chem. Heterocycl. Compd. (Engl. Transl.)* **10, 516 (1974)**. 7-4Ba
74CHE699	K. M. Dyumaev, E. P. Popova, I. F. Mikhailova, L. I. Shibaeva, and L. D. Smirnov, *Chem. Heterocycl. Compd. (Engl. Transl.)* **10**, 699 (1973). 11-4A
74CHE930	L. G. Yudin, A. N. Kost, E. Ya. Zuichenko, and A. G. Zhigulin, *Chem. Heterocycl. Compd. (Engl. Transl.)* **10**, 930 (1974). 8-2Ab(iii)
74CHE953	A. K. Sheinkman, M. M. Mestechkin, A. P. Kucherenko, N. A. Klyuev, V. N. Poltaveta, G. A. Maltseva, L. A. Palagushkina, and Yu. B. Vysotskii, *Chem. Heterocycl. Compd. (Engl. Transl.)* **10**, 953 (1974). Table 11-6
74CHE1397	N. N. Zatsepina and I. F. Tupitsyn, *Chem. Heterocycl. Compd. (Engl. Transl.)* **10**, 1397 (1974). 9-3; Table 10-1; 11-3, Table 2
74CJC451	O. S. Tee and S. Banerjee, *Can. J. Chem.* **52**, 451 (1974). 10-5B
74CJC3960	A. Fischer and J. N. Ramsay, *Can. J. Chem.* **52**, 3960 (1974). 3-2F
74CPB21	M. Hori, T. Kataoka, and C. F. Hsu, *Chem. Pharm. Bull.* **22**, 21 (1974). 11-4C
74CPB27	M. Hori, T. Kataoka, C. F. Hsu, Y. Asahi, and E. Mizuta, *Chem. Pharm. Bull.* **22**, 27 (1974). 11-4C
74CPB2359	M. Masui, K. Suda, M. Inoue, K. Izukura and M. Yamauchi, *Chem. Pharm. Bull.* **22**, 2359 (1974). 7-6
74CS(5)217	S. Gronowitz, B. Cederlund, and A-B. Hornfeldt, *Chem. Scr.* **5**, 217 (1974). 4-7A; 6-8A
74FRP2193823	Upjohn Co., Fr. Pat. 2,193,823 (1974). 4–6; 7-5B
74GEP2354786	R. Maksimovic, D. Dumanovic, N. Radovic, and P. Bubic, Ger. Pat. 2,354,786 (1974). 7-4Ad
74GEP2457082	G. Skipka and A. Vogel, Ger. Pat. 2,457,082 (1974). 8-3Ab(ii)
74HC(14,1-4)	R. A. Abramovitch *et al.*, *Chem. Heterocycl. Compd.* **14**, Parts I–IV (1974). 9-1
74JA6908	G. A. Olah, H. C. Lin, and D. A. Forsyth, *J. Am. Chem. Soc.* **96**, 6908 (1974). 2-1A
74JCS(P1)1751	R. G. Coombes and L. W. Russell, *J. S. Perkin Trans. 1*, 1751 (1974). 11-4A
74JCS(P1)2095	D. H. R. Barton, W. A. Bubb, R. H. Hesse, and M. M. Pechet, *J.C.S. Perkin Trans. 1*, 2095 (1974). 10-5B
74JCS(P2)332	F. Fringuelli, G. Marino, A. Taticchi, and G. Grandolini, *J.C.S. Perkin Trans. 2*, 332 (1974). 6-10A

74JCS(P2)382	A. G. Burton, M. Dereli, A. R. Katritzky, and H. O. Tarhan, *J.C.S. Perkin Trans. 2*, 382 (1974). 7-4B*c*
74JCS(P2)389	A. G. Burton, A. R. Katritzky, M. Kouya, and H. O. Tarhan, *J.C.S. Perkin Trans. 2*, 389 (1974). 7-4B*c*, Scheme 11
74JCS(P2)394	J. Banger, C. D. Johnson, A. R. Katritzky, and B. R. O'Neill, *J.C.S. Perkin Trans. 2*, 394 (1974). 2-1B*a*, Fe(i)
74JCS(P2)399	S. Clementi, P. P. Forsythe, C. D. Johnson, A. R. Katritzky, and B. Terem, *J.C.S. Perkin Trans. 2*, 399 (1974). 2-1G*c*(i); 7-2A,B, Table 1
74JCS(P2)1294	S. Clementi, C. D. Johnson, and A. R. Katritzky, *J.C.S. Perkin Trans. 2*, 1294 (1974). 9-8; 10-2, 4
74JCS(P2)1610	P. Linda, A. Lucarelli, G. Marino, and G. Savelli, *J.C.S. Perkin Trans. 2*, 1610 (1974). 4-3E*a,b;* 6-7A, Table 6.6
74JCS(P2)1893	M. H. Palmer and S. M. F. Kennedy, *J.C.S. Perkin Trans. 2*, 1893 (1974). Table 8-1
74JHC205	L. H. Klemm, R. E. Merrill, F. H. W. Lee, and C. E. Klopfenstein, *J. Heterocycl. Chem.* **11**, 205 (1974). 8-D*b*(iv)
74JHC355	L. H. Klemm and R. E. Merrill, *J. Heterocycl. Chem.* **11**, 355 (1974). 4-2A; 8-D*b*(iii)
74JHC459	Y. Tamura, H. Hayashi, E. Sacki, J.-H. Kim, and M. Ikeda, *J. Heterocycl. Chem.* **11**, 459 (1974). 8-5C
74JHC813	K. Pilgram and M. Zupan, *J. Heterocycl. Chem.* **11**, 813 (1974). 4–6; 8-C*b*(iv)
74JHC1013	J. Arriau, O. Chalvet, A. Dargelos, and G. Maury, *J. Heterocycl. Chem.* **11**, 1013 (1974). 8-D*a*
74JHC1017	C. D'Erba, G. Garbarino, and G. Guanti, *J. Heterocycl. Chem.* **11**, 1017 (1974). 6-3A
74JOC587	C. W. Whitehead, C. A. Whitesitt, and A. R. Thompson, *J. Org. Chem.* **39**, 587 (1974). 4-3A; 10-6E
74JOC591	C. W. Whitehead and C. A. Whitesitt, *J. Org. Chem.* **39**, 591 (1974). 10-6E
74JOC1157	C. K. Bradsher, L. L. Braun, J. D. Turner, and G. L. Walker, *J. Org. Chem.* **39**, 1157 (1974). 11-4A
74JOC1192	G. Knaus and A. I. Meyers, *J. Org. Chem.* **39**, 1192 (1974). 4-2A; 7-8B
74JOC2398	J. L. Wong and J. H. Keck, *J. Org. Chem.* **39**, 2398 (1974). 7-2E,3B
74JOC2828	D. S. Noyce and D. A. Forsyth, *J. Org. Chem.* **39**, 2828 (1974). 8-2A*b*(vii)
74JOC2934	J. D. Vaughan, E. C. Wu, and C. T. Huang, *J. Org. Chem.* **39**, 2934 (1974). 7-3B
74JOC3481	H. M. Gilow and J. H. Ridd, *J. Org. Chem.* **39**, 3481 (1974). 4–6; 9-5A
74JOC3598	G. P. Rizzi, *J. Org. Chem.* **39**, 3598 (1974). 4-2A; 10-6F
74JOU2489	A. V. Eltsov, V. P. Martynova, E. R. Zakhs, and L. P. Shustova, *J. Org. Chem. USSR (Engl. Transl.)* **10**, 2489 (1974). 8-D*b*(iv)
74MI1	R. Taylor, *Aromat. Heteroaromat. Chem.* **2**, 219, 222 (1974). 6-10C; 9-7B*a*
74MI2	A. A. Krasheninnikov and Yu. A. Panteleev, *Teor. Eksp. Khim.* **10**, 335 (1974). 11-8A

74NKK1708	T. Keumi, H. Hashio, Y. Maegawa, and Y. Oshima, *Nippon Kaguka Kaishi*, 1708 (1974)[*CA* **82**, 43114 (1974)]. 4-3E*b*; 8-3A*b*(iv), Scheme 20
74RC2145	J. Mlochowski, *Rocz. Chem.* **48**, 2145 (1974). 11-5B
74T2123	T. Hino, M. Tonozuka, and M. Nakagawa, *Tetrahedron* **30**, 2123 (1974). 4–6; 8-2A*b*(iv)
74TL853	H. C. Bell, J. R. Kalman, J. T. Pinhey, and S. Sternhell, *Tetrahedron Lett.*, 853 (1974). 4-2D; 6-8A
74URP437763	B. A. Tertov, V. V. Burykin, and A. S. Morkovnik, USSR Pat. 437,763 (1974). 7-4A*d*
74USP3792017	F. E. Arnold, F. L. Hedberg, and R. F. Kovar, U. S. Pat., 3,792,017 (1974). 8-3A*b*(ii)
75ACS457	S. Gronowitz and A. Maltessa, *Acta Chem. Scand.* **29**, 457 (1975). 10-4E; 11-4D
75BSF2334	B.-P. Roques, M.-C. Fournie-Zaluski, and R. Oberlin, *Bull. Soc. Chim. Fr.*, 2334 (1975). 6-4A
75CA(82)16654	Yu. G. Yurev, V. L. Ivasenko, and V. P. Lopatinskii, *Chem. Abstr.* **82**, 16654 (1975). 4-3E*b*; 8-3A*b*(iv)
75CB3762	H. Reimlinge, J. J. M. Vanderwalle, R. Merenyi, and W. R. F. Lingier, *Chem. Ber.* **108**, 3762 (1975). 8-3B
75CC875	E. Baciocchi, S. Clementi, and G. V. Sebastiani, *J.C.S. Chem. Commun.*, 875 (1975). 6-4B
75CC956	C. Wakselman and M. Tordeux, *J.C.S. Chem. Commun.*, 956 (1975). 6-7D
75CCC1163	V. Machecek and V. Sterba, *Collect. Czech. Chem. Commun.* **40**, 1163 (1975). 9-3
75CHE352	Sh. V. Abdullaev, Yu. V. Kurbatov, O. S. Otroshenenko, A. S. Sadyov, and M. F. Khatumova, *Chem. Heterocycl. Compd. (Engl. Transl.)* **11**, 352 (1975). 9-4C
75CHE478	L. D. Smirnov, V. S. Zhurarlev, E. E. Merzon, B. E. Zaitsev, V. P. Lezina, and K. M. Dyumaev, *Chem. Heterocycl. Compd. (Engl. Transl.)* **11**, 478 (1966). 4–6; 9-5B
75CHE571	G. P. Sharnin, I. F. Falyakov, and D. N. Butovetskii, *Chem. Heterocycl. Compd. (Engl. Transl.)* **11**, 571 (1975). 6-3B
75CHE643	S. D. Sokolov and I. M. Yudintseva, *Chem. Heterocycl. Compd. (Engl. Transl.)* **11**, 643 (1975). 7-5B
75CHE745	K. M. Dyumaev and R. E. Lokhov, *Chem. Heterocycl. Compd. (Engl. Transl.)* **11**, 745 (1975). 4-5B; 9-6D
75CJC1	D. R. Arnold and B. M. Clarke, *Can. J. Chem.* **53**, 1 (1975). 4-3E*b*; 6-7A
75CJC119	B. M. Lynch, M. A. Khan, S. C. Sharma, and H. C. Teo, *Can. J. Chem.* **53**, 119 (1975). 8-D*b*(iv)
75CPB923	S. Kamiya and M. Tanno, *Chem. Pharm. Bull.* **23**, 923 (1975). 4–6; 10-4A,5A,6E
75CPB1879	S. Kamiya and M. Tanno, *Chem. Pharm Bull.* **23**, 1879 (1975). 10-4A
75CPB2990	T. Hino, T. Nakamura, and M. Nakagawa, *Chem. Pharm. Bull.* **23**, 2990 (1975). 8-2A*b*(iv)
75G539	S. Clementi, A. El-Anani, A. R. Katritzky, and B. R. O'Neill, *Gazz. Chim. Ital.* **105**, 539 (1975). 2-1G*c*(i)

75JA760	M. J. Cook, N. L. Dessanayake, C. D. Johnson, A. R. Katritzky, and T. W. Toone, *J. Am. Chem. Soc.* **97**, 760 (1975). 2-1G*c*(i); 3-4C
75JCS(P2)277	R. Taylor, *J.C.S. Perkin Trans. 2*, 277 (1975). 9-7A*b*, 9D, Table 17; 10-1; 11-4A
75JCS(P2)366	E. Silla, J. Bertran, and J. I. Fernandez-Alonso, *J.C.S. Perkin Trans. 2*, 366 (1975). Table 8-1
75JCS(P2)551	G. T. Bruce, A. R. Cooksey, and K. J. Morgan, *J.C.S. Perkin Trans. 2*, 551 (1975). 6-9A
75JCS(P2)648	J. W. Barnett, R. B. Moodie, K. Schofield, and J. B. Weston, *J.C.S. Perkin Trans. 2*, 648 (1975). 3-2B
75JCS(P2)1316	D. M. Muir and M. C. Whiting, *J.C.S. Perkin Trans. 2*, 1316 (1975). 2-1E*a;* 6-1D
75JCS(P2)1463	E. Glyde and R. Taylor, *J.C.S. Perkin Trans. 2*, 1463 (1975). 5.
75JCS(P2)1600	A. R. Katritzky, B. Terem, E. V. Scriven, S. Clementi, and H. O. Tarhan, *J.C.S. Perkin Trans. 2*, 1600 (1975). 2-1G*c*(iii); 3-4A,C,D,E,H,5A,B,C,D; 6-3A; 7-4B*a,c*, Schemes 9,11; 9-4F; 10-4; 11-4E
75JCS(P2)1609	M. Dereli, A. R. Katritzky, and H. O. Tarhan, *J.C.S. Perkin Trans. 2*, 1609 (1975). 7-4B*c*, Scheme 11
75JCS(P2)1614	A. R. Katritzky, C. Ögretir, H. O. Tarhan, H. M. Dou, and J. V. Metzger, *J.C.S. Perkin Trans. 2*, 1614 (1975). 7-4A*b*, Scheme 8
75JCS(P2)1620	A. R. Katritzky, H. O. Tarhan, and B. Terem, *J.C.S. Perkin Trans. 2*, 1620 (1975). 7-4B*b*, Scheme 10
75JCS(P2)1624	A. R. Katritzky, S. Clementi, and H. O. Tarhan, *J.C.S. Perkin Trans. 2*, 1624 (1975). 9-4F; 10-4
75JCS(P2)1627	A. R. Katritzky, M. Konya, H. O. Tarhan, and A. G. Burton, *J.C.S. Perkin Trans. 2*, 1627 (1975). 7-4B*a*, Scheme 9
75JCS(P2)1632	A. R. Katritzky, H. O. Tarhan, and B. Terem, *J.C.S. Perkin Trans. 2*, 1632 (1975). 7-4B*c*, Scheme 11
75JCS(P2)1783)	E. Glyde and R. Taylor, *J.C.S. Perkin Trans. 2*, 1783 (1975). 10-1; 11-7,8A,B, Table 6
75JHC195	B. P. Roques, D. Florentin, and M. Callanquin, *J. Heterocyl. Chem.* **12**, 195 (1975). 4-1B; 6-8A,C
75JHC379	O. Fuentes and W. W. Paudler, *J. Heterocycl. Chem.* **12**, 379 (1975). 4-3E*a;* 8-B*a*,D*b*(iv)
75JHC597	D. M. Mulvey and H. Jones, *J. Heterocycl. Chem.* **12**, 597 (1975). 4-5D; 7-7
75JHC705	J. W. McFarland, W. A. Essary, L. Cienti, W. Cozart, and P. E. McFarland, *J. Heterocycl. Chem.* **12**, 705 (1975). 8-D*b*(iv)
75JHC861	M. F. DePompei and W. W. Paudler, *J. Heterocycl. Chem.* **12**, 861 (1975). 8-D*b*(ii),(iv)
75JHC1091	R. Weber and M. Renson, *J. Heterocycl. Chem.* **12**, 1091 (1975). 8-C*b*(iii)
75JOC3373	E. Abushanab, A. P. Bindra, D. Y. Lee, and L. Goodman, *J. Org. Chem.* **40**, 3373 (1975). 8-D*b*(iv)
75JOC3381	D. S. Noyce and B. B. Sandel, *J. Org. Chem.* **40**, 3381 (1975). 7-9A, Scheme 12

75JOU412	L. I. Belenkii, A. P. Yakubov, and Ya. L. Goldfarb, *J. Org. Chem. USSR (Engl. Transl)* **11**, 412 (1975). 4-3E*b*; 6-7A
75JOU889,902	M. G. Voronkov *et al.*, *J. Org. Chem. USSR (Engl. Transl.)* **11**, 889, 902 (1975). 4–6 8-C*b*(iv)
75JOU1883	V. D. Pokhodenko, V. A. Khizhnyi, V. G. Koshechko, and O. I. Shkrebtii, *J. Org. Chem. USSR (Engl. Transl.)* **11**, 1883 (1975). 3-2D
75JOU2691	B. A. Korolev and M. A. Maltseva, *J. Org. Chem. USSR (Engl. Transl.)* **11**, 2691 (1975). 10-2D
75MI1	A. Ohta, T. Watanabe, Y. Akita, and T. Kurihara, *Hukusokan Kagaku Toronkai Koen Yoshishu, 8th*, 84 (1975) [*CA* **84**, 164723 (1976)]. 10-4, 4C
75MI2	R. Taylor, *Aromat. Heteroaromat. Chem.* **3**, 222 (1975). 7-9B
75MI3	R. Taylor, *Aromat. Heteroaromat. Chem.* **3**, 246 (1975). 3-2F; 11-4A
75RCR823	K. M. Dyumaev, *Russ. Chem. Rev. (Engl. Transl.)* **44**, 823 (1975). 9-1
75TL435	R. Taylor, *Tetrahedron Lett.*, 435 (1975). 2-1E*b*
75TL1395	S. Clementi, A. R. Katritzky, and H. O. Tarhan, *Tetrahedron Lett.*, 1395 (1975). 10-4
75ZPK2241	K. K. Preobrazhenskii, S. N. Kharkov, V. I. Shlyakhov, and E. N. Smyalkovskaya, *Zh. Prikl. Khim.* **48**, 2241 (1975). 4-5B; 8-c*b*(vi)
76ACR287	R. B. Moodie and K. Schofield, *Acc. Chem. Res.* **9**, 287 (1976). 3-2F
76ACS(B)605	F. Fringuelli, S. Gronowitz, A.-B. Hornfeldt, I. Johnson, and A. Taticchi, *Acta Chem. Scand., Ser. B* **B 30**, 605 (1976). 4-2A; 6-8B
76AHC(20)1	P. Tomasik, and C. D. Johnson, *Adv. Heterocycl. Chem.* **20**, 1 (1976). 11-4A
76BAU2609	V. I. Dronov and R. F. Nigmatullina, *Bull. Acad. Sci. USSR, Div. Chem. Sci. (Engl. Transl.)* 2609 (1976). 4-3B; 8-3A*b*(iv)
76CA(84)89931	N. R. Radzhabov, I. M. Nasyrov, and I. U. Numanov, *Chem. Abstr.* **84**, 89931 (1976). 4-5C; 8-2A*b*(vi)
76CA(84)135510	B. Iteke Fefe, L. Christiaens, and M. Renson, *Chem. Abstr.* **84**, 135510 (1976). 4-3E*b*, 6;8-4B
76CHE64	L. M. Alekseeva, G. G. Dvoryantseva, I. V. Persianova, Y. N. Sheinker, R. M. Palei, and P. M. Kochergin, *Chem. Heterocycl. Compd. (Engl. Transl.)* **12**, 64 (1976). 8-4B
76CHE955	L. D. Smirnov and K. M. Dyumaev, *Chem. Heterocycl. Compd. (Engl. Transl.)* **12**, 955 (1976). 9-1
76CHE1397	V. I. Zaionts and O. V. Maksimova, *Chem. Heterocycl. Compd. (Engl. Transl.)* **12**, 1397 (1976). 9-1
76CHE1399	B. A. Tertov, A. Morkovnik, and Yu. G. Borgachev, *Chem. Heterocycl. Compd. (Engl. Transl.)* **12**, 1399 (1976). 8-C*b*(v)
76CI(M)220	A. DaSettimo, V. Santerini, G. Primofirre, and G. Biagi, *Chim. Ind. (Milan)* **58**, 220 (1976). 8-2A*b*(iv)
76CI(M)880	A. De Munno, V. Bertini, and F. Lucchesini, *Chim. Ind. (Milan)* **58**, 880 (1976). 7-4B*a*; 5B
76CS(10)165	S. Gronowitz, C. Westerlund, and A. B. Hornfeldt, *Chem. Scr.* **10**, 165 (1976). 8-5A

76JCS(P1)2355	N. B. Chapman, K. Clarke, and J. M. Willis, *J.C.S. Perkin Trans. 1*, 2355 (1976). 6-4A
76JCS(P2)266	E. Baciocchi, S. Clementi, and G. V. Sebastiani, *J.C.S. Perkin Trans. 2*, 266 (1976). 4–6;82Ab(iv)
76JCS(P2)388	D. M. Muir and M. C. Whiting. *J.C.S. Perkin Trans. 2*, 388 (1976). 6-1D;8-2Ab(i)
76JCS(P2)559	M. M. J. LeGuen and R. Taylor, *J.C.S. Perkin Trans. 2*, 559 (1976). 2-1C
76JCS(P2)696	R. S. Alexander and A. R. Butler, *J.C.S. Perkin Trans. 2*, 696 (1976). 4-3A; 6-8D
76JCS(P2)925	C. Eaborn and G. Seconi, *J.C.S. Perkin Trans. 2*, 925 (1976). 4-1C; 8-2Ab(vi)
76JCS(P2)1135	S. Alunni, S. Clementi, and L. H. Klemm, *J.C.S. Perkin Trans. 2*, 1135 (1976). 3-3A
76JHC393	J. Iriarte, E. Martinez, and J. M. Muchowski, *J. Heterocycl. Chem.* **13**, 393 (1976). 4–6; 6-4A
76JHC581	S. W. Schneller, F. W. Clough, and P. N. Skancke, *J. Heterocycl. Chem.* **13**, 581 (1976). 8-Da
76JHC961	H. C. van der Plas and M. Wozniak, *J. Heterocycl. Chem.* **13**, 961 (1976). 11-5B
76JHC1021	O. Attanasi, G. Bartoli, and P. E. Todesco, *J. Heterocycl. Chem.* **13**, 1021 (1976). 8-Cb(ii)
76JHC1141	W. E. Hymans, *J. Heterocycl. Chem.* **13**, 1141 (1976). 10-4B
76JHC1265	D. Florentin, M. C. Fournie-Zaluski, M. Callanquin, and B. P. Roques, *J. Heterocycl. Chem.* **13**, 1265 (1976). 6-8C
76JHC1297	F. E. Herkes and T. A. Blazer, *J. Heterocycl. Chem.* **13**, 1297 (1976). 7-5A
76JOC93	T. J. Kress, L. L. Moore, and S. M. Costantino, *J. Org. Chem.* **41**, 93 (1976). 4–6; 9-5A
76JOC351	R. Buchan, M. Fraser, and C. Shand, *J. Org. Chem.* **41**, 351 (1976).8-Db(iv)
76JOC838	O. S. Tee and G. V. Patil, *J. Org. Chem.* **41**, 838 (1976). 11-5B
76JOC3549	E. S. Hand and W. W. Paudler, *J. Org. Chem.* **41**, 3549 (1976). 8-Db(iv)
76JOC4004	S. Banerjee and O. S. Tee, *J. Org. Chem.* **41**, 4004 (1976). 4–6; 10-5D
76JOU1550	I. Ya. Kvitko, N. B. Sokolova, S. P. Fradkina, and A. V. Eltsov, *J. Org. Chem. USSR (Engl. Transl.)* **12**, 1550 (1976). 8-5C
76MI1	D. N. Kursanov, V. N. Setkina, Yu. D. Konovalov, M. N. Nefedova, N. K. Baranetskaya, G. A. Panosyan, and F. I. Adyrkhaeva, *Doekl. Chem. (Engl. Transl.)* **227**, 310 (1976). 6-1A
76MI2	R. Taylor, *Aromat. Heteroaromat. Chem.* **4**, 259 (1976). 8-2Ab(vi)
76T399	O. Attanasi, G. Bartoli, and P. E. Todesco, *Tetrahedron* **32**, 399 (1976). 8-Db(ii)
76T1403	S. Gronowitz and I. Ander, *Tetrahedron* **32**, 1403 (1976). 6-4A
76T1767	M. P. Carmody, M. J. Cook, N. L. Dassanayake, A. R. Katritzky, P. Linda, and R. D. Tack, *Tetrahedron* **32**, 1767 (1976). 6-10A

76T2595	A. Cipiciani, S. Clementi, P. Linda, G. Savelli, and G. V. Sebastiani, *Tetrahedron* **32**, 2595 (1976). 4-3E*b;* 8-2A*b*(v), 3A*b*(iv)
76TL771	R. G. Coombes and J. G. Golding, *Tetrahedron Lett.*, 771 (1976). 3-2C
76URP388556	USSR Pat. 388,556 (1976) [*CA* **85**, 123957 (1976)]. 10-5A.
77CA(82)57518	L. N. Yakhontov *et al.*, *Chem. Abstr.* **82**, 57518 (1977). 8-D*b*(iv).
77CA(87)151948	N. I. Baranova and V. I. Shishkina, *Chem. Abstr.* **87**, 151948 (1977). 4-5E; 8-3A*b*(vi)
77CA(86)171800	B. M. Kotlyarevskaya and I. I. Gubenko, *Chem. Abstr.* **86**, 171800 (1977). 4-3E*a;* 8-2A*b*(v)
77CC301	K. Fujiwara, J. C. Giffney, and J. H. Ridd, *J.C.S. Chem. Commun.*, 301 (1977). 3-2F
77CCC2694	D. Cech, H. Beerbaum, and A. Holy, *Collect. Czech. Chem. Commun.* **42**, 2694 (1977). 4–6; 10-5B
77CHE1110	M. A. Iradyan, A. G. Torosyan, R. G. Mirzoyan, and A. A. Aroyan, *Chem. Heterocycl. Compd. (Engl. Transl.)* **13**, 1110 (1977). 7-4A*d*
77CHE1235	A. I. Belyashova, N. N. Zatsepina, E. N. Malysheva, A. F. Pozharskii, L. P. Smirnova, and I. F. Tupitsyn, *Chem. Heterocycl. Compd. (Engl. Transl.)* **13**, 1235 (1977). 8-C*b*(ii), 3B; 11-3
77CS(11)87	S. Clementi and G. Marino, *Chem. Scr.* **11**, 87 (1977). 6-10C
77CS(12)1	S. Gronowitz, C. Westerlund, and A. B. Hornfeldt, *Chem. Scr.* **12**, 1 (1977). 4–6; 8-5C
77CS(12)97	S. Gronowitz and T. Dahlgren, *Chem. Scr.* **12**, 97 (1977). 4–6; 8-4A,B
77G339	S. Clementi, F. Fringnelli, P. Linda, G. Marino, G. Savelli, A. Taticchi, and J. L. Piette, *Gazz. Chim. Ital.* **107**, 339 (1977). 4-3E*a,b;* 6-7B,C,D,9B,D
77G359	O. Attanasi, *Gazz. Chim. Ital.* **107**, 359 (1977). 8-2A*b*(vii), C*b*(ii)
77H929	D. T. Hurst and J. A. Saldanha, *Heterocycles* **6**, 929 (1977). 8-D*b*(iv)
77HC(30)1	J. P. Paolini, *Chem. Heterocycl. Compd.* **30**, 1 (1977). 4-4B,6; 8-5C
77IJC(B)1058,1061	K. A. Thakar, D. D. Goswami, and B. M. Bhawal, *Indian J. Chem. Sect. B* **15B**, 1058, 1061 (1977). 8-C*b*(iii)
77IJC(B)1063	M. S. Shingare, D. B. Ingle, *Indian J. Chem., Sect. B* **15B**, 1063 (1977). 7-7
77JA5516	C. L. Perrin, *J. Am. Chem. Soc.* **99**, 5516 (1977). 3-2D
77JCS(P1)672	I. J. Ferguson, M. R. Grimmett, and K. Schofield, *J.C.S. Perkin Trans. 1*, 672 (1977). 7-4A*d*
77JCS(P1)887	D. J. Chadwick and C. Willbe, *J.C.S. Perkin Trans. 1*, 887 (1977). 4-2A; 6-8A,C; 8-2A*b*(vi)
77JCS(P1)1862	D. Lloyd, H. McNab, and K. S. Tucker, *J.C.S. Perkin Trans. 1*, 1862 (1977). 10-5B
77JCS(P1)1985	D. T. Hurst and M. L. Wong, *J.C.S. Perkin Trans. 1*, 1985 (1977). 10-6A
77JCS(P2)47	G. Bianchi, L. Casotti, D. Passadore, and N. Stabile, *J.C.S. Perkin Trans. 2*, 47 (1977), 8-C*b*(iii)

77JCS(P2)248	J. W. Barnett, R. B. Moodie, K. Schofield, J. B. Weston, R. G. Coombes, J. C. Golding, and G. D. Tobin, *J.C.S. Perkin Trans. 2*, 248 (1977). 6-3A
77JCS(P2)678	E. Glyde and R. Taylor, *J.C.S. Perkin Tans. 2*, 678 (1977). 5; 6-1D
77JCS(P2)845	P. G. Traverso, N. C. Marziano, and R. C. Passerini, *J.C.S. Perkin Trans. 2*, 845 (1977). 3-4C
77JCS(P2)866	H. V. Ansell and R. Taylor, *J.C.S. Perkin Trans. 2*, 866 (1977). 8-3Ac(ii)
77JCS(P2)1284	A. Cipiciani, S. Clementi, P. Linda, G. Marino, and G. Savelli, *J.C.S. Perkin Trans. 2*, 1284 (1977). 4-3Eb; 6-7D; 8-2Ab(v)
77JCS(P2)1361	N. C. Marziano, R. Passerini, J. H. Rees, and J. H. Ridd, *J.C.S. Perkin Trans. 2*, 1361 (1977). 3-2C
77JCS(P2)1452	A. R. Butler, P. Pogorzelec, and P. T. Shepherd, *J.C.S. Perkin Trans. 2*, 1452 (1977). 4-4B; 6-8D
77JCS(P2)1693	R. B. Moodie, P. N. Thomas, and K. Schofield, *J.C.S. Perkin Trans. 2*, 1693 (1977). 3-2B,4C, E; 6-3A
77JCS(P2)1998	R. S. Alexander and A. R. Butler, *J.C.S. Perkin Trans. 2*, 1998 (1977). 6-1A
77JHC95	O. Attanasi, *J. Heterocycl. Chem.* **14**, 95 (1977). 8-2Ab(ii)
77JHC359	E. Baciocchi, S. Clementi, and G. V. Sebastiani, *J. Heterocycl. Chem.* **14**, 359 (1977). 8-2Ab(iv)
77JHC517	R. J. Sundberg, *J. Heterocycl. Chem.* **14**, 517 (1977). 4-2A; 7-5A,8B
77JHC627	E. D. Weiler, R. B. Petigara, M. H. Wolfersberger, and G. A. Miller, *J. Heterocycl. Chem.* **14**, 627 (1977). 7-5B
77JHC725	E. D. Weiler and G. A. Miller, *J. Heterocycl. Chem.* **14**, 725 (1977). 7-5B
77JHC893	S. Gronowitz, C. Roos, E. Sandberg, and S. Clementi, *J. Heterocycl. Chem.* **14**, 893 (1971). 2-1Gc(iii); 8-Db(i)
77JIC1151	J. N. Chatterjee and R. S. Gandhi, *J. Indian Chem. Soc.* **54**, 1151 (1977). 8-3Ab(iv)
77JOC897	S. N. Balasubrahmanyam, A. S. Radhakrishna, A. J. Boulton, and T. Kan-Woon, *J. Org. Chem.* **42**, 897 (1977). 8-Cb(iii)
77JOC2448	R. Buchan, M. Fraser, and C. Shand, *J. Org. Chem.* **42**, 2448 (1977). 8-Db(iv)
77JOC2511	N. C. Marziano, A. Zingales, and V. Ferlito, *J. Org. Chem.* **42**, 2511 (1977). 3-3C, 4C
77JOC3498	B. T. Keen, R. J. Radel, and W. W. Paudler, *J. Org. Chem.* **42**, 3498 (1977). 10-5D
77JOC3670	S. Banerjee, O. S. Tee, and K. D. Wood, *J. Org. Chem.* **42**, 3670 (1977). 10-5B
77JOC3821	G. F. Huang and P. F. Torrence, *J. Org. Chem.* **42**, 3821 (1977). 10-4B
77JOC4197	J. Bradac *et al.*, *J. Org. Chem.* **42**, 4197 (1977). 8-Db(iv)
77JOU329	L. I. Belenkii, A. P. Yakubov, and A. I. Bessonova, *J. Org. Chem. USSR (Engl. Transl.)* **13**, 329 (1977). 6-5
77JOU1192	M. S. Pevzner, N. V. Gladkova, G. A. Lopukhova, M. P. Bedin, and V. Yu. Dolmatov, *J. Org. Chem. USSR (Engl. Transl.)* **13**, 1192 (1977). 8-Cb(iii)

77LA145	E. Regel and K.-H. Büchel, *Justus Liebigs Ann. Chem.*, 145 (1977). 4-3E*b*; 7-6; 8-C
77LA159	E. Regel, *Justus Liebigs Ann. Chem.*, 159 (1977). 4-3E*b*; 7-6
77MI1	M. L. Gandhi, *Curr. Sci.* **46**, 291 (1977). 11-4A
77NKK1518	T. Keumi, S. Shimakawa, and Y. Oshima, *Nippon Kagaku Kaishi*, 1518 (1977). 4-3E*b*; 8-3A*b*(iv), Table 17, Scheme 18
77TL389	T. Kauffmann, J. König, D. Körber, H. Lexy, H.-J. Streitberger, A. Vahrenhorst, and A. Woltermann, *Tetrahedron Lett.*, 389 (1977). 4-3E*b*; 6-7A,D
77ZN(B)1331	N. Soundarajan and P. Shanmugan, *Z. Naturforsh., B: Anorg. Chem., Org. Chem.* **32**, 1331 (1977). 8-3B
78CA(88)152349	N. I. Baranova, L. N. Pushkina, and V. I. Shishkina, *Chem. Abstr.* **88**, 152349 (1978). 4-5E; 8-3A*b*(vi)
78CB1006	I. W. Southon and W, Pfleiderer, *Chem. Ber.* **111**, 1006 (1978). 10-4B
78CHE1132	V. F. Sedova, A. S. Lisitsyn, and V. P. Mamaev, *Chem. Heterocycl. Compd. (Engl. Transl.)* **14**, 1132 (1978). 10-5B
78CI(M)348	L. Forlani, M. Magagni, and P. E. Todesco, *Chim. Ind. (Milan)* **60**, 348 (1978). 7-3B
78CJC1970	D. R. Arnold and C. P. Hadjiantoniou, *Can. J. Chem.* **56**, 1970 (1978). 6-7A
78CJC2970	O. S. Tee, D. C. Thackray, and C. G. Berks, *Can J. Chem.* **56**, 2970 (1978). 10-5B
78CPB3498	H. Uno and M. Kurokawa, *Chem. Pharm. Bull.* **26**, 3498 (1978). 8-C*b*(iii)
78CPB3884	S. Kamiya, M. Miyahara, S. Sueyoshi, I. Suzuki, and S. Odashima, *Chem. Pharm. Bull.* **26**, 3884 (1978). 4–6; 10-5A
78GEP2749235	A. Blank. Ger. Pat. 2,749,235 (1978). 6-4A
78H247	H. Fuchs and W. Pfleiderer, *Heterocycles* **11**, 247 (1978). 4-4A; 10-6B
78JCR(S)10	P. D. Clark, K. Clarke, R. M. Scrowston, and T. M. Sutton, *J. Chem. Res., Synop*, 10 (1978). 4-3E*b*; 8-2A*b*(iii),(iv),(v)
78JCR(S)133	J. Pankiewicz, B. Decroix, C. Fugier, J. Morel, and P. Pastour, *J. Chem. Res., Synop.*, 133 (1978). 10-4
78JCS(P2)72	G. P. Bean and T. J. Wilkinson, *J.C.S. Perkin Trans.* 2, 72 (1978). 6-1D
78JCS(P2)613	A. R. Katritzky, S. Clementi, G. Milleti, and G. V. Sebastiani, *J.C.S. Perkin Trans.* 2, 613 (1978). 2-1G*e*; 9-9I; 10-2B
78JCS(P2)632	M. W. Austin, *J.C.S. Perkin Trans.* 2, 632 (1978). 8-C*b*(iii)
78JCS(P2)751	H. V. Ansell and R. Taylor, *J.C.S. Perkin Trans.* 2, 751 (1978). 2-1G*c*(iii)
78JCS(P2)861	S. Clementi, S. Lepri, G. V. Sebastiani, S. Gronowitz, C. Westerlund, and A. B. Hörnfeldt, *J.C.S. Perkin Trans.* 2, 861 (1978). 2-1G*d*; 8-D*b*(i)
78JCS(P2)865	D. J. Evans, H. F. Thimm, and B. A. W. Coller, *J.C.S. Perkin Trans.* 2, 865 (1978). 7-1B,5A,B; 8-C*a,b*(iv)
78JCS(P2)1053	H. B. Amin and R. Taylor, *J.C.S. Perkin Trans.* 2, 1053 (1978). 4-3E*b*; 8-2A*a,b*(v),(vii),*c*(i)
78JHC123	A. Tanaka, K. Yakushijin, and S. Yoshina, *J. Heterocycl. Chem.* **15**, 123 (1978) 4-3E*b*,6; 8-4B

78JHC665	N. Sato *J. Heterocycl. Chem.* **15**, 665 (1978). 10-5C	
78JOC2514	R. J. Radel, J. L. Atwood, and W. W. Paudler, *J. Org. Chem.* **43**, 2514 (1978). 10-5D	
78JOC2639	M. DeRosa and J. L. Triana Alonso, *J. Org. Chem.* **43**, 2639 (1978). 8-2A*b*(iv)	
78JOC3565	Y. Takeuchi, H. J. C. Yeh, K. L. Kirk, and L. A. Cohen, *J. Org. Chem.* **43**, 3565 (1978). 7-3A,C	
78JOC3570	Y. Takeuchi, K. L. Kirk, and L. A. Cohen, *J. Org. Chem.* **43**, 3570 (1978). 7-3C, Scheme 6	
78JPR172	V. Denes and R. Chira, *J. Prakt. Chem.* **320**, 172 (1978). 11-5B	
78PHA419	U. Wrzeciono, E. Linkowska, and W. Felinska, *Pharmazie* **33**, 419 (1978). 8-C*b*(iii)	
78RTC151	C. G. Kruse, P. B. M. W. M. Timmermans, C. Van der Laken, and A. Van der Gen, *Recl. Trav. Chim. Pays-Bas* **97**, 151 [5~(1978). 8-D*b*(iv)	
78S675	L. A. M. Bastiaansen and E. F. Godefroi, *Synthesis*, 675 (1978). 4-3E*b*; 7-6	
78TL267	H. B. Amin and R. Taylor, *Tetrahedron Lett.*, 267 (1978). 5.	
78TL2537	A. J. Ashe, W.-T. Chan, and T. W. Smith, *Tetrahedron Lett.*, 2537 (1978). 4-3E*b*; 9-6B	
79AHC(25)147	B. J. Wakefield and D. J. Wright, *Adv. Heterocycl. Chem.* **25**, 147 (1979). 7-2B	
79AHC(24)151	J. W. Barton, *Adv. Heterocycl. Chem.* **24**, 151 (1979). 11-5B	
79AJC1727	K.-C. Chang, M. R. Grimmett, D. D. Ward, and R. T. Weavers, *Aust. J. Chem.* **32**, 1727 (1979). 7-4B*c*	
79AJC2049	W. B. Cowden and N. W. Jacobsen, *Aust. J. Chem.* **32**, 2049 (1979). 10-4B	
79BAU633	S. B. Gashev and L. D. Smirnov, *Bull. Acad. Sci. USSR, Div. Chem. Sci. (Engl. Transl.)* **28**, 633 (1979). 4-3C; 10-6A,E	
79BAU1446	M. S. Shvartsberg, L. N. Bizhan, A. N. Sinyakov, and R. N. Myasnikova, *Bull. Acad. Sci. USSR, Div. Chem. Sci. (Engl. Transl.)* 1446 (1979). 7-5A	
79CA(91)73755	F. Wudl, A. A. Kruger, and G. A. Thomas, *Chem. Abstr.* **91**, 73755 (1979). 4-2C; 8-B*a*	
79CA(91)175261	Yu. M. Yutilov and I. A. Svertilova, *Chem. Abstr.* **91**, 175261 (1979). 8-D*b*(iv)	
79CHE695	A. A. Prokopov and L. N. Yakhontov, *Chem. Heterocycl. Compd. (Engl. Transl.)* **15**, 695 (1979). 8-D*b*(iv)	
79CHE1195	V. A. Azimov, A. A. Prokopov, I. S. Zhivotovskaya, M. K. Polievktov, and L. N. Yakhontov, *Chem. Heterocycl. Compd. (Engl. Transl.)* **15**, 1195 (1979). 8-D*b*(iv)	
79CI(L)28	M. W. Austin, *Chem Ind. (London)*, 28 (1979). 11-4A	
79CJC626	O. S. Tee and S. Banerjee, *Can. J. Chem.* **57**, 626 (1979). 10-5B	
79CJC937	J. Llinares, J.-P. Galy, R. Faure, and E.-J. Vincent, *Can. J. Chem.* **57**, 937 (1979). 8-C*b*(iii)	
79CPB2627	A. Ohta, T. Kurihara, H. Ichimura, and T. Watanabe, *Chem. Pharm. Bull.* **27**, 2627 (1979). 11-4A	
79H475	H. Saito and M. Hamana, *Heterocycles* **12**, 475 (1979). 11-5A	

79H745	W. D. Guither, M. D. Coburn, and R. N. Castle, *Heterocycles* **12**, 745 (1979). 10-4
79IJC(B)342	K. S. Sharma, R. Parshad, and V. Singh, *Indian J. Chem., Sect. B* **17A**, 342 (1979). 8-3B
79JCS(P1)1503	J. W. Barton and D. J. Lapham, *J.C.S. Perkin Trans. 1*, 1503 (1979). 4–6; 11-5B
79JCS(P1)2334	D. H. Reid, R. G. Webster, and S. McKenzie, *J.C.S. Perkin Trans. 1*, 2334 (1979). 8-5C
79JCS(P2)224	T. A. Kortekaas and M. Cerfontain, *J.C.S. Perkin Trans. 2*, 224 (1979). 4-5B; 8-3A*b*(vi)
79JCS(P2)228	H. B. Amin and R. Taylor, *J.C.S. Perkin Trans. 2*, 228 (1979). 9-7A*a* 9.H.
79JCS(P2)312	L. Greci and J. H. Ridd, *J.C.S. Perkin Trans. 2*, 312 (1979). 8-B*a*
79JCS(P2)381	H. V. Ansell, P. J. Sheppard, C. F. Simpson, M. A. Stroud, and R. Taylor, *J.C.S. Perkin Trans. 2*, 381 (1979). 8-2A*b*(i),(iii),(v), C*b*(iii), 5A
79JCS(P2)624	H. B. Amin and R. Taylor, *J.C.S. Perkin Trans. 2*, 624 (1979). 5; 9-8
79JCS(P2)1145	L. Forlani, M. Magagni, and P. E. Todesco, *J.C.S. Perkin Trans. 2*, 1145 (1979). Scheme 7-3(b)
79JHC393	H. S. Kuo, S. Yoshina, and Y.-C. Tung, *J. Heterocycl. Chem.* **16**, 393 (1979). 8-B*b*
79JHC1029	M. Hannoun, N. Blazevic, D. Kolbah, A. Sabljic, N. Trinajstic, A. Sega, A. Lisini, F. Kajfez, and V. Sunjic, *J. Heterocycl. Chem.* **16**, 1029 (1979). 4-3E*b*; 8-2A*b*(v)
79JHC1153	J. S. Amato, V. J. Grenada, T. M. H. Liu, and E. J. J. Grabowski, *J. Heterocycl. Chem.* **16**, 1153 (1979). 7-4A*d*
79JOC3256	O. S. Tee and S. Banerjee, *J. Org. Chem.* **44**, 3256 (1979). 4–6; 10-5B
79JOC4240	Y. Takeuchi, K. L. Kirk, and L. A. Cohen, *J. Org. Chem.* **44**, 4240 (1979). 7-2E, Scheme 2
79JOC4385	H. Taguchi and S. Y. Wang, *J. Org. Chem.* **44**, 4385 (1979). 10-5B
79JOU357	I. V. Vigalok, Yu. A. Fedotov, A. A. Vigalok, and Ya. A. Levin, *J. Org. Chem. USSR (Engl. Transl.)* **15**, 357 (1979). 4–6; 10-5B
79JOU528	A. G. Zhigulin, N. B. Librovich, G. F. Burya, and M. I. Vinnik, *J. Org. Chem. USSR. (Engl. Transl.)* **15**, 528 (1979). 8-2A*b*(iii)
79PS(5)305	O. Attanasi, P. Battistoni, and G. Fava, *Phosphorus Sulfur* **5**, 305 (1979). 8-2A*b*(ii),C*b*(ii)
79T2895	L. W. Deady, M. R. Grimmett, and C. H. Potts, *Tetrahedron* **35**, 2895 (1979). 9-4A
80CA(92)215176	Kh. Yu. Yuldashev, *Chem. Abstr.* **92**, 215176 (1980). 4-3E*b*; 8-2A*b*(v)
80CA(92)215319	N. Saldabols and J. Popelis, *Chem. Abstr.* **92**, 215319 (1980). 8-D*b*(iv)
80CB3675	R. Kreher and P. H. Wagner, *Chem. Ber.* **113**, 3675 (1980). 8-2A*b*(v)

80CCC2949	A. Krutosikova, J. Kovac, M. Chudobova, and D. Ilavsky, *Collect. Czech. Chem. Commun.* **45**, 2949 (1980). 8-5
80CHE142	I. A. Abronin, V. P. Litvinov, G. M. Zhidomirov, A. Z. Dzhumanazarova, and Ya. L. Goldfarb, *Chem. Heterocycl. Compd. (Engl. Transl.)* **16**, 142 (1980). 8-5
80CHE230	G. N. Friedlin, N. A. Kuraeva, and K. A. Solop, *Chem. Heterocycl. Compd. (Engl. Transl.)* **16**, 230 (1980). Table 6.6
80CHE272	G. P. Sharnin, I. F. Falyakhov, and F. G. Khairutchinov, *Chem. Heterocycl. Compd. (Engl. Transl.)* **16**, 272 (1980). 9-4A
80CHE339	S. S. Mochalov, F. M. Abdelrazak, T. P. Surikova, and Yu. S. Shabarov, *Chem. Heterocycl. Compd. (Engl. Transl.)* **16**, 339 (1980). 6-7A
80CPB1909	T. Uno, K. Takagi, and M. Tomoeda, *Chem. Pharm. Bull.* **28**, 1909 (1980). 8-Cb(iii)
80CS(15)20	S. Gronowitz and I. Ander, *Chem. Scr.* **15**, 20 (1980). 6-3A
80CS(15)102	S. Liljefors and S. Gronowitz, *Chem. Scr.* **15**, 102 (1980). 4–6; 7-5B
80CS(15)206	S. Gronowitz, A. Konar, and V. Litvinov, *Chem. Scr.* **15**, 206 (1980). 4-3Ea,b,6; 8-5C
80H1753	D. T. Hurst, A. D. Stacey, and D. K. Weerasinghe, *Heterocycles* **14**, 1753 **(1980)**. 10-6A
80JCR(S)197	K. Clarke, B. Gleadhill, and R. M. Scrowston, *J. Chem. Res., Synop.*, 197 (1980). 4–6; 8-Cb(iii),(iv)
80JCR(S)201	N. Soundararajan, R. Palaniappan, S. Nagarajam, T. K. Raja, and P. Shanmugam, *J. Chem. Res., Synop.*, 201 (1980). 4–6; 8-3B
80JCS(P1)959	E. E. Glover and L. W. Peck, *J.C.S. Perkin Trans. 1*, 959 (1980). 8-Db(iv)
80JCS(P2)110	R. S. Alexander and A. R. Butler, *J.C.S. Perkin Trans. 2*, 110 (1980). 6-1D
80JCS(P2)773	D. L. Brydan, A. J. G. Sagar, and D. M. Smith, *J.C.S. Perkin Trans. 2*, 773 (1980). 7-4Ca
80JGU618	A. I. Razumov, P. A. Gurevich, S. A. Muslimov, T. V. Komina, T. V. Zykov, and R. A. Salakhutdinov, *J. Gen. Chem. USSR (Engl. Transl.)* **50**, 618 (1980). 4-4C; 8-2Ab(vi)
80JHC143	N. Sato, *J. Heterocycl. Chem.* **17**, 143 (1980). 10-5C
80JCH1019	J. Bourguignon, M. Lemarchand and G. Queguiner, *J. Heterocycl. Chem.* **17**, 1019 (1980). 4-2A; 8-Db(iii),(iv)
80JHC1399	H. El-Kashef, S. Rault, M. Cugnon de Sevricourt, P. Touzot and M. Robba, *J. Heterocycl. Chem.* **17**, 1399 (1980). 4–6; 8-3B
80JOC76	J. F. Hansen, Y. I. Kim, L. J. Griswold, G. W. Hoelle, D. L. Taylor, and D. E. Vietti, *J. Org. Chem.* **45**, 76 (1980). 7-5B
80JOC830	O. S. Tee and C. G. Berks, *J. Org. Chem.* **45**, 830 (1980). 10-5B, Table 5
80JOC2072	O. S. Tee and M. Paventi, *J. Org. Chem.* **45**, 2072 (1980). 10-5B
80JOC3108	J. D. Vaughan, V. L. Vaughan, S. S. Daly, and W. A. Smith, *J. Org. Chem.* **45**, 3108 (1980). 7-5A
80JOU391	A. Barudi, A. B. Kudryavtsev, A. Ya. Zheltov, and B. I. Stepa-

nov, *J. Org. Chem. USSR (Engl. Transl.)* **16**, 391 (1980). 8-4B

80MI1 K. Schofield, "Aromatic Nitration," pp. 1, 36, 109, 147, 148, 151, 153, 198. Cambridge Univ. Press, Cambridge, England, 1980. 3,3-2D,F,4A,C

80S139 T. Keumi, R. Taniguchi, and H. Kitajima, *Synthesis*, 139 (1980). 4-3E*b*; 8-3A*b*(iv)

80S800 A. R. Katritzky, D. Winwood, and N. E. Grzeskowiak, *Synthesis*, 800 (1980). 4-2A; 7-8B

80URP707916 T. K. Khanina, USSR Pat. 707,916 (1980). 4-5B; 6-8A

81AHC(29)10 R. K. Smalley, *Adv. Heterocycl. Chem.* **29**, 10 (1981). 8-C*b*(iii)

81BAU1089 S. Gronowitz, A. Konar, and V. P. Litvinov, *Bull. Acad. Sci. USSR, Div. Chem. Sci. (Engl. Transl.)* **30**, 1089 (1981). 8-5,5C

81CA(94)156679 D. K. Chae, W. K. Chung, M. W. Chun, and B. Y. Lee, *Chem. Abstr.* **94**, 156679 (1981). 8-2A*b*(iv)

81CA(95)80688 G. I. Migachev and N. G. Grekhova, *Chem. Abstr.* **95**, 80688 (1981). 11-4A

81CHE152 S. S. Mochalov, T. P. Surikova, F. M. Abdelrazak, and V. D. Zakharova, *Chem. Heterocycl. Compd. (Engl. Transl.)* **17**, 152 (1981). 6-8A

81CHE1217 N. S. Prostakov, G. Datta Rai, N. D. Sergeeva, and V. I. Kuznetsov, *Chem. Heterocycl. Compd. (Engl. Transl.)* **17**, 1217 (1981). 11-4A

81CJC1022 N. H. Werstiuk and G. Timmins, *Can. J. Chem.* **59**, 1022 (1981). 9-3

81HC(381)32 S. F. Dyke and R. G. Kinsman, *Chem. Heterocycl. Compd.* **38** (Part 1), 32 (1981). 11-1B

81IJC(19A)1183 V. Kannappan, M. J. Nanjan, and R. Ganesan, *Indian J. Chem., Sect. A* **19A**, 1183 (1981). 4–6, 6.D.1

81JCR(S)104 J. Bourguignon, C. Becue, G. Queguiner, *J. Chem. Res., Synop.*, 104 (1981). 10-4

81JCS(P2)628 M. Colonna, L. Greci, and M. Poloni, *J.C.S. Perkin Trans. 2*, 628 (1981). 8-2A*b*(iii)

81JCS(P2)931 G. Seconi and C. Eaborn, *J.C.S. Perkin Trans. 2*, 931 (1981). 4-1C; 6-8A

81JCS(P2)1153 W. J. Archer and R. Taylor, *J.C.S. Perkin Trans. 2*, 1153 (1981). 11-1B, 2A, 4A,7

81JFC67 J. Mirek and A. Haas, *J. Fluorine Chem.* **19**, 67 (1981). 4-5D; 8-B*a*

81JHC885 W. J. Hammar and M. A. Rustad, *J. Heterocycl. Chem.* **18**, 885 (1981). 7-4A*a*

81JHC1081 R. C. Boruah, J. S. Sandhu, and G. Thyagarajan, *J. Heterocycl. Chem.* **18**, 1081 (1981). 8-C*b*(iii)

81JHC1639 Z. Swistun and H. C. van der Plas, *J. Heterocycl. Chem.* **18**, 1639 (1981). 10-6A

81JMC959 I. T. Barnish, P. E. Cross, R. P. Dickinson, M. J. Parry, and M. J. Randall, *J. Med. Chem.* **24**, 959 (1981). 4-5C; 6-8A

81JOC881 A. J. Ashe, W.-T. Chan, T. W. Smith, and K. M. Taba, *J. Org. Chem.* **46**, 881 (1981). 9-2G,4E

81JOC1646	H. C. Brown, M. Periasamy, and K.-T. Liu, *J. Org. Chem.* **46**, 1646 (1981). 8-Cb(iv)
81JOC2221	H. M. Gilow and D. E. Burton, *J. Org. Chem.* **46**, 2221 (1981). 4–6; 6-4B
81JOC3056	J. V. Crivello, *J. Org. Chem.* **46**, 3056 (1981). 8-Cb(iii)
81JOC4172	O. S. Tee and M. Paventi, *J. Org. Chem.* **46**, 4172 (1981). 4–6; 10-5B
81JOM(204)153	G. Seconi, C. Eaborn, and J. G. Stamper, *J. Organomet. Chem.* **204**, 153 (1981). 4-1C; 6-8A
81JOU2183	V. V. Mezheritskii, D. M. Elisevich, and G. N. Dorofeenko, *J. Org. Chem. USSR (Engl. Transl.)* **17**, 2183 (1981). 4-3Ea; 11-6C
81PS111	R. J. Cremlyn, K. H. Goulding, F. J. Swinbourne, and K-M. Yung. *Phosphorus Sulphur*, 111 (1981). 4-5C; 6-8A
81RTC267	C. N. M. Bakker and F. M. Kaspersen, *Recl. Trav. Chim. Pays-Bas* **100**, 267 (1981). 10-5B
81S701	G. Szilagyi and H. Wamhoff, *Synthesis*, 701 (1981). 10-5B
81USP4288445	T. Kusumi and K. Nakanishi, U.S. Pat. 4,288,445 (1981). 7-4Ba
82AJC1761	J. C. Teulade, R. Escale, J. C. Rossi, J. P. Chapat, G. Grassy, and M. Payard, *Aust. J. Chem.* **35**, 1761 (1982). 8-Da,b(iv)
82AJC2025	L. W. Deady, O. L. Korytsky, and J. E. Rowe, *Aust. J. Chem.* **35**, 2025 (1982). 9-4A
82AJC2035	L. W. Deady and O. L. Korytsky, *Aust. J. Chem.* **35**, 2035 (1982). 9.D.1
82BAU2104	Ya. L. Goldfarb, A. A. Dudinov, and V. P. Litvinov, *Bull. Acad. Sci. USSR, Div. Chem. Sci. (Engl. Transl.)* 2104 (1982). 4–6; 6-4A
82BCJ629	T. Keumi, H. Yamada, H. Takahashi, and H. Kitajima, *Bull. Chem. Soc. Jpn.* **55**, 629 (1982). 8-3Ab(ii)
82CA(96)68737	V. G. Poludnenko and V. I. Enya, *Chem. Abstr.* **96**, 68737 (1982). 8-3Ab(ii)
82CHE127	G. N. Friedlin, A. A. Glushkova, and K. A. Salop, *Chem. Heterocycl. Compd. (Engl. Transl.)* **18**, 127 (1982). 6-3A
82CHE297	S. B. Gashev and L. D. Smirnov, *Chem. Heterocycl. Compd. (Engl. Transl.)* **18**, 297 (1982). 4-3C; 10-6A,E
82CHE539	V. A. Bakulev, V. S. Mokrushin, and Z. V. Pushkareva, *Chem. Heterocycl. Compd. (Engl. Transl.)* **18**, 539 (1982). 7-5A
82CI(L)57	M. W. Austin, *Chem. Ind. (London)*, 57 (1982). 7-4Bc
82CJC2668	J. Bourguignon, S. Chapelle, P. Granger, and G. Queguiner, *Can. J. Chem.* **60**, 2668 (1982). 10-4
82CPB3392	T. Itaya, C. Shioyama, and S. Kagatani, *Chem. Pharm. Bull.* **30**, 3392 (1982). 4-4A; 10-6B
82CS(20)208	I. A. Abronin, A. Z. Djumanazarova, V. P. Litvinov, and A. Konar, *Chem. Scr.* **20**, 208 (1982). 8-5
82HC(382)447	G. Jones and D. J. Baty, *Chem. Heterocycl. Compd.* **38** (Part 2), 447 (1982). 11-4A
82IJC(A)417	M. R. Nair, *Indian J. Chem., Sect. A* **21A**, 417 (1982). 4–6; 6-4A, Table 6.4
82JA4142	O. S. Tee and M. Paventi, *J. Am. Chem. Soc.* **104**, 4142 (1982). 4–6; 9-5B

82JA7084	G. Angelini, C. Sparapani, and M. Speranza, *J. Am. Chem. Soc.* **104**, 7084 (1982). 6-5
82JA7091	G. Angelini, G. Lilla, and M. Speranza, *J. Am. Chem. Soc.* **104**, 7091 (1982). 6-5
82JCS(P2)181	W. J. Archer, M. A. Hossaini, and R. Taylor, *J.C.S. Perkin Trans. 2*, 181 (1982). 5.
82JCS(P2)187	M. A. Hossanini and R. Taylor, *J.C.S. Perkin Trans. 2*, 187 (1982). 5.
82JCS(P2)295	W. J. Archer and R. Taylor, *J.C.S. Perkin Trans. 2*, 295 (1982). 6-1,1A,10A; 8-5A, 8.F.1.
82JCS(P2)301	W. J. Archer and R. Taylor, *J.C.S. Perkin Trans. 2*, 301 (1982). 6-10A; 8-6A
82JCS(P2)1489	H. B. Amin, A. A. Awad, W. J. Archer, and R. Taylor, *J.C.S. Perkin Trans. 2*, 1489 (1982). 8-2Aa,b(i),(ii), Table 1, Scheme 3
82JGU2291	Yu. A. Manaev, M. A. Andreeva, V. P. Perevalov, B. I. Stepanov, V. S. Dubrovskaya, and V. I. Seraya, *J. Gen. Chem. USSR (Engl. Transl.)* **52**, 2291 (1982). 7-4Bc,5B
82JHC279	J.-M. Clavel, J. Guillaumel, P. Demerseman, and R. Royer, *J. Heterocycl. Chem.* **19**, 279 (1982). 4-3Eb; 8-2Ab(v)
82JHC665	C. Galvez and P. Viladoms, *J. Heterocycl. Chem.* **19**, 665 (1982). 8-Db(iv)
82JHC673	N. Sato, *J. Heterocycl. Chem.* **19**, 673 (1982). 10-5C
82MI1	A. R. Katritzky and C. Ögretir, *Chim. Acta Turc.* **10**, 137 (1982). 7-4Cc
82S1096	V. Bocchi and G. Palla, *Synthesis*, 1096 (1982). 8-2Ab(iv)
82T3693	J. Catalan, P. Perez, and M. Yanez, *Tetrahedron* **38**, 3693 (1982). 8-2Aa, Table 1
83AJC1227	R. W. Read, R. J. Spear, and W. P. Norris, *Aust. J. Chem.* **36**, 1227 (1983). 8Cb(iii)
83AJC1659	D. T. Hurst, *Aust. J. Chem.* **36**, 1659 (1983). 4-4B; 10-6A
83CCC2676	J. Farkas, *Collect. Czech. Chem. Commun.* **48**, 2676 (1983). 10-4D
83CHE514	G. P. Sharnin, I. Sh. Saitfullin, I. F. Falyakov, F. G. Khairutchinov, T. G. Bolshakova, and V. U. Zverev, *Chem. Heterocycl. Compd. (Engl. Transl.)* **19**, 514 (1983). 9-4A
83CHE871	D. O. Kadzhrishvili, Sh. A. Samsoniya, E. H. Gordeev, L. N. Kurkovskaya, V. E. Zhigachev, and N. N. Suvorov, *Chem. Heterocycl. Compd. (Engl. Transl.)* **19**, 871 (1983). 4-3Eb; 8-4B
83CHE1003	S. B. Gashev, V. P. Lezina, and L. D. Smirnov, *Chem. Heterocycl. Compd. (Engl. Transl.)* **19**, 1003 (1983). 10-2B
83CHE1008	S. B. Gashev, V. F. Sedova, L. D. Smirnov, and V. P. Mamaev, *Chem. Heterocycl. Compd. (Engl. Transl.)* **19**, 1008 (1983). 4-3C; 10-5B,6E
83CHE1012	S. B. Gashev, B. A. Korolev, L. A. Osmolovskaya, and L. D. Smirnov, *Chem. Heterocycl. Compd. (Engl. Transl.)* **19**, 1012 (1983). 10-5B
83CJC2287	J. Einhorn, P. Demerseman, and R. Royer, *Can. J. Chem.* **61**, 2287 (1983). 8-2Ab(iii)

83CJC2556	O. S. Tee and M. Paventi, *Can. J. Chem.* **61**, 2556 (1983). 4–6; 9-5B
83CS(22)22	A. Konar and V. P. Litvinov, *Chem. Scr.* **22**, 22 (1983). 8-5
83JCS(P2)813	W. J. Archer, R. Cook, and R. Taylor, *J.C.S. Perkin Trans. 2*, 813 (1983). 6-10A; 8-4A
83JCS(P2)1491	A. Margonelli and M. Speranza, *J.C.S. Perkin Trans. 2*, 1491 (1983). 6-5
83JHC61	J.-M. Webert, D. Cagniant, P. Cagniant, G. Kirsch, and J.-V. Weber, *J. Heterocycl. Chem.* **20**, 61 (1983). 4-3Eb;8-4B
83JOC1064	W. W. Paudler and M. V. Jovanovic, *J. Org. Chem.* **48**, 1064 (1983). 4–6; 10-5A,B,C
83JOM159	P. Jutzi and U. Gilge, *J. Organomet. Chem.* **246**, 159 (1983). 8-Cb(v)
83MI1	G. Chuchani and R. M. Dominguez, *Int. J. Chem. Kinet.* **15**, 1275 (1983). 9-7Ab
83PHA83	H. A. H. El-Sherief, A. E. Abdul-Raman, and A. M. Mahmoud, *Pharmazie* **38**, 83 (1983). 11-5A
83S987	A. J. Guildford, M. A. Tometzki, and R. W. Turner, *Synthesis*, 987 (1983). 4-2A; 8-Db(iii)
83T1777	E. Bisagni, N. C. Hung, and J. M. Lhoste, *Tetrahedron* **39**, 1777 (1983). 4-2A; 8-Db(iii)
83T2851	J. Catalan, O. Mo, P. Perez, and M. Yanez, *Tetrahedron* **39**, 2851 (1983). 8-Da
84AHC(36)394	C. R. Hardy, *Adv. Heterocycl. Chem.* **36**, 394 (1984). 8-Db(iv)
84BAU2469	V. P. Lezina, S. B. Gashev, M. M. Boranov, L. D. Smirnov, and K. M. Dyumaev, *Bull. Acad. Sci. USSR, Div. Chem. Sci. (Engl. Transl.)*, 2469 (1984). 10-2B
84CHE687	V. T. Grachev et al., *Chem. Heterocycl. Compd. (Engl. Transl.)* **20**, 687 (1984). 8-3B
84H241	Y. Murakami, M. Tani, K. Tanaka, and Y. Yokoyama, *Heterocycles* **22**, 241 (1984). 4-3Eb; 8-2Ab(v)
84H1195	M. V. Jovanovic, *Heterocycles* **22**, 1195 (1984). 10-5C
84JA37	G. Angelini, G. Laguzzi, C. Sparapani, and M. Speranza, *J. Am. Chem. Soc.* **106**, 37 (1984). 6-1D
84JCR(S)390	T. Nishiwaki and N. Kunishige, *J. Chem. Res., Synop.*, 390 (1984). 4-3Eb,6; 8-4B
84JCS(P1)2839	C. D. Buttery, D. W. Knight, and A. P. Nott, *J. C. S. Perkin Trans. 1*, 2839 (1984). 4-2A; 8-2Ab(vi)
84JCS(P2)165	M. Colonna, L. Greci, and M. Poloni, *J. C. S. Perkin Trans. 2*, 165 (1984). 4-4A; 8-2Ab(iii),Ba
84JCS(P2)1179	J. White and G. McGillivray, *J. C. S. Perkin Trans. 2*, 1179 (1984). 4-3Ea; 6-7D
84JCS(P2)1607	J. White, *J. C. S. Perkin Trans. 2*, 1607 (1984). 4-3Ea; 6-7D
84JCS(P2)1659,1667	A. H. Clems and J. H. Ridd, *J. C. S. Perkin Trans. 2*, 1659,1667 (1984). 3-2D
84JHC177	J. P. Bachelet, J. M. Clavel, P. Demerseman, and R. Royer, *J. Heterocycl. Chem.* **21**, 177 (1984). 4-3Eb; 8-2Ab(v)
84JHC725	S. Shiotani, H. Morita, M. Inoue, T. Ishida, Y. Iitaka, and A. Itai, *J. Heterocycl. Chem.* **21**, 725 (1984). 8-Db(iv)

84JHC785	L. H. Klemm and J. N. Louris, *J. Heterocycl. Chem.* **21**, 785 (1984). 4-2A; 8-Db(iii),(iv)
84JHC1485	R. Deschner and U. Pindur, *J. Heterocycl. Chem.* **21**, 1485 (1984). 4-3A; 8-2Ab(v)
84JOC3401	P. S. Waalwijk, P. Cohen-Fernandes, and C. L. Habraken, *J. Org. Chem.* **49**, 3401 (1984). 8-Cb(iv)
84JOC4409	D. Dauzonne, I. A. O'Neil, and A. Renaud, *J. Org. Chem.* **49**, 4409 (1984). 4-7C; 8-2Ab(vi)
84MI1	M. Tisler and B. Stanovnik, *Compr. Heterocycl. Chem.* **3**, 20 (1984). 10-4A
84MI2	H. Neunhoeffer, *Compr. Heterocycl. Chem.* **3**, 369 (1984). 10-4D
84MI3	A. R. Katritzky and J. M. Lagowski, *Compr. Heterocycl. Chem.* **5**, 57 (1984). 7-2B
84S252	T. Sakamoto, Y. Kondo, and H. Yamanaka, *Synthesis*, 252 (1984). 4–6; 10-5B
84T1857	J. B. Kyziol and Z. Daszkiewicz, *Tetrahedron* **40**, 1857 (1984). 8-3Ab(ii)
84TL3325	H. Ikehira, T. Matsuura, and I. Saito, *Tetrahedron Lett.* **25**, 3325 (1984). 10-5B
85CC396	J. L. Morris, C. W. Rees, and D. J. Rigg, *J. C. S. Chem. Commun.*, 396 (1985). 12
85CC398	R. Jones, J. L. Morris, A. W. Potts, C. W. Rees, D. J. Rigg, H. S. Rzepa, and D. J. Williams, *J. C. S. Chem. Commun.*, 398 (1985). 12
85CS(25)295	L. I. Belenkii, G. P. Gromova, M. A. Cheskis, and Ya. L. Goldfarb, *Chem. Scr.* **25**, 295 (1985). 7-4Aa,5A, Table 4
85H295	H. Chikashita and K. Itoh, *Heterocycles* **23**, 295 (1985). 8-Cb(v)
85JCR(S)318	R. Taylor, *J. Chem. Res., Synop.*, 318 (1985). 6-1D
85JCS(P2)97	R. Lazzaroni, J. P. Boutique, J. Riga, J. J. Verbist, J. G. Fripiat, and J. Delhalle, *J. C. S. Perkin Trans. 2*, 97 (1985). 8-5,5A
85JCS(P2)1227	A. H. Clemens, J. H. Ridd, and J. P. B. Sandall, *J. C. S. Perkin Trans. 2*, 1227 (1985). 3-2D
85JOC1324	R. Buchan, M. Fraser, and P. V. S. Kong Thoo Lin, *J. Org. Chem.* **50**, 1324 (1985). 8-Db(i)
85MI1	T. L. Gilchrist, "Heterocyclic Chemistry," p. 277. Putman, London, 1981. 11-1B
86CSR1	J. L. Morris and C. W. Rees, *Chem. Soc. Rev.* **15**, 1 (1986). 12
86HC(44,2)1	R. Taylor, *Chem. Heterocycl. Compd.* **44**, (Part 2), p. 1. Wiley, London, 1986. 4-7c; 6,6-1A,D,7C,8A,10A
86JCS(P2)1581	N. Al-Awadi and R. Taylor, *J. C. S. Perkin Trans. 2*, 1581 (1986). 5-Bb
86JCS(P2)1265	R. August, C. Davis, and R. Taylor, *J. C. S. Perkin Trans. 2*, 1265 (1986). 7-9A, Scheme 13
86PAC197	J. L. Morris and C. W. Rees, *Pure Appl. Chem.* **58**, 197 (1986). 12
86PC1	S. Clementi and P. Linda, personal communication (1986).
86UP1	G. Marino and R. Taylor, unpublished work (1986). 6-9B

87JCS(P2)591	A. Laws and R. Taylor, *J.C.S. Perkin Trans. 2*, 591 (1987). 8-B,B*a,b*,D*b*(iv),4A,5A, Table 13
89JCS(P2)	A. Laws and R. Taylor, *J.C.S. Perkin Trans. 2*, 1911 (1989). 12
90MI1	R. Taylor, "Electrophilic Aromatic Substitution," p. 481. Wiley, Chichester, England, 1990. 6-10C